Handbook of Algal Technologies and Phytochemicals

Volume I: Food, Health and Nutraceutical Applications

Handbook of Algal Technologies and Phytochemicals

Volume I: Food, Health and Nutraceutical Applications

Edited by
Gokare A. Ravishankar and Ambati Ranga Rao

CRC Press
Taylor & Francis Group
Boca Raton London New York

CRC Press is an imprint of the
Taylor & Francis Group, an **informa** business

CRC Press
Taylor & Francis Group
6000 Broken Sound Parkway NW, Suite 300
Boca Raton, FL 33487-2742

© 2020 by Taylor & Francis Group, LLC
CRC Press is an imprint of Taylor & Francis Group, an Informa business

No claim to original U.S. Government works

Printed on acid-free paper

International Standard Book Number-13: 978-0-367-14979-6 (Hardback)

This book contains information obtained from authentic and highly regarded sources. Reasonable efforts have been made to publish reliable data and information, but the author and publisher cannot assume responsibility for the validity of all materials or the consequences of their use. The authors and publishers have attempted to trace the copyright holders of all material reproduced in this publication and apologize to copyright holders if permission to publish in this form has not been obtained. If any copyright material has not been acknowledged please write and let us know so we may rectify in any future reprint.

Except as permitted under U.S. Copyright Law, no part of this book may be reprinted, reproduced, transmitted, or utilized in any form by any electronic, mechanical, or other means, now known or hereafter invented, including photocopying, microfilming, and recording, or in any information storage or retrieval system, without written permission from the publishers.

For permission to photocopy or use material electronically from this work, please access www.copyright.com (http://www.copyright.com/) or contact the Copyright Clearance Center, Inc. (CCC), 222 Rosewood Drive, Danvers, MA 01923, 978-750-8400. CCC is a not-for-profit organization that provides licenses and registration for a variety of users. For organizations that have been granted a photocopy license by the CCC, a separate system of payment has been arranged.

Trademark Notice: Product or corporate names may be trademarks or registered trademarks, and are used only for identification and explanation without intent to infringe.

Visit the Taylor & Francis Web site at
http://www.taylorandfrancis.com

and the CRC Press Web site at
http://www.crcpress.com

Dedication

Lalgudi Vaidyanthan Venkataraman

(1938–2017)

The editors respectfully dedicate this volume to their beloved former colleague and mentor Dr. Lalgudi Vaidyanthan Venkataraman, who, along with Dr. E. Wolfgang Becker of Tuebingen, Germany, was responsible for starting algal research at the Central Food Technological Research Institute, Mysore, India. Dr. Venkataraman was fondly called LV by his friends and colleagues. LV was a dedicated scientist and a great human being. LV's passionate association with algal technology resulted in seminal studies in the field of applied phycology, leading to the industrial production of Spirulina in India. One of the editors of this volume—GAR—was associated with LV in the saga of the industrialization of Spirulina in India, and they were both jointly honoured by the Department of Biotechnology, Government of India, with a National Technology Day Award in the year 2003.

LV's research with algae resulted in developing low-cost technology for Spirulina cultivation which became popular in India for rural masses for both nutrition and livelihood. Further, LV was responsible for the partnership of Thapar Group and Ballarpur Industries Ltd. to develop industrial technology for large-scale production of Spirulina; this culminated in (at the time) Asia's largest production unit at Nanjangud, near Mysore, in the year 1990.

Dr. Lalgudi Vaidyanathan Venkataraman was born on March 31, 1938, in Tiruchirapalli. He graduated with a BSc degree from St. Joseph's College Tiruchirapalli in 1957 and earned his MSc degree in 1959, from Banaras Hindu University. After a 10-year career teaching at St. Joseph's College, he joined Bowling Green State University, Ohio, in 1969, and earned his PhD at that university in 1972. On his return to India, in 1973, he joined the Central Food Technological Research Institute (CFTRI), where he served in various capacities from scientist to the level of deputy director, before he voluntarily retired in 1994. LV had a distinguished scientific career with immense contributions to protein chemistry, algal biotechnology, and plant secondary metabolites. He was the founding Head of the Autotrophic Cell Culture Discipline, which was later named the Plant Cell Biotechnology Department, at CFTRI. Subsequently, he founded a consultancy firm, Elvees Biotech. He was also a director of South-East Agro Industries, Delhi, and an advisor to the South China Biotech Exploitation Centre, PR China. He served as FAO consultant to Cuba and Thailand.

LV developed integrated systems for recycling effluents with microalgae. His primary contribution has been in the outdoor mass culture of Spirulina, using simple agro fertilizer-based medium and developing harvesting of biomass and drying techniques. He was the first to introduce the algae in the tablet form as a nutraceutical. Several algae-based formulations for feed and food uses were made by him and his students. LV's published work includes 6 books, 14 book chapters, 190 research papers, including numerous reviews and popular articles. He had four patents to his credit. He mentored more than 15 PhD students, 50 master's students, and 8 visiting scientists who occupy high positions globally.

LV received several awards and recognitions, viz. Fellow of Indian Botanical Society; Fellow of National Academy of Sciences; Fellow of Association of Food Scientists and Technologists of India (AFSTI); Prof. J.V Bhat-Eurka Forbes Award; Prof. Vyas Memorial Award of Association of Microbiologists of India; Lalji Godhoo Smarak Nidhi Award of AFSTI; National Technology Day Award of Government of India; Distinguished Member of Plant Tissue Culture Association of India. He served as editor of Journal of Green Vegetation Research, Journal of Food Science and Technology, *and several others. He also served on the committees of departments of science and technology, departments of biotechnology, technology development boards, and in boards of studies of several universities in India. He served as national president of AFSTI. He was a Rotarian who rendered service to society for the handicapped and for female empowerment at Ayikudi, Tamil Nadu, India.*

LV was ably supported by his wife, the late Dr. M.K Krishnakumari (Krishna), a scientist also at CFTRI, who was a pioneer in research on rodent control and pesticide toxicology. LV and Krishna had two children, daughter Sujatha and son Sanjeev, and four grandchildren.

LV will be remembered for his kindhearted, friendly nature, oratory skills, prolific writing, and passionate scientific pursuits, mainly as Spirulina Man of India. He breathed his last on January 18, 2017.

Contents

Preface xi
Acknowledgments xiii
Editors xv
Contributors xvii

SECTION I Algal Constituents for Food, Health and Disease Applications

Chapter 1 Algal Polysaccharides: Innovative Extraction Technologies, Health Benefits and Industrial Applications 3

M. Clemente-Carazo, V. Sanchez, S. Condon-Abanto, and Marco García-Vaquero

Chapter 2 Macroalgal Fucoidan for Biomedical Applications 13

Jayachandran Venkatesan, Sukumaran Anil, Sneha Rao, and Se-Kwon Kim

Chapter 3 Bioprocess Parameters of Production of Cyanobacterial Exopolysaccharide: Biomass Production and Product Recovery 25

Onkar Nath Tiwari, Sagnik Chakraborty, Indrama Devi, Abhijit Mondal, Biswanath Bhunia, and Shen Boxiong

Chapter 4 Production, Extraction and Characterization of Alginates from Seaweeds 33

Faiez Hentati, Alina V. Ursu, Guillaume Pierre, Cedric Delattre, Bogdan Trica, Slim Abdelkafi, Gholamreza Djelveh, Tanase Dobre, and Philippe Michaud

Chapter 5 Champion Microalgal Forms for Food and Health Applications: Spirulina *and* Chlorella 43

Chiara Toniolo, Marcello Nicoletti, Paola Del Serrone, Ambati Ranga Rao, and Gokare A. Ravishankar

Chapter 6 Seaweed Antimicrobials: Present Status and Future Perspectives 61

María José Pérez, Elena Falqué, and Herminia Domínguez

Chapter 7 Marine-Algal Bioactive Compounds: A Comprehensive Appraisal 71

Iahtisham-Ul-Haq, Masood Sadiq Butt, Natasha Amjad, Iqra Yasmin, and Hafiz Ansar Rasul Suleria

Chapter 8 Ultrasound-Assisted Extraction of Bioactive Compounds from Microalgae 81

Mona Ahmed J. Alzahrani and Conrad O. Perera

Chapter 9 Biogeneration of Volatile Organic Compounds in Microalgae-Based Systems 89

Pricila Nass Pinheiro, Karem Rodrigues Vieira, Andriéli Borges Santos, Eduardo Jacob-Lopes, and Leila Queiroz Zepka

Chapter 10 Antidiabetic Properties of Brown Seaweeds (*Sargassum polycystum* C.Ag) 99

Suhaila Mohamed and Mahsa Motshakeri

Chapter 11 Biologically Active Vitamin B_{12} from Edible Seaweeds 105

Tomohiro Bito, Fei Teng and Fumio Watanabe

Chapter 12 Potentials and Challenges in the Production of Microalgal Pigments with Reference to Carotenoids, Chlorophylls, and Phycobiliproteins 111

Delia B. Rodriguez-Amaya and Iriani R. Maldonade

Chapter 13 Potential Health and Nutraceutical Applications of Astaxanthin and Astaxanthin Esters from Microalgae 121

Ambati Ranga Rao and Gokare A. Ravishankar

Chapter 14 *Dunaliella salina*: Sustainable Source of β-Carotene 139

J. Paniagua-Michel

Chapter 15 Exploring the Potential of Using Micro- and Macroalgae in Cosmetics 149

W.A.J.P. Wijesinghe and N.E. Wedamulla

Chapter 16 Microalgae for Human Nutrition: Perspectives for the Future 161

Mariana F.G. Assuncao, Ana Paula Batista, Raquel Amaral, and Lília M.A. Santos

Chapter 17 Nutraceutical Aspects of Microalgae: Will Our Future Space Foods Be Microalgae Based? 177

Ceren Gurlek, Cagla Yarkent, Izel Oral, Ayse Kose, and Suphi S. Oncel

Chapter 18 Microalgae and Cyanobacteria as a Potential Source of Anticancer Compounds 185

Wan-Loy Chu and Siew-Moi Phang

Chapter 19 Macroalgae and Microalgae: Novel Sources of Functional Food and Feed 207

Shama Aumeerun, Joyce Soulange-Govinden, Marie Francoise Driver, Ambati Ranga Rao, Gokare A. Ravishankar, and Hudaa Neetoo

Chapter 20 Angiogenic Actions of Anionic Polysaccharides from Seaweed 221

Celina Maria Pinto Guerra Dore, Monique G. das Chagas Faustino Alves, Luiza Sheyla Evenni P. Will Castro, Luciana Guimaraes Alves Figueira, and Edda Lisboa Leite

Chapter 21 Platform Molecules from Algae by Using Supercritical CO_2 and Subcritical Water Extraction 229

Nidhi Hans, S.N. Naik and Anushree Malik

Chapter 22 Biological Activities and Safety Aspects of Fucoxanthin 245

Naveen Jayapal, Madan Kumar Perumal, and Baskaran Vallikannan

SECTION II Algal Genomics and Metabolomics

Chapter 23 Functional Omics and Big Data Analysis in Microalgae: The Repertoire of Molecular Tools in Algal Technologies 261

Chetan Paliwal, Tonmoy Ghosh, Asha A. Nesamma, and Pavan P. Jutur

Chapter 24 Bioactive Metabolites: Genetic Regulation and Potential Market Implications 273

Ayse Kose, Claire Remacle, Young-Woo Kim, Suphi S. Oncel, and Murat Elibol

Chapter 25 Leveraging Genome Sequencing Strategies for Basic and Applied Algal Research, Exemplified by Case Studies 281

Ariana A. Vasconcelos and Vitor H. Pomin

Index 293

Preface

Algae are the most significant group of autotrophic organisms which support life forms in aquatic, semi aquatic and terrestrial environments. They are also the earliest organisms on this planet which shaped the ecosystem making it fit for an enormous number of diverse life forms including humans. Ironically, on one hand their biodiversity has still not been fully explored by mankind, but on the other hand we are unfortunately losing them from our ecosystem due to human activities. Their destruction is bound to create imbalances in our planet causing irreversible damages to the ecosystem and food chain.

Human ingenuity to utilize the macroalgae or the seaweeds has been well known for several centuries. Whereas the advancement of science and technology has laid emphasis on the identification of micro and macro algal forms, their cultivation and utilization for various intended uses, their utility as sources of food, and as phytochemical factories, is beginning to open up immense opportunities for industrial production. They also have enormous abilities for environmental cleanup of polluted ecosystems. The shift in consumer preferences for nature derived, natural and nature identical molecules has provided a new impetus into research and development to explore the potential of micro and macro algae for food, health, energy and environmental needs.

Realizing the need to provide a focus on this important group of organisms with unlimited potential for multifaceted utilities and industrial applications, we embarked on bringing out two comprehensive volumes for the benefit of researchers, industrialists, entrepreneurs and consumers. The contents of the two volumes offer an overview of the organisms, their distribution in various parts of the globe, the methods to culture them, scale-up technologies, downstream processing, biological activities, regulatory considerations, utilization of biomass in processed foods, feed, health and pharma products, and other novel applications including fuel and environmental aspects.

These aspects have been described in the two volumes and the following points provide the focus at a glance.

VOLUME I

This volume relates to the use of micro and macro algae in food, health and nutraceutical applications.

Applications of algae for polysaccharides such as alginate, agar and most recent trends in utility of fucoidan in food, nutraceuticals and health care products have been highlighted.

Utility of *Spirulina, Chlorella* which are the earliest cultivated microalgal forms have been discussed. *Spirulina* has been termed a "super food" by the WHO and is expected to rule the microalgal market in the future as well.

Antimicrobials from seaweeds are finding utility in various applications in food preservation, cosmetics and health formulations.

The emerging opportunities of volatiles of algae add newer dimensions to the potentials of micro and macro algae.

The extraction methods using ultrasound technology, supercritical carbon dioxide and subcritical water affords recovery of metabolite without the loss of biological activities during downstream processing.

Applications of seaweed for treatment of diabetes is interesting as it opens up newer options of usage as nutraceutical or functional food.

For the vegetarians, the sources of B12 being limited, here is a detailed report from a novel seaweed source, with a possibility of utilization in vegan foods.

Carotenoids such as β-carotene, astaxanthin which are industrially produced pigments from microalgae *Dunaliella* and *Haematococcus* have been presented in detail with newer perspectives of future utility as nutraceutical, cosmeceutical. Fucoxanthin from seaweeds has been dealt with extensively bringing to focus the health applications. Thus, the emerging applications of carotenoids from algal sources in the treatment of diseases such as cancer, ulcers and many more, provide future directions.

A wide range of utilities of micro and macroalgae with special reference to marine forms has been discussed with examples of human trials on the basis of several leads obtained through drug discovery programs leading to the evaluation of bio-efficacies. Their utility in the health foods and nutraceuticals segments is beginning to expand. The market share of algae-based products has been dealt with in a few of the chapters providing global perspectives in bio-business.

The future lies in the exploitation of omics tools to elucidate genetic regulation of the phytochemicals of microalgae for augmented production; also to produce the molecules of interest in newer hosts, which is expected to open up production of novel products and processes.

VOLUME II

This volume relates to the use of micro- and macroalgae in bioremediation, biofuels and global biomass production for commercial applications.

Phycoremediation is receiving attention due to its potential to clean up heavy metals, pollutants and atmospheric carbon dioxide, acting as a sequestration agent and in treatment strategies. The eco-friendly nature and efficiency of the process has been discussed. Also, the utilization of algal biomass in open cultivation, or closed cultivation through photobioreactor systems, for the treatment of contaminants is highlighted.

Algae-based biofuels are being explored globally as a source of alternate fuels. However, the challenges to overcome low productivity are still a big bottleneck. However, emerging strategies to enhance the yields have been discussed. Integration of algal biofuel production with wastewater treatment, carbon-dioxide sequestration and genetic regulation of lipid production pathways has been recommended as a sustainable production strategy. Photobioreactor applications for biofuels would provide adoption of technology independent of environmental conditions and also usage of flue gases in effective manner. The biorefinery approach to produce multiple products is envisaged to provide economically viable technologies.

Application of seaweed as a source of fertilizer, health foods and cosmetics, coupled to their cultivation methodologies for the production of value-added biomass for gainful employment is promising.

Detailed treatise on the cyanobacterial diversity and potential for future applications has thrown open innumerable industrial possibilities.

The scale-up technologies are constantly being innovated for open pond production or closed cultivation in photobioreactors. The current research on the design of reactors for mass cultivation and downstream processing strategies for varied applications have been presented in a few chapters.

Global perspectives of algal biomass production with a few of the major industries across the globe has been highlighted. Case studies with reference to industries in Thailand and Indonesia have been detailed.

The overall objective of these two volumes is to provide up-to-date information and projected future possibilities based on the advancing research and innovations in the diversified facets of phycotechnologies.

The chapters are written by experts from 28 countries. This collective effort from scientists from all over the world represents global perspectives on the topics, which are discussed in a focused manner to bring up to date information, and is largely based on the authors' own experience in working with various systems.

The organization of the chapters in each of the volumes is under the following heads:

Algal Constituents for Food, Health and Disease Applications (Vol I: Chapters 1 through 22)

Algal Genomics and Metabolomics (Vol I: Chapter 23 through 25)

Phycoremediation Applications (Vol II: Chapters 1 through 3)

Algal Biofuels (Vol II: Chapters 4 through 9)

Other Products of Economic Value (Vol II: Chapters 10 through 14)

Mass Production of Microalgae (Vol II: Chapters 15 through 20)

Production of Algal Biomass and Products Worldwide (Vol II: Chapters 21 through 23)

These two volumes will be a treasure trove of information to students and researchers of plant sciences, biological sciences, agricultural sciences, foods and nutrition sciences, health sciences and environmental sciences. However, its application value will impact professionals such as agricultural scientists, food experts, biotechnologists, ecologists, environmentalists and biomass specialists. Its global relevance and outreach have economic implications in industries dealing with foods, nutraceuticals, pharmaceuticals, cosmeceuticals, bioenergy, health care products and bioenergy.

Gokare A. Ravishankar
Ambati Ranga Rao

Acknowledgments

At the outset the editors are extremely thankful to the contributors for their dedication in providing the material in a comprehensive manner for the benefit of readers across the globe. Their kind cooperation in every facet of publication is gratefully acknowledged. The quality of these volumes is attributable to the contributors' commitment to share their knowledge and experience with all those interested in the topic pertaining to basic and applied aspects of phycotechnologies.

We are grateful to Alice Oven, Lara Spieker, Jennifer Blaise, and their team at CRC Press for their diligent efforts in publishing these volumes in an elegant manner.

We are thankful to our families who have extended wholehearted support and encouraged us to take up this task, even though it involved a lot of time away from them. G.A.R. thanks his wife Shyla, son Prashanth, daughter-in-law Vasudha, and daughter Apoorva. A.R.R. thanks his wife Deepika, daughter Jesvisree, parents Venkateswaralu and Tulasidevi, brothers, sisters-in-law, sisters and brothers-in-law.

G.A.R. is thankful to Dr. Premachandra Sagar, Vice Chairman, Dayananda Sagar Institutions, Bengaluru, for granting permission to take this additional responsibility.

A.R.R. is thankful to Dr. L. Rathaiah, Chairman; Mr. L. Sri Krishnadevarayalu, Vice-Chairman; Prof. Dr. K. Ramamurthy Naidu, Chancellor; Dr. M.Y.S. Prasad, Vice-Chancellor; Dr. Madhusudhan Rao, Director, Engineering and Management; Dean Academics, Dean R&D; and Head, Biotechnology Department of Vignan's Foundation for Science, Technology and Research University for providing facility and support to fulfill this new assignment.

We thank the institutions of the Government of India, Department of Biotechnology, Department of Science and Technology and Indian Council of Medical Research, for the grant of financial support to our studies on algal biotechnology done at CSIR-Central Food Technological Research Institute (CFTRI), Mysuru, India.

We thank the staff and students of the Plant Cell Biotechnology Department and collaborating departments at CSIR-CFTRI for the research done by our team, especially, the late Dr. L.V. Venkararaman, Dr. R. Sarada, Dr. M. Mahadevaswamy, Dr. KSMS Raghava Rao, Dr. V. Baskaran, Dr. Shylaja Dharmesh, T.R. Shamala, Dr. K. Udaya Sankar and many non-technical staff, including K. Shivanna, C.V. Venkatesh, H.S. Jaya in many of our algal projects. Also, thanks are due to a number of research students and associates of G.A.R. for their dedication to research and authorship in a large number of publications.

G.A.R. is proud to have associated with doyens of Algal Biotechnology, Late Dr. L.V. Venkataraman (LVV) of CFTRI, Mysore and Late Dr. E. Wolfgang Becker of Eberhard-Karls-Universitat, Tubingen, Germany, who started the Algal Projects at CFTRI, with the support of GTZ (Gesellschaft für Technische Zusammenarbeit), Germany. This laboratory at CFTRI gained prominence as world renowned center for algal biotechnology research.

G.A.R. is also thankful to the Department of Biotechnology Government of India for the National Technology Day Award conferred on him and LVV on May 11, 2003 for commercialization of *Spirulina* Technology in India.

A.R.R. is thankful to IUFoST (Canada) and TWAS-CAS for being honored as a young affiliate for his work on Astaxanthin from *Haematococcus*.

Gokare A. Ravishankar
Ambati Ranga Rao

Editors

Gokare A. Ravishankar is presently the vice president of research and development (R&D) in life sciences and biotechnology at Dayananda Sagar Institutions, Bengaluru, India; he is a professor of biotechnology. Earlier, he had a distinguished research career of more than 30 years working at the Central Food Technological Research Institute (CFTRI), Mysuru, India, and in the institutions of government of India. He served as chairman of the board of studies in biotechnology at the Visvesvaraya Technological University, Belgavi, India, and as academic council member of Dayananda Sagar University. He has also been member of the boards of eight universities. He is an internationally recognized expert in the areas of plant biotechnology, algal biotechnology, food biotechnology and postharvest technologies, plant secondary metabolites, functional foods, herbal products, genetic engineering, and biofuels, and served as visiting professor to universities in Japan, Korea, Taiwan, and Russia.

Dr. Ravishankar earned a master's and PhD degree from Maharaja Sayajirao University of Baroda, India. He mentored more than 40 PhD students, 62 master's students, 7 postdocs, and 8 international guest scientists; he has authored more than 260 peer-reviewed research papers in international and national journals, 48 reviews, 55 patents in India and abroad, edited 3 books, with a h-index of 59. He has presented more than 200 lectures in various scientific meetings in India and abroad, including visits to more than 25 countries.

Dr. Ravishankar has received several coveted honors and awards: Young Scientist award (Botany) by the then prime minister of India in 1992; National Technology Day Award of Government of India in 2003; Laljee Godhoo Smarak Nidhi Award for food biotechnology R&D of industrial relevance; the prestigious Professor V. Subramanyan Food-Industrial Achievement Award; Professor S.S. Katiyar Endowment Lecture Award in New Biology by Indian Science Congress; Professor Vyas Memorial Award of Association of Microbiologists of India; Professor V.N. Raja Rao Endowment Lecture Award in Applied Botany, University of Madras, India; Lifetime Achievement Award by the Society of Applied Biotechnologists; Dr. Diwaker Patel Memorial Award by Anand Agricultural University, India; Prof C.S. Paulose Memorial Oration Award by Society for Biotechnologists of India.

He is honored as fellow of several national organizations in India, viz. National Academy of Sciences; National Academy of Agricultural Sciences; Association of Microbiologists of India; Society of Agricultural Biochemists; Society for Applied Biotechnology; Indian Botanical Society; and the Association of Food Scientists and Technologists of India. He has held honorary positions of President of the Society of Biological Chemists, Mysore Chapter, and President of Association of Microbiologists of India, Mysore and Bangalore Chapters.

Several premier international bodies have also honored him with fellowships, viz., the International Academy of Food Science and Technology (Canada); the Institute of Food Technologists (USA); the Institute of Food Science and Technology (UK); and the Certified Food Scientists of the United States. Dr. Ravishankar is a very active member of the task forces of several organizations of the government of India and has served an expert in the selection committees for the appointment of professors, scientists, and research students in the universities, as well as R&D institutions. He has also served as advisor and resource at international conferences, seminars, workshops, and short courses; he has convened national and international seminars in biology, biotechnology, and food science and technology. He is an associate editor and reviewer of a large number of reputed research journals.

****

Ambati Ranga Rao is a scientist and assistant professor in the Department of Biotechnology at Vignan's Foundation for Science, Technology, and Research (Deemed to be University), Andhra Pradesh, India. Dr. Ambati earned bachelor's and master's degrees from Acharya Nagarjuna University, Andhra Pradesh, India, and a PhD degree from University of Mysore, India. He started his research career in 2004 as a research assistant at the Department of Plant Cell Biotechnology, Council of Scientific and Industrial Research (CSIR) Central Food Technological Research Institute (CFTRI), Mysuru, India, under the supervision of Dr. G. A. Ravishankar and Dr. R. Sarada. He was awarded Senior Research Fellow of Indian Council of Medical Research (ICMR), New Delhi, in the year 2007. His PhD work at CFTRI focused on the production of astaxanthin from cultured green alga *Haematococcus pluvialis*, and its biological activities. He worked extensively on process optimization of algal biomass production; mass culture of various algal species in raceway ponds and photobioreactors; downstream processing of algal metabolites and evaluation of their possible nutraceutical applications in *in vitro* and *in vivo* models. Further, Dr. Ambati worked as lead scientist in Algal Technologies, Carot Labs Pvt Ltd, India; postdoctoral research associate in Laboratory of Algal Research and Biotechnology, Arizona State University, under the supervision of Prof. Milton Sommerfeld and Qiang Hu; visiting assistant professor in Food Science and Technology Programme, Beijing Normal University–Hong Kong Baptist University, United International College, China, under the supervision of Prof. Bo Lei; visiting senior research fellow in the Institute of Ocean and Earth Sciences, University of Malaya, Malaysia, under the guidance of Prof. Phang Siew Moi. He is the author of more than 30 peer-reviewed publications, 36 international/national conference papers (including invited talks), 5 reviews, and 9 chapters in books. His research citations exceed 1,600, with h-Index (12), and i-index (14) as per Google Scholar. He has attended international and national conferences/symposia in the United States, Canada, Brazil, China, Malaysia, Indonesia, and Oman. Based on his research accomplishments, he is honored TWAS-Young Affiliate (2014) by Regional Office of East South-East Asia and the Pacific, Chinese Academy of Sciences (CAS), China; Carl Storm International Diversity Fellowship Award (2010) by Gordon Research Conferences. He is a fellow of the Society for Applied Biotechnology (2013), India. He has received research grants and travel grant fellowships as both international and national awards, under young scientist schemes. He also serves as editorial board member and reviewer for reputed international and national journals.

****

Contributors

Slim Abdelkafi
Department of Biological Engineering
University of Sfax
Sfax, Tunisia

Mona Ahmed J. Alzahrani
School of Chemical Sciences, Food Science
　Programme
University of Auckland
Auckland, New Zealand

Monique G. das Chagas Faustino Alves
Department of Biochemistry
Federal University of Rio Grande do Norte
Natal, Brazil

Raquel Amaral
Coimbra Collection of Algae
Department of Life Sciences
University of Coimbra
Coimbra, Portugal

Natasha Amjad
Department of Diet and Nutritional Sciences
Faculty of Health and Allied Sciences
Imperial College of Business Studies
Lahore, Pakistan

Sukumaran Anil
Department of Dentistry
Hamad Medical Corporation
Qatar

Mariana F.G. Assuncao
Coimbra Collection of Algae
Department of Life Sciences
University of Coimbra
Coimbra, Portugal

Shama Aumeerun
Department of Agriculture and Food Sciences
Faculty of Agriculture
University of Mauritius
Moka, Mauritius

Ana Paula Batista
UMR 1208 IATE Agro-Polymer Engineering and
　Emerging Technologies
INRA
Montpellier, France

Biswanath Bhunia
Department of Bioengineering
National Institute of Technology
Agartala, India

Tomohiro Bito
Department of Agricultural, Life and Environmental
　Sciences
Faculty of Agriculture
Tottori University
Tottori, Japan

Shen Boxiong
School of Energy and Environmental Engineering
Hebei University of Technology
Tianjin, China

Masood Sadiq Butt
National Institute of Food Science and Technology
Faculty of Food, Nutrition and Home Sciences
University of Agriculture
Faisalabad, Pakistan

Luiza Sheyla Evenni P. Will Castro
Department of Biochemistry
Federal University of Rio Grande do Norte
Natal, Brazil

Sagnik Chakraborty
School of Energy and Environmental Engineering
Hebei University of Technology
Tianjin, China

Wan-Loy Chu
School of Postgraduate Studies
International Medical University
Kuala Lumpur, Malaysia

Clemente-Carazo M
Departamento de Ingeniería de Alimentos y del
　Equipamiento Agrícola
Instituto de Biotecnología Vegetal
Universidad Politécnica de Cartagena
Spain

Condon-Abanto S
Grupo de Nuevas Technologías de Conservación de
　Alimentos
Facultad de Veterinaria, Universidad de Zaragoza
Spain

Cedric Delattre
Institut Pascal, SIGMA Clermont
Clermont Auvergne University
Clermont-Ferrand, France

Paola Del Serrone
Council for Agricultural Research and Economics CREA
Research Centre for Zootechnic and Aquaculture CREA ZA
Rome, Italy

Indrama Devi
DBT Institutes of Bioresources and Sustainable Development
Manipur, India

Gholamreza Djelveh
Institut Pascal, SIGMA Clermont
Clermont Auvergne University
Clermont-Ferrand, France

Tanase Dobre
Politehnica University of Bucharest
Bucharest, Romania

Herminia Domínguez
Departamento de Enxeñería Química
Universidade de Vigo
Ourense, Spain

Celina Maria Pinto Guerra Dore
Department of Biochemistry
Federal University of Rio Grande do Norte
Natal, Brazil

Marie Francoise Driver
Department of Agricultural Production and Systems
Faculty of Agriculture
University of Mauritius
Moka, Mauritius

Murat Elibol
Ege University
Department of Bioengineering
Izmir, Turkey

Elena Falqué
Departamento de Química Analítica y Alimentaria
Universidade de Vigo
Ourense, Spain

Luciana Guimaraes Alves Figueira
Department of Biochemistry
Federal University of Rio Grande do Norte
Natal, Brazil

Tonmoy Ghosh
Omics of Algae
International Centre for Genetic Engineering and Biotechnology
New Delhi, India

Joyce Soulange-Govinden
Department of Agriculture and Food Sciences
Faculty of Agriculture
University of Mauritius
Moka, Mauritius

Ceren Gurlek
Ege University
Department of Bioengineering
İzmir, Turkey

Nidhi Hans
Centre of Rural Development and Technology
Indian Institute of Technology
New Delhi, India

Faiez Hentati
Institut Pascal, SIGMA Clermont
Clermont Auvergne University
Clermont-Ferrand, France

Eduardo Jacob-Lopes
Food Science and Technology Department
Federal University of Santa Maria
Santa Maria, Brazil

Naveen Jayapal
Department of Biochemistry
CSIR-CFTRI
Mysore, India

Pavan P. Jutur
Omics of Algae
International Centre for Genetic Engineering and Biotechnology
New Delhi, India

Se-Kwon Kim
Department of Marine Life Sciences
Korean Maritime and Ocean University
Korea

Contributors

Young-Woo Kim
College of Oriental Medicine
Daegu Haany University
Daegu, South Korea

Ayse Kose
Ege University
Department of Bioengineering
İzmir, Turkey

Edda Lisboa Leite
Department of Biochemistry
Federal University of Rio Grande do Norte
Natal, Brazil

Iriani R. Maldonade
Laboratorio de Ciência e Tecnologia de Alimentos
Embrapa Hortaliças
Brasilia, Brazil

Marco Garcia-Vaquero
School of Veterinary Medicine
University College Dublin
Ireland

Anushree Malik
Centre of Rural Development and Technology
Indian Institute of Technology
New Delhi, India

Philippe Michaud
Institut Pascal, SIGMA Clermont
Clermont Auvergne University
Clermont-Ferrand, France

Suhaila Mohamed
Institute of Bioscience
Universiti Putra Malaysia
Selangor, Malaysia.

Abhijit Mondal
Department of Chemical Engineering
National Institute of Technology
Agartala, India

Satya Narayan Naik
Centre of Rural Development and Technology
Indian Institute of Technology
New Delhi, India

Asha A. Nesamma
Omics of Algae
International Centre for Genetic Engineering and Biotechnology
New Delhi, India

Marcello Nicoletti
Department of Environmental Biology
University Sapienza of Rome, Italy

Hudaa Neetoo
Department of Agriculture and Food Sciences
Faculty of Agriculture, University of Mauritius
Moka, Mauritius

J. Paniagua-Michel
Department of Marine Biotechnology
Centro de Investigación Científica y de Educación Superior de Ensenada
Ensenada, Mexico

Suphi S. Oncel
Ege University
Department of Bioengineering
İzmir, Turkey

Izel Oral
Ege University
Department of Bioengineering
İzmir, Turkey

Chetan Paliwal
Omics of Algae
International Centre for Genetic Engineering and Biotechnology
New Delhi, India

Siew-Moi Phang
Institute of Biological Sciences
University of Malaya
Kuala Lumpur, Malaysia

and

Institute of Ocean and Earth Sciences
University of Malaya
Kuala Lumpur, Malaysia

Conrad O. Perera
School of Chemical Sciences
Food Science Programme
University of Auckland
Auckland, New Zealand

María José Pérez
Departamento de Biología Funcional y Ciencias de la Salud
Universidade de Vigo
Ourense, Spain

Madan Kumar Perumal
Department of Biochemistry
CSIR-CFTRI
Mysore, India

Guillaume Pierre
Institut Pascal, SIGMA Clermont
Clermont Auvergne University
Clermont-Ferrand, France

Pricila Nass Pinheiro
Food Science and Technology Department
Federal University of Santa Maria
Santa Maria, Brazil

Vitor H. Pomin
Department of BioMolecular Sciences
Research Institute of Pharmaceutical Sciences
School of Pharmacy
University of Mississippi
University, Mississippi

Sneha Rao
Yenepoya Research Center
Yenepoya University
Deralakatte, Mangalore
India

Claire Remacle
Laboratory of Genetics and Physiology of Microalgae,
Institute of Botany
University of Liège
Liège, Belgium

Delia B. Rodriguez-Amaya
Faculty of Food Engineering,
University of Campinas,
Sao Paulo, Brazil

Sanchez V
Grupo de Análisis y Simulación de Procesos Alimentarios
Departamento de Tecnología de los Alimentos
Universitat Politècnica de València
Spain

Andriéli Borges Santos
Food Science and Technology Department
Federal University of Santa Maria
Santa Maria, Brazil

Lília M.A. Santos
Coimbra Collection of Algae
Department of Life Sciences
University of Coimbra
Coimbra, Portugal

Hafiz Ansar Rasul Suleria
Department of Agriculture and Food Systems
The University of Melbourne
Victoria, Australia

Fei Teng
Department of Food Quality and Safety
College of Food Science
Northeast Agricultural University
Heilongjiang, China

Bogdan Trica
Politehnica University of Bucharest
Bucharest, Romania

Chiara Toniolo
Department of Environmental Biology
University Sapienza of Rome, Italy

Onkar Nath Tiwari
Centre for Conservation and Utilisation of Blue Green Algae
Division of Microbiology, ICAR-Indian Agricultural Research Institute
New Delhi, India

Iahtisham-Ul-Haq
Department of Diet and Nutritional Sciences
Faculty of Health and Allied Sciences
Imperial College of Business Studies
Lahore, Pakistan

Alina V. Ursu
Institut Pascal, SIGMA Clermont
Clermont Auvergne University
Clermont-Ferrand, France

Baskaran Vallikannan
Department of Biochemistry
CSIR-CFTRI
Mysore, India

Ariana A. Vasconcelos
Program of Glycobiology
Institute of Medical Biochemistry Leopoldo de Meis
Federal University of Rio de Janeiro
Rio de Janeiro, Brazil

Jayachandran Venkatesan
Yenepoya Research Center
Yenepoya University
Deralakatte, Mangalore
India

Karem Rodrigues Vieira
Food Science and Technology Department,
Federal University of Santa Maria
Santa Maria, Brazil

Fumio Watanabe
Department of Agricultural, Life and Environmental Sciences
Faculty of Agriculture
Tottori University
Tottori, Japan

Wedamulla N.E.
Department of Export Agriculture
Faculty of Animal Science and Export Agriculture
Uva Wellassa University
Badulla, Sri Lanka.

W.A.J.P Wijesinghe
Department of Export Agriculture
Faculty of Animal Science and Export Agriculture
Uva Wellassa University
Badulla, Sri Lanka.

Cagla Yarkent
Ege University
Department of Bioengineering
İzmir, Turkey

Iqra Yasmin
Department of Diet and Nutritional Sciences
Faculty of Health and Allied Sciences
Imperial College of Business Studies
Lahore, Pakistan

Leila Queiroz Zepka
Food Science and Technology Department
Federal University of Santa Maria
Santa Maria, Brazil

Section I

Algal Constituents for Food, Health and Disease Applications

1 Algal Polysaccharides
Innovative Extraction Technologies, Health Benefits and Industrial Applications

M. Clemente-Carazo, V. Sanchez, S. Condon-Abanto, and Marco García-Vaquero

CONTENTS

Abbreviations 3
Introduction 3
Extraction of Microalgal Polysaccharides 4
   Extraction of Extra-Cellular Polysaccharides 5
   Extraction of Cellular Polysaccharides 6
Purification of Microalgal Polysaccharides 7
Industrial Applications and Prospects of Microalgal Polysaccharides 7
Acknowledgments 9
References 9

BOX 1.1 SALIENT FEATURES

Microalgae offer certain advantages for the production and commercialisation of carbohydrates over other traditional sources such as the possibility to excrete extra-cellular polymers to the culture media together with the production of other cell wall and intracellular compounds. Due to the wide heterogeneity and industrial potential of carbohydrates, the extraction and purification strategies used to obtain these molecules from marine organisms have recently gained a great deal of attention in the scientific community. This chapter summarises the current methodological approaches (traditional or innovative technologies) used to extract microalgal carbohydrates together with the main biological activities described in the literature for both extra-cellular and cellular polysaccharides. The purification of carbohydrates for future commercialisation is also covered together with key aspects of the current market of microalgal compounds and the future prospects of this industry.

ABBREVIATIONS

$CaCl_2$: calcium chloride
DHA: docosahexaenoic acid
EPA: eicosapentaenoic acid:
PUFA: polyunsaturated fatty acids
NaCl: sodium chloride
NaOH: sodium hydroxide

INTRODUCTION

There is an increased interest in high-value products from algae due to the wide variety of compounds (i.e. proteins, carbohydrates and pigments) with potential applications in food/feed, cosmeceutical and pharmaceutical industries (Deniz, García-Vaquero and Imamoglu 2018, García-Vaquero 2019b, García-Vaquero, Lopez-Alonso and Hayes 2017, Hayes et al. 2019, Miranda, Lopez-Alonso and García-Vaquero 2017). Microalgae offer certain advantages for the production and commercialisation of carbohydrates over other traditional sources of these compounds. Microalgae are capable of excreting extra-cellular polymers to the culture media at different stages during their biological cycle (i.e. sulphated and non-sulphated extra-cellular polysaccharides). Moreover, the microalgal cell wall is also rich in polysaccharides (i.e. sulphated galactan hetero-polysaccharides) and the cells produce intra-cellular carbohydrates involved in different cellular processes (i.e. starch – amylose and amylopectin – and other sulphated compounds) (Arad and Levy-Ontman 2010, Singh, Kate and Banerjee 2005, Williams and Laurens 2010). The chemical composition of microalgal polysaccharides (degree of sulphation, monosaccharide composition and linkages, type and number of chains) is influenced by the algae species and strain (de Jesus Raposo, de Morais and de Morais 2014a). The chemical structure of the building blocks conforming polysaccharides in *Porphyridium* sp. are presented in Figure 1.1. Also, the culture conditions of microalgae (i.e. nutrients, light and temperature conditions) could be optimised to increase the production of the polysaccharides and other high-value compounds (du Plooy et al. 2015, Guihéneuf and Stengel 2015). For instance, the addition of glyoxylate to the media increased the cells' metabolism and excretion of intra- and extra-cellular polysaccharides (Liu et al. 2010).

Due to the wide heterogeneity and industrial potential of carbohydrates, the extraction and purification strategies used to obtain these compounds from marine organisms, such as seaweed and microalgae, have recently gained a great deal of attention in the scientific community, aiming to achieve high yields of compounds while preserving intact their biological properties (García-Vaquero et al. 2018, García-Vaquero and Hayes 2016, García-Vaquero, et al., 2017, Michalak and Chojnacka 2014). This chapter summarises the current methodological approaches used to extract and purify carbohydrates from microalgae and the prospects for the industrial exploitation of microalgal polysaccharides.

EXTRACTION OF MICROALGAL POLYSACCHARIDES

Extraction techniques of carbohydrates should ideally achieve high yields of compounds, preserve the nature of the products and co-products, minimise energy consumption, generate minimum waste and be suitable for scaling-up the process to industry (García-Vaquero et al. 2018, Michalak and Chojnacka 2014). Microalgae are capable of excreting extra-cellular polymers and accumulate cellular carbohydrates (cell wall and intracellular compounds). There are similarities, but also substantial differences, between the extraction methodologies described for extra-cellular and cellular compounds. Extra-cellular polysaccharides are released to the media

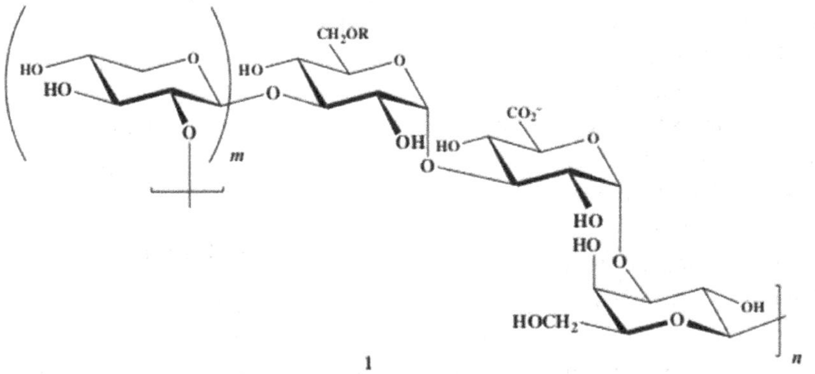

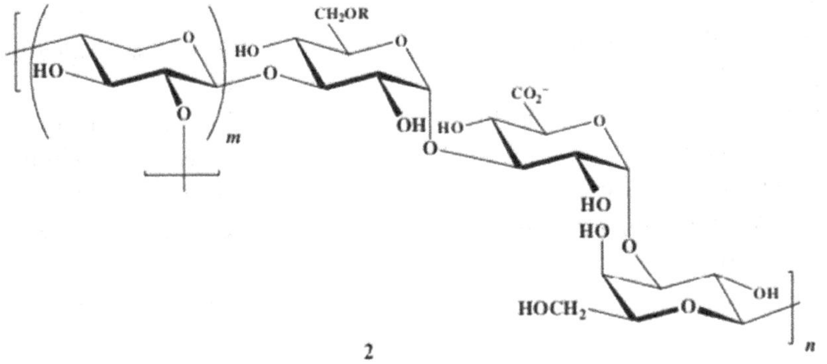

FIGURE 1.1 Image from Geresh et al. (2009) showing the chemical structure of a building block of polysaccharides from *Porphyridium* sp. Image reprinted with permission from Elsevier.

during the culture of algae, and thus, the first steps in the extraction process include the separation or harvesting of the algal biomass followed by the extraction techniques of choice. The remaining polysaccharides (cell wall and intracellular polysaccharides) require further extraction steps to achieve the desired algal extracts such as preparation and pretreatments of the microalgal biomass, cell wall disruption techniques and precipitation of the carbohydrates of interest.

EXTRACTION OF EXTRA-CELLULAR POLYSACCHARIDES

Extra-cellular polysaccharides show great potential due to the wide variety of applications and the easy and environmentally friendly production of these compounds (Öner 2013). The production of extra-cellular polysaccharides was reported in several organisms including fungi, yeasts (Duan et al. 2008, Zou, Sun and Guo 2006) and several microalgae species such as *Porphyridium* and *Chlorella* (Guzman-Murillo and Ascencio 2000).

The methods used to obtain extra-cellular polysaccharides from microalgae in the recent literature are summarised in Table 1.1.

In general, most studies include one or two steps of centrifugation and/or filtration to remove the algal biomass from the media containing the polysaccharides of interest. The wide majority of researches in the recent literature applied solvents at low temperatures to concentrate

TABLE 1.1
Summary of the Extraction Methods Used to Obtain Extra-Cellular Polysaccharides from Microalgae and the Potential Use or Bioactivity of These Compounds

Microalgae sp.	Extraction Method	Potential Use or Bioactivity	References
Porphyridium cruentum, *Chaetoceros* sp., *Chlorella autotrophica*, *Chlorella capsulate*, *Chlorella* sp., *Dunalliela tertiolecta*, *Isochrysis galbana*, *Isochrysis* sp., *Nannochloropsis oculata*, *Phaeodactylum tricornutum*, *Rodhosorusmarinus*, *Tetraselmis* sp., *Tetraselmis suecica*, *Botryococcussudeticus*, *Botryococcus braunii*; *Chlamydomonas Mexicana*, *Chlorococcum oleofaciens*, *Dysmorphococcusglobosus*, *Hormotilopsisgelatinosa*, *Neochloris oleoabundans* and *Ochromonasdanica*	Centrifugation followed by filtration (0.45 μm), heating (40°C, 1h) and precipitation with cetylpyridinium chloride. The precipitate collected by centrifugation was precipitated with $CaCl_2$, and the pellet washed with ethanol, ethanol:ether and ether.	Antibacterial (inhibition of the cytoadhesion process of *Helicobacter pylori* to HeLa S3 cells).	Guzman-Murillo and Ascencio (2000)
Gyrodinium impudicum KG03	Centrifugation followed by ethanol precipitation, cetyltrimethylammonium bromide re-precipitation. The pellet collected was re-suspended in NaCl, precipitated with ethanol and dialysed.	Antiviral (encephalomyocarditis virus) and antitumour.	Bae et al. (2006), Yim et al. (2004)
Porphyridium sp.	Centrifugation followed by precipitation of the supernatant with NaOH and different solvents (HCl, NaCl, ethanol, cetyltrimethylammonium bromide and acetone).	–	Geresh et al. (2009)
Amphora sp., *Ankistrodesmus angustus* and *Phaeodactylum tricornutum*	Centrifugation to separate the cells followed by filtration, crossed-flow ultrafiltration and stirred cell diafiltration of the supernatant.	–	Chen et al. (2011)
Porphyridium cruentum	Diafiltration (tangential flow filtration with 300 kDa membrane).	–	Patel et al. (2013)
Porphyridium cruentum	2 step concentration-diafiltration with different molecular weight cut-off membranes.	–	Marcati et al. (2014)
Trebouxia sp. from lichen *Ramalina farinacea*	Centrifugation followed by ethanol precipitation 2 times of the supernatant.	–	Casano et al. (2015)

the carbohydrates without damaging their properties (de Jesus Raposo, de Morais and de Morais 2014b, Guzman-Murillo and Ascencio 2000, Parikh and Madamwar 2006), followed by one or several precipitation steps with cationic surfactants including cetyltrimethylammonium bromide (Bae et al. 2006, Geresh et al. 2009, Yim et al. 2004) or cetylpyridinium chloride (Guzman-Murillo and Ascencio 2000); organic solvents such as ethanol (Bae et al. 2006, Casano et al. 2015, de Jesus Raposo, de Morais and de Morais 2014b, Díaz Bayona and Garcés 2014, Magaletti et al. 2004, Pletikapić et al. 2011, Urbani et al. 2005, Yim et al. 2004, Yim et al. 2007), methanol (Mishra and Jha 2009), acetone (Parikh and Madamwar 2006) ether (Geresh et al. 2009); or a combination of several of these solvents in different proportions (Geresh et al. 2009, Guzman-Murillo and Ascencio 2000). After these solvent precipitation steps, the extracts containing polysaccharides were dialysed to remove salts or small components followed by a lyophilisation step to preserve the compounds (see multiple references in Table 1.1).

Novel technologies are being used for extracting or fractionating carbohydrates. Membrane filtration techniques have been successfully used to fractionate multiple compounds from milk (Brans et al. 2004), bacteria (Delattre et al. 2005) or plants (Wan, Prudente and Sathivel 2012). More recently, membrane techniques have been used to extract and purify extra-cellular polysaccharides from a wide variety of microalgae species such as *Amphora* sp., *Ankistrodesmus angustus*, *Phaeodactylum tricornutum*, *Graesiella emersonii*, *Graesiella vacuolata* and *Porphyridium cruentum* (Chen et al. 2011, Marcati et al. 2014, Mezhoud et al. 2014, Patel et al. 2013, Zhang and Santschi 2009).

Filtration techniques seem well suited for initial extraction and purification steps at industrial scale as they can be automated; and permit to treat large volumes of samples (Patel et al. 2013). However, the widespread application of membranes has been hindered due to excessive membrane fouling which could result in reduced performance, severe flux decline, high energy consumption and frequent membrane cleaning or replacement (Feng et al. 2009). Recent studies focused on achieving a better understanding of anti-fouling agents (Feng et al. 2009) and other strategies to reduce fouling. For example the addition of an initial high-molecular weight cut-off membrane before the ultrafiltration step reduces the fouling of the membranes when extracting bacterial oligosaccharides (Mellal et al. 2008).

In the case of extra-cellular polysaccharides from *Porphyridium cruentum*, the tangential flow filtration (diafiltration) with a 300 kDa membrane showed better extraction, purification and desalting efficiency of high-molecular weight extra-cellular polysaccharides than other methods such as dialysis and solvent precipitation with methanol, ethanol and isopropanol (Patel et al. 2013). However, a significant proportion of cell-attached polymers were not extracted by this method. The application of a two-step membrane process with ultrafiltration and diafiltration through 300 kDa molecular weight cut-off, followed by ultrafiltration and diafiltration with a 10 kDa membrane, showed to be highly efficient for extracting high and low molecular weight extra-cellular polysaccharides and other molecules such as phycoerythrin (Marcati et al. 2014).

Other novel extraction techniques to generate polysaccharide extracts, such as ultrasonication and microwave technologies are used to breakdown extra-cellular polysaccharides, but not with the purposes of extraction (Sun et al. 2009, Sun, Wang and Zhou 2012). Ultrasonication and microwave technologies could be important tools during the development of pharmaceutical products from polysaccharides. Most products from marine origin with pharmaceutical and cosmeceutical properties currently on the market and approved by the Food and Drug Administration or the European Medicines Agency are modifications of the natural molecules obtained during the processes of optimisation and drug development (Martins et al. 2014).

Extraction of Cellular Polysaccharides

Less information is available concerning the extraction of cellular polysaccharides from microalgae. Traditionally these polymers were not extracted and purified to obtain high-value products. The majority of the scientific literature related to cellular polysaccharides focuses on general extraction of carbohydrates or fermentation of the biomass from oil extraction to produce bioethanol (Behera et al. 2014). Due to the low lignin and hemicellulose contents, microalgae have been considered a suitable source of polysaccharides for bioethanol production as an alternative to conventional crops such as corn and soy bean (Behera et al. 2014, Chaudhary et al. 2014). Several microalgae species have shown promising results in this field. For example, *Chlorococcum* produced high yields of bioethanol (Harun et al. 2011, Harun and Danquah 2011), at levels comparable or even higher than those obtained from traditional crops like rice straw or sugar cane (Behera et al. 2014).

The downstream processing steps required to produce microalgal cellular carbohydrates include energy and cost intensive steps such as harvesting and concentration of the biomass such as filtration, flocculation, centrifugation and sonication (Liang 2015, Vandamme, Foubert and Muylaert 2013) followed by other processes to dry the biomass using spray-, drum-, freeze- and sun-drying techniques (Behera et al. 2014). Some have avoided the extra energy consumption to dry the biomass if there are no preservation issues or the extraction processes are not affected by moisture. The extraction process normally starts with different pretreatments to eliminate pigments, lipids or proteins (Casano et al. 2015, Cheng, Labavitch and VanderGheynst 2015, Sadovskaya et al. 2014) followed by different methods designed to breakdown the cell walls and precipitate the polysaccharides of interest for further purification and characterisation (see Table 1.2).

The breakdown of the cell walls is normally achieved by conventional strategies such as solubilisation of biomass in water (Casano et al. 2015, Lee et al. 1998, Lee et al. 2000, Mader et al. 2016) or solutions of $CaCl_2$ (Sadovskaya et al. 2014) and ethanol (Balavigneswaran et al. 2013, Pugh et al. 2001). Recently, new technologies have been used for this purpose such as sonication (alone or in combination with heat treatment) (Guzmán et al. 2003, Jo et al. 2010, Sun et al. 2014, Sun et al. 2016) and subcritical water extraction (Chakraborty et al. 2012). Other innovative extraction processes include the use of enzymes – i.e. proteases (ProtamexTM) (Jo et al. 2010) and α-amylase and amyloglucosidase (Cheng et al. 2011, Cheng, Labavitch and VanderGheynst 2015, Fu et al. 2010). Although promising, the enzymatic treatments do not have industrial applications to date due to the high cost and time required by these processes (Michalak and Chojnacka 2014).

After breaking down the cell walls, the dissolved polymers can be precipitated by using membrane filtration systems (Guzmán et al. 2003, Jo et al. 2010) and different solvents such as ethanol (Balavigneswaran et al. 2013, Casano et al. 2015, Chakraborty et al. 2012, Sun et al. 2014, Sun et al. 2016), methanol (Jo et al. 2010), acetone or ether (Sun et al. 2014). Independently of the use or not of pretreatments, some authors also describe techniques to remove undesired compounds (mainly lipids and proteins) at this stage, i.e. trichloroacetic acid (Lee et al. 1998, Lee et al. 2000, Mader et al. 2016), chloroform and butanol (Sadovskaya et al. 2014) and octanol in chloroform (Sun et al. 2014).

PURIFICATION OF MICROALGAL POLYSACCHARIDES

After the extraction procedures, extra-cellular or cellular polysaccharides could go through one or several purification steps that do not differ to those described for other carbohydrates. The purification techniques commonly used in microalgal polysaccharides include the application of different solvents (hexane, ethyl acetate, acetone, ethanol, petroleum ether, methanol and water) to separate the molecules with respect to their polarity (Challouf et al. 2011, Jo et al. 2010). Other chromatographic techniques include ion-exchange chromatography (Guzmán et al. 2003, Huheihel et al. 2002, Huleihel et al. 2001, Sadovskaya et al. 2014, Sun et al. 2014) and gel permeation chromatography (Bae et al. 2006, Casano et al. 2015, Feng et al. 2009, Guzmán et al. 2003, Huleihel et al. 2001, Magaletti et al. 2004, Yim et al. 2004). The use of chromatographic and non-chromatographic techniques to purify and characterise carbohydrates was recently reviewed in detail by García-Vaquero (2019a).

INDUSTRIAL APPLICATIONS AND PROSPECTS OF MICROALGAL POLYSACCHARIDES

Microalgae are traditionally commercialised as full dried biomass mainly for human food and animal feed applications in the form of powders, tablets, capsules or liquids (Spolaore et al. 2006). The main microalgae species commercialised as whole dried biomass are *Spirulina* sp. and *Chlorella* sp. with an approximate turnover of 80 million USD per year (Vigani et al. 2015). Other high-value products obtained from microalgae are currently in the market, with important applications in the food, pharmaceutical, cosmeceutical and energy industries (Ventura et al. 2018). The high-value compounds commercialised from microalgae include pigments (i.e. asthaxanthin, ß-carotene), PUFA [mainly docosahexaenoic acid (DHA) and eicosapentaenoic acid (EPA)] and proteins (phycobiliproteins) (Ventura et al. 2018).

Microalgae are also a promising source of other valuable compounds such as polysaccharides. For instance, extra-cellular and sulphated polysaccharides extracted from *Porphyridium*, *Chlorella* and *Spirulina* sp. namely alguronic acid are used as cosmeceutical, nutraceutical and pharmaceutical (Borowitzka 2013, Laurienzo 2010, Raposo, de Morais and Bernardo de Morais 2013). Recently, microalgal polysaccharides have dragged the attention of the scientific community due to the wide range of biological properties of these compounds (see Tables 1.1 and 1.2). Several bioactivities from microalgal polysaccharides include antiviral, antibacterial, antioxidant, anti-inflammatory, immunomodulatory, antitumour, anti-lipidemic, anti-glycemic, anticoagulant, anti-thrombotic, bio-lubricant and anti-adhesive properties (de Jesus Raposo, de Morais and de Morais 2014a). Moreover, extra-cellular polysaccharides produced by cyanobacteria have also been used as soil conditioners, improving the water

TABLE 1.2
Summary of the Extraction Methods Used to Obtain Cellular Polysaccharides from Microalgae and the Potential Use or Bioactivity of These Compounds

Microalgae sp.	Pretreatment	Extraction Method	Potential Use or Bioactivity	References
Tetraselmis suecica	–	Dried algae was suspended in water, sonicated and incubated with a protease enzyme (50°C, 3h). After centrifugation, the supernatant was ultrafiltrated.	Anti-inflammatory.	Jo et al. (2010)
Chlorella sorokiniana	–	Subcritical water extraction (2 times at 160°C and 180°C) followed by ethanol precipitation and lyophilisation of supernatant.	–	Chakraborty et al. (2012)
Isocrysis galbana	Dried biomass with 70% ethanol (85°C, 2h, 3 times)	Ethanol precipitation, filtration, acetone:chloroform washes (3 times), hot water extraction and ethanol precipitation.	Antioxidant (hydrogen peroxide, DPPH and superoxide anion).	Balavigneswaran et al. (2013)
Chlorella vulgaris, Chlorella sorokiniana, Chlorella minutissima and *Chlorella variabilis*	Dried biomass treated with chloroform:methanol (3 times). Precipitate washed in NaCl to remove proteins followed by acetone (85°C, 3 times) to remove starch.	Pretreated pellet was incubated with α-amylase and amyloglucosidase (37°C, overnight) and the cell walls were collected.	–	Cheng, Labavitch, and VanderGheynst (2015), Cheng et al. (2011)
Isochrysis galbana	–	Dried algae were re-suspended in water, sonicated and heated (70°C, 180 min). Supernatant was concentrated with ethanol. Pellet was washed with ethanol, acetone and ether and treated with Sevag method (remove proteins) and dialysis.	Antioxidant.	Sun et al. (2014)
Trebouxia sp. from lichen *Ramalina farinacea*	Dried biomass was pressure-treated (1200 psi).	Pellet was re-dissolved and extracted with ethanol, chloroform:methanol and acetone (different times and temperatures) followed by boiling water and ethanol precipitation.	–	Casano et al. (2015)
Pavlova viridis	–	Dried algae re-suspended in water, sonicated (65°C). Supernatant was concentrated, treated with trichloroacetic acid (precipitate proteins) and supernatant dialysed and precipitated with ethanol several times.	Immunomodulation and antitumour.	Sun et al. (2016)
Isochrysis galbana	Dried algae mixed with ethanol (different % and temperatures, 2 times).	Pretreated pellet was extracted with $CaCl_2$ (boiling temperature, 3 times). Pooled supernatants were ethanol precipitated and pellet washed with chloroform (remove proteins and lipophilic materials) and dialysed.	Antitumour activity.	Sadovskaya et al. (2014)

holding capacity of the soil and the detoxification of heavy metals/radionuclides and removal of solid matter from contaminated water (Bender and Phillips 2004). Some extra-cellular polysaccharides produced by marine microorganisms are currently in the cosmeceutical market with great success. For example, extra-cellular polysaccharides from *Alteromonas macleodii* (Abyssine®) and glycoproteins from *Pseudoalteromonas* sp. (SeaCode®) (Martins et al. 2014). However, the main producers of extra-cellular polysaccharides currently in the market are bacteria from the species *Xanthomonas*, *Leuconostoc*, *Sphingomonas* and *Alcaligenes* that produce xanthan, dextran, gellan and curdlan (Öner 2013).

Despite the great potential of microalgae for the generation of high-value compounds, there are certain challenges that this industry has to address to increase the presence of microalgal products in the market, such as the strong presence in the market of chemically synthesised compounds. The successful establishment of microalgal products will depend not only on the aptitude of the different microalgae species to produce high-value compounds, but in reducing the cost of production and downstream processing of the biomass, designing new products and creating new markets (de la Jara et al. 2016).

ACKNOWLEDGMENTS

Marta Clemente-Carazo received funding from the Ministerio de Economia, Industria y Competitividad (MINECO), Agencia Estatal de Investigacion (AEI) and Fondo Europeo de Desarrollo Regional (FEDER) (code 30.05.18.80.79.541A642.10).

REFERENCES

Arad, S.M., and Levy-Ontman, O. 2010. "Red microalgal cell-wall polysaccharides: Biotechnological aspects." *Current Opinion in Biotechnology* 21 (3):358–364.

Bae, S.Y., Yim, J.H., Lee, H.K., and Pyo, S. 2006. "Activation of murine peritoneal macrophages by sulfated exopolysaccharide from marine microalga *Gyrodinium impudicum* (strain KG03): Involvement of the NF-kappa B and JNK pathway." *International Immunopharmacology* 6 (3):473–484.

Balavigneswaran, C.K., Kumar, T.S.J., Moses Packiaraj, R., Veeraraj, A., and Prakash, S. 2013. "Anti-oxidant activity of polysaccharides extracted from *Isocrysis galbana* using RSM optimized conditions." *International Journal of Biological Macromolecules* 60:100–108.

Behera, S., Singh, R., Arora, R., Sharma, N.K., Shukla, M., and Kumar, S. 2014. "Scope of algae as third generation biofuels." *Frontiers in Bioengineering and Biotechnology* 2:90.

Bender, J., and Phillips, P. 2004. "Microbial mats for multiple applications in aquaculture and bioremediation." *Bioresource Technology* 94 (3):229–238.

Borowitzka, M.A. 2013. "High-value products from microalgae—Their development and commercialisation." *Journal of Applied Phycology* 25 (3):743–756.

Brans, G.B.P.W., Schroën, C.G.P.H., Van der Sman, R.G.M., and Boom, R.M. 2004. "Membrane fractionation of milk: State of the art and challenges." *Journal of Membrane Science* 243 (1–2):263–272.

Casano, L.M., Braga, M.R., Álvarez, R., Del Campo, E.M., and Barreno, E. 2015. "Differences in the cell walls and extracellular polymers of the two *Trebouxia* microalgae coexisting in the lichen *Ramalina farinacea* are consistent with their distinct capacity to immobilize extracellular Pb." *Plant Science* 236:195–204.

Chakraborty, M., Miao, C., McDonald, A., and Chen, S. 2012. "Concomitant extraction of bio-oil and value added polysaccharides from *Chlorella sorokiniana* using a unique sequential hydrothermal extraction technology." *Fuel* 95:63–70.

Challouf, R., Trabelsi, L., Ben Dhieb, R., El Abed, O., Yahia, A., Ghozzi, K., Ben Ammar, J., Omran, H., and Ben Ouada, H. 2011. "Evaluation of cytotoxicity and biological activities in extracellular polysaccharides released by cyanobacterium *Arthrospira platensis*." *Brazilian Archives of Biology and Technology* 54 (4):831–838.

Chaudhary, L., Pradhan, P., Soni, N., Singh, P., and Tiwari, A. 2014. "Algae as a feedstock for bioethanol production: New entrance in biofuel world." *International Journal of ChemTech Research* 6:1381–1389.

Chen, C.S., Anaya, J.M., Zhang, S., Spurgin, J., Chuang, C.Y., Xu, C., Miao, A.J., Chen, E.Y., Schwehr, K.A., Jiang, Y., Quigg, A., Santschi, P.H., and Chin, W.C. 2011. "Effects of engineered nanoparticles on the assembly of exopolymeric substances from phytoplankton." *PLOS ONE* 6 (7):e21865.

Cheng, Y.S., Labavitch, J.M., and VanderGheynst, J.S. 2015. "Elevated CO2 concentration impacts cell wall polysaccharide composition of green microalgae of the genus *Chlorella*." *Letters in Applied Microbiology* 60 (1):1–7.

Cheng, Y.S., Zheng, Y., Labavitch, J.M., and VanderGheynst, J.S. 2011. "The impact of cell wall carbohydrate composition on the chitosan flocculation of *Chlorella*." *Process Biochemistry* 46 (10):1927–1933.

de Jesus Raposo, M.F., de Morais, A.M., and de Morais, R.M. 2014a. "Bioactivity and applications of polysaccharides from marine microalgae." In: *Polysachharides: Bioactivity and Biotechnology*. Cham, Switzerland: Springer International.

de Jesus Raposo, M.F., de Morais, A.M., and de Morais, R.M. 2014b. "Influence of sulphate on the composition and antibacterial and antiviral properties of the exopolysaccharide from *Porphyridium cruentum*." *Life Sciences* 101 (1-2):56–63.

de la Jara, A., Assunção, P., Portillo, E., Freijanes, K., and Mendoza, H. 2016. "Evolution of microalgal biotechnology: A survey of the European Patent Office database." *Journal of Applied Phycology* 28 (5):2727–2740.

Delattre, C., Michaud, P., Courtois, B., and Courtois, J. 2005. "Oligosaccharides engineering from plants and algae: Applications in biotechnology and therapeutics." *Minerva Biotecnologica* 17 (3):107.

Deniz, I., García-Vaquero, M., and Imamoglu, E. 2018. "Trends in red biotechnology: Microalgae for pharmaceutical applications." In: *Microalgae-Based Biofuels and Bioproducts*, edited by R. Muñoz and C. Gonzalez-Fernandez. Elsevier.

Díaz Bayona, K.C., and Garcés, L.A. 2014. "Effect of different media on exopolysaccharide and biomass production by the green microalga *Botryococcus braunii*." *Journal of Applied Phycology* 26 (5):2087–2095.

Duan, X., Chi, Z., Wang, L., and Wang, X. 2008. "Influence of different sugars on pullulan production and activities of alpha-phosphoglucose mutase, UDPG-pyrophosphorylase and glucosyltransferase involved in pullulan synthesis in *Aureobasidium pullulans* Y68." *Carbohydrate Polymers* 73 (4):587–593.

du Plooy, S.J., Perissinotto, R., Smit, A.J., and Muir, D.G. 2015. "Role of salinity, nitrogen fixation and nutrient assimilation in prolonged bloom persistence of Cyanothece sp. in Lake St Lucia, South Africa." *Aquatic Microbial Ecology* 74 (1):73.

Feng, L., Li, X., Du, G., and Chen, J. 2009. "Characterization and fouling properties of exopolysaccharide produced by *Klebsiella oxytoca*." *Bioresource Technology* 100 (13):3387–3394.

Fu, C.C., Hung, T.C., Chen, J.Y., Su, C.H., and Wu, W.T. 2010. "Hydrolysis of microalgae cell walls for production of reducing sugar and lipid extraction." *Bioresource Technology* 101 (22):8750–8754.

García-Vaquero, M. 2019a. "Analytical methods and advances to evaluate dietary fiber." In: *Dietary Fiber: Properties, Recovery & Applications* edited by C. Galanakis (ed.). Elsevier.

García-Vaquero, M. 2019b. "Seaweed proteins and applications in animal feed." In: *Novel Proteins for Food, Pharmaceuticals and Agriculture: Sources, Applications and Advances*, 139–161.

García-Vaquero, M., and Hayes, M. 2016. "Red and green macroalgae for fish and animal feed and human functional food development." *Food Reviews International* 32 (1):15–45.

García-Vaquero, M., Lopez-Alonso, M., and Hayes, M. 2017. "Assessment of the functional properties of protein extracted from the brown seaweed *Himanthalia elongata* (Linnaeus) SF Gray." *Food Research International* 99 (3):971–978.

García-Vaquero, M., Rajauria, G., O'Doherty, J.V., and Sweeney, T. 2017. "Polysaccharides from macroalgae: Recent advances, innovative technologies and challenges in extraction and purification." *Food Research International* 99 (3):1011–1020.

García-Vaquero, M., Rajauria, G., Tiwari, B., Sweeney, T., and O'Doherty, J. 2018. "Extraction and yield optimisation of fucose, glucans and associated antioxidant activities from *Laminaria digitata* by applying response surface methodology to high intensity ultrasound-assisted extraction." *Marine Drugs* 16 (8):257.

Geresh, S., Arad, S.M., Levy-Ontman, O., Zhang, W., Tekoah, Y., and Glaser, R. 2009. "Isolation and characterization of poly- and oligosaccharides from the red microalga *Porphyridium* sp." *Carbohydrate Research* 344 (3):343–349.

Guihéneuf, F., and Stengel, D.B. 2015. "Towards the biorefinery concept: Interaction of light, temperature and nitrogen for optimizing the co-production of high-value compounds in *Porphyridium purpureum*." *Algal Research* 10:152–163.

Guzmán, S., Gato, A., Lamela, M., Freire-Garabal, M., and Calleja, J.M. 2003. "Anti-inflammatory and immunomodulatory activities of polysaccharide from *Chlorella stigmatophora* and *Phaeodactylum tricornutum*." *Phytotherapy Research* 17 (6):665–670.

Guzman-Murillo, M.A., and Ascencio, F. 2000. "Anti-adhesive activity of sulphated exopolysaccharides of microalgae on attachment of red sore disease-associated bacteria and *Helicobacter pylori* to tissue culture cells." *Letters in Applied Microbiology* 30 (6):473–478.

Harun, R., and Danquah, M.K. 2011. "Influence of acid pretreatment on microalgal biomass for bioethanol production." *Process Biochemistry* 46 (1):304–309.

Harun, R., Jason, W.S.Y., Cherrington, T., and Danquah, M.K. 2011. "Exploring alkaline pre-treatment of microalgal biomass for bioethanol production." *Applied Energy* 88 (10):3464–3467.

Hayes, M., Bastiaens, L., Gouveia, L., Gkelis, S., Skomedal, H., Skjanes, K., Murray, P., García-Vaquero, M., Hosoglu, M.I., and Dodd, J. 2019. "Microalgal bioactive compounds including protein, peptides, and pigments: Applications, opportunities, and challenges during biorefinery processes." In: *Novel Proteins for Food, Pharmaceuticals and Agriculture: Sources, Applications and Advances* edited by M. Hayes (ed.). Wiley.

Huheihel, M., Ishanu, V., Tal, J., and Arad, S.M. 2002. "Activity of *Porphyridium* sp. polysaccharide against *Herpes simplex* viruses in vitro and in vivo." *Journal of Biochemical and Biophysical Methods* 50 (2–3):189–200.

Huleihel, M., Ishanu, V., Tal, J., and Arad, S.M. 2001. "Antiviral effect of red microalgal polysaccharides on *Herpes simplex* and *Varicella zoster* viruses." *Journal of Applied Phycology* 13 (2):127–134.

Jo, W.S., Choi, Y.J., Kim, H.J., Nam, B.H., Hong, S.H., Lee, G.A., Lee, S.W., Seo, S.Y., and Jeong, M.H. 2010. "Anti-inflammatory effect of microalgal extracts from *Tetraselmis suecica*." *Food Science and Biotechnology* 19 (6):1519–1528.

Laurienzo, P. 2010. "Marine polysaccharides in pharmaceutical applications: An overview." *Marine Drugs* 8 (9):2435–2465.

Lee, J.B., Hayashi, T., Hayashi, K., and Sankawa, U. 2000. "Structural analysis of calcium spirulan (Ca-SP)-derived oligosaccharides using electrospray ionization mass spectrometry." *Journal of Natural Products* 63 (1):136–138.

Lee, J.B., Hayashi, T., Hayashi, K., Sankawa, U., Maeda, M., Nemoto, T., and Nakanishi, H. 1998. "Further purification and structural analysis of calcium spirulan from *Spirulina platensis*." *Journal of Natural Products* 61 (9):1101–1104.

Liang, H. 2015. *Harvesting micro algae*. Google Patents, US9464268B2.

Liu, Y., Wang, W., Zhang, M., Xing, P., and Yang, Z. 2010. "PSII-efficiency, polysaccharide production, and phenotypic plasticity of *Scenedesmus obliquus* in response to changes in metabolic carbon flux." *Biochemical Systematics and Ecology* 38 (3):292–299.

Mader, J., Gallo, A., Schommartz, T., Handke, W., Nagel, C.H., Günther, P., Brune, W., and Reich, K. 2016. "Calcium spirulan derived from *Spirulina platensis* inhibits herpes simplex virus 1 attachment to human keratinocytes and protects against *Herpes labialis*." *The Journal of Allergy and Clinical Immunology* 137 (1):197.e3–203.e3.

Magaletti, E., Urbani, R., Sist, P., Ferrari, C.R., and Cicero, A.M. 2004. "Abundance and chemical characterization of extracellular carbohydrates released by the marine diatom *Cylindrotheca fusiformis* under N- and P-limitation." *European Journal of Phycology* 39 (2):133–142.

Marcati, A., Ursu, A.V., Laroche, C., Soanen, N., Marchal, L., Jubeau, S., Djelveh, G., and Michaud, P. 2014. "Extraction and fractionation of polysaccharides and B-phycoerythrin from the microalga *Porphyridium cruentum* by membrane technology." *Algal Research* 5:258–263.

Martins, A., Vieira, H., Gaspar, H., and Santos, S. 2014. "Marketed marine natural products in the pharmaceutical and cosmeceutical industries: Tips for success." *Marine Drugs* 12 (2):1066–1101.

Mellal, M., Jaffrin, M.Y., Ding, L.H., Delattre, C., Michaud, P., and Courtois, J. 2008. "Separation of oligoglucuronans of low degrees of polymerization by using a high shear rotating disk filtration module." *Separation and Purification Technology* 60 (1):22–29.

Mezhoud, N., Zili, F., Bouzidi, N., Helaoui, F., Ammar, J., and Ouada, H.B. 2014. "The effects of temperature and light intensity on growth, reproduction and EPS synthesis of a thermophilic strain related to the genus *Graesiella*." *Bioprocess and Biosystems Engineering* 37 (11):2271–2280.

Michalak, I., and Chojnacka, K. 2014. "Algal extracts: Technology and advances." *Engineering in Life Sciences* 14 (6):581–591.

Miranda, M., Lopez-Alonso, M., and García-Vaquero, M. 2017. "Macroalgae for functional feed development: Applications in aquaculture, ruminant and swine feed industries." In: *Seaweeds: Biodiversity, Environmental Chemistry and Ecological Impacts*, edited by P. Newton. NOVA Publishers.

Mishra, A., and Jha, B. 2009. "Isolation and characterization of extracellular polymeric substances from microalgae *Dunaliellasalina* under salt stress." *Bioresource Technology* 100 (13):3382–3386.

Öner, T.E. 2013. "Microbial production of extracellular polysaccharides from biomass." In: *Pretreatment Techniques for Biofuels and Biorefineries*, edited by Z. Fang, 35–56. Berlin, Heidelberg: Springer Berlin Heidelberg.

Parikh, A., and Madamwar, D. 2006. "Partial characterization of extracellular polysaccharides from cyanobacteria." *Bioresource Technology* 97 (15):1822–1827.

Patel, A.K., Laroche, C., Marcati, A., Ursu, A.V., Jubeau, S., Marchal, L., Petit, E., Djelveh, G., and Michaud, P. 2013. "Separation and fractionation of exopolysaccharides from *Porphyridium cruentum*." *Bioresource Technology* 145:345–350.

Pletikapić, G., Radić, T.M., Zimmermann, A.H., Svetličić, V., Pfannkuchen, M., Marić, D., Godrijan, J., and Zutić, V. 2011. "AFM imaging of extracellular polymer release by marine diatom *Cylindrotheca closterium* (Ehrenberg) Reiman & J.C. Lewin." *Journal of Molecular Recognition* 24 (3):436–445.

Pugh, N., Ross, S.A., ElSohly, H.N., ElSohly, M.A., and Pasco, D.S. 2001. "Isolation of three high molecular weight polysaccharide preparations with potent immunostimulatory activity from *Spirulina platensis*, Aphanizomenon flos-aquae and *Chlorella pyrenoidosa*." *Planta Medica* 67 (8):737–742.

Raposo, M., de Morais, R., and Bernardo de Morais, A. 2013. "Bioactivity and applications of sulphated polysaccharides from marine microalgae." *Marine Drugs* 11 (1):233–252.

Sadovskaya, I., Souissi, A., Souissi, S., Grard, T., Lencel, P., Greene, C.M., Duin, S., Dmitrenok, P.S., Chizhov, A.O., Shashkov, A.S., and Usov, A.I. 2014. "Chemical structure and biological activity of a highly branched (1 --> 3,1 --> 6)-beta-D-glucan from *Isochrysis galbana*." *Carbohydrate Polymers* 111:139–148.

Singh, S., Kate, B.N., and Banerjee, U.C. 2005. "Bioactive compounds from cyanobacteria and microalgae: An overview." *Critical Reviews in Biotechnology* 25 (3):73–95.

Spolaore, P., Joannis-Cassan, C., Duran, E., and Isambert, A. 2006. "Commercial applications of microalgae." *Journal of Bioscience and Bioengineering* 101 (2):87–96.

Sun, L., Chu, J., Sun, Z., and Chen, L. 2016. "Physicochemical properties, immunomodulation and antitumor activities of polysaccharide from *Pavlova viridis*." *Life Sciences* 144:156–161.

Sun, L., Wang, C., Shi, Q., and Ma, C. 2009. "Preparation of different molecular weight polysaccharides from *Porphyridium cruentum* and their antioxidant activities." *International Journal of Biological Macromolecules* 45 (1):42–47.

Sun, Y., Wang, H., Guo, G., Pu, Y., and Yan, B. 2014. "The isolation and antioxidant activity of polysaccharides from the marine microalgae *Isochrysis galbana*." *Carbohydrate Polymers* 113:22–31.

Sun, L., Wang, L., and Zhou, Y. 2012. "Immunomodulation and antitumor activities of different-molecular-weight polysaccharides from *Porphyridium cruentum*." *Carbohydrate Polymers* 87 (2):1206–1210.

Urbani, R., Magaletti, E., Sist, P., and Cicero, A.M. 2005. "Extracellular carbohydrates released by the marine diatoms *Cylindrotheca closterium*, *Thalassiosira pseudonana* and *Skeletonema costatum*: Effect of P-depletion and growth status." *The Science of the Total Environment* 353 (1–3):300–306.

Vandamme, D., Foubert, I., and Muylaert, K. 2013. "Flocculation as a low-cost method for harvesting microalgae for bulk biomass production." *Trends in Biotechnology* 31 (4):233–239.

Ventura, S., Nobre, B., Ertekin, F., Hayes, M., García-Vaquero, M., Vieira, F., and Palavra, A. 2018. "Extraction of added-value compounds from microalgae." In: *Microalgae-Based Biofuels and Bioproducts* edited by Muñoz, R. and Gonzalez-Fernandez, C. (eds). Elsevier.

Vigani, M., Parisi, C., Rodríguez-Cerezo, E., Barbosa, M.J., Sijtsma, L., Ploeg, M., and Enzing, C. 2015. "Food and feed products from micro-algae: Market opportunities and challenges for the EU." *Trends in Food Science and Technology* 42 (1):81–92.

Wan, Y., Prudente, A., and Sathivel, S. 2012. "Purification of soluble rice bran fiber using ultrafiltration technology." *LWT – Food Science and Technology* 46 (2):574–579.

Williams, P.J.B., and Laurens, L.M.L. 2010. "Microalgae as biodiesel & biomass feedstocks: Review & analysis of the biochemistry, energetics & economics." *Energy and Environmental Science* 3 (5):554–590.

Yim, J.H., Kim, S.J., Ahn, S.H., Lee, C.K., Rhie, K.T., and Lee, H.K. 2004. "Antiviral effects of sulfated exopolysaccharide from the marine microalga *Gyrodinium impudicum* strain KG03." *Marine Biotechnology* 6 (1):17–25.

Yim, J.H., Kim, S.J., Ahn, S.H., and Lee, H.K. 2007. "Characterization of a novel bioflocculant, p-KG03, from a marine dinoflagellate, *Gyrodinium impudicum* KG03." *Bioresource Technology* 98 (2):361–367.

Zhang, S., and Santschi, P.H. 2009. "Application of cross-flow ultrafiltration for isolating exopolymeric substances from a marine diatom (*Amphora* sp.)." *Limnology and Oceanography: Methods* 7 (6):419–429.

Zou, X., Sun, M., and Guo, X. 2006. "Quantitative response of cell growth and polysaccharide biosynthesis by the medicinal mushroom *Phellinus linteus* to NaCl in the medium." *World Journal of Microbiology and Biotechnology* 22 (11):1129–1133.

2 Macroalgal Fucoidan for Biomedical Applications

Jayachandran Venkatesan, Sukumaran Anil, Sneha Rao, and Se-Kwon Kim

CONTENTS

Abbreviations 13
Introduction 13
Sulfated Polysaccharides 14
Fucoidan 14
Isolation of Fucoidan from Marine Macroalgae 14
Characterization of Fucoidan 14
Biomedical Applications of Fucoidan 14
    Tissue Engineering 15
    Fucoidan in Drug Delivery 17
    Fucoidan for Wound Dressing 20
Conclusion 20
Acknowledgment 20
References 20

BOX 2.1 SALIENT FEATURES

Fucoidan is considered as "gold from the sea" and it has enormous application in the field of biological and biomedical applications. Fucoidan is a sulfated polysaccharide, and commonly observed in marine brown algae. Fucoidan has the ability to differentiate from stem cells to osteoblastic cells. Fucoidan can be converted into nanoparticles, hydrogels, nanofibers, microsphere and scaffolds for tissue engineering, drug delivery and wound dressing applications. Fucoidan structure and activity can vary depending on the brown seaweed source and geographical area.

ABBREVIATIONS

BMP-2: Bone Morphogenetic Protein 2
VEGF: Vascular Endothelial Growth Factor
HIV: Human Immunodeficiency Virus
PCL: Polycaprolactone
DCFH-DA: Dichlorofluorescindiacetate

INTRODUCTION

Over 71% of earth is covered by ocean which is an abode of a range of living forms including microorganisms, plants and animals. Among the photosynthetic forms, micro and macro algae are the predominant forms in the marine ecosystem. Marine macroalgae have the ability to adopt to a wide range of temperatures, salinity and produce different kinds of secondary metabolites including polysaccharides, proteins, peptides, organic compounds, lipids etc. (Garcia-Vaquero et al., 2017). Polysaccharides such as alginate (Lee and Mooney, 2012), carrageenan (Prajapati et al., 2014), agar (Fuse and Goto, 1971), fucoidan (Li et al., 2008), ulvan (Lahaye and Robic, 2007) are abundantly produced by the seaweeds. Moreover, these polysaccharides have excellent biological and biomedical applications impacting future bio-economy. These polysaccharides are widely utilized in pharmaceuticals (Selinger et al., 1998), cosmeceutical (Fitton et al., 2007; Varco et al., 1990) and nutraceutical applications Ruocco et al., 2016).

SULFATED POLYSACCHARIDES

Sulfated polysaccharides are commonly found in structural elements of plants and animals (Venkatesan et al., 2017). Carrageenan (Prajapati et al., 2014), fucoidan (Ale et al., 2011), ulvan (Lahaye and Robic, 2007) are commonly available sulfated polysaccharides which are found in different kinds of algae viz., red, green and brown (McCandless and Craigie, 1979). These sulfated polysaccharides are extensively studied in tissue engineering (Jeong et al., 2013; Puvaneswary et al., 2015; Santo et al., 2009), drug delivery (Alves et al., 2012b; Li et al., 2014) and wound dressing applications (Boateng et al., 2013; Kikionis et al., 2015; Sezer et al., 2007) due to their excellent chemical, physical and biological properties.

FUCOIDAN

Fucoidan is a sulfated polysaccharide and is mainly composed of a rich amount of fucose moieties with differently substituted other sugar units and it is usually found in brown seaweeds. However, the structure of fucoidan varies among seaweed species and geographic condition. For example, the *sargassum* species consists of sulfated fucogalacturonan. Recent research results suggest that fucoidan has excellent pharmaceutical properties and antitumor (Zhuang et al., 1995), anticancer (Kwak, 2014), immunomodulatory (Raghavendran et al., 2011), antioxidant (Wang et al., 2010), anti-inflammatory (Cumashi et al., 2007), suppression of fat (Kim et al., 2014), anti-HIV (Thuy et al., 2015) and anticoagulant (Düriget et al., 1997) activities. These biological activities are mainly related to the structure of fucoidan, its monosaccharides unit structure, sulfated condition and position of this group. Soft isolation procedure might be required to get biologically active fucoidan molecules (Ale and Meyer, 2013).

ISOLATION OF FUCOIDAN FROM MARINE MACROALGAE

Fucoidan can be extracted from seaweed using different approaches (Lim and Aida, 2017) and is one of the most widely used techniques to isolate fucoidan through ethanol precipitation (Rani et al., 2017). Some of the other interesting methods utilized to isolate the fucoidan from macroalgae include microwave-assisted extraction (Rodriguez-Jasso et al., 2011), ultrasound-assisted extraction (Flórez-Fernández et al., 2017), enzymatic extraction techniques (Hahn et al., 2012), pressurized liquid extraction (Saravana et al., 2016), subcritical water extraction (Saravana et al., 2018). Figure 2.1 shows the isolation procedure for fucoidan from brown seaweed and deodorization method for fucoidan isolation.

For fucoidan isolation from seaweed, dry material is finely powered through either pulverisation to get the micrometer size particulate matter. Further, sufficient amount of ethanol should be added into the seaweed powder to remove the pigments, proteins and other dust particles. Subsequently, seaweed power is washed with acetone and centrifuged and dried. A small amount of seaweed (approximately 5g) is treated with 100 ml of distilled water and stirred for an hour, and this process can be repeated two to three times to extract the maximum amount of fucoidan. The fractions are combined and centrifuged, and the supernatant treated with 1% (W/V) $CaCl_2$ solution to get the alginate precipitation. Further, the complete solution is kept overnight at 4°C for total alginic acid precipitation, followed by centrifugation to spate the alginate (Rani et al., 2017; Yang et al., 2008). To get the fucoidan, the solution is mixed firstly with 30% ethanol and kept for 4 h at 4°C and centrifuged to remove the impurities. The supernatant is made to 70% of ethanol to precipitate the fucoidan and placed at 4°C overnight for total yield. Precipitated fucoidan solution is centrifuged and washed with ethanol and acetone to get the crude fucoidan and dried at room temperature (Rani et al., 2017). Figure 2.1 (A) shows the similar isolation procedure while Synytsya et al. (2010) have used mild acid (HCl) for the isolation of fucoidan. The yield of fucoidan is calculated as follows,

$$\text{Yield}(\%) = \frac{\text{Weight of the obtained fucoidan in gram}}{\text{Weight of dried seaweed in gram}} \times 100$$

CHARACTERIZATION OF FUCOIDAN

Fourier transform infrared spectroscopy is a widely utilized technique to characterize the isolated fucoidan from seaweed. The main peaks in the FT-IR spectrum shows at 843.78 cm^{-1} and 842 cm^{-1} corresponding to sulfate stretching frequencies (Rani et al., 2017). See Figure 2.2.

Further chemical characterization of isolated fucoidan were done using nuclear magnetic resonance spectroscopy and molecular weight determination using mass spectral studies (Bilan et al., 2002; Clément et al., 2010; Daniel et al., 2007).

BIOMEDICAL APPLICATIONS OF FUCOIDAN

Biomedical research on fucoidan is increasingly focused on tissue engineering, drug delivery and biosensor

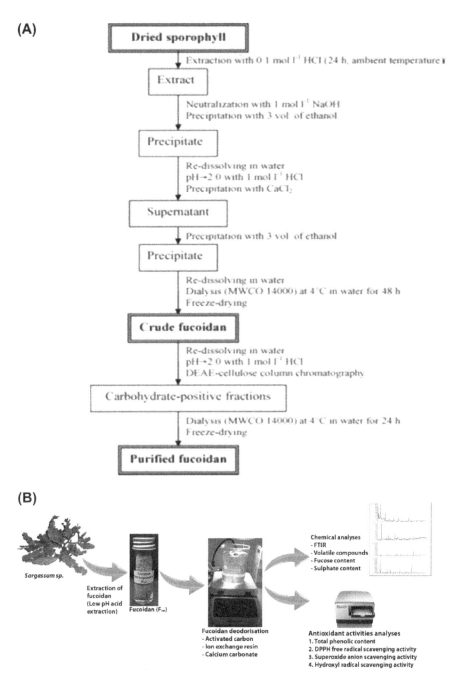

FIGURE 2.1 (A) Isolation and purification of fucoidan from Miyeokgui. Figure adapted with permission from (Synytsya et al., 2010), (B) Isolation and deodorisation of fucoidan from brown seaweed *sargassum* sp. and their antioxidant property. (Figure adapted with permission from Khalafu et al., 2017.)

application due to its unique properties such as biocompatibility, biodegradable nature and biological activities.

Tissue Engineering

Tissue Engineering is an interdisciplinary field of research with an aim to repair and regenerate the defective and damaged tissue. Tissue engineering studies involve disciplines such as materials science, life sciences, medical science, computer science, electrical engineering and biomedical engineering. The basic paradigm of tissue engineering is the combination of materials viz., polymeric, ceramics, protein, peptides, lipids, small molecules in the form of scaffolds, hydrogels, nanofibers, nanoparticles; cells (primary cells, stem cells etc.); and growth factors (bone morphogenetic protein 2 (BMP-2), platelet rich plasma (PRP), vascular endothelial growth factor (VEGF) etc. Scaffold plays a

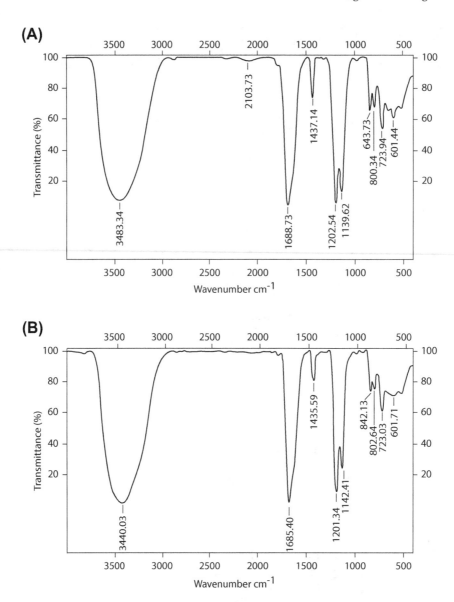

FIGURE 2.2 Fourier transform infrared spectrum of fucoidan which is isolated from brown seaweed *P. tetrastromatica* and *S. oligocystum*. (Figure adapted with permission from Rani et al., 2017.)

major role in the tissue engineering application due to its mimicking properties with extra-cellular matrix, and also helps in the cell adhesion and differentiation process (Lee et al., 2012). The ideal scaffold for the tissue engineering application should be biocompatible, biodegradable and should render natural tissue mechanical strength. Furthermore, it should be the least toxicity and without inflammatory effects. Importantly, it should promote cell adhesion and proliferation as well as integration with the native tissues. Presently both synthetic polymeric substances and natural polymeric materials are widely utilized in the tissue engineering application. Seaweed derived polysaccharides substances, including polysaccharides find utility in tissue engineering applications. For example, alginate (Venkatesan et al., 2015), agarose (Zarrintaj et al., 2018), fucoidan (Lowe et al., 2016), carrageenan (Mihaila et al., 2014), chitosan, ulvan (Alves et al., 2012a) are widely utilized natural polysaccharides for tissue engineering (Malafaya et al., 2007).

It is reported that fucoidan has the osteo-conductive property which facilitates bone tissue engineering (Changotade et al., 2008). These authors used Lubboc, a commercially available bone biomaterial, which was substituted and impregnated with low molecular weight fucoidan molecules to study the osteoblastic cell behavior. They observed an increase in the collagen formation and alkaline phosphatase activity. Similar to this study, Cho et al. (2009) conducted a research on fucoidan with MG-63 osteoblastic like cells which induced the calcium formation and showed increased BMP-2 production.

Hwang et al. (2016) also checked the low molecular weight fucoidan for the bone tissue engineering application. Sargassum derived fucoidan influenced a significant level of cell viability around 150.33 ± 6.50% compared to normal fucoidan molecules. Electrospun nanofibers were developed using fucoidan with the combination of polycaprolactone (PCL) and proposed for the tissue engineering application. The tensile strength of nanofibers enhanced with the combination of fucoidan when compared to PCL alone (Lee et al., 2012) (Figure 2.3).

Venkatesan et al. (2014) developed chitosan-alginate fucoidan composites for bone tissue engineering. A simple polyelectrolyte complexation was utilized to obtain the composite scaffold using the lyophilization method. Firstly, chitosan and alginate were combined to form a hydrogel, and further addition of fucoidan enhanced the viscosity properties and mechanical strength. Water uptake and retention ability were not altered (Figure 2.4). Further, protein absorption and excellent mineralization were observed in the scaffold containing fucoidan with chitosan and alginate. Also many other reports also indicated osteogenic properties of fucoidan; furthermore these kinds of scaffold have been recommended for bone tissue engineering (Hwang et al., 2016; Jin and Kim, 2012; Kim and Cho, 2011; Park et al., 2012; Puvaneswary et al., 2015; Wang et al., 2017; Young et al., 2016).

Fucoidan in Drug Delivery

Fucoidan is an ionic polysaccharide and it is easy to combine with positively charged polysaccharides (Sezer and Cevher, 2011). Due to the formation of polyelectrolyte complexation between cationic materials and fucoidan, different kinds of composites can be obtained. Materials such as scaffold, nanoparticles, nanofibers, hydrogels, microsphere, nanosphere, nanogels are formed using chitosan with fucoidan for drug delivery applications (Cardoso et al., 2016; Cunha and Grenha, 2016; Manivasagan et al., 2016). An amine group of chitosan and sulfate from fucoidan forms an instant polyelectrolyte with a proper stoichiometric ratio providing desired nanoparticles for drug delivery applications (Lee and Lim, 2016). Huang and coworkers developed a series of chitosan-fucoidan based nanoparticles with the polyelectrolyte complexation method and utilized them for delivery of different agents such as curcumin, SDF-1,

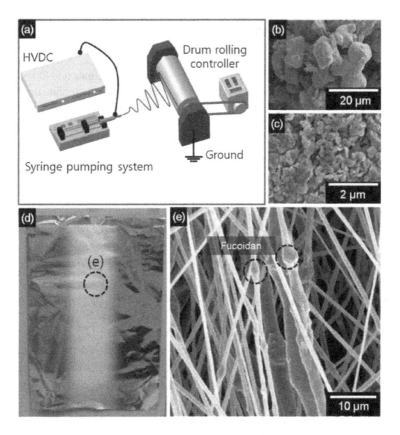

FIGURE 2.3 (a) Schematic of an electrospinning process, (b and c) SEM images of poly-caprolactone and fucoidan powders, respectively, (d) Electrospun P/F3 on aluminum foil, (e) Magnified scanning electron microcopy image of (d). (Figure adapted with permission from J. S. Lee et al., 2012.)

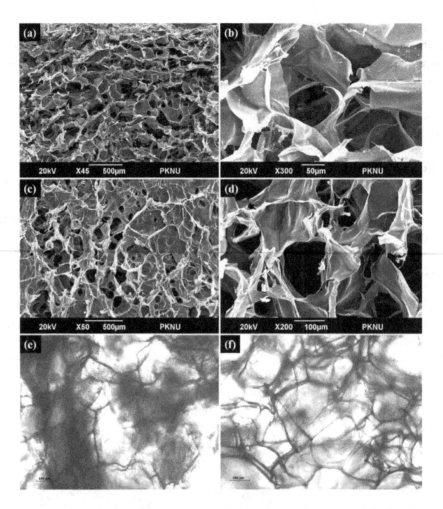

FIGURE 2.4 High and low magnification SEM micrographs of (a, b) Chitosan-Alginate and (c, d) Chitosan-Alginate-fucoidan; and optical microscopy images of (e) Chitosan-alginate and (f) Chitosan-alginate-fucoidan. Images show the interconnected pore size found in chitosan-alginate-fucoidan scaffold. (Figure adapted with permission from Venkatesan et al., 2014.)

proteins and peptides (Huang et al., 2016; Huang and Kuo, 2016; Huang and Li, 2014; Huang and Liu, 2012; Huang and Yang, 2016; Huang et al., 2014). Chitosan-fucoidan nanoparticles were also utilized for antioxidant drug delivery. The developed nanoparticles were spherical, around 230–250 nm. The ratio between chitosan and fucoidan was 5/1 to form a desirable nanoparticle. Gentamicin was used as a model drug, whereas the 1,1-diphenyl-2-picrylhydrazyl (DPPH), 2′,7′-dichlorofluorescin diacetate (DCFH-DA), Nitric Oxide assay, Interleukin-6 assay were utilized to check the antioxidant properties of the nanoparticles (Huang and Li, 2014). In addition to lack of toxicity, the chitosan-fucoidan nanoparticles delivered the drug, gentamicin, in a sustainable manner with an efficacy of 99% delivery, in 72 h. Chitosan-fucoidan-tripolyphosphate nanoparticles were utilized for stem cell derived factor 1 (SDF-1) delivery which is an important agent in the stem cell differentiation process. Chitosan-fucoidan nanoparticles were utilized to prevent the degradation of SDF-1 by enzymes and also to increase the half-life period to retain its activity for a long time (Huang and Liu, 2012) (Figure 2.5).

pH sensitive o-carboxymethyl chitosan-fucoidan nanoparticles were utilized for curcumin drug delivery. The particle size of the developed nanoparticles was around 270 nm which effectively encapsulated the curcumin (92.8%). The developed nanoparticles was stable at pH 2.5 and disrupted at pH 7.4. Nanoparticles showed lesser cytotoxicity against mouse fibroblast cells (L929) than free curcumin. As a conclusion of the study, o-carboxymethyl chitosan/fucoidan nanoparticles effectively increased the curcumin uptake which can be potentially used for the oral drug delivery system (Huang and Kuo, 2016). In another study, layer-by-layer of chitosan and fucoidan nanocapsules were developed on polystyrene nanoparticles. Morphology of hallow multilayer were measured through scanning electron and transmission electron microscopy (Pinheiro et al., 2015). Nano carrier

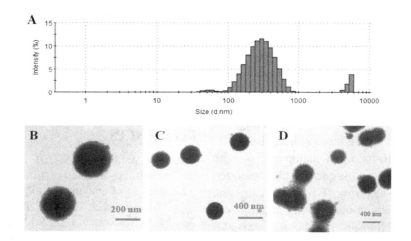

FIGURE 2.5 (A) Size distribution of chitosan-fucoidan-tripolyphosphate nanoparticles, (B) different magnification of chitosan-fucoidan-tripolyphosphate nanoparticles. (Figure adapted with permission from Y. C. Huang and Liu, 2012.)

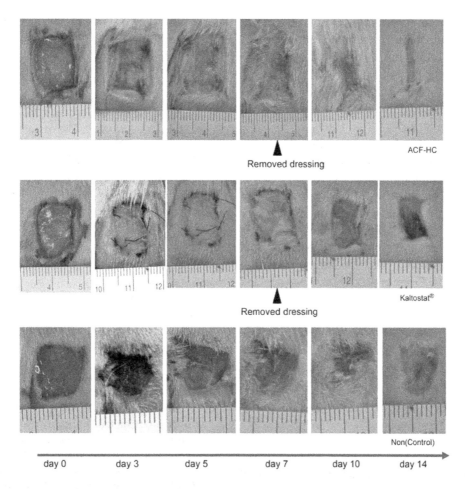

FIGURE 2.6 (**See color insert.**) Photographic findings of wounds covered with chitin/chitosan, alginate, fucoidan hydrogel sheet or Kaltostat, and controls. Each wound on the indicated day is representative of eight wounds (four rats) covered with ACF-HS or Kaltostat, or not covered (control). (Figure adapted with permission Murakami et al., 2010b.)

with chitosan and fucoidan were also used to deliver berberine and subsequently characterized through FT-IR to ensure the chemical interaction between the molecules. Release of berberine from the materials at simulated intestinal fluid at pH 7.5 was fast, whereas the release was slow at simulated gastric fluid at pH 2.0 (Wu et al., 2014). Protamine-fucoidan nanoparticles were prepared for the delivery of Doxorubicin, an anticancer drug (Lu et al., 2017). Recently, chitosan-fucoidan nanoparticles with silver were developed and subsequently characterized through dynamic light scattering, FT-IR, transmission electron microscopy, antimicrobial activity and anticancer activity. Silver in the chitosan-fucoidan nanoparticles significantly inhibited the microbial growth and cancer cells growth (Venkatesan et al., 2018).

Fucoidan for Wound Dressing

Typical wound dressing materials should be biocompatible, non-toxic, should have water retention properties and should prevent infection. To mimic these properties, alginate, anionic polysaccharides from seaweed, has been extensively used as a wound dressing material (Chandika et al., 2015; Cole and Nelson, 1993; Mogoşanu and Grumezescu, 2014; Paul and Sharma, 2004). Murakami et al. (2010a) have developed hydrogel which contains chitin/chitosan, fucoidan and alginate for wound dressing application (Figure 2.6).

In another study, fucoidan with transforming growth factor (TGF-β1) were used for treating wounds (O'Leary et al., 2004). Chitosan containing fucoidan was developed for wound dressing to treat dermal burns, which was extensively evaluated by *in vitro* and *in vivo* studies (Sezer et al., 2007; Sezer et al., 2008). New Zealand rabbits were utilized in the study to check the wound healing capacity of the composite scaffold with periodic observation up to 3 weeks through macroscopically and histopathological staining. Complete closure of the wound was achieved by using chitosan-fucoidan composite and suggesting the efficacy of these biomaterials in wound-healing applications.

CONCLUSION

There is enormous scope for advancing the application of fucoidan for health care. Different kinds of composites scaffold were developed with fucoidan in the form of nanoparticles, scaffolds, nanofibers and hydrogels to aid tissue engineering, drug delivery and wound healing application. The technology of developing fucoidan based biomaterials holds tremendous promise in a number of biomedical applications in the future.

ACKNOWLEDGMENT

Authors are thankful to Yenepoya Research Center, Yenepoya (Deemed to be University) for their support to write this chapter.

REFERENCES

Ale, M. T., & Meyer, A. S. (2013). Fucoidans from brown seaweeds: An update on structures, extraction techniques and use of enzymes as tools for structural elucidation. *RSC Advances*, 3(22), pp. 8131–8141.

Ale, M. T., Mikkelsen, J. D., & Meyer, A. S. (2011). Important determinants for fucoidan bioactivity: A critical review of structure-function relations and extraction methods for fucose-containing sulfated polysaccharides from brown seaweeds. *Marine Drugs*, 9(10), pp. 2106–2130.

Alves, A., Duarte, A. R. C., Mano, J. F., Sousa, R. A., & Reis, R. L. (2012a). PDLLA enriched with ulvan particles as a novel 3D porous scaffold targeted for bone engineering. *The Journal of Supercritical Fluids*, 65, pp. 32–38.

Alves, A., Pinho, E. D., Neves, N. M., Sousa, R. A., & Reis, R. L. (2012b). Processing ulvan into 2D structures: Cross-linked ulvan membranes as new biomaterials for drug delivery applications. *International Journal of Pharmaceutics*, 426(1–2), pp. 76–81.

Bilan, M. I., Grachev, A. A., Ustuzhanina, N. E., Shashkov, A. S., Nifantiev, N. E., & Usov, A. I. (2002). Structure of a fucoidan from the brown seaweed *Fucus evanescens* C. Ag. *Carbohydrate Research*, 337(8), pp. 719–730.

Boateng, J. S., Pawar, H. V., & Tetteh, J. (2013). Polyox and carrageenan based composite film dressing containing anti-microbial and anti-inflammatory drugs for effective wound healing. *International Journal of Pharmaceutics*, 441(1–2), pp. 181–191.

Cardoso, M. J., Costa, R. R., & Mano, J. F. (2016). Marine origin polysaccharides in drug delivery systems. *Marine Drugs*, 14(2), p. 34.

Chandika, P., Ko, S. C., & Jung, W. K. (2015). Marine-derived biological macromolecule-based biomaterials for wound healing and skin tissue regeneration. *International Journal of Biological Macromolecules*, 77, pp. 24–35.

Changotade, S. I. T., Korb, G., Bassil, J., Barroukh, B., Willig, C., Colliec-Jouault, S., Durand, P., Godeau, G., & Senni, K. (2008). Potential effects of a low-molecular-weight fucoidan extracted from brown algae on bone biomaterial osteoconductive properties. *Journal of Biomedical Materials Research Part A*, 87(3), pp. 666–675.

Cho, Y.-S., Jung, W.-K., Kim, J.-A., Choi, I.-W., & Kim, S.-K. (2009). Beneficial effects of fucoidan on osteoblastic MG-63 cell differentiation. *Food Chemistry*, 116(4), pp. 990–994.

Clément, M. J., Tissot, B., Chevolot, L., Adjadj, E., Du, Y., Curmi, P. A., & Daniel, R. (2010). NMR characterization and molecular modeling of fucoidan showing the importance of oligosaccharide branching in its anticomplementary activity. *Glycobiology*, 20(7), pp. 883–894.

Cole, S. M., & Nelson, D. L. (1993). Alginate wound dressing of good integrity. Google Patents.

Cumashi, A., Ushakova, N. A., Preobrazhenskaya, M. E., D'incecco, A., Piccoli, A., Totani, L., Tinari, N., Morozevich, G. E., Berman, A. E., Bilan, M. I., & Usov, A. I. (2007). A comparative study of the anti-inflammatory, anticoagulant, antiangiogenic, and antiadhesive activities of nine different fucoidans from brown seaweeds. *Glycobiology*, 17(5), pp. 541–552.

Cunha, L., & Grenha, A. (2016). Sulfated seaweed polysaccharides as multifunctional materials in drug delivery applications. *Marine Drugs*, 14(3), p. 42.

Daniel, R., Chevolot, L., Carrascal, M., Tissot, B., Mourão, P. A., & Abian, J. (2007). Electrospray ionization mass spectrometry of oligosaccharides derived from fucoidan of Ascophyllumnodosum. *Carbohydrate Research*, 342(6), pp. 826–834.

Dürig, J., Bruhn, T., Zurborn, K. H., Gutensohn, K., Bruhn, H. D., & Béress, L. (1997). Anticoagulant fucoidan fractions from *Fucus vesiculosus* induce platelet activation in vitro. *Thrombosis Research*, 85(6), pp. 479–491.

Fitton, J. H., Irhimeh, M., & Falk, N. (2007). Macroalgal-fucoidan extracts: A new opportunity for marine cosmetics. *Cosmetics and Toiletries*, 122(8), p. 55.

Flórez-Fernández, N., López-García, M., González-Muñoz, M. J., Vilariño, J. M. L., & Domínguez, H. (2017). Ultrasound-assisted extraction of fucoidan from Sargassummuticum. *Journal of Applied Phycology*, 29(3), pp. 1553–1561.

Fuse, T., & Goto, F. (1971). Studies on utilization of agar: Part X. Some properties of agarose and agaropectin isolated from various mucilaginous substances of red seaweeds. *Agricultural and Biological Chemistry*, 35(6), pp. 799–804.

Garcia-Vaquero, M., Rajauria, G., O'Doherty, J. V., & Sweeney, T. (2017). Polysaccharides from macroalgae: Recent advances, innovative technologies and challenges in extraction and purification. *Food Research International*, 99(3), pp. 1011–1020.

Hahn, T., Lang, S., Ulber, R., & Muffler, K. (2012). Novel procedures for the extraction of fucoidan from brown algae. *Process Biochemistry*, 47(12), pp. 1691–1698.

Huang, Y. C., Chen, J. K., Lam, U. I., & Chen, S. Y. (2014). Preparing, characterizing, and evaluating chitosan/fucoidan nanoparticles as oral delivery carriers. *Journal of Polymer Research*, 21(5), p. 415.

Huang, Y. C., & Kuo, T. H. (2016). O-carboxymethyl chitosan/fucoidan nanoparticles increase cellular curcumin uptake. *Food Hydrocolloids*, 53, pp. 261–269.

Huang, Y. C., & Li, R. Y. (2014). Preparation and characterization of antioxidant nanoparticles composed of chitosan and fucoidan for antibiotics delivery. *Marine Drugs*, 12(8), pp. 4379–4398.

Huang, Y. C., Li, R. Y., Chen, J. Y., & Chen, J. K. (2016). Biphasic release of gentamicin from chitosan/fucoidan nanoparticles for pulmonary delivery. *Carbohydrate Polymers*, 138, pp. 114–122.

Huang, Y. C., & Liu, T. J. (2012). Mobilization of mesenchymal stem cells by stromal cell-derived factor-1 released from chitosan/tripolyphosphate/fucoidan nanoparticles. *Acta Biomaterialia*, 8(3), pp. 1048–1056.

Huang, Y. C., & Yang, Y. T. (2016). Effect of basic fibroblast growth factor released from chitosan-fucoidan nanoparticles on neurite extension. *Journal of Tissue Engineering and Regenerative Medicine*, 10(5), pp. 418–427.

Hwang, P. A., Hung, Y. L., Phan, N. N., Hieu, B. T., Chang, P. M., Li, K. L., & Lin, Y. C. (2016). The in vitro and in vivo effects of the low molecular weight fucoidan on the bone osteogenic differentiation properties. *Cytotechnology*, 68(4), pp. 1349–1359.

Jeong, H. S., Venkatesan, J., & Kim, S. K. (2013). Hydroxyapatite-fucoidannanocomposites for bone tissue engineering. *International Journal of Biological Macromolecules*, 57, pp. 138–141.

Jin, G., & Kim, G. (2012). Multi-layered polycaprolactone–alginate–fucoidanbiocomposites supplemented with controlled release of fucoidan for bone tissue regeneration: Fabrication, physical properties, and cellular activities. *Soft Matter*, 8(23), pp. 6264–6272.

Khalafu, S. H. S., Aida, W. M. W., Lim, S. J., & Maskat, M. Y. (2017). Effects of deodorisation methods on volatile compounds, chemical properties and antioxidant activities of fucoidan isolated from brown seaweed (Sargassum sp.). *Algal Research*, 25, pp. 507–515.

Kikionis, S., Ioannou, E., Toskas, G., & Roussis, V. (2015). Electrospun biocomposite nanofibers of ulvan/PCL and ulvan/PEO. *Journal of Applied Polymer Science*, 132(26).

Kim, B. S., Kang, H. J., Park, J. Y., & Lee, J. (2015). Fucoidan promotes osteoblast differentiation via JNK-and ERK-dependent BMP2–Smad 1/5/8 signaling in human mesenchymal stem cells. *Experimental and Molecular Medicine*, 47(1), p. e128.

Kim, B. S., Yang, S. S., You, H. K., Shin, H. I., & Lee, J. (2018). Fucoidan-induced osteogenic differentiation promotes angiogenesis by inducing vascular endothelial growth factor secretion and accelerates bone repair. *Journal of Tissue Engineering and Regenerative Medicine*, 12(3), pp. e1311–e1324.

Kim, M. J., Jeon, J., & Lee, J. S. (2014). Fucoidan prevents high-fat diet-induced obesity in animals by suppression of fat accumulation. *Phytotherapy Research*, 28(1), pp. 137–143.

Kim, S.-K., & Cho, Y.-S. (2011). Pharmaceutical compositions containing fucoidan for stimulating and activating osteogenesis. Google Patents.

Kwak, J. Y. (2014). Fucoidan as a marine anticancer agent in preclinical development. *Marine Drugs*, 12(2), pp. 851–870.

Lahaye, M., & Robic, A. (2007). Structure and functional properties of ulvan, a polysaccharide from green seaweeds. *Biomacromolecules*, 8(6), pp. 1765–1774.

Lee, E. J., & Lim, K. H. (2016). Formation of chitosan-fucoidan nanoparticles and their electrostatic interactions: Quantitative analysis. *Journal of Bioscience and Bioengineering*, 121(1), pp. 73–83.

Lee, J. S., Jin, G. H., Yeo, M. G., Jang, C. H., Lee, H., & Kim, G. H. (2012). Fabrication of electrospun biocomposites comprising polycaprolactone/fucoidan for tissue regeneration. *Carbohydrate Polymers*, 90(1), pp. 181–188.

Lee, K. Y., & Mooney, D. J. (2012). Alginate: Properties and biomedical applications. *Progress in Polymer Science*, 37(1), pp. 106–126.

Li, B., Lu, F., Wei, X., & Zhao, R. (2008). Fucoidan: Structure and bioactivity. *Molecules*, 13(8), pp. 1671–1695.

Li, L., Ni, R., Shao, Y., & Mao, S. (2014). Carrageenan and its applications in drug delivery. *Carbohydrate Polymers*, 103, pp. 1–11.

Lim, S. J., & Aida, W. M. W. (2017). Extraction of sulfated polysaccharides (fucoidan) from brown seaweed. In: *Seaweed Polysaccharides* (pp. 27–46): Elsevier.

Lowe, B., Venkatesan, J., Anil, S., Shim, M. S., & Kim, S. K. (2016). Preparation and characterization of chitosan-natural nano hydroxyapatite-fucoidan nanocomposites for bone tissue engineering. *International Journal of Biological Macromolecules*, 93(B), pp. 1479–1487.

Lu, K. Y., Li, R., Hsu, C. H., Lin, C. W., Chou, S. C., Tsai, M. L., & Mi, F. L. (2017). Development of a new type of multifunctional fucoidan-based nanoparticles for anticancer drug delivery. *Carbohydrate Polymers*, 165, pp. 410–420.

Malafaya, P. B., Silva, G. A., & Reis, R. L. (2007). Natural–origin polymers as carriers and scaffolds for biomolecules and cell delivery in tissue engineering applications. *Advanced Drug Delivery Reviews*, 59(4–5), pp. 207–233.

Manivasagan, P., Bharathiraja, S., Bui, N. Q., Jang, B., Oh, Y. O., Lim, I. G., & Oh, J. (2016). Doxorubicin-loaded fucoidan capped gold nanoparticles for drug delivery and photoacoustic imaging. *International Journal of Biological Macromolecules*, 91, pp. 578–588.

McCandless, E. L., & Craigie, J. S. (1979). Sulfated polysaccharides in red and brown algae. *Annual Review of Plant Physiology*, 30(1), pp. 41–53.

Mihaila, S. M., Popa, E. G., Reis, R. L., Marques, A. P., & Gomes, M. E. (2014). Fabrication of endothelial cell-laden carrageenan microfibers for microvascularized bone tissue engineering applications. *Biomacromolecules*, 15(8), pp. 2849–2860.

Mogoşanu, G. D., & Grumezescu, A. M. (2014). Natural and synthetic polymers for wounds and burns dressing. *International Journal of Pharmaceutics*, 463(2), pp. 127–136.

Murakami, K., Aoki, H., Nakamura, S., Nakamura, S. I., Takikawa, M., Hanzawa, M., Kishimoto, S., Hattori, H., Tanaka, Y., Kiyosawa, T., & Sato, Y. (2010a). Hydrogel blends of chitin/chitosan, fucoidan and alginate as healing-impaired wound dressings. *Biomaterials*, 31(1), pp. 83–90.

Murakami, K., Ishihara, M., Aoki, H., Nakamura, S., Nakamura, S. I., Yanagibayashi, S., Takikawa, M., Kishimoto, S., Yokoe, H., Kiyosawa, T., & Sato, Y. (2010b). Enhanced healing of mitomycin C-treated healing-impaired wounds in rats with hydrosheets composed of chitin/chitosan, fucoidan, and alginate as wound dressings. *Wound Repair and Regeneration*, 18(5), pp. 478–485.

O'Leary, R., Rerek, M., & Wood, E. J. (2004). Fucoidan modulates the effect of transforming growth factor (TGF)-β1 on fibroblast proliferation and wound repopulation in in vitro models of dermal wound repair. *Biological and Pharmaceutical Bulletin*, 27(2), pp. 266–270.

Park, S. J., Lee, K. W., Lim, D. S., & Lee, S. (2012). The sulfated polysaccharide fucoidan stimulates osteogenic differentiation of human adipose-derived stem cells. *Stem Cells and Development*, 21(12), pp. 2204–2211.

Paul, W., & Sharma, C. P. (2004). Chitosan and alginate wound dressings: A short review. *Trends in Biomaterials and Artificial Organs*, 18(1), pp. 18–23.

Pinheiro, A. C., Bourbon, A. I., Cerqueira, M. A., Maricato, É., Nunes, C., Coimbra, M. A., & Vicente, A. A. (2015). Chitosan/fucoidan multilayer nanocapsules as a vehicle for controlled release of bioactive compounds. *Carbohydrate Polymers*, 115, pp. 1–9.

Prajapati, V. D., Maheriya, P. M., Jani, G. K., & Solanki, H. K. (2014). Carrageenan: A natural seaweed polysaccharide and its applications. *Carbohydrate Polymers*, 105, pp. 97–112.

Puvaneswary, S., Talebian, S., Raghavendran, H. B., Murali, M. R., Mehrali, M., Afifi, A. M., Kasim, N. H. B. A., & Kamarul, T. (2015). Fabrication and in vitro biological activity of βTCP-Chitosan-Fucoidan composite for bone tissue engineering. *Carbohydrate Polymers*, 134, pp. 799–807.

Raghavendran, H. R. B., Srinivasan, P., & Rekha, S. (2011). Immunomodulatory activity of fucoidan against aspirin-induced gastric mucosal damage in rats. *International Immunopharmacology*, 11(2), pp. 157–163.

Rani, V., Shakila, R., Jawahar, P., & Srinivasan, A. (2017). Influence of species, geographic location, seasonal variation and extraction method on the fucoidan yield of the brown seaweeds of Gulf of Mannar, India. *Indian Journal of Pharmaceutical Sciences*, 79(1), pp. 65–71.

Rodriguez-Jasso, R. M., Mussatto, S. I., Pastrana, L., Aguilar, C. N., & Teixeira, J. A. (2011). Microwave-assisted extraction of sulfated polysaccharides (fucoidan) from brown seaweed. *Carbohydrate Polymers*, 86(3), pp. 1137–1144.

Ruocco, N., Costantini, S., Guariniello, S., & Costantini, M. (2016). Polysaccharides from the marine environment with pharmacological, cosmeceutical and nutraceutical potential. *Molecules*, 21(5), p. 551.

Santo, V. E., Frias, A. M., Carida, M., Cancedda, R., Gomes, M. E., Mano, J. F., & Reis, R. L. (2009). Carrageenan-based hydrogels for the controlled delivery of PDGF-BB in bone tissue engineering applications. *Biomacromolecules*, 10(6), pp. 1392–1401.

Saravana, P. S., Cho, Y. J., Park, Y. B., Woo, H. C., & Chun, B. S. (2016). Structural, antioxidant, and emulsifying activities of fucoidan from Saccharina japonica using pressurized liquid extraction. *Carbohydrate Polymers*, 153, pp. 518–525.

Saravana, P. S., Tilahun, A., Gerenew, C., Tri, V. D., Kim, N. H., Kim, G.-D., Woo, H.-C., & Chun, B.-S. (2018). Subcritical water extraction of fucoidan from Saccharina japonica: Optimization, characterization and biological studies. *Journal of Applied Phycology*, 30(1), pp. 579–590.

Selinger, E., Dell, S. M., Colliopoulos, J. A., & Reilly Jr, W. J. (1998). MCC: alginate pharmaceutical suspensions. Google Patents.

Sezer, A. D., & Cevher, E. (2011). Fucoidan: A versatile biopolymer for biomedical applications. In: *Active Implants and Scaffolds for Tissue Regeneration* (pp. 377–406): Springer.

Sezer, A. D., Cevher, E., Hatıpoğlu, F., Oğurtan, Z., Baş, A. L., & Akbuğa, J. (2008). Preparation of fucoidan-chitosan hydrogel and its application as burn healing accelerator on rabbits. *Biological and Pharmaceutical Bulletin*, 31(12), pp. 2326–2333.

Sezer, A. D., Hatipoğlu, F., Cevher, E., Oğurtan, Z., Baş, A. L., & Akbuğa, J. (2007). Chitosan film containing fucoidan as a wound dressing for dermal burn healing: Preparation and in vitro/in vivo evaluation. *AAPS PharmSciTech*, 8(2), pp. E94–E101.

Synytsya, A., Kim, W.-J., Kim, S.-M., Pohl, R., Synytsya, A., Kvasnička, F., Čopíková, J., & Park, Y. I. (2010). Structure and antitumour activity of fucoidan isolated from sporophyll of Korean brown seaweed *Undaria pinnatifida*. *Carbohydrate Polymers*, 81(1), pp. 41–48.

Thuy, T. T. T., Ly, B. M., Van, T. T. T., Van Quang, N., Tu, H. C., Zheng, Y., Seguin-Devaux, C., Mi, B., & Ai, U. (2015). Anti-HIV activity of fucoidans from three brown seaweed species. *Carbohydrate Polymers*, 115, pp. 122–128.

Varco, J. J., Wis-Surel, G. M., & Jachowicz, J. (1990). Aerosol hair setting composition containing an alginate. Google Patents.

Venkatesan, J., Anil, S., & Kim, S.-K. (2017). *Seaweed Polysaccharides: Isolation, Biological and Biomedical Applications*. Elsevier.

Venkatesan, J., Bhatnagar, I., & Kim, S. K. (2014). Chitosan-alginate biocomposite containing fucoidan for bone tissue engineering. *Marine Drugs*, 12(1), pp. 300–316.

Venkatesan, J., Bhatnagar, I., Manivasagan, P., Kang, K. H., & Kim, S. K. (2015). Alginate composites for bone tissue engineering: A review. *International Journal of Biological Macromolecules*, 72, pp. 269–281.

Venkatesan, J., Singh, S. K., Anil, S., Kim, S. K., & Shim, M. S. (2018). Preparation, characterization and biological applications of biosynthesized silver nanoparticles with chitosan-fucoidan coating. *Molecules*, 23(6), p. 1429.

Wang, F., Schmidt, H., Pavleska, D., Wermann, T., Seekamp, A., & Fuchs, S. (2017). Crude fucoidan extracts impair angiogenesis in models relevant for bone regeneration and osteosarcoma via reduction of VEGF and SDF-1. *Marine Drugs*, 15(6), p. 186.

Wang, J., Zhang, Q., Zhang, Z., Song, H., & Li, P. (2010). Potential antioxidant and anticoagulant capacity of low molecular weight fucoidan fractions extracted from Laminaria japonica. *International Journal of Biological Macromolecules*, 46(1), pp. 6–12.

Wu, S. J., Don, T. M., Lin, C. W., & Mi, F. L. (2014). Delivery of berberine using chitosan/fucoidan-taurine conjugate nanoparticles for treatment of defective intestinal epithelial tight junction barrier. *Marine Drugs*, 12(11), pp. 5677–5697.

Yang, C., Chung, D., & You, S. (2008). Determination of physicochemical properties of sulphated fucans from sporophyll of *Undaria pinnatifida* using light scattering technique. *Food Chemistry*, 111(2), pp. 503–507.

Young, A. T., Kang, J. H., Kang, D. J., Venkatesan, J., Chang, H. K., Bhatnagar, I., Chang, K. Y., Hwang, J. H., Salameh, Z., Kim, S. K., & Kim, H. T. (2016). Interaction of stem cells with nano hydroxyapatite-fucoidan bionanocomposites for bone tissue regeneration. *International Journal of Biological Macromolecules*, 93(B), pp. 1488–1491.

Zarrintaj, P., Manouchehri, S., Ahmadi, Z., Saeb, M. R., Urbanska, A. M., Kaplan, D. L., & Mozafari, M. (2018). Agarose-based biomaterials for tissue engineering. *Carbohydrate Polymers*, 187, pp. 66–84.

Zhuang, C., Itoh, H., Mizuno, T., & Ito, H. (1995). Antitumor active fucoidan from the brown seaweed, umitoranoo (*Sargassum thunbergii*). *Bioscience, Biotechnology, and Biochemistry*, 59(4), pp. 563–567.

3 Bioprocess Parameters of Production of Cyanobacterial Exopolysaccharide
Biomass Production and Product Recovery

Onkar Nath Tiwari, Sagnik Chakraborty, Indrama Devi, Abhijit Mondal, Biswanath Bhunia, and Shen Boxiong

CONTENTS

Introduction 26
Production of Exopolysaccharide 26
    Commercial-Scale Algae Cultivation 26
        Open Cultivation Systems 26
        Shallow Ponds 26
        Cultivation Tanks 26
        Circular and Raceway Ponds..27
        Sloped Open Systems 27
        Closed Reactor Systems 27
        Tubular Reactors 27
    Laboratory-Scale Bioreactors 28
        Optimization Strategies for Yield Improvement 28
        Algae Cultivation Regimes 28
Extraction and Purification of EPS 28
Treatments for the Extraction of Cell-bound EPS 28
    Isolation and Extraction..29
    Traditional Extraction Route via Alcoholic Precipitation 29
    Extraction Technique Applying Peripheral Ultrafiltration 30
    Ultrasound or Microwave-Assisted Extraction Technique 30
Conclusion/Inference 30
References 30

BOX 3.1 SALIENT FEATURES

The exopolysaccharides (EPSs) produced by cyanobacteria are important constituents for the development of biofilm through association of microbial communities in various habitats. Discoveries in the field of cyanobacterial exopolysaccharide synthesis have opened up new opportunities for bioprocess engineers towards the production of biopolymers which is suitable for various industrial applications. Such engineering approaches are able to produce efficient exopolysaccharide as well as modified polymers exhibiting unique functional properties for specific interest. Therefore, a low-cost production system along with improving the quality of exopolysaccharide becomes economically viable, which is only possible through the development of an optimized production system. This book chapter aims to demonstrate the most important studies on the bioprocess engineering approaches of native producer for enhanced production of EPS available in literature to date. The chapter also discusses technical concerns about the application of various reactors and their optimization for high yield cyanobacterial exopolysaccharide production. In addition, progress toward purification of exopolysaccharide using suitable unit operation is highlighted extensively in this chapter.

INTRODUCTION

The marine environment is a vast and complex source of microorganisms. It envisages that one milliliter of seawater contains more than 10^5 bacterial cells. Marine bacteria encompass many different phyla, including actinobacteria, proteobacteria, and cyanobacteria (Watson et al. 1977). Cyanobacteria are water harboring microorganisms which sustain their livelihood by gaining energy from sunlight (2Bhunia et al. 2017; Sardar et al. 2018). They occur in bodies of water such as marine lakes and estuaries. The cyanobacteria play a pivotal role in controlling the biogeochemical cycle of the atmosphere (Suthers & Rissik 2009). It is evident that marine cyanobacteria yield numerous unique nutraceuticals and natural products.

The potential of cyanobacteria for the production of enormous quantity of exopolysaccharide is well recognized (De Philippis et al. 2011). The ubiquitous polysaccharides are abundantly available within microorganisms, plants, and animals such as starch, glycogen, and cellulose. They find applications as exopolysaccharide in the preparation of adhesives, bioflocculants, soil booster including biosorbents, etc. (Table 3.1).

Hence, this comprehensive review encompasses and envisions the significantly influencing factors chiefly related to the yield of exopolysaccharide and its extraction from nutrient growth medium as documented in published literature.

PRODUCTION OF EXOPOLYSACCHARIDE

It has been demonstrated that controlled growth and harvesting of cyanobacterial biomass would enrich the productivity and production of value-added compounds in an economically viable manner. Bioreactors are being specifically designed for maximizing the yield of biomass and its significant products. They could be grown in open or closed systems depending on the ease of cultivation and scale-up requirements.

COMMERCIAL-SCALE ALGAE CULTIVATION

Open Cultivation Systems

In open systems, the cells are exposed directly to the atmosphere. Open systems are the oldest cultivation types and are still use widely (Figure 3.1A). In fact, the majority of commercial cyanobacterial production adopts open pond cultivation (Lee 1997). Even though many constraints of these systems hinder productivity, the open systems are still significantly cheap in comparison to closed bioreactors (Carvalho et al. 2006). Due to contamination issues, sustained production in open ponds has been successful only for a small number of organisms which thrive in extreme environments including high salinity or high pH (Parmar et al. 2011).

Shallow Ponds

The simplest systems for cultivation of cyanobacteria are shallow ponds (Richmond 2008). The production limitations, due to insufficient mixing, poor light utilization and high evaporation rates, result in a nutrient imbalance including carbon delivery which is quite inefficient. To achieve high productivity one needs to ensure easy access to water, nutrients, and sunlight for the most of the year. Moreover, the temperature should be stable, rainfall should be small, and land costs should be low. For these reasons, shallow ponds are not applicable in countries where land is expensive, and climate conditions vary to a significant extent.

Cultivation Tanks

Cultivation tanks possess a similar design to shallow ponds, but they are typically of much smaller volume, which allows better control of environmental conditions. Abysmal light utilization efficiency usually limits the productivity. Also, the gas transfer is typically insufficient, and thus possibilities for successful scale-up are minimal (Borowitzka 1999).

TABLE 3.1
The Application of Cyanobacterial Exopolysaccharide

Cyanobacteria	Application of EPS	References
Anabaena sp. BTA992	Bioflocculants property	(Khangembam et al. 2016)
Sargassum thunbergii	Antitumor agent	(Itoh et al. 1993)
Cyanothece sp. ATCC 51142	Adsorbent for heavy metals, organic pollutants such as dyes and pesticides	(Aksu 2005; Shah et al. 2000)
Microcoleus sp.	Soil conditioner	(Mazor et al. 1996)
M. vaginatus	Biological soil crusts inducer	(Wang et al. 2009)
Arthrospira sp.	Thickening agent	(Chentir et al. 2017; Velasco et al. 2009)
Anabaena sp. ATCC 33047	Biopolymer	(Moreno et al. 1998)

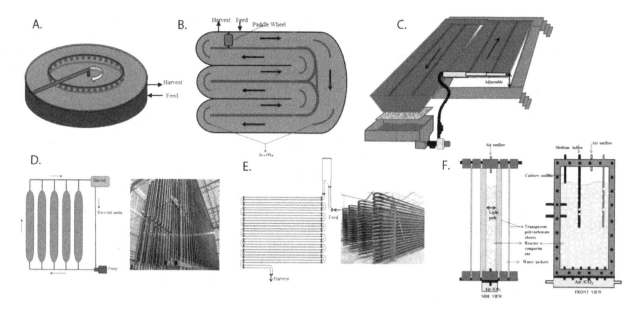

FIGURE 3.1 **(See color insert.)** A Circular pond, B Raceway pond, C Sloped open bioreactor, D Vertical tubular reactor, E Horizontal tubular reactor, F Flat panel laboratory reactor (Chisti 2007; Olivieri et al. 2014; Parmar et al. 2011; Posten 2009; Sforza et al. 2014).

Circular and Raceway Ponds

Culture reservoirs with constant circulation represent improved systems which overcome some of the constraints mentioned previously. Mixing is often secured by a rotating arm (circular ponds, Figure 3.1) or paddle wheel (raceway ponds, Figure 3.1B) (Chisti 2007). The culture depth is required to stay at the minimal level of approximately 15 cm as a prevention of flow reduction and formation of turbulence. Such depth is far from sufficient for light penetration, and typically low biomass concentrations are achieved in both raceway and circular ponds (Sheehan et al. 1998). Circular basins are slightly more expensive constructions and have significantly higher energy requirements for mixing; however, they are adopted widely. The raceway ponds are the most utilized industrial plants for outdoor cultivation.

Sloped Open Systems

Light availability is effectively managed using sloped outdoor cultivation units (Figure 3.1C) with culture depth reduced to approximately 1 cm. The liquid pump secures culture circulation and carbon dioxide incorporated into the system through its suction part. By connecting multiple cultivation panels, culture mixing is significantly improved, and thus the algae population can reach higher densities when compared to other open systems. With denser culture, harvesting costs can be reduced. However, even in sloped systems, the productivity is potentially limited by contamination and environmental variations, and the biomass productivity is far from the maxima achieved in closed cultivation systems (Olivieri et al. 2014).

Closed Reactor Systems

In closed systems, culture separates from the outside environment. Besides the possibility of avoiding contamination, closed systems offer broader options of cultivation conditions control, which gives them at least partial independence on geographical location. With increased cell densities, harvesting costs are further reduced. Moreover, they can be grown vertically in reactors reducing the land required for production.

However, the initial investments for reactors' setup are much higher when compared to open systems, which is probably the main reason for the still limited commercial use of these systems. It is opined that capital investments required per liter volume of culture, as well as production costs for a kilogram of dried biomass, were the highest for closed reactor systems, in comparison with raceway ponds and even heterotrophic cultivation in a closed fermenter (Davis et al. 2011).

Tubular Reactors

Tubular reactors are made of flat transparent tubes connected by U-shape or L-shape bands to form flat loops which can be further arranged either vertically (Figure 3.1D) or horizontally (horizontal tubular reactor, (Figure 3.1E). Gas exchange, as well as nutrient addition, is typically carried out in a separate vessel, with a pump used for culture circulation. Photobioreactor tubes transmit photosynthetically active irradiation and are reliable and stable in both mechanical and transparency properties. The tubes are usually of diameters between 10–60 mm, and length of up to several hundred meters (Posten 2009). Cyanobacteria can potentially attach to the internal tube walls. Besides the systems mentioned previously, a cylindrical airlift bioreactor can also be considered as a particular type of tubular reactor; with typically smaller volumes, vertical orientation, and mixing secured by gas injection (Cozma & Gavrilescu 2012). In the cylindrical bioreactor with a fiber-optic light transmission system, the highest real production rates have been reported (Olivieri et al. 2014).

Laboratory-Scale Bioreactors

The most commercially successful cyanobacterial biotechnological applications start in the laboratory with screening experiments, identification of growth, and production optima or limitations, as well as hypotheses validations which all typically take place in a controlled laboratory environment. The cultivation of cyanobacteria can take place in theory in any vessel that secures a stable environment – thus, even the basic setups which utilize Erlenmeyer flasks or test tubes for cultivation are still very popular in research laboratories (Figure 3.1F). They are used to develop innocula for further scale-up in bioreactors described previously. Various designs of laboratory-scale bioreactors have been demonstrated for the cultivation of algal biomass (Sforza et al. 2014).

Optimization Strategies for Yield Improvement

The high productivity of the cyanobacteria is possible via boosting cellular production on a metabolic state by recombinant DNA technology as well as via designing the best-suited photoreactor encompassing the cultivation methods. The influencing parameters include water quality, temperature, light, pH, nourishments (macro/micro), and a suitable amount of salts as well as ions concentration including gaseous exchange. The cyanobacterial strains sustain an exceptional range of conditions ranging from subzero to an elevated temperature of around 70°C generally present in naturally occurring hot springs (Seckbach 2007). Likewise, extremophiles thrive in extreme pH, light, or salinity conditions. However, the parameters of the physical and chemical environment is to be optimized for maximized growth of biomass.

Algae Cultivation Regimes

Cyanobacteria is often cultured as batch, semi-continuous, or in continuous regime. Each of these modes have advantages and constraints. The simplest is the batch mode, where resources are finite, and cell concentration continually increases until some factor becomes limiting (typically some nutrient is exhausted). Potential products also increase their intensity in the medium over time. For growth restoration, the limiting factors need to be replenished. The batch growth is a highly dynamic process with culture density increasing as the typical sigmoid growth curve. Individual phases of batch growth can be categorized into lag, acceleration, exponential, retardation ("linear"), stationary, and decline period (Finkel 2006).

Another cultivation regime is the continuous mode, where cell density is maintained at a defined level by the constant addition of fresh cultivation medium. The medium inflow rate is proportional to culture growth rate and the culture removal rate. Typically, during the continuous cultivation, cells are maintained in an exponential phase, to reach maximal growth rates and biomass production. Variation of the continuous mode is the semi-continuous regime, where culture medium is replaced periodically within a defined period. A special type, standing between continuous and semi-continuous growth, is the "quasi-continuous" cultivation regime, with typically small dilution range, based on monitoring of specific culture parameters like turbidity. In the so-called turbidostat case, dilution is typically automatically controlled; it starts after the culture reach defined upper-density level and stops when the culture is diluted to a defined lower density level. Besides turbidostat, other quasi-continuous regimes have been developed, including pH-stat, physiostat, or luminostat.

EXTRACTION AND PURIFICATION OF EPS

The majority of the extraction techniques relate to a dissolvable liberated polysaccharide (DLP) within the nutrient media chiefly via unicellular cyanobacteria. The methods of extraction is often customized for enhanced recovery (Helm et al. 2000; Li et al. 2001). The common protocol adapted for extraction and purification of exopolysaccharide is shown in Figure 3.2.

TREATMENTS FOR THE EXTRACTION OF CELL-BOUND EPS

It is a general observation that a fraction of EPS is retained as adhered slime coat over the microalgal cell surface (Arad & Levy-Ontman 2010). Methods employing the use of formaldehyde (FA), ethylenediaminetetraacetic acid (EDTA), sodium hydroxide (NaOH), coupled with sonication, heating, cell cleansing in aqueous media, adoption of ionic resin are followed with suitable modification tailored to the needs to extract the

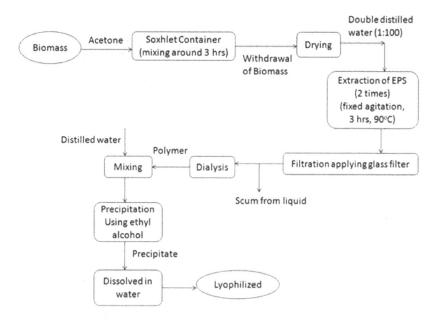

FIGURE 3.2 Schematic diagram of extraction and purification of EPS.

adhering EPS (Pierre et al. 2014; Takahashi et al. 2009). However, another research elucidated the application of FA along with glutaraldehyde (GTA) as a fixative for shielding microalgal cell during segregation of the EPS. An additional choice or route is to rinse the microalgae by applying water for further extraction of cell adhered EPS. Depending on the class of microalgae, warm water (30–95°C), range of pH, the extent of treatment modification (1–4 hr) is optimized for subjecting the biomass for release and dissolution of EPS. The small molecular mass (SMM) intracellular carbohydrate impurity can be removed from EPS extracts using overnight selective alcoholic precipitation with cold and absolute ethanol (−20°C) at the final concentration of 75% (in the mixture). This strategy led to the sole recovery of EPS of high molecular weights (insoluble ethanol fraction) since LMW (ethanol soluble fraction) did not precipitate.

ISOLATION AND EXTRACTION

Several genera of cyanobacteria can produce EPSs. They include *Anabaena variabilis* (Bhatnagar et al. 2012), *Nostoc calcicola* (Singh & Das 2011), *Limnothrix redekei* PUPCCC116 (Khattar et al. 2010) and many more.

There are several species reviewed by Pengfu et al. (Li et al. 2001) about exploring the possibilities for the production of cyanobacterial EPS. The physical and chemical methods are generally used for the separation of soluble and bound EPS. Though the soluble EPS can be released from cyanobacteria through the centrifugation process, additional chemical treatment is required for separation of the bound EPS. Comte et al. (Comte et al. 2006) reported that the extraction process determines the EPS yield as well as the chemical composition of EPS. The ultrasonication, cation exchange resin, heating, high-speed centrifugation are adopted under physical process and alkaline reagents, ethylene diamine tetraacetic acid (EDTA), aldehyde solutions are some of the chemical process steps.

TRADITIONAL EXTRACTION ROUTE VIA ALCOHOLIC PRECIPITATION

Ramus (Ramus 1972) reported segregation of the "encapsulating polysaccharide" from *Porphyridium* biomass. Red microalgae produces sulfated polysaccharides dissolvable within the nutrient media. The removal of color from biomass was achieved by applying the acetone along with ethanol, further solubilization of exopolysaccharide glue from porphyridium cells was achieved using warm water. Another study also reports the use of absolute alcohol or isopropanol (Liu et al. 2015; Patel et al. 2013). The yield of polysaccharide is influenced by temperature of precipitation and polarity of alcohol. This technique robustly yields EPS. The method possesses specific merits like further recycling the alcohol via distillation to modify extremely viscid solution. Patel et al. (Patel et al. 2013) delineated the extraction and salting out of EPS from *Porphyridium cruentum* applying the alcoholic precipitation, separation via membrane method. Furthermore, they have inferred the application of diafiltration employing 300 kDa membrane as a highly effective technique. Occasionally, an additional refinement phase is necessary to remove the undesired molecules like proteins, pigments, salts and another compounds via trichloroacetic acid modification, peripheral ultrafiltration, or precipitation by using alcohol (Li et al. 2011; Patel et al. 2013).

EXTRACTION TECHNIQUE APPLYING PERIPHERAL ULTRAFILTRATION

The peripheral ultrafiltration method is projected as a substitute to the traditional extraction of EPS applying ethanol precipitation (Charcosset 2006). The salting out method is studied by applying the diafiltration process to facilitate the substitution of the solvent via a new buffer (Charcosset 2006). The ultrafiltration approach significantly diminishes the generation of "sieve bar" and passes the suitable concentration of the biomolecules reasonably to the final edge of the filtration process. The efficacy of the membrane system is solely reliant on the following points, (i) the thickness and concentration of EPS; (ii) the orifice mass allocation along with the architecture of membrane, and (iii) the speed of the flow and the pressure of the transmembrane.

Li et al. (Li et al. 2011) established a precommercial extraction and refinement technique of functional biological EPS via the nutrient medium of numerous cyanobacteria along with microalgae like *Chaetocerous mueleri, Chlorella pyrenoidosa, spirulina platensis, Haematococcus pluvialis, Nostoc commune* along with *Nostoc sphaeroides*. The expertise is solely reliant on the microfiltration approach (polypropylene membrane) to separate microalgae along with a peripheral flow ultrafiltration system using a polyethersulfone membrane (MWCO, 5000 Da) to accumulate the EPS (20 to 40 times). Furthermore, adoption of two additional phases of the membrane process was incorporated by Maracati et al. (Marcati et al. 2014) to segregate polysaccharide along with B-PE from *Porphyridium* sp. accompanying polyethersulfone membrane possessing MWCOs around 300 kDa to 10k Da (Patel et al. 2013).

ULTRASOUND OR MICROWAVE-ASSISTED EXTRACTION TECHNIQUE

An unconventional route like ultrasound-assisted extraction (UAE) along with microwave-assisted extraction (MAE) is also recommended for the advancement of extraction of biomolecules chiefly from macroalgae along with microalgae (Budarin et al. 2012; Kadam et al. 2013). The UAE approach has been efficiently applied to extract polysaccharides from *Spirulina platensis* (Kurd & Samavati 2015). Whereas MAE portrays the extraction procedure of intracellular metabolite as originated via microalgae like carotenoid from *Dunaliella tertiolecta* (Pasquet et al. 2011) or phycobiliproteins from *Porphyridium purpureum* (Juin et al. 2015).

CONCLUSION/INFERENCE

The exopolysaccharides from microalgae are gaining importance as sources of bioactive molecules of therapeutic importance. Several commercially available exopolysaccharides are obtained from cyanobacterial sources. Development of biotechnological processes for the production of biomass and downstream processing for the production of high-quality EPS products is being considered. However the process innovation is centric to the type of organism and conditions favoring the production of biomass and EPS. Accordingly the innovative interventions are being adopted for maximizing the yield for commercial feasibility.

REFERENCES

Aksu, Z. Application of biosorption for the removal of organic pollutants: A review. *Process Biochemistry*. 2005. 40: 997–1026.

Arad, S. M., Levy-Ontman, O. Red microalgal cell-wall polysaccharides: Biotechnological aspects. *Current Opinion in Biotechnology*. 2010. 21: 358–364.

Bhatnagar, M., Pareek, S., Ganguly, J., Bhatnagar, A. Rheology and composition of a multi-utility exopolymer from a desert borne cyanobacterium *Anabaena variabilis*. *Journal of Applied Phycology*. 2012. 24: 1387–1394.

Bhunia, B., Uday, U. S. P., Oinam, G., Mondal, A., Bandyopadhyay, T. K., Tiwari, O. N. Characterization, genetic regulation and production of cyanobacterial exopolysaccharides and its applicability for heavy metal removal. *Carbohydrate Polymers*. 2017. 179: 228–243.

Borowitzka, M. A. Commercial production of microalgae: Ponds, tanks, tubes and fermenters. *Journal of Biotechnology*. 1999. 70: 313–321.

Budarin, V., Ross, A. B., Biller, P., Riley, R., Clark, J. H., Jones, J. M., Gilmour, D. J., Zimmerman, W. Microalgae biorefinery concept based on hydrothermal microwave pyrolysis. *Green Chemistry*. 2012. 14: 3251–3254.

Carvalho, A. P., Meireles, L. A., Malcata, F. X. Microalgal reactors: A review of enclosed system designs and performances. *Biotechnology Progress*. 2006. 22: 1490–1506.

Charcosset, C. Membrane processes in biotechnology: An overview. *Biotechnology Advances*. 2006. 24: 482–492.

Chentir, I., Hamdi, M., Doumandji, A., Sadok, A. H., Ouada, H. B., Nasri, M., Jridi, M. Enhancement of extracellular polymeric substances (EPS) production in Spirulina

(Arthrospira sp.) by two-step cultivation process and partial characterization of their polysaccharidic moiety. *International Journal of Biological Macromolecules.* 2017. 105: 1412–1420.

Chisti, Y. Biodiesel from microalgae. *Biotechnology Advances.* 2007. 25: 294–306.

Comte, S., Guibaud, G., Baudu, M. Relations between extraction protocols for activated sludge extracellular polymeric substances (EPS) and EPS complexation properties: Part I. Comparison of the efficiency of eight EPS extraction methods. *Enzyme and Microbial Technology.* 2006. 38: 237–245.

Cozma, P., Gavrilescu, M. Airlift reactors: Applications in wastewater treatment. *Environmental Engineering and Management Journal.* 2012. 11: 1505–1515.

Davis, R., Aden, A., Pienkos, P. T. Techno-economic analysis of autotrophic microalgae for fuel production. *Applied Energy.* 2011. 88: 3524–3531.

De Philippis, R., Colica, G., Micheletti, E. Exopolysaccharide-producing cyanobacteria in heavy metal removal from water: Molecular basis and practical applicability of the biosorption process. *Applied Microbiology and Biotechnology.* 2011. 92: 697–708.

Finkel, S. E. Long-term survival during stationary phase: Evolution and the GASP phenotype. *Nature Reviews Microbiology.* 2006. 4: 113–120.

Helm, R. F., Huang, Z., Edwards, D., Leeson, H., Peery, W., Potts, M. Structural characterization of the released polysaccharide of desiccation-tolerant Nostoc commune DRH-1. *Journal of Bacteriology.* 2000. 182: 974–982.

Itoh, H., Noda, H., Amano, H., Zhuaug, C., Mizuno, T., Ito, H. Antitumor activity and immunological properties of marine algal polysaccharides, especially fucoidan, prepared from *Sargassum thunbergii* of Phaeophyceae. *Anticancer Research.* 1993. 13: 2045–2052.

Juin, C., Chérouvrier, J. R., Thiéry, V., Gagez, A. L., Bérard, J. B., Joguet, N., Kaas, R., Cadoret, J. P., Picot, L. Microwave-assisted extraction of phycobiliproteins from Porphyridium purpureum. *Applied Biochemistry and Biotechnology.* 2015. 175: 1–15.

Kadam, S. U., Tiwari, B. K., O'Donnell, C. P. Application of novel extraction technologies for bioactives from marine algae. *Journal of Agricultural and Food Chemistry.* 2013. 61: 4667–4675.

Khangembam, R., Tiwari, O., Kalita, M. Production of exopolysaccharides by the cyanobacterium Anabaena sp. BTA992 and application as bioflocculants. *Journal of Applied Biology and Biotechnology.* 2016. 4: 008–011.

Khattar, J. I. S., Singh, D. P., Jindal, N., Kaur, N., Singh, Y., Rahi, P., Gulati, A. Isolation and characterization of exopolysaccharides produced by the cyanobacterium Limnothrix redekei PUPCCC 116. *Applied Biochemistry and Biotechnology.* 2010. 162: 1327–1338.

Kurd, F., Samavati, V. Water soluble polysaccharides from *Spirulina platensis*: Extraction and in vitro anticancer activity. *International Journal of Biological Macromolecules.* 2015. 74: 498–506.

Lee, Y.-K. Commercial production of microalgae in the Asia-Pacific rim. *Journal of Applied Phycology.* 1997. 9: 403–411.

Li, H., Li, Z., Xiong, S., Zhang, H., Li, N., Zhou, S., Liu, Y., Huang, Z. Pilot-scale isolation of bioactive extracellular polymeric substances from cell-free media of mass microalgal cultures using tangential-flow ultrafiltration. *Process Biochemistry.* 2011. 46: 1104–1109.

Li, P., Harding, S. E., Liu, Z. Cyanobacterial exopolysaccharides: Their nature and potential biotechnological applications. *Biotechnology and Genetic Engineering Reviews.* 2001. 18: 375–404.

Liu, B., Sun, Z., Ma, X., Yang, B., Jiang, Y., Wei, D., Chen, F. Mutation breeding of extracellular polysaccharide-producing microalga *Crypthecodinium cohnii* by a novel mutagenesis with atmospheric and room temperature plasma. *International Journal of Molecular Sciences.* 2015. 16: 8201–8212.

Marcati, A., Ursu, A. V., Laroche, C., Soanen, N., Marchal, L., Jubeau, S., Djelveh, G., Michaud, P. Extraction and fractionation of polysaccharides and B-phycoerythrin from the microalga *Porphyridium cruentum* by membrane technology. *Algal Research.* 2014. 5: 258–263.

Mazor, G., Kidron, G. J., Vonshak, A., Abeliovich, A. The role of cyanobacterial exopolysaccharides in structuring desert microbial crusts. *FEMS Microbiology Ecology.* 1996. 21: 121–130.

Moreno, J., Vargas, M. A., Olivares, H., Rivas, J., Guerrero, M. G. Exopolysaccharide production by the cyanobacterium Anabaena sp. ATCC 33047 in batch and continuous culture. *Journal of Biotechnology.* 1998. 60: 175–182.

Olivieri, G., Salatino, P., Marzocchella, A. Advances in photobioreactors for intensive microalgal production: Configurations, operating strategies and applications. *Journal of Chemical Technology and Biotechnology.* 2014. 89: 178–195.

Parmar, A., Singh, N. K., Pandey, A., Gnansounou, E., Madamwar, D. Cyanobacteria and microalgae: A positive prospect for biofuels. *Bioresource Technology.* 2011. 102: 10163–10172.

Pasquet, V., Chérouvrier, J.-R., Farhat, F., Thiéry, V., Piot, J.-M., Bérard, J.-B., Kaas, R., Serive, B., Patrice, T., Cadoret, J.-P., Picot, L. Study on the microalgal pigments extraction process: Performance of microwave assisted extraction. *Process Biochemistry.* 2011. 46: 59–67.

Patel, A. K., Laroche, C., Marcati, A., Ursu, A. V., Jubeau, S., Marchal, L., Petit, E., Djelveh, G., Michaud, P. Separation and fractionation of exopolysaccharides from *Porphyridium cruentum*. *Bioresource Technology.* 2013. 145: 345–350.

Pierre, G., Zhao, J.-M., Orvain, F., Dupuy, C., Klein, G. L., Graber, M., Maugard, T. Seasonal dynamics of extracellular polymeric substances (EPS) in surface sediments of a diatom-dominated intertidal mudflat (Marennes–Oléron, France). *Journal of Sea Research.* 2014. 92: 26–35.

Posten, C. Design principles of photo-bioreactors for cultivation of microalgae. *Engineering in Life Sciences.* 2009. 9: 165–177.

Ramus, J. The production of extracellular polysaccharide by the unicellular red alga *Porphyridium aerugineum*. *Journal of Phycology*. 1972. 8: 97–111.

Richmond, A. 2008. *Handbook of Microalgal Culture: Biotechnology and Applied Phycology*. John Wiley & Sons.

Sardar, U. R., Bhargavi, E., Devi, I., Bhunia, B., Tiwari, O. N. Advances in exopolysaccharides based bioremediation of heavy metals in soil and water: A critical review. *Carbohydrate Polymers*. 2018. 199: 353–364.

Seckbach, J. 2007. *Algae and Cyanobacteria in Extreme Environments*. Springer Science & Business Media.

Sforza, E., Enzo, M., Bertucco, A. Design of microalgal biomass production in a continuous photobioreactor: An integrated experimental and modeling approach. *Chemical Engineering Research and Design*. 2014. 92: 1153–1162.

Shah, V., Ray, A., Garg, N., Madamwar, D. Characterization of the extracellular polysaccharide produced by a marine cyanobacterium, Cyanothece sp. ATCC 51142, and its exploitation toward metal removal from solutions. *Current Microbiology*. 2000. 40: 274–278.

Sheehan, J., Dunahay, T., Benemann, J., Roessler, P. 1998. *A Look Back at the US Department of Energy's Aquatic Species Program: Biodiesel From Algae*. National Renewable Energy Laboratory. p. 328.

Singh, S., Das, S. Screening, production, optimization and characterization of cyanobacterial polysaccharide. *World Journal of Microbiology and Biotechnology*. 2011. 27: 1971–1980.

Suthers, I. M., Rissik, D. 2009. *Plankton: A Guide to Their Ecology and Monitoring for Water Quality*. CSIRO Publishing.

Takahashi, E., Ledauphin, J., Goux, D., Orvain, F. Optimising extraction of extracellular polymeric substances (EPS) from benthic diatoms: Comparison of the efficiency of six EPS extraction methods. *Marine and Freshwater Research*. 2009. 60: 1201–1210.

Velasco, S. E., Areizaga, J., Irastorza, A., Dueñas, M. T., Santamaria, A., Muñoz, M. E. Chemical and rheological properties of the β-glucan produced by *Pediococcus parvulus* 2.6. *Journal of Agricultural and Food Chemistry*. 2009. 57: 1827–1834.

Wang, W., Liu, Y., Li, D., Hu, C., Rao, B. Feasibility of cyanobacterial inoculation for biological soil crusts formation in desert area. *Soil Biology and Biochemistry*. 2009. 41: 926–929.

Watson, S. W., Novitsky, T. J., Quinby, H. L., Valois, F. W. Determination of bacterial number and biomass in the marine environment. *Applied and Environmental Microbiology*. 1977. 33: 940–946.

4 Production, Extraction and Characterization of Alginates from Seaweeds

Faiez Hentati, Alina V. Ursu, Guillaume Pierre, Cedric Delattre, Bogdan Trica, Slim Abdelkafi, Gholamreza Djelveh, Tanase Dobre, and Philippe Michaud

CONTENTS

Introduction 34
Structural Characterization of Alginates 35
　Global Composition of Alginates 35
　Footprints and Main Functional Groups Identification by Fourier Transformed Infrared
　(FT-IR) Spectroscopy 36
　Determination of M/G Ratios by Chromatography 36
　Determination of Sequence and Blocks Distribution by Spectroscopic Methods 36
Physicochemical Characterization of Alginates 38
　Determination of Macromolecular Magnitudes by SEC/MALLS 38
　Determination of the Critical Overlap Concentration (C*) 38
Rheological Measurements 38
　Steady Shear Measurements 38
　Dynamic Oscillatory Measurements 39
Alginates Gel Properties 39
Biological Properties of Alginates 39
Conclusion 40
References 40

BOX 4.1 SALIENT FEATURES

Alginates are a group of unbranched anionic copolymers composed of uronic acids. They are mainly derived from the cell walls of brown marine macroalgae (Phaeophyceae) and are composed of (1,4) linked β-D-mannuronic (M) and α-L-guluronic acids (G) arranged in homogeneous (MM or GG) and heterogeneous (MG or GM) blocks. The gelling properties of alginates are related to their GG blocks which have the ability to bind divalent cations leading to hydrogel formation by cross-linking processes called the "egg-box system". These seaweed polysaccharides are authorized as a food additive by major regulatory agencies including the European Commission and U.S. Food and Drug Administration (FDA) and their market is driven by food applications notably for high G type alginate. The main species exploited for alginate production are *Macrocystis pyrifera* (South Africa), *Fucus* species (Australia, California, Chile and Mexico), *Ascophyllum nodosum*, *Ecklonia* and *Durvillea* (Europe) and *Laminaria* (Europe, Korea) and the annual production ranges from 30 to 40,000 t.yr^{-1}. After their extraction using acid and alkaline treatments alginates are characterized using biochemical assays, chromatographic and spectroscopic analyses to obtain their fine structure identification and their potential physicochemical but also biological properties. This book chapter focuses on the extraction, characterization and applications of alginates from seaweeds.

INTRODUCTION

"Alginate" or rather "alginates" are old polysaccharides which have been largely investigated. In the years 2015–2018 alone, 12,863 research articles, 411 reviews and 15322 patents can be found in the literature using the keyword "alginate" (SciFinder portal). The term alginate is employed to design alginic acids and their salts which are a group of unbranched anionic copolymers composed of uronic acids. They are mainly derived from the cell walls of brown marine macroalgae (Phaeophyceae) even if some of them may have a bacterial origin (Rinaudo 2007). Their existence in a mixer of salt forms (Ca^{2+}, Na^+ or Mg^{2+}) in cell walls of brown algae makes the tissues flexible and strong. They are composed of (1,4) linked β-D-mannuronic (M) with $^4C_1$ ring conformation and α-L-guluronic acids (G) with $^1C_4$ ring conformation, the two uronic acids being in pyranosic conformation. These two monosaccharides are arranged in homogeneous (MM or GG) and heterogeneous (MG or GM) blocks leading to a large diversity of structures, molecular weights and physicochemical properties (Figure 4.1) (Matsumoto et al. 1992).

Indeed, M blocks give linearity and flexibility to alginic backbone whereas G-block segments provide rigid and folded structural conformations responsible for the stiffness of this macromolecule. So alginates with high GG blocks content have higher viscosity. The gelling properties of alginate are also related to their GG blocks where selective alkaline earth metal multivalent cations contents interacting with divalent ions (the most commonly used is Ca^{2+}) take place by chelation to form stronger gels. Bacterial alginates are produced as exopolysaccharides by some *Azetobacter* and *Pseudomonas* species and are acetylated on C-2 and/or C-3 of M residues. Even if numerous authors have investigated the production and properties of bacterial alginates to meet the demand for different applications, none of them are currently on the market. The sourcing and characterization are of primary importance before considering the exploitation of seaweeds alginates by industry. Basically, alginates are characterized by their molecular weights, M/G ratio, monad values (F_M and F_G) and diad frequencies (F_{GG}, F_{MM}, F_{MG} and F_{GM}) which give structural information easily correlated with their rheological properties in solution or as cationic or acid gels (Khajouei et al. 2018). Industrial alginates are extracted from wild marine macroalgae with yields and quality depending on algae species, water temperature and season of harvesting. The main species exploited for alginate production have alginate content up to 20–40% of their dry weight (Rinaudo 2007). They are *Macrocystis pyrifera* (South Africa), *Fucus* species (Australia, California, Chile and Mexico), *Ascophyllum nodosum*, *Ecklonia* and *Durvillea* (Europe) and *Laminaria* species (Europe, Korea). The annual production ranges from 30 to 40,000 t.yr^{-1} (Hay et al. 2013). Various grades of seaweed alginates are currently on the market and classified depending on their distribution pattern of M- and G-blocks, molecular weight, purity and composition. The extraction of alginate can be summarized in 5 steps (Figure 4.2).

First, the washed and grinded seaweeds were extracted with a mineral acid leading to insoluble alginic acids, easily separated to other contaminating glycans such as laminarans and fucoidans by centrifugation or filtration. The insoluble residue is then treated by alkaline solution to convert insoluble alginic acid into sodium alginate. After another separation step, the soluble sodium alginate is precipitated using alcohol, acids or calcium chloride. Alginates are widely used in textiles, food (E400-405), printing and pharmaceutical industries for their viscosifying and gelling properties (Rinaudo 2007). Recently, alginates

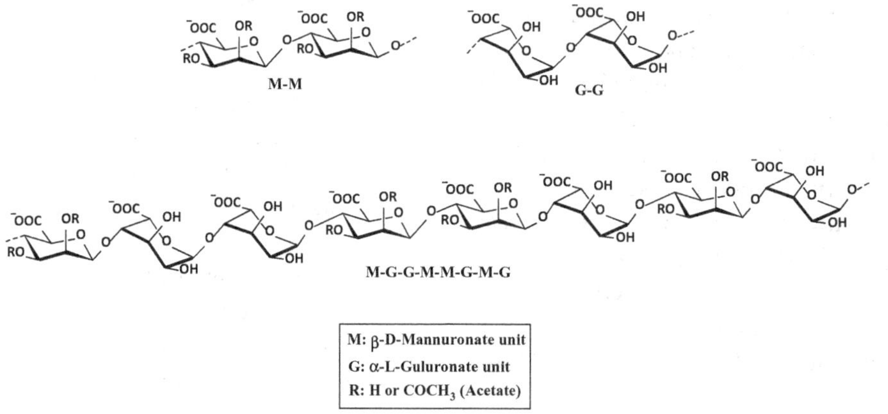

M: β-D-Mannuronate unit
G: α-L-Guluronate unit
R: H or COCH$_3$ (Acetate)

FIGURE 4.1 Main structures of alginates.

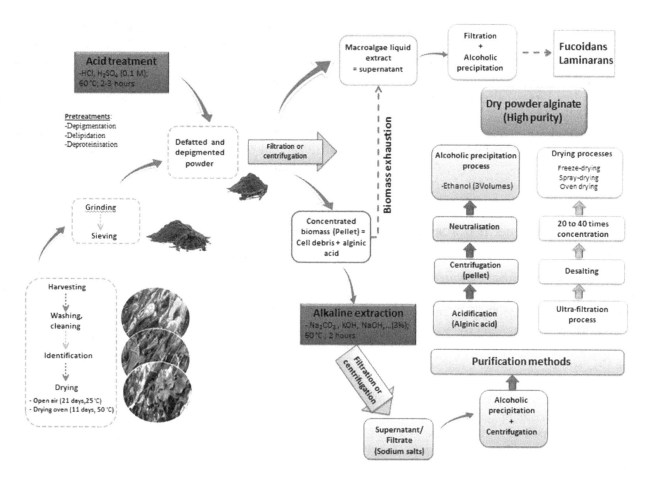

FIGURE 4.2 (See color insert.) Flowchart of the main processes for the extraction and purification of alginates.

which were approved by the FDA, became one of the most important biomaterials used as hydrogel for applications in regeneration medicine (Balakrishnan and Jayakrishnan 2005). The recent developments on structural and physicochemical characterizations of alginates as well as their biological properties are discussed in this chapter.

STRUCTURAL CHARACTERIZATION OF ALGINATES

The structural characterization of alginate includes elementary biochemical assays but also more complex analyses leading to fine structure identification.

GLOBAL COMPOSITION OF ALGINATES

The phenol-sulfuric acid assay (Dubois et al. 1956) improved by Albalasmeh et al. (2013) remains the most reliable method to obtain a response of each monosaccharide in the alginate sample. Under the action of concentrated and hot mineral acids, the hydrolyzed saccharides undergo extensive internal dehydration, followed by a cyclization resulting in the formation of furfural and hydroxymethylfurfural derivatives, reacting with phenol. Finally, the formation of a colored (yellow-red) complex makes it possible to follow the total sugar concentration by measuring the absorbance at λ_{485}. D-glucose is generally used as standard.

Since alginates are polyuronides, their uronic acid contents can be quantified by the meta-hydroxydiphenyl (MHDP) assay described by Blumenkrantz and Asboe-Hansen (1973) and modified by Filisetti-Cozzi and Carpita (1991) using D-glucuronic acid as standard. Uronic acids react with MHDP to form an absorbent pink chromophore at λ_{525}. The comparison of these uronic acids contents with those of neutral sugars is the signature of their purity. Neutral sugars content can be evaluated using the resorcinol/sulfuric acid method of Monsigny et al. (1988) based on the dehydration of hydrolyzed saccharides to furfural derivatives which form colored complexes with resorcinol reagent. The results are expressed in equivalent of D-glucose and the absorbance is measured at λ_{510}. The correction factors of sugar concentrations are calculated by the formula proposed by Montreuil et al. (1963) to eliminate the interferences due to the neutral monosaccharides in particular in the MHDP assay.

Other assays are expected to quantify both functional groups and non-sugar contaminants co-extracted with seaweed alginates (Hentati et al. 2018) such as phenolic compounds (Singleton et al. 1999), proteins (Lowry et al. 1951; Bradford 1976; Smith et al. 1985) and sulfated groups (Dodgson and Price 1962; Jaques et al. 1968; Craigie et al. 1984).

FOOTPRINTS AND MAIN FUNCTIONAL GROUPS IDENTIFICATION BY FOURIER TRANSFORMED INFRARED (FT-IR) SPECTROSCOPY

FT-IR spectroscopy is frequently used to partially characterize polysaccharides by studying four regions of absorption spectra: (i) 3500–2900 cm^{-1}, (ii) 1600–1150 cm^{-1}, (iii) 1150–950 cm^{-1} and (iv) 950–700 cm^{-1} corresponding respectively to O-H and C-H stretching vibrations, C-O stretching vibration, alginate fingerprint and anomeric region of uronic acids (Hentati et al. 2018). The signals observed at 1420–1400 cm^{-1} and 1610–1590 cm^{-1} are characteristic of symmetrical and asymmetrical stretching vibrations of carboxylate groups (COO^-). The deformation of O-C-H groups can be identified at 1320–1310 cm^{-1} while C-O-C elongation vibrations can be detected at 1130–1115 cm^{-1}. The anomeric region at 950–750 cm^{-1} presents several characteristic signals which can correspond to C-O stretching vibrations of alginate backbone. The absence of a broad absorption band at 1260–1210 cm^{-1} confirms the lack of sulfates contamination. Finally, two specific peaks in the 1100–1070 cm^{-1} and 1030–1020 cm^{-1} anomeric regions are specific of units of M and G, respectively (Hentati et al. 2018). The detection of proteins and nucleic acids contaminants can also be revealed through the presence of characteristic bands at 1710–1590 cm^{-1} (amide I) and 1585–1480 cm^{-1} (amide II) and at 1356–1191 cm^{-1}, respectively (Dammak et al. 2017).

DETERMINATION OF M/G RATIOS BY CHROMATOGRAPHY

High Performance Anion Exchange Chromatography with Pulsed Amperometric Detection (HPAEC-PAD) has been widely used for determination of M/G ratios of alginates after their complete acid hydrolysis using formic acid (90% v/v) (Hentati et al. 2018). Separations can be performed using a stationary phase composed of quaternary ammoniums (CarboPac PA-1 column) eluted by a linear gradient of 100 mM NaOH/1 M NaOAc (eluent B) in 100 mM NaOH (eluent A). M and G are separated based on their pKa values. With this chromatographic method 60% of uronic acids are lost during the analysis, 43–44% of which is due to acidic hydrolysis and neutralization and 15% by the alkaline elution. Haug et al. (1962) proposed a procedure multiplying the proportion between M and G by 0.66 to correct the difference in breakdown rates of these two uronic acids. HPAEC-PAD is also a powerful method to detect neutral monosaccharides. The detection of Rha, Xyl, Man, Ara, Gal, Fuc and Glc using isocratic elution of CarboPac PA1 column with 18 mM NaOH can be then correlated with the presence of fucoidans co-extracted with alginates (Rhein-Knudsen et al. 2017).

Gaz Chromatography coupled with Mass Spectrometry and Electron Ionization (GC/MS-EI) is another method for determining M/G ratios of hydrolyzed alginates (trifluoroacetic acid 2 M, 90 min, 120°C) after derivatization of their constitutive uronic acids (Hentati et al. 2018). Silylation using BSTFA (Bis (trimethylsilyl) trifluoroacetamide) and TMCS (trimethylchlorosilane) reagents (99:1–90:10) is the most common method used at various temperatures (from ambient temperature to 80°C) and reactions times (30 to 240 min). Trimethylsilyl-O-glycosides into dichloromethane can be separated for example into a OPTIMA-1MS column eluted with a helium flow rate (Hentati et al. 2018).

DETERMINATION OF SEQUENCE AND BLOCKS DISTRIBUTION BY SPECTROSCOPIC METHODS

Even if the sequential structure and acid monomer distribution along the alginate backbone can be investigated using circular dichroism (Morris et al. 1980) and FT-IR (Hentati et al. 2018), the structural composition of alginates is generally efficiently achieved by ^{1}H- and ^{13}C-NMR experiments (Hadj Ammar et al. 2018; Hentati et al. 2018). This analysis is generally performed using 5–10 mg of purified alginate in D_2O. ^{1}H-NMR spectra are achieved at 85°C on 300–600 MHz spectrometers. The anomeric regions of the ^{1}H-NMR spectra of alginates give access to anomeric proton H_G-1 at 4.9–5.20 ppm (signal A), H_G-5 in the triad GG-5G at 4.30–4.40 ppm (signal C) and H_M-1+H_{GM}-5 overlap at 4.45–4.60 ppm (signal B) (Figure 4.3).

According to Grasdalen (1983), the two monad frequencies (F_M, F_G) and the four diads (F_{GG}, F_{MM}, F_{MG} and F_{GM}) are calculated using the areas of A, B and C signals as following:

$$F_G = \frac{A}{B+C}; F_{GG} = \frac{C}{B+C}; F_G + F_M = 1$$

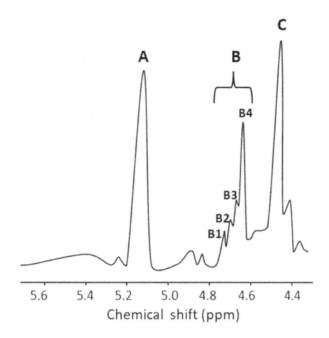

FIGURE 4.3 ^{1}H-NMR spectrum (anomeric region) of an alginate from brown seaweed.

$$F_{GM} = F_{MG} = F_G - F_{GG}; F_{MM} = F_M - F_{MG};$$

$$F_{MM} + F_{GG} + F_{GM} + F_{MG} = 1$$

^{1}H spectra give also information about triads (Grasdalen et al. 1979). Signal B2 at 4.72–4.74 ppm, B1 (4.74–4.78 ppm) and signal at 4.42–4.44 ppm refer to the H-5 of the central G in MGM, GGM or MGG triads, respectively. The H-1 of M neighboring M (signal B4) and G (signal B3), respectively, are presented at 4.66–4.68 (MM) and 4.68–4.70 (MG) ppm. F_{GGG}, F_{MGG}, F_{GGM} and F_{MGM} can be calculated based on the following relationships:

$$F_G = \frac{G}{M+G}; F_M = \frac{M}{M+G}; F_G + F_M = 1$$

$$F_{GG} = \frac{GG}{M+G}; F_{MM} = \frac{MM}{M+G}; F_{GM} = F_{MG} = \frac{MG}{M+G};$$

$$F_{MM} + F_{GG} + F_{GM} + F_{MG} = 1$$

$$F_{GGG} = \frac{GGG}{M+G}; F_{MGM} = \frac{MGM}{M+G};$$

$$F_{MGG} = F_{GGM} = \frac{MGG}{M+G}$$

$$F_{MG} = F_{GM} = F_{GGM} + F_{MGM};$$

$$F_G = F_{GGG} + F_{MGG} + F_{GGM} + F_{MGM}$$

With $\quad G = 0.5(A+C+0.5(B1+B2+B3));$
$M = B4 + 0.5(B1+B2+B3)$

$GG = 0.5(A+C-0.5(B1+B2+B3)); MM = B4$

The values of M/G ratio, the number of average G-block length larger than one ($N_{G>1}$), the parameter η (showing the abundance of homopolymeric or heteropolymeric diads) and the degree of polymerization (DP_n) are defined by the following equations:

$$\frac{M}{G} = \frac{F_M}{F_G}; N_{G>1} = \frac{F_G - F_{MGM}}{F_{GGM}}; \eta = \frac{F_{GM}}{F_M \times F_G};$$

$$DP_n = \frac{M + G + \text{red } \beta + \text{red } \alpha}{\text{red } \alpha + \text{red } \beta}$$

Where the signals at around (5.15–5.25 ppm) and (4.85–4.95 ppm) represent αH-1 (red-α) and βH-1 (red-β) of terminal reducing end respectively (Rhein-Knudsen et al. 2017).

The spectral region 50–200 ppm of the ^{13}C-NMR spectra of alginates shows three major signal families; the ring carbons in the range of 60–90 ppm, the anomeric carbons at 90–110 ppm and finally the carboxylic groups between 170 and 180 ppm. As previously reported by Llanes et al. (1997) only the signals at 71, 76, 65, 68 and 82 ppm (pyranose rings) are useful for the determination of the M/G ratio according to the following equation:

$$\frac{M}{G} = \frac{P_{71} + P_{76}}{P_{65} + P_{68} + P_{82}}.$$

Where Px are the areas of each signal.

PHYSICOCHEMICAL CHARACTERIZATION OF ALGINATES

Among the salts of alginic acid, sodium alginates present the greatest industrial importance for the food and pharmaceutical industries. Indeed, alginate salts are multivalent additives: binder and thickening agent, emulsifier, gelling agent, stabilizer for suspensions and foams, texture-improver or hydration agents. As for the other hydrocolloids, it is mandatory to know: (i) the molecular weight and (ii) the intrinsic viscosity to relate viscosity to molecular weight, polymer concentration and shear rate to provide the potential use of alginates. The physicochemical properties of alginates depend on their molecular weights, compositions and sequences (Draget et al. 2006).

DETERMINATION OF MACROMOLECULAR MAGNITUDES BY SEC/MALLS

Size exclusion chromatography, coupled with: refractometer and/or Multi-Angle Light Scattering analysis (SEC-MALLS) and/or osmometer and/or viscometer are used to estimate the weight-average molecular weight (M_w), the number-average molecular weight (M_n), the radius of gyration (R_g), the hydrodynamic radius (R_h) and the polydispersity index ($Đ = M_w/M_n$) of alginates (Draget et al. 2006). M_w of sodium alginates were evaluated between 100 and 553 kg/mol depending on brown seaweeds species and extraction processes (Belalia and Djelali 2014; Khajouei et al. 2018; Hentati et al. 2018). As related by Draget et al. (2006) and confirmed by Belalia and Djelali (2014), alginates possessed a Đ ranging from 1.5 to 3. Đ values represent the size distribution of a particle population in polysaccharide solution. Recently, Morris et al. (2014) presented some hydrodynamic methodologies used to investigate the main shapes of polysaccharides in a dilute regime. As mentioned by authors, from some specific coefficients (power law) and parameters (Wales and Van Holde 1954) obtained by using SEC-MALLS in combination with viscometer, it is possible to give a well-established idea of the conformation, shape and flexibility of polysaccharides in solution. Low molecular weight alginates (M_w around 100 kg/mol) had R_h and R_g of 17.3–17.7 and 26.2 nm respectively (Khajouei et al. 2018; Hentati et al. 2018). As related by Morris et al. (2014), diverse sodium alginates with M_w ranging from 15–2700 kg/mol had R_g and R_h ranging from 40–190 nm and 25–55 nm respectively. Nevertheless, the structure factor (or shape factor) $\rho = R_g/R_h$ provides better information regarding the polysaccharides shapes according to the specific conformation in dilute regime with: $\rho = 2$ (rigids rods), $\rho = 1.7$ (random coils) and $\rho = 0.78$ (hard spheres) (Burchard 1988).

DETERMINATION OF THE CRITICAL OVERLAP CONCENTRATION (C*)

The critical overlap concentration (C*) represents the limit between dilute and semi-dilute regimes for polysaccharides solution and C** represents the last transition to the concentrated regimes (Bouldin et al. 1988). The plot of log (η_{sp}) *versus* log (C) leads to a slope break indicating the experimental C* value which is the signature of the beginning of the macromolecules entanglement (Bouldin et al. 1988). C* could be also calculated using intrinsic viscosity. Then, C* is an inverse function of the intrinsic viscosity (C* = k/[η]) where the constant k (values between 1 and 4) is specific for each type of polysaccharides (Norton et al. 1984). Concerning alginate solutions, it was clearly established that for random coil polysaccharide solutions, the critical concentration C* value is around 4/[η] for a zero shear specific viscosity (η_{sp}) value near to 10 (Morris et al. 1980).

RHEOLOGICAL MEASUREMENTS

The most important rheological characteristics of polysaccharides are the flow behavior, thixotropy and dynamical viscoelastic properties (Ma et al. 2014). The factors affecting the rheology of alginates are species and structure dependent. Solutions of alginate in water can exhibit oppositional behavior at the same concentrations, i.e. Newtonian or non-Newtonian fluids. Khajouei et al. (2018) stated that solutions of sodium alginate (1.0 to 5.0%, w/v) from *Nizimuddinia zanardini* exhibited almost Newtonian or very low shear-thinning behavior whereas Ma et al. (2014) found that, above a critical shear rate, solutions of commercial G-rich sodium alginate (1.0 to 3.0%, w/v) exhibited non-Newtonian shear thinning one.

STEADY SHEAR MEASUREMENTS

At the concentrations used in industrial applications, alginates are non-Newtonian pseudoplastic fluids since the viscosity decreases as the rate of shear increases (McHugh 1987). The addition of NaCl to sodium alginate solutions induced a significant increase of alginate solutions viscosity as Na$^+$ ions favor inter-chain associations (Draget et al. 2006). This effect becomes more pronounced when the G content increases (Ma et al.

2014). Seale (1982) proposed three modes of binding of univalent cations in addition to general polyelectrolyte effects: weak chelation by M and isolated G residues, specific site-binding to adjacent G and cooperative "egg-box" binding. The viscosifying properties depend on the molecular conformation of the polymer and can be influenced by the ionic strength of the solvent. This sensitivity of viscosity to ionic strength is typical of polyelectrolytes which normally exhibit high viscosities at low ionic strength (due to an expanded chain conformation caused by charge-charge repulsions) and lower viscosities at higher ionic strength (due to a more compacted conformation) (Lapasin and Pricl 1995). The viscosity of an alginate solution will significantly decrease with the temperature increasing (McHugh 1987). Ma et al. (2014) published the temperature-dependent behavior of sodium alginate solution and a good correlation to the Arrhenius model. Low concentration of Ca^{2+} in an alginate solution resulted, for example, in an increase of its viscosity due to the formation of inter-chain interactions (McHugh 1987).

Dynamic Oscillatory Measurements

The oscillatory rheology is generally used to quantify the viscous and elastic responses at different time scales of viscoelastic systems such as solutions of alginates. Parameters that can be evaluated by oscillatory analysis are: (i) the storage modulus (G'), (ii) the loss modulus (G'') and (iii) the dynamical mechanical loss tangent ($tan\ \delta = G''/G'$). Oscillatory frequency sweep is useful for evaluating gelation kinetics and can provide data on the structure and energy consumption (Ma et al. 2014). Dynamic oscillatory measurements are sensitive to molecular structure and interactions in solutions. In a recent work, Ma et al. (2014) showed that, under small deformation conditions, G' and G'' of a solution at 2.5% (w/v) of sodium alginate increased with the angular frequency and decreased with the increasing of temperatures from 5 to 35°C. Moreover, G'' values were slightly larger than G' values and at high frequency both moduli tended to approach each other. Therefore, the majority of the energy was dissipated by viscous flow and the alginate solution had a fluid-like viscoelastic behavior. This behavior was confirmed by the values of tan δ bigger than unity (Ma et al. 2014). The dynamical oscillatory test realized by Belalia and Djelali (2014) with a sodium alginate solution at 2.0% (w/v) showed higher differences between G' and G'' and higher values for G'' than for G'. Authors concluded that the both moduli are frequency dependent in the case of sodium alginate solutions.

ALGINATES GEL PROPERTIES

Some applications of alginates are based on their ability to bind divalent cations leading to hydrogel formation by cross-linking processes called "egg-box system" (Draget et al. 2006). The selective binding ability of alginates towards divalent ions increases in the following order: $Mg^{2+}<Mn^{2+}<Ca^{2+}<Sr^{2+}Ba^{2+}<Cu^{2+}<Pb^{2+}$.

Alginates have a higher affinity for divalent ions than for monovalent ones. Therefore, sodium alginate in the presence of calcium ions can forms strong gels. The gelation mechanism is well described in the literature. Due to the higher affinity for Ca^{2+} ions, a continuous swollen network where Ca^{2+} ions hold together polymer chains is developed. The affinity for a specific ion increases with its concentration. The gelling process is highly dependent on the structure of alginate, the degree of conversion into calcium alginate (that is 100% when the molar ration between Na^+ and Ca^{2+} is 0.5), the type of calcium salt, the diffusion of gelling ions into the alginate network and the specific method of preparation (Mancini et al. 1999).

Alginate gels are obtained by two fundamental methods: the dialysis/diffusion setting method and the internal setting method (gelation is mediated by a change in pH in order to cross-link cations from insolubility of chelated form) (Smidsrød and Draget 1996). Gurnikov and Smirnova (2018) presented three more nonconventional methods: (i) cryotopic gelation, (ii) non-solvent induce phase separation and (iii) carbon dioxide induce gelation. Regarding the literature, alginate gels obtained by the diffusion setting procedure can exhibit an inhomogeneous distribution of alginate chains where the highest concentration is found at the surface and gradually decreases towards the center of the gel (Draget and Taylor 2011).

BIOLOGICAL PROPERTIES OF ALGINATES

Alginates have been extensively studied for their numerous biological properties, which are affected by their structures and molecular weights. New knowledges point towards more advanced biological and biomedical devices as well as pharmaceutical properties of alginates including antitumor, antioxidant and antibacterial activities, inhibition of the both vascular smooth muscle cell proliferation and histamine release from mast cells, enhancement of cholesterol excretion and glucose tolerance, *in vitro* inhibition of nuclear factor-κB activation and *in vivo* inhibition of histidine decarboxylase (Kimura et al. 1996; Asada et al. 1997; Logeart et al. 1997; Hu et al. 2005; Jeong et al. 2006; Hentati et al.

2018). Recently, Hadj Ammar et al. (2018) showed an interesting gastroprotective effect for alginates extracted from three *genus* of Phaeophyceae and having M/G ratios greater than 1. Other reports showed the potential applications of alginate polymers as anti-diabetic and anti-obesity agents (Brownlee et al. 2005; Sari-Chmayssem et al. 2016). Low molecular weight alginates constituted of MM blocks and MG blocks have been reported to stimulate monocytes to produce tumor necrosis factor (TNF-α), interleukin (IL) 1 and 6, whereas G-blocks failed to induce cytokine production (Ueno and Oda 2014). Oligo-alginate obtained using enzymatic degradation or physical treatments including high temperatures, high pressure and sonication had various biological activities such as stimulations of RAW264.7 macrophage cells proliferation and nitric oxide production, human keratinocyte growth, endothelial cells growth and migration (Rostami et al. 2017). They also inhibited Th2 development and IgE secretion through IL-12 secretion and fibroblast multiplication and collagen synthesis in human skin (Ueno and Oda 2014). Since these alginate oligomers have fairly weak viscosity in water even at great concentration and possess no gelling characteristic in the presence of divalent ions, it is suggested that they are more applicable for *in vivo* systems (Kurachi et al. 2005). Indeed, it has been reported that intraperitoneal administration of hydrolyzed alginate oligomers to mice showed a significant rise in serum levels of several cytokines such as granulocyte colony-stimulating factor and TNF-α (Yamamoto et al. 2007). Interestingly, alginates seem also to have some biological effects on plants as several reports presented their effects on elongation of barley and carrot roots and on the growth of rice (Ueno and Oda 2014).

CONCLUSION

Depending on their origins, alginates have various structures and molecular weights leading to specific biological and rheological properties at the origin of various grades for specific uses. So the correlation of their structures with their physicochemical properties is of primary importance to develop applications. *Sargassum* species usually give low viscosity alginates whereas *Macrocystis* genus give medium-viscosity products and *Laminaria* species medium strength or strong gels. These are some of the reasons why alginate producers try to possess a huge variety of seaweed sources. The global alginate market is mainly driven by the food applications where the gelifying properties of these seaweed polysaccharides authorized as additives by major regulatory agencies including the European Commission and FDA are valorized. The production of alginates with several grades is furnished by several suppliers such as ISP alginate (UK), FMC biopolymer (US), Degussa texturants systems (Germany), Danisco (Denmark), Kimica Corporation (Japan), Algisa (Chile), Ceamsea (Spain) and Algaia (France). Consumers in the food and beverage industries exhibit a higher demand for high G type alginate having superior gelling performance. The demand for high M type product will probably increase in the future, driven by their use in paper and textile manufacturing industries. New applications of calcium alginates in the pharmaceutical industry account for high consumption volumes in wound care and wound dressing areas. In addition, calcium alginate based fibers are used to accelerate wound healing, for wounds exhibiting drainage or fluid excretion. Despite the abundance of Phaeophyceae, notably in Chile and Norway, the prices of alginates are quietly expensive due to the consumer demand for high quality and purity but also because the supply of brown seaweeds from Chile can be curtailed by El Niño events.

REFERENCES

Albalasmeh, A. A., Berhe, A. A., Ghezzehei, T. A. A new method for rapid determination of carbohydrate and total carbon concentrations using UV spectrophotometry. *Carbohyd Polym*. 2013. 97: 253–261.

Ammar, H. H., Lajili, S., Sakly, N., Cherif, D., Rihouey, C., Le Cerf, D., Bouraoui, A., Majdoub, H. Influence of the uronic acid composition on the gastroprotective activity of alginates from three different genus of Tunisian brown algae. *Food Chem*. 2018. 239: 165–171.

Asada, M., Sugie, M., Inoue, M., Nakagomi, K., Hongo, S., Murata, K., Irie, S., Takeuchi, T., Tomizuka, N., Oka, S. Inhibitory effect of alginic acids on hyaluronidase and on histamine release from mast cells. *Biosci Biotechnol Biochem*. 1997. 61: 1030–1032.

Balakrishnan, B., Jayakrishnan, A. Self-cross-linking biopolymers as injectable *in situ* forming biodegradable scaffolds. *Biomaterials* 2005. 26: 3941–3951.

Belalia, F., Djelali, N. E. Rheological properties of sodium alginate solutions. *Rev Roum Chim*. 2014. 59: 135–145.

Blumenkrantz, N., Asboe-Hansen, G. New method for quantitative determination of uronic acids. *Anal Biochem*. 1973. 54: 484–489.

Bouldin, M., Kulicke, W. M., Kehler, H. Prediction of the non-Newtonian viscosity and shear stability of polymer solutions. *Colloid Polym Sci*. 1988. 266: 793–805.

Bradford, H. M. A rapid and sensitive method for the quantification of microgram quantities of protein utilizing the principle of protein–dye binding. *Anal Biochem*. 1976. 72: 248–254.

Brownlee, I. A., Allen, A., Pearson, J. P., Dettmar, P. W., Havler, M. E., Atherton, M. R., Onsøyen, E. Alginate as a source of dietary fiber. *Crit Rev Food Sci Nutr.* 2005. 45: 497–510.

Burchard, W. Polymer characterization: Quasi-elastic and elastic light scattering. In *Makromolekulare Chemie. Macromolecular Symposia.* 1988. (Vol. 18, No. 1, pp. 1–35). Basel: Hüthig & Wepf Verlag.

Craigie, J. S., Wen, Z. C., Van der Meer, J. P. Interspecific, intraspecific and nutritionally-determined variations in the composition of agars from Gracilaria spp. *Bot Mar.* 1984. 27: 55–62.

Dammak, M., Hadrich, B., Miladi, R., Barkallah, M., Hentati, F., Hachicha, R., Laroche, C., Michaud, P., Fendri, I., Abdelkafi, S. Effects of nutritional conditions on growth and biochemical composition of *Tetraselmis* sp. *Lipids Health Dis.* 2017. 16: 41.

Dodgson, K. S., Price, R. G. A note on the determination of the ester sulphate content of sulphated polysaccharides. *Biochem J.* 1962. 84: 106–110.

Draget, K. I., Taylor, C. Chemical, physical and biological properties of alginates and their biomedical implications. *Food Hydrocolloid.* 2011. 25(2): 251–256.

Draget, K. I., Moe, S. T., Skjak-Braek, G., Alginates, S. O. *Food Polysaccharides and Their Applications* (2nd edition, pp. 289–334), edited by Stephen A. M., Phillips G.O., & Williams P.A. CRC Press, Boca Raton, FL, 2006.

Dubois, M., Gilles, K. A., Hamilton, J. K., Rebers, P. A., Smith, F. Colorimetric method for determination of sugars and related substances. *Anal Chem.* 1956. 28: 350–356.

Filisetti-Cozzi, T. M., Carpita, N. C. Measurement of uronic acids without interference from neutral sugars. *Anal Biochem.* 1991. 197: 157–162.

Grasdalen, H. High-field, ^{1}H-n.m.r. Spectroscopy of alginate: Sequential structure and linkage conformations. *Carbohyd Res.* 1983. 118: 255–260.

Grasdalen, H., Larsen, B., Smidsrød, O. An NMR study of the composition and sequence of uronate residues in alginates. *Carbohyd Res.* 1983. 68(1): 23–31.

Gurikov, A., Smirnova, I. Amorphization of drugs by adsorptive precipitation from supercritical solutions: A review. *J Supercrit Fluid.* 2018. 132: 105–125.

Haug, A., Larsen, B., Fykse, O., Block-Bolten, A., Toguri, J. M., Flood, H. Quantitative determination of the uronic acid composition of alginates. *Acta Chem Scand.* 1962. 16: 1908–1918.

Hay, I. D., Rehman, Z. U., Moradali, M. F., Wang, Y., Rehm, B. H. A. Microbial alginate production, modification and its applications. *Microb Biotechnol.* 2013. 6: 637–650.

Hentati, F., Delattre, C., Ursu, A. V., Desbrières, J., Le Cerf, D., Gardarin, C., Abdelkafi, S., Michaud, P., Pierre, G. Structural characterization and antioxidant activity of water-soluble polysaccharides from the Tunisian brown seaweed *Cystoseira compressa*. *Carbohyd Polym.* 2018. 198: 589–600.

Hu, X., Jiang, X., Gong, J., Hwang, H., Liu, Y., Guan, H. Antibacterial activity of lyase-depolymerized products of alginate. *J Appl Phycol.* 2005. 17: 57–60.

Jaques, L. B., Ballieux, R. E., Dietrich, C. P., Kavanagh, L. W. A microelectrophoresis method for heparin. *Can J Physiol Pharmacol.* 1968. 46: 351–360.

Jeong, H. J., Lee, S. A., Moon, P. D., Na, H. J., Park, R. K., Um, J. Y., Kim, H. M., Hong, S. H. Alginic acid has anti-anaphylactic effects and inhibits inflammatory cytokine expression via suppression of nuclear factor-κB activation. *Clin Exp Allergy* 2006. 36: 785–794.

Khajouei, R. A., Keramat, J., Hamdami, N., Ursu, A. V., Delattre, C., Laroche, C., Gardarin, C., Lecerf, D., Desbrières, J., Djelveh, G., Michaud, P. Extraction and characterization of an alginate from the Iranian brown seaweed *Nizimuddinia zanardini*. *Int J Biol Macromol.* 2018. 118: 1073–1081.

Kimura, Y., Watanabe, K., Okuda, H. Effects of soluble sodium alginate on cholesterol excretion and glucose tolerance in rats. *J Ethnopharmacol.* 1996. 54: 47–54.

Kurachi, M., Nakashima, T., Miyajima, C., Iwamoto, Y., Muramatsu, T., Yamaguchi, K., Oda, T. Comparison of the activities of various alginates to induce TNF-alpha secretion in RAW264.7 cells. *J Infect Chemother.* 2005. 11: 199–203.

Lapasin, R., Pricl, S. *Rheology of Industrial Polysaccharides: Theory and Applications.* Springer Science and Business Media, Dordrecht, 1995.

Llanes, F., Sauriol, F., Morin, F. G., Perlin, A. S. An examination of sodium alginate from *Sargassum* by NMR spectroscopy. *Can J Chem.* 1997. 75: 585–590.

Logeart, D., Prigent-Richard, S., Jozefonvicz, J., Letourneur, D. Fucans, sulfated polysaccharides extracted from brown seaweeds, inhibit vascular smooth muscle cell proliferation. I. Comparison with heparin for antiproliferative activity, binding and internalization. *Eur J Cell Biol.* 1997. 74: 376–384.

Lowry, O. H., Rosebrough, N. J., Farr, A. L., Randall, R. J. Protein measurement with the Folin phenol reagent. *J Biol Chem.* 1951. 193: 265–275.

Ma, J., Lin, Y., Chen, X., Zhao, B., Zhang, J. Flow behavior, thixotropy and dynamical viscoelasticity of sodium alginate aqueous solutions. *Food Hydrocol.* 2014. 38: 119–128.

Mancini, M., Moresi, M., Rancini, R. Mechanical properties of alginate gels: Empirical characterization. *J Food Eng.* 1999. 39: 369–378.

Matsumoto, T., Kawai, M., Masuda, T. Influence of concentration and mannuronate/guluronate ratio on steady flow properties of alginate aqueous systems. *Biorheology* 1992. 29: 411–417.

McHugh, D. J. Production, properties and uses of alginates. In: *Production and Utilization of Products From Commercial Seaweeds*, edited by McHugh D.J. FAO Fisheries Technical Papers, Rome, 1987.

Monsigny, M., Petit, C., Roche, A. C. Colorimetric determination of neutral sugars by a resorcinol sulfuric acid micromethod. *Anal Biochem.* 1988. 175: 525–530.

Montreuil, J., Spik, G., Chosson, A., Segard, E., Scheppler, N. Methods of study of the structure of glycoproteins. *J Pharm Belg.* 1963. 18: 529–546.

Morris, E. R., Rees, D. A., Thom, D. Characterisation of alginate composition and block-structure by circular dichroism. *Carbohyd Res.* 1980. 81(2): 305–314.

Morris, G. A., Adams, G. G., Harding, S. E. On hydrodynamic methods for the analysis of the sizes and shapes of polysaccharides in dilute solution: A short review. *Food Hydrocoll.* 2014. 42: 318–334.

Norton, I. T., Morris, E. R., Rees, D. A. Lyotropic effects of simple anions on the conformation and interactions of kappa-carrageenan. *Carbohyd Res.* 1984. 134: 89–101.

Rhein-Knudsen, N., Ale, M. T., Ajalloueian, F., Meyer, A. S. Characterization of alginates from Ghanaian brown seaweeds: *Sargassum* spp. and *Padina* spp. *Food Hydrocol.* 2017. 71: 236–244.

Rinaudo, M. Seaweed polysaccharides. In: *Comprehensive Glycoscience, 2.2: Polysaccharide Functional Properties* (pp. 691–773), edited by Kamerling J. P. Elsevier, Amsterdam, the Netherlands, 2007.

Rostami, Z., Tabarsa, M., You, S., Rezaei, M. Relationship between molecular weights and biological properties of alginates extracted under different methods from *Colpomenia peregrina*. *Proc Biochem.* 2017. 58: 289–297.

Sari-Chmayssem, N. S., Taha, S., Mawlawi, H., Guégan, J. P., Jeftić, J., Benvegnu, T. Extracted and depolymerized alginates from brown algae *Sargassum vulgare* of Lebanese origin: Chemical, rheological, and antioxidant properties. *J Appl Phycol.* 2016. 28: 1915–1929.

Seale, R., Morris, E. R., Rees, D. A. Interactions of alginates with univalent cations. *Carbohyd Polym.* 1982. 111(1): 101–112.

Smidsrød, O., Draget, K. L. Chemistry and physical properties of alginates. *Carbohyd Europ.* 1996. 14: 6–13.

Smith, P. K., Krohn, R. I., Hermanson, G. T., Mallia, A. K., Gartner, F. H., Provenzano, M. D., Fujimoto, E. K., Goeke, N. M., Olson, B. J., Klenk, D. C. Measurement of protein using bicinchoninic acid. *Anal Biochem.* 1985. 150: 76–85.

Singleton, V. L., Orthofer, R., Lamuela-Raventós, R. M. Analysis of total phenols and other oxidation substrates and antioxidants by means of folin-ciocalteu reagent. *Method Enzymol.* 1999. 299: 152–178.

Ueno, M., Oda, T. Biological activities of alginate. *Adv Food Nutr Res.* 2014. 72: 95–112.

Wales, M., Van Holde, K. E. The concentration dependence of the sedimentation constants of flexible macromolecules. *J Polym Sci.* 1954. 14: 81–86.

Yamamoto, Y., Kurachi, M., Yamaguchi, K., Oda, T. Stimulation of multiple cytokines production in mice by alginate oligosaccharides following intraperitoneal administration. *Carbohydr Res.* 2007. 342: 1133–1137.

5 Champion Microalgal Forms for Food and Health Applications
Spirulina *and* Chlorella

*Chiara Toniolo, Marcello Nicoletti, Paola Del Serrone,
Ambati Ranga Rao, and Gokare A. Ravishankar*

CONTENTS

Abbreviations 43
Introduction 44
The Microalgae Market 44
   Spirulina 45
Commercial Production of *Spirulina* Biomass 45
   Chlorella 46
Commercial Production of *Chlorella* Biomass 47
Industrial Potential of *Spirulina* and *Chlorella* 49
Antimicrobial Activity of Microalgae 53
Market Value for *Chlorella* and *Spirulina* Products 53
Safety Regulations for Algal Products 53
Conclusions 54
Acknowledgment 54
References 54

BOX 5.1 SALIENT FEATURES

In recent years, considerable interest has been focused on the potential of algae mainly due to the identification of several substances synthesized by these microorganisms. Algae are of great importance, both biologically and economically. Their economic importance is related to the wide range of microalgae applications all over the world, from the food industry to feed, medicine, biofuels, cosmetics, and agriculture. The biodiversity of microalgae and consequent high variability in their biochemical composition, combined with the use of genetic improvement and the establishment of large-scale cultivation technology has allowed certain species to be commercially available. The commercial cultivation of algal species such as *Spirulina*, *Dunaliella*, *Chlorella* and *Haematococcus* has been established aiming at the production of biomass for both food manufacturing and also for obtaining natural compounds with high added value. In view of this, the current book chapter is focused on health food applications of *Spirulina* and *Chlorella* culture, commercial biomass productivities of *Spirulina* and *Chlorella* in raceway and photobioreactors by various authors, and also biomass producing companies are discussed. Further, bioactive molecules such as proteins, aminoacids, minerals, enzymes, pigments, essential fatty acids, vitamins, polysaccharides, phenolics, sterols, lipids, carbohydrate in both algal species, and their use in food, feed, nutraceutical, pharmaceutical, and cosmetic industry are discussed. Current market and safety regulations of algal products are provided.

ABBREVIATIONS

A. platensis: *Arthrospira platensis*
S. plantensis: *Spirulina platensis*

PUFA:	Polyunsaturated Fatty Acid
IIMSAM:	Inter-Governmental Institution for the use of Micro-Algae *Spirulina* Against Malnutrition
WHO:	World Health Organization
NASA:	National Aeronautics and Space Administration
ESA:	European Space Agency
PBR:	Photobioreactor
RNA:	Ribonucleic Acid
DNA:	Deoxyribonucleic
GLA:	Gamma-Linolenic Acid
EPA:	Eicosapentaenoic Acid
DHA:	Docosahexaenoic Acid
EFSA:	European Food Safety Authority
FD&C:	Federal Food Drug and Cosmetic Act
NDA:	Dietetic Products Nutrition Allergies
FDC:	Food, Drug and Cosmetic
CFSAN:	Centre for Food Safety and Applied Nutrition
USFDA:	United States Food and Drug Administration
GRAS:	Generally Recognized As Safe
CVM:	Center for Veterinary Medicine
DSHEA:	Dietary Supplement Health and Education Act

INTRODUCTION

For a long period algae were used in food, feed, and fertilizer applications (Da Silva Vaz et al. 2016). Of late microalgae have gained tremendous impetus in areas such as food, nutraceutical, and biofuel. Microalgae are increasingly used as a source of proteins, vitamins, hydrocarbons, and PUFA (Becker 2004; Christaki et al. 2011; Ambati et al. 2018, Ranga Rao et al. 2018). Nutraceuticals, phytochemicals, food supplements, botanicals, are among the many terms used to market and commercialize health food products (Eskin and Tanir 2006; Kent et al. 2015). Some of the algal products have been termed as super foods as well (Moorhead et al. 2001). Though the boom for algal products appears to be increasing for its utilization as nutraceutical, scientific validations are continuing (Nicoletti 2012). Although the species of microalgae actually in use are limited in number, the possibilities of adulteration are encountered. Sporadically toxic species can gain entry if algae is not cultivated in strict conditions (Hikmet et al. 2004; Miller et al. 2010). Therefore, safety evaluation is a continuous necessity for maintenance of quality requirements (Chacon-Lee and Gonzalez-Marino 2010).

THE MICROALGAE MARKET

Two species, i.e. *Spirulina* and *Chlorella*, are dominating the microalgae market and therefore their products are carefully reported and analysed here (Figure 5.1A and 5.1B). *Spirulina* is easily cultivated and utilized as raw material for food and nutraceuticals. *Spirulina* belongs to *Cyanobacteria* and is a prokaryote. On the contrary, *Chlorella* is a eukaryotic organism. Actually, *Cyanobacteria* were grouped under algae until 1962, when they were included in the prokaryote kingdom (Monera), based on the understanding of its cellular structure and genetics (Guiry 2012; Komarek et al. 2014; Nadis 2003; Singh and Dhar 2011). Algae are a very large and diverse group of eukaryotic organisms, included in the uneven kingdom Protista. In any case, all of them possess characteristic photosynthetic capability, but with different additional pigments. The cyanobacteria were the most ancient of all living organisms. For at least 3.5 billion years, cyanobacteria have provided oxygen, changing the primitive atmosphere and nutrients to the myriad varieties of life forms. For the first 2 billion years of the history of life on our planet, they were the dominating forms specially constituting the major part of the aquatic biomass and also now they are fundamental for our planet's equilibrium. In comparison with the apparent simplicity of the cell organization, their metabolism is complex (Juneja et al. 2013).

Considering the importance of algal forms for mainly food uses, R&D efforts have focused on

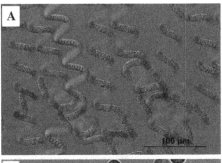
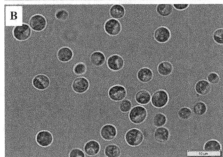

FIGURE 5.1 (See color insert.) (A) *Spirulina* (https://sagdb.uni-goettingen.de/detailedList.php?str_number=85.79), (B) *Chlorella*.

various commercial species, such as *Spirulina, Chlorella, Scenedesmus, Dunaliella,* and *Haematococcus* (Borowitzka 1999; Borowitzka and Moheimani 2013; Pulz and Gross 2004).

Spirulina

Marketed *Spirulina* products are currently derived from three species: *Arthrospira platensis* Goomont, *A. fusiformis* Voronich, and *A. maxima* Geitl. They are the most intensively used and extensively investigated, as those species are edible with high nutritional value (Belay 2008; Gershwin and Belay 2007). In the chapter we have used the term *Spirulina* for the species of the genus, as referred to by a large number of authors. The first information about the use of *Spirulina* as food can be found in Mexico during the Aztec civilization, as early as over 400 years ago, where *Spirulina* growing in Lake Texcoco was eaten by the Mayas, Toltecs, and Kanembu (Del Castillo 1928). According to the reports of conquistadores, *Spirulina* algae were harvested, dried, and used to make a sort of cake. In Central Africa, *Spirulina* has been harvested from the Lake Kossorom (Chad) over centuries for the Chadians to prepare cake or broths, as meals, and sold on the market (Abdulqader et al. 2000). It also occurs naturally in high-salt alkaline water reservoirs in subtropical and tropical areas including America, Mexico, Asia, and Central Africa, as well as cultivated in several countries in similar environmental conditions. This microalga is also currently grown in bioreactors, where environmental conditions, like pH and temperature can be controlled, as well as any contamination is avoided (Mata et al. 2010; Torzillo et al. 1986).

Since the middle of the 1980s *Spirulina* has been used extensively as food. Presently its popularity as functional food or nutraceutical is well known (Khan et al. 2005). It has been reported that consumption of *Spirulina* as a diet supplement could afford health benefits in preventing or managing hypercholesterolemia, hyperglycerolemia, certain inflammatory diseases, allergies, cancer, viral infections, diabetes, cardiovascular disease; they also possess hypolipidemia, antioxidant, and anti-inflammatory activities (Gershwin and Belay 2007). *Spirulina* was reported by the IIMSAM, as early as the mid-70s, as highly nutritional food to fight starvation and malnutrition in the world (Siva Kiran et al. 2015). In consideration of ease of cultivation coupled with nutritional value such as its high content of proteins and aminoacids (Table 5.1), the National Aeronautics and Space Administration (NASA) and the European Space Agency (ESA) promoted *Spirulina* as one of the primary foods during long-term space missions.

COMMERCIAL PRODUCTION OF *SPIRULINA* BIOMASS

Among the commercial algal species, *Spirulina* has gained commercial popularity because of its ease of cultivation in raceway and photobioreactors coupled with a wide range of nutritional and health applications. *Spirulina* possesses high protein and vitamin content by virtue of which is used in feed and food formulations (Lu et al. 2011; Soni et al. 2017). The production of *Spirulina* is estimated around 8,000 metric tons per year (Vonshak et al. 2014). The major production of *Spirulina* has been in China followed by others in North America, India, and the Asia-Pacific region (Belay 2013). Production of *Spirulina* depends on various factors such as light intensity, pH, temperature, salinity, nutrients, medium composition, and mode of cultivation in raceway ponds or photobioreactors (Belay 2008). The growth rate, photosynthetic efficiency, and chlorophyll content were reduced during the peak hours of high light intensity due to photo-inhibition (Vonshak et al. 2014). The culture temperature between 35–37 °C was found to be optimum for biomass productivity while increase in temperature was observed to hinder the growth rate (Chaiklahan et al. 2007; Richmond 1988). Various cultivation techniques and adoption of adequate culture conditions have enhanced *Spirulina* biomass productivity through several innovative approaches (Table 5.2). One way to economically produce *Spirulina* biomass is by growing under outdoor conditions using natural solar radiation at ambient temperatures, with reduction in cost of cultivation set up, coupled with optimised farming practices and harvesting methods (Vonshak 1997).

Outdoor cultivation of algae is a very complex system which is affected by solar light, photoperiod, salinity, pH, and temperature (Borowitzka 1999). Despite these challenges, successful outdoor cultivation of *Spirulina* has been achieved at an industrial scale by various companies (Figure 5.2).

Vonshak and Richmond (1988) discussed improvements and constraints of outdoor mass production of *Spirulina*. Large-scale production of biomass of *Spirulina* is achieved using raceway ponds in outdoor conditions (Belay 1997; Hidasi and Belay 2018; Jimenez et al. 2003; Olgulin et al. 2003; Pushparaj et al. 1997; Richmond et al.

TABLE 5.1
Chemical Composition of *Spirulina**

General Composition	(%)	Vitamins	(mg/kg)
Moisture	3–7	Provitamin A	2,330,000 IU/kg*
Protein	55–70	Vitamin E	100
Fat (Lipids)	6–8	Thiamin B_1	35
Carbohydrates	15–25	Riboflavin B_2	40
Minerals (Ash)	7–13	Niacin B_3	140
Fiber	8–10	Vitamin B_6	8
		Vitamin B_{12}	3.2
Amino acids	**(g/Kg)**	Inositol	640
Alanine	47	Folic acid	0.1
Arginine	43	Biotin	0.05
Aspartic acid	61	Pantothenic acid	1.0
Cysteine	6	Vitamin K_1	22
Glutamic acid	91		
Glycine	32	**Minerals**	**(mg/kg)**
Histidine	10	Calcium	7000
Isoleucine	35	Chromium	2.8
Leucine	54	Copper	12
Lysine	29	Iron	1000
Methionine	14	Magnesium	4000
Phenylalanine	28	Manganese	50
Proline	27	Phosphorus	8000
Serine	32	Potassium	14000
Threonine	32	Sodium	9000
Tryptophan	9	Zinc	30
Tyrosine	30		
Valine	40	**Pigments**	**(g/kg)**
		Carotenoids	3.7
Physical properties		Chlorophyll	10
Appearance	Fine power	Phycocyanin	140
Color	Dark blue-green	**Essential fatty acids**	**(g/kg)**
Odor and taste	Mild like seaweed	Linoleic acid	8
Bulk Density	0.35–0.55/kg	γ-linolenic acid	10
Particle size	64 mesh through		
		Enzymes (Units/kg)	**(Units/kg)**
		Superoxide dismutase	1500000

(Source: Belay 1997); *Chemical composition in *Spirulina* culture may vary upon culture conditions.

1990). The highest biomass productivity of *S. maxima* was recorded as 37.67 g/m²/d in the batch mode cultivation using raceway ponds under greenhouse conditions. In another study conducted by Richmond et al. (1990) on biomass production from *S. platensis* using raceway (750 L) ponds, the productivity was reported as 15–27 g/m²/d. Whereas a study on culture of *S. platensis* in raceway ponds (135000 L) in Spain reported a yield of 2–17 g/m²/d (Jimenez et al. 2003).

CHLORELLA

Among the species of *Chlorella*, mainly *C. vulgaris* is cultivated worldwide. *C. vulgaris* Beyerinck is a green, roughly spherical, single celled fresh and ocean water microalga, belonging to the phylum Chlorophyta. Chlorella cells are many times smaller than those of other microalgae. *Chlorella* sps. are widely used in industries producing health foods, feed and food supplements, as well as in pharmaceutical and cosmetics. They

TABLE 5.2
Data on Outdoor Cultivation of *Spirulina* Reported in Literature – Some Examples

Culture Conditions	Productivity (g/m²/d)	Volume of Culture (L)	Mode of Cultivation & Location	References
Spirulina platensis	15–27	750	Raceway, Israel	Richmond et al. (1990)
Spirulina platensis	8.2	NR	Raceway, USA	Belay (1997)
Spirulina platensis	14.47	282	Raceway, Italy	Puspharaj et al. (1997)
Spirulina sp.	2–17	135,000	Raceway, Spain	Jimenez et al. (2003)
Spirulina sp.	9–13	NR	Raceway, Mexico	Olgulin et al. (2003)
Spirulina maxima	5.68–37.67	10.000–150,000	Semi-open raceway pond	Kim et al. (2018)

NR, not reported.

FIGURE 5.2 (See color insert.) Production of biomass from *Spirulina* by various companies (Sources: http://www.algaeindustrymagazine.com). (A) Earthrise-California, (B) Cyanotech-Hawaii, (C) Boonsom, Thailand, (D) Parry-India, (E) Hainan-China, and (F) Yaeyama-Japan.

are rich in proteins, amino acids, lipids, vitamins, and minerals (Table 5.3). Chlorella is marketed as a nutrient dense food supplement administered for boosting human health (Nakashima et al. 2009).

COMMERCIAL PRODUCTION OF *CHLORELLA* BIOMASS

Chlorella biomass production has reached to 2,000 tons annually (Table 5.4). The main producers of *Chlorella* are Taiwan, Japan, and Germany. *Chlorella* grows rapidly compared to other algal species. *C. vulgaris* is an ideal strain which can grow in a wide range of cultural conditions. Open cultivation is adopted for biomass production in commercial scale, using raceway ponds (Jorquera et al. 2010). However, exposure to unfavorable culture conditions such as nitrogen and phosphorus limitation; high light and temperature; carbon-dioxide concentration and high iron concentration, enhanced the production of lipids with a decrease in biomass

TABLE 5.3
Chemical Composition of *Chlorella**

[a]General Composition	%	[d]Vitamins	(mg/100g)
Moisture	5.83	Thiamine B_1	1.5
Protein	51.45	Riboflavin B_2	4.8
Fat (Lipids)	12.18	Niacin B_3	23.8
Carbohydrates	11.86	Pantothenic acid B_5	1.3
Ash	9.50	Pyridoxine B_6	1.7
Fiber	9.18	Biotin B_7	191.6
		Folic acid B_9	26.9
[b]Amino acids	(g/100g)	Cobalamin B_{12}	125.9
Alanine	10.90	Ascorbic acid C	15.6
Arginine	7.38		
Aspartic acid	10.94	[e]Minerals	(g/100g)
Cysteine	0.19	Calcium	0.59
Glutamic acid	9.08	Magnesium	0.34
Glycine	8.60	Phosphorus	1.76
Histidine	1.25	Potassium	0.05
Isoleucine	0.09	Sodium	1.35
Leucine	7.49	Zinc	30
Lysine	6.83	Fe	0.26
Methionine	1.30	Manganese	Tr
Phenylalanine	5.81	Se	Tr
Proline	2.97	Chromium	Tr
Serine	7.77	Copper	Tr
Threonine	6.09	Zinc	Tr
Tryptophan	2.21		
Tyrosine	8.44	[f]Fatty acids and phytonutrients	(mg/100g)
Valine	3.09	Unsaturated fatty acids	1377
Ornithine	0.13	Saturated fatty acids	256
[c]Pigments	(µg/g)	GLA	6
Astaxanthin	550,000	Coenzyme Q9	14
Canthaxanthin	362,000	RNA	2950
Lutein	52–3830	DNA	280
Chlorophyll-a	250–9630		
Chlorophyll-b	72–5770		
Pheophytin-a	2310–5640		
Violoxanthin	10–37		

tr, traces; Source from [a]Mohamed et al. (2013); [b]Safi et al. (2013); [c]Safi et al. (2014); [d]Panahi et al. (2012); [e]Tokuşoglu and Unal (2003); [f]Abeilled'Or Corporation Sdn Bhd., *Chemical composition in *Chlorella* culture may vary upon culture conditions.

productivity (Coverti et al. 2009; Liu et al. 2008; Lv et al. 2010; Widjaja et al. 2009).

Raceway ponds have limitations because the open cultivation is vulnerable to contamination, zooplankton, pollution, water evaporation, bacteria, and other competing algal species (Ranga Rao et al. 2012). Temperature, carbon-dioxide, and light vary due to environmental changes, posing challenges to the efficient production of biomass. To circumvent these problems several industries are adopting closed photobioreactor systems. These reactors can also be used for algal strains which are sensitive to environmental conditions. The photobioreactor tubes are maintained 20 cm in diameter, the thickness of their transparent wall is of a few millimeters, allowing proper light absorption. Column photobioreactors, tubular photobioreactor and flat plate photobioreactor are tested for algae and for biomass production (Kojima and Zhang 1999; Molina Grima

TABLE 5.4
Annual Production of *Spirulina* and *Chlorella* with Other Commercial Algal Species

Algal Species Name	Dry Weight (tons)	Producers Name and Country	Applications
Spirulina (*Arthrospira*)	3000	Earthrise Farms, USA, Myanma Microaglae-Myanmar, Cyanotech Corporation, USA. Hainan DIC Microalgae Co., Ltd China, Neotech Food Co., Ltd., Thailand, Genix, Cuba, E.I.D Parry Ltd, India, Hydrolina Biotech Pvt Ltd., Nutrex Hawaii Inc., China, India, USA, Japan, Myanmar, Chile, Cuba, Thailand.	Food (beverages, pasta, liquid extracts, chips), animal feed, cosmetics, pet foods; pharmaceutical (tablets, powder, extracts), nutraceutical
Chlorella	2000	Sun Chlorella Corporation, Yaeyama Shokusan Co. Ltd., Maypro Industries Inco., Taiwan Chlorella Manufacturing Co Ltd., Far East Microaglae Ind Co., Ltd., Roquette Klotze GmbH and Co.KG, Taiwan, Germany, Japan, USA.	Food (noodles), human nutrition, aquaculture, pharmaceutical (powder, extracts, tablets) nutraceutical, cosmetics
Dunaliella salina	2000	Aquacarotene Ltd., Cognis Nutrition and health, Nikken Sohonsha Corporation, Tianjin Norland Biotech Co Ltd., E.I.D Parry Ltd., Cyanotech Corporation, Australia, Israel, USA, Chi Corporation, Tianjin Norland Biotech Co Ltd., E.I.D Parry Ltd., Cyanotech Corporation, USA, Australia, Japan, China, India.	Human nutrition, cosmetics, β-carotene, Vitamin A precursor, antioxidant, food color, malnutrition, prevent macular degeneration, skin burn from UV rays, depression, asthma, infertility, psoriasis, high blood pressure, pharmaceutical, cosmetics, animal feed, fish feed, pet foods
Haematococcus pluvialis	1200	Algae Technologies, Mera pharmaceuticals, Cyanotech Corporation, Valensa International, Fuji Chemicals Industry Co Ltd., Jingzhou Natural Astaxanthin Inc., *Cyanotech Corporation, USA, India, Israel, Japan, Sweden, China.*	Aquaculture, astaxanthin, antioxidants, protect from UV rays, bioavailability, food colorant, animal feed, enhance immune functions, improve eye/skin health, anti-aging property, pharmaceutical, nutraceutical, cosmeceutical

(Source: Spolaore et al. (2006); Ambati et al. (2018); Ranga Rao et al. (2018); www.oilgae.com)

et al. 2003; Qiang and Richmond 1996; Zhang et al. 2001). Degen et al. (2001) reported that the biomass productivity of 0.11 g/L/h was achieved in *Chlorella* in a flat panel air-lift photobioreactor under continuous illumination (980E/m²/s). The biomass productivity of *C. vulgaris* and *C. sorokiniana* cultivated in flat panel reactors was reported as 17 g/m²/d and 7.7 g/m²/h respectively (Cuaresma et al. 2009; Pruvost et al. 2011). *Chlorella* was grown in a stirred tank reactor or bioreactor under heterotrophic culture conditions wherein the biomass productivity up to 0.25 g/L/d was obtained and high accumulation of lipids 22–54 mg/L/d was achieved (Liang et al. 2009; Ogawa and Aiba 1981). *C. vulgaris* was cultivated in mixotrophic conditions with reported biomass yield of 2–5 g/L/d (Yeh and Chang 2012). The advantages of this technique have been to limit the loss of biomass during dark respiration, that reduced the organic substances produced during the light regime. Biomass yield of *Chlorella* and its productivities varied depending on the mode of cultivation system reported by various researchers (Table 5.5). Despite the challenges just described *Chlorella* is successfully cultivated at an industrial scale by many companies using various cultivation methods (Figure 5.3).

INDUSTRIAL POTENTIAL OF *SPIRULINA* AND *CHLORELLA*

Spirulina and *Chlorella* are well known to produce high value molecules such as proteins, lipids, fatty acids, polyunsaturated fatty acids, phenolic, amino acids, volatile compounds, pigments, vitamins, sterols, and polysaccharides for health benefits. These bioactive compounds are used in food, feed, nutraceutical, cosmeceutical, and pharmaceutical applications (Ranga Rao et al. 2018).

TABLE 5.5
Reported *Chlorella* Biomass and Its Productivities by Various Authors*

Species and Strain Names	Mode of Cultivation	Biomass (g/L)	Biomass Productivity (g/L/d)	References
Chlorella protothecoides	Bioreactors	15.5	nd	Li et al. (2007)
Chlorella protothecoides	PBR	15.5	nd	Xu et al. (2006)
Chlorella protothecoides	Bioreactors	14.4	nd	Li et al. (2007)
Chlorella protothecoides	Bioreactors	12.8	nd	Li et al. (2007)
Chlorella minutissima (UTEX 2341)	PBR	8.3	0.01	Oh et al. (2010)
Chlorella protothecoides UTEX 25	Tubular PBR	6.3	0.41	Wawrik and Harriman (2010)
Chlorella minutissima (UTEX 2341)	PBR	3.4	0.006	Oh et al. (2010)
Chlorella sp	Transparent chambers	3.745	0.34	Hsieh and Wu (2009)
Chlorella vulgaris(13.86)	Bubble column PBR	3.79	0.6	Hulatt and Thomass (2011)
Chlorella vulgaris(INETI 58)	Race way ponds	3	0.18	Gouveia and Oliveira (2009)
Chlorella sp. TISTR 8990	Bioreactor	2.6	0.15	Wawrik and Harriman (2010)
Chlorella vulgaris (CPCC90)	Bubble column PBR	2.01	0.24	Cristiane et al. (2011)
Chlorella vulgaris (259)	Polycarbonate bottles	1.6	0.25	Liang et al. (2009)
Chlorella sp.	Airlift photobioreactor	1.484	0.32	Fulke et al. (2010)
Chlorella emersonii Shihira and Karus *(CCAP 211/11N)*	PBR	1.11	0.46	Illman et al. (2000)
Chlorella vulgaris (FACHB1068)	Column PBR	0.89	0.22	Fend et al. (2011)
Chlorella minutissima	Air-lift PBR	–	0.66	Dineshkumar et al. (2016)
Chlorella vulgaris	Flat panel PBR	–	0.45	Degen et al. (2001)
Chlorella vulgaris	Bubble column PBR	–	0.27	Degen et al. (2001)

*Biomass and its productivities may vary upon culture conditions and mode of cultivation system.

Spirulina and *Chlorella* have shown potential biological activities such as antioxidant, anti-inflammatory, anticancer, immune-stimulant, antimicrobial, antidiabetic, antihypertensive, anti-hepatoprotective, neuroprotecion, anti-anaemic, anti-leucopenic, and tissue engineering (Table 5.6).

Long-chain polyunsaturated fatty acids, especially of the ω-6 family such as γ-linolenic acid and arachidonic acid, and the ω-3 family such as eicosapentaenoic acid and docosahexaenoic acid are known to be useful in human nutrition. Omega-3 and 6 families cannot be synthesized *de novo* by humans; and therefore, must be received through food (Handayania et al. 2012). Omega-3 and 6 play a major role in maintaining human health and disease prevention. Phenolic compounds such as caffeic acid, ferulic acid, and and *p*-coumaric acid produced in the biomass of *Chlorella* are potent antioxidants (Zakaria et al. 2017). Similarly, the phenylpropanoid compounds such as pinostrobin, gallic acid, cinnamic acid, *p*-hydroxy-benzoic acid, chlorogenic acid, and vanillic acid were extracted from *S. maxima* (EI-Baky et al. 2009); whereas phenolic compounds- 3,4-dihydroxybenzaldehyde, *p*-hydroxybenzoic acid, protocatechuic acid, vanillic, syringic, caffeic acid, chlorogenic acid, and 4-hydroxybenzaldehyde in *S. platensis* was reported (Klejdus et al. 2009). Phloroglucinol from *Chlorella* proved to be effective against gastrointestinal disorders and autophagy in leukemia cells (Ruela-de-Sousa et al. 2010). Both *p*-coumaric acid and ferulic acid from *Spirulina* showed potential antioxidant activities against hepato-toxicity in *in vitro* models (EI-Baky et al. 2009).

One of the most important properties of foods is the flavour. *Chlorella* and *Spirulina* had great potential in food products due to their rich composition of volatile molecules. Volatile compounds (hydrocarbons, acids, alcohols esters, aldehydes, and ketones) were reported in *C. vulgaris* (Abdel-Baky et al. 2002) and *Spirulina* (Milovanović et al. 2015) and they were applied to various food products. Sterols play a fundamental role in the membrane integrity of microalgae (Patterson 1974). Microalgal sterols have several potential nutraceutical and pharmaceutical applications. The sterol composition in *Chlorella* sps. were detailed by Volkman (2016). The biological activity of phytosterol derived from algae showed immunomodulatory, anti-inflammatory, anti-hypercholestrolemic, antioxidant, anticancer, and antidiabetic properties (Luo et al. 2015). Ergosterol, 7-dehydroporiferasterol peroxide,

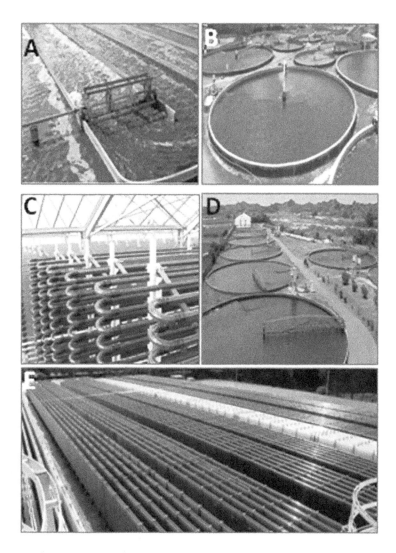

FIGURE 5.3 **(See color insert.)** Cultivation of *Chlorella* in various production systems for biomass production; (A) SAB, Singapore; (B) Nutriphys, Belgium; (C) Chlorella-echlorial (https://chlorella-echlorial.com); (D) Sun Chlorella, USA; and (E) Allmicroalgae-natural products, Lisbon, Portugal.

7-dehydroporiferasterol, ergosterol peroxide, and 7-oxocholesterol were identified in *Chlorella*, which showed potential effects in the prevention of inflammatory disorders and cancer (Yasukawa et al. 1996).

Spirulina contains phycobiliproteins, which have shown potential hepatoprotective, anti-inflammatory, anticancer, antioxidant, and immunomodulatory properties (Stengel et al. 2011). These proteins are used as labels in antibodies, receptors, and immune-blotting, fluorescence diagnosis (Waghmare et al. 2016). Proteins in *Chlorella* sps. were isolated and evaluated for their significant influence in improving the immune system, reducing blood pressure, accelerating wound healing, and lowering cholesterol levels (Halperin et al. 2003; Lisboa et al. 2014).

Algal proteins are a rich source of essential amino acids, which have received a great deal of attention by the food industry (Becker 2007). *Chlorella* and *Spirulina* contain ~50% protein content on the dry wt. basis (Becker 2004). Microalgae as a source of essential amino acids, lysine, leucine, isoleucine, tyrosine, tryptophan, and valine, constitute up to 35% of the total amino acids present in them (Waghmare et al. 2016). Algal aminoacids are being used in nutritional supplements as nutraceuticals. Peptides from *Chlorella* protect DNA against oxidative damage of cells. Peptides are used in the prevention of coronary, cancer, and atherosclerosis diseases (Dewapriya and Kim 2014; Romero García et al. 2012).

Spirulina is a rich source of Vitamin-B_1, B_7, B_{12}, and E. Vitamin-E protects the membrane lipids from oxidative damage, and also prevents atherosclerosis and coronary disease, whereas B_1 presents anti-inflammatory properties, while B_7 maintains nails, hair, and skin

TABLE 5.6
Bioactive Molecules in *Spirulina* and *Chlorella* for Industrial Applications

Compounds	Chlorella	Spirulina	Applications	References
Phenolic compounds	Phloroglucinol, p-coumaric acid, ferulic acid, apigenin	3,4-dihydroxybenzaldehyde, p-hydroxy benzoic acid, protocatechuic acid, vanillic, syringic caffeic acid, chlorogenic acid, and 4-hydroxybenzaldehyde	Antioxidant Properties, pharmaceutical applications, gastrointestinal disorders	Klejdus et al. (2009); Zakaria et al. (2017); El-Baky et al. (2009); Ruela-de-Sousa et al. (2010)
Volatile compounds	Hydrocarbons, acids, alcohols, esters, aldehydes ketones	Hydrocarbons, acids, alcohols, esters, ketones	Food supplements; Food products; Phytotoxic effects	Abdel-Baky et al. (2002); Milovanović et al. (2015)
Sterols	Brassicasterol, ergosterol, poriferasterol, clionaasterol, chondrillasterol, 7-dehydroporiferasterol peroxide, 7-dehydroporiferasterol, 7-oxocholesterol,	β-sitosterol, campesterol, stigmasterol,	Reduce coronary heart disease, immunomodulatory, anti-inflammatory, anti-hypercholesterolemic, antioxidant, anticancer and antidiabetic, antimicrobial property.	Volkman (2016); Patterson (1974); Yasukawa et al. (1996); Martinez Nadal (1971)
Proteins and aminoacids	Peptides, phycobiliproteins, isoleucine, leucine, phenylalanine	isoleucine, leucine, phenylalanine	Dietary supplements, anti-inflammatory, immunomodulatory, anticancer and antioxidant, labels for antibodies, receptors, and other biological moleculesreducing blood pressure, lowering cholesterol and glycaemia levels	Becker (2004); Waghmare et al. (2016); Halperin et al. (2003); Lisboa et al. (2014); Manirafasha et al. (2016)
Vitamins		Vitamin-A, Vitamin-C, Vitamin-E (alpha-tocopherol), Vitamin-K (phylloquinogfine), Vitamin-B_1; Vitamin-B_2, Vitamin-B_3, Vitamin-B_5, Vitamin-B_6, Vitamin-B_9, Vitamin-B_{12}	Nourish the body, detoxify and normalize intestinal function, stimulate immune system, protect lipids from oxidative damage, prevent atherosclerosis, coronary disease	Watanabe et al. (2002); Brown et al. (1999); Solomons et al. (2012)
Polysaccharides	Arabinose, galactose, rhamnose, glucose, 3-methyl-arabinose, 3-methyl-mannose, xylose, 4-methyl-arabinose, mannose, ribose, 2,4-dimethyl-arabinose, 3-methyl-galactose, 3-methyl-xylose, 3-methyl-rhamnose, 3,5-dimethyl-hexose, 6-methyl-galactose, glycerol, 2-keto-3-deoxy-octulonosic acid, 2,3,6-trimethyl-mannose, 3,6-dimethyl-mannose, 2-mehtyl-galactose, N-acetyl-galactoglucosamine, N-acetyl-glucosamine, amino sugar.	Rhamnose, glucuronic acid, fructose, galactose, 2-methyl-rhamnose, xylose, 3-methyl-rhamnose, 2-methyl-xylose, 4-methyl-rhamnose, glucose, mannose, galacturonic acid, 3-methyl-galactose, arabinose, amino sugar, 2-3-dimethyl-fucose, N-acetyl-glucosamine, 2-methyl-glucose, glycerol	antioxidant properties, tumor activities, and immunomodulatory, antiviral activity	Sheng et al. (2007); Pugh et al. (2001); Majdoub et al. (2009)
Pigments and carotenoids	β-carotene, lutein, antheraxanthin, zeaxanthin, violaxanthin, canthaxanthin, astaxanthin, chlorophyll-a &b, pheophytin-a & b	Lutein, zeaxanthin, β-carotene	Antioxidant property, animal feed, food colorant enhance immune function, improve skin health, antioperty, pharmaceutical, nutraceutical, cosmeceutical	Zhang et al. (2017); Safi et al. (2014); Liu et al. (2014, 2013); Cha et al. (2008); Gouveia et al. (2006); Ambati et al. (2018); Ranga Rao et al. (2018, 2013, 2010a, 2010b)

health properties. Vitamin-A from *Chlorella* is involved in vision, cellular communication, reproduction, and immune function in humans and animals (Solomons et al. 2012). Vitamin B_3 is important for the metabolism of fats, cholesterol synthesis, DNA synthesis, regulation of glucose, reduction in cholesterol, and cardiovascular diseases (Johnson and Russell 2012). *Chlorella* and *Spirulina* contain vitamin B_{12} in reasonably high concentrations. Vitamin B_{12} in *Chlorella* has shown better bioavailability compared to that from *Spirulina* (Watanabe et al. 2002). Both *Chlorella* and *Spirulina* have a good quantity of Vitamin B_9, which is important for the formation of cells and maintenance of metabolism (Brown et al. 1999).

Polysaccharides are used in food industries to develop a range of products. They are also used in nutraceuticals, cosmetics, and pharmaceutical industries (Plaza et al. 2009). Polysaccharides from *S. platensis* and *C. pyrenoidosa* have shown antitumor, immunomodulatory, and antioxidant properties (Sheng et al. 2007). The polysaccharides- immulina and immurella were isolated from *S. platensis* and *C. pyrenoidosa*, respectively (Pugh et al. 2001). Rhamnose was found as the main compound in *Spirulina* (Lee et al. 1998). Two polysaccharides (658Da) and (406Da) that were produced by *C. pyrenoidosa* strains were characterized (Becker 2007).

Specific algal pigments such as astaxanthin, lutein, and β-carotene are of commercial importance (Ranga Rao et al. 2014). These pigments are used in the formulation of food ingredients, fortified foods, dietary supplements, and also animal feed (Ranga Rao et al. 2018). Other class of pigments namely, phycobiliproteins are phycoerythrin, phycocyanin and allophycocyanin. Algal pigments can act as antioxidant, anti-carcinogen, anti-inflammatory, anti-obesity, antiangiogenic, neuroprotective agent, antitumor, immune-modulating, radical scavenging, antiviral, and antifungal agents (Manirafasha et al. 2016; Ranga Rao et al. 2018). Phycobiliproteins are used in the production of fermented milk products, ice creams, soft drinks, sweet cake decoration, milk shakes (Sonani et al. 2016).

ANTIMICROBIAL ACTIVITY OF MICROALGAE

The emergence of antibiotic resistance has laid the emphasis on the investigations of a range of natural products as a source of antibiotics. Over the last few years, several interesting and promising compounds were isolated from marine microorganisms and phytoplankton including green, brown, and red algae. Cholesterol and β-sitosterol found in *Spirulina maxima* were evaluated for their antimicrobial activity (Nadal 1971). Many microalgae have been shown to have antibacterial activities (Dineshkumar et al. 2017; Ranga Rao et al. 2010). The exact mechanism of the action of microalgal PUFAs exhibiting antimicrobial activity is not yet clarified. They may act towards multiple cellular targets, even though cell membranes are the most probable targets causing reduction of nutrient uptake, besides inhibiting cellular respiration; whereas Desbois et al. (2009) claimed a peroxidative process. Pratt (1942) reported that growth of *C. vulgaris* was suppressed by chlorellin, a compound produced and excreted into the medium.

MARKET VALUE FOR *CHLORELLA* AND *SPIRULINA* PRODUCTS

Market value for *Chlorella* and its products has been increasing over the last few years. Most consumers are looking for health foods from natural resources, which may prevent various diseases and also promote better health (Yim et al. 2007). Therefore, the use of algal-based health products are in full demand (Price Waterhouse Coopers 2009). Globally (*Asia-Pacific, Europe, North America, Middle East, Africa and Latin America*), algal products are projected to reach $3,318 million by 2022. The North American market was found to dominate global algae products with a specific focus on the nutraceutical market. *Spirulina* products have the largest market with more than 12,000 tons of *Spirulina* biomass produced every year. 70% of the production of *Spirulina* biomass comes from China, followed by India and Taiwan. The production of *Chlorella* was around 5,000 tons per annum. In the year 2015, the global market for functional foods, ingredients, personal care, pharmaceutical, and dietary supplements was valued at $182.60 billion, which is expected to reach $278.96 billion by 2021 (Gouveia et al. 2008). The market for *Chlorella* ingredients is expected to reach $700 million by 2022. The demand for *Spirulina* and *Chlorella* based products is projected to increase to $238 million by 2022.

SAFETY REGULATIONS FOR ALGAL PRODUCTS

Most consumers or suppliers pay a great deal of attention to the safety regulations on algal products (http://ec.europa.eu). Safety is an important issue and requires due attention when the algae produced in raceway ponds may lead to contamination with other microorganisms. The European Community Regulation on Food Safety (EC 178/2002) was published in 2002 in the Official Journal of the European Communities (1.2.2002 EN L 31/1) which applies to food products using whole cell algal biomass, or products. This regulation provides production, processing, distribution of food and feed. This regulation is applicable for *Chlorella* as it is used as a whole cell biomass or in combination with other ingredients. This regulation applies to novel foods and food ingredients that were not on the European market before May 15, 1997. The EPA- and DHA-rich microalgal oils have a significant history of consumption. According to this regulation, companies have to provide proof of the safety of the food product to the EFSA before the commercialization of the product. Labeling of novel food ingredients are added to the general European requirements on food labelling and has been provided by this regulation (Barsanti and Gualtieri 2018). This regulation states that health claims on food/feed products will be based on and substantiated by accepted scientific evidence. The EFSA panel on Dietetic Products Nutrition and Allergies (NDA) provides the scientific opinions on all health claims made in food/feed products. Two laws are applied to algal products in the United States when algal products are on the market: 1. The Federal Food, Drug and Cosmetic Act (FD&C), introduced in 1938 (http://www.ecolex.org/ecolex/ledge/view), regulates foods and food additives; 2. The Dietary Supplement Health and Education Act (DSHEA), introduced in 1994, regulates dietary ingredients and supplements. Both the FDC and DSHEA are controlled by USFDA. The Centre for Food Safety and Applied Nutrition (CFSAN) is responsible for regulating food ingredients and ensuring that those ingredients are safe. The authorization of feed products comes under the FDA Center for Veterinary Medicine (CVM).

For the FDA, any substance which has been added intentionally to food is a food additive and is subject to premarket review and approval by the FDA, except when the substance is generally recognized, among qualified experts, as safe (GRAS) under the conditions of its intended use. As per the center for food safety and applied nutrition, *Spirulina, Dunaliella, Schizochytrium, P. cruentum Chlorella* are used as food sources falling under Generally Recognized As Safe (GRAS). Some other algal products granted GRAS status are oils from *Schyzochitrium* and *Ulkenia*, whole algal protein powder, and a lipid ingredient derived from *Chlorella* (FDA, 2010).

CONCLUSIONS

Utility of edible algal forms as a source of proteins, vitamins, and minerals has now expanded to the utilities of nutraceutical value in recent years (Nicoletti 2016; Ramakrishnan et al. 2008). Some of the forms are now considered as superfoods (Cachon-Lee et al. 2010; Gouveia and Oliveira 2009; Luo et al. 2015; Vigani et al. 2015; Wells et al. 2017). For example, *Spirulina* is now added to pasta, chocolate, beverages and others, in order to increase and improve digestion or metabolic functions (Gantar and Svirčev 2008) and antioxidant properties (Romay et al. 2003). Several forms are useful in controlling bacterial growth (Raja et al. 2016; Romay et al. 1998). However, they have a positive influence on the gut microbiome. Furthermore, considering the nutraceuticals effects, the prebiotic effects (Gupta et al. 2017, 2011; Parada et al. 1998), including immune-stimulating properties (Suarez et al. 2005), several algal forms including *S. platensis* (Parada et al. 1998) and *Chlorella pyrenoidosa* (Liu and Chen 2016) have been utilized in health food products. The utility of both *Spirulina* and *Chlorella* is bound to expand in the years to come and is here to stay forever.

ACKNOWLEDGMENT

ARR acknowledge Vignan's Foundation for Science, Technology and Research University for providing the facility for preparing this manuscript.

REFERENCES

Abdel-Baky, H.H., Shallan, M.A., El-Baroty, G., and El-Baz, F.K. Volatile compounds of the microalgae *Chlorella vulgaris* and their phytotoxic effect. *Pak. J. Biol. Sci.* 2002. 5: 61–65.

Abdulqader, G., Barsanti, L., and Tredici, M.R. Harvest of *Arthrospira platensis* from Lake Kossorom (Chad) and its household usage among the Kanembu. *J. Appl. Phycol.* 2000. 12: 493–498.

Ambati, R.R., Gogisetty, D., Aswathnarayana Gokare, R., Ravi, S., Bikkina, P.N., Su, Y., and Lei, B. *Botryococcus* as an alternative source of carotenoids and its possible applications-an overview. *Crit. Rev. Biotechnol.* 2018. 38: 541–558.

Barsanti, L., and Gualtieri, P. Is exploitation of microalgae economically and energetically sustainable? *Algal Res.* 2018. 31: 107–115.

Becker, E.W. Micro-algae as a source of proteins. *Biotechnol. Adv.* 2007. 25: 207–210.

Becker, E.W. Microalgae in human and animal nutrition. In: Richmond, A., (Ed.), *Handbook of Microalgal Culture: Biotechnology and Applied Phycology*, 1st edition. Blackwell Science, London, UK, 2004, pp. 312–351.

Belay, A. Biology and industrial production of *Arthrospira* (*Spirulina*). In: Richmond, A., and Hu, Q., (Eds.), *Handbook of Microalgal Culture: Applied Phycology and Biotechnology*. Blackwell Publishing Ltd., Oxford, London, 2013, pp. 339–358.

Belay, A. Mass culture of *Spirulina* outdoors: The earthrise farms experience. In: Vonshak, A., (Ed.), *Spirulina platensis (Arthrospira), Physiology, Cell Biology and Biotechnology*. Taylor & Francis, London, 1997, pp. 131–158.

Belay, A. *Spirulina (Arthrospira)*: Production and quality assurance. In: Gershwin, M.E., and Belay, A., (Eds.), *Spirulina in Human Nutrition and Health*. CRC Press, Taylor and Francis Group, Boca Raton, FL, 2008, pp. 1–25.

Borowitzka, M.A. Commercial production of microalgae: Ponds, tanks, tubes and fermentors. *J. Biotechnol.* 1999. 70: 313–321.

Borowitzka, M.A., and Moheimani, N.R. Open pond culture systems. In: *Algae for Biofuels and Energy*. Springer, Dordrecht, 2013, pp. 133–152.

Brown, M.R., Mular, M., Miller, I., Farmer, C., and Trenerry, C. The vitamin content of microalgae used in aquaculture. *J. App. Phycol.* 1999. 11: 247–255.

Cha, K.H., Koo, S.Y., and Lee, D.U. Anti-proliferative effects of carotenoids extracted from *Chlorella ellipsoidea* and *Chlorella vulgaris* on human colon cancer cells. *J. Agric. Food Chem.* 2008. 56: 10521–10526.

Chacon-Lee, T.L., and Gonzalez-Marino, G.E. Microalgae for healthy foods possibilities and challenges. *Com. Rev. Foods Sci. Food Saf.* 2010. 9: 655–675.

Chaiklahan, R., Khonsarn, N., Chirasuwan, N., Ruengjitchatchawalya, M., Bunnag, B., and Tanticharoen, M. Response of *Spirulina platensis* C1 to high temperature and high light intensity. *Kasetsart J. Nat. Sci.* 2007. 41: 123–129.

Christaki, E., Florou-Paneri, P., and Bonos, E. Microalgae: A novel ingredient in nutrition. *Int. J. Food Sci. Nutr.* 2011. 62: 794–799.

Da Silva Vaz, B., Moreira, J.B., de Morais, M.G., and Costa, J.A.V. Microalgae as a new source of bioactive compounds in food supplements. *Curr. Opin. Food Sci.* 2016. 7: 73–77.

Degen, J., Uebele, A., Retze, A., Schmid-Staiger, U., and Trösch, W.A. Novel air-lift photobioreactor with baffles for improved light utilization through the flashing light effect. *J. Biotechnol.* 2001. 92: 89–94.

Del Castillo, B. *The Discovery and Conquest of Mexico, 1517–1521*. Routledge, London, UK, 1928, p. 300.

Desbois, A.P., Mearns-Spragg, A., and Smith, V.J. A fatty acid from the diatom *Phaeodactylum tricornutum* is antibacterial against diverse bacteria including multi-resistant *Staphylococcus aureus* (MRSA). *Mar. Biotechnol.* 2009. 11: 45–52.

Dewapriya, P., and Kim, S.K. Marine microorganisms: An emerging avenue in modern nutraceuticals and functional foods. *Food Res. Int.* 2014. 56: 115–125.

Dineshkumar, R., Narendran, R., Jayasingam, P., and Sampathkumar, P. Centre of advanced study in marine cultivation and chemical composition of microalgae *Chlorella vulgaris* and its antibacterial activity against human pathogens. *J. Aquac. Mar. Biol.* 2017. 5(3): 119.

Dineshkumar, R., Subramanian, G., Dash, S.K., and Sen, R. Development of an optimal light-feeding strategy coupled with semi-continuous reactor operation for simultaneous improvement of microalgal photosynthetic efficiency, lutein production and CO2 sequestration. *Biochem. Eng. J.* 2016. 113: 47–56.

El-Baky, H.H.A., El-Baz, F.K., and El-Baroty, G.S. Production of phenolic compounds from *Spirulina maxima* microalgae and its protective effects *in vitro* toward hepatotoxicity model. *Afr. J. Pharm. Pharmacol.* 2009. 3: 133–139.

Eskin, N.A.M., and Tamir, S. *Dictionary of Nutraceuticals and Functional Foods*. CRC Press, Boca Raton, FL, 2006, p. 520.

FDA. Federal Food, Drug, and Cosmetic Act (FD&C Act). 2010 http://www.fda.gov/regulatoryinformation/legislation/federalfooddrugandcosmeticactfdcact/default.htm.

Fulke, A.B., Mudliar, S.N., Yadav, R., Shekh, A., Srinivasan, N., Ramanan, R., Krishnamurthi, K., Devi, S.S., and Chakrabarti, T. Bio-mitigation of CO2, calcite formation and simultaneous biodiesel precursors production using *Chlorella* sp. *Bioresour. Technol.* 2010. 101: 8473–8476.

Gantar, M., and Svirčev, Z. Microalgae and cyanobacteria: Food for thought. *J. Phycol.* 2008. 44: 260–268.

Gershwin, M.E., and Belay, A. *Spirulina in Human Nutrition and Health*, 1st edition. CRC Press, Boca Raton, FL, 2007, p. 328. ISBN 9781420052565.

Gouveia, L., Batista, A.P., Sousa, I., Raymundo, A., and Bandarra, N.M. *Food Chemistry Research Developments*, 1st edition. Nova Science Publishers, New York, 2008, pp. 75–111.

Gouveia, L., and Oliveira, A.C. Microalgae as a raw material for biofuels production. *J. Ind. Microbiol. Biotechnol.* 2009. 36: 269–274.

Gouveia, L., Raymundo, A., Batista, A.P., Sousa, I., and Empis, J. *Chlorella vulgaris* and *Haematococcus pluvialis* biomass as colouring and antioxidant in food emulsions. *Eur. Food Res. Technol.* 2006. 222: 362–367.

Guiry, M.D. How many species of algae are there? *J. Phycol.* 2012. 48: 1057–1063.

Gupta, S., Gupta, C., Garg, A.P., and Prakash, D. Prebiotic efficiency of blue green algae on probiotics microorganisms. *J. Microbiol. Exp.* 2017. 4: 1–4.

Halperin, S.A., Smith, B., Nolan, C., Shay, J., and Kralovec, J. Safety and immune-enhancing effect of a *Chlorella*-derived dietary supplement in healthy adults undergoing influenza vaccination: Randomized, double-blind, placebo-controlled trial. *Can. Med. Assoc. J.* 2003. 169: 111–117.

Handayania, N.A., Ariyantib, D., and Hadiyanto, H. Potential production of polyunsaturated fatty acids from microalgae. *Int. J. Eng. Sci.* 2012. 02: 13–16.

Hidasi, N., and Belay, A. Diurnal variation of various culture and biochemical parameters of *Arthrospira platensis* in large-scale outdoor raceway ponds. *Algal Res.* 2018. 29: 121–129.

Hikmet, K., Akın, B.S., and Atici, A.T. Microalgal toxin(s): Characteristics and importance. *Afr. J. Biotechnol.* 2004. 3: 667–674.

Hobuss, C.B., Rosales, P.F., Venzke, D., Souza, P.O., Gobbi, P.C., Gouvea, L.P., Santos, M.A.Z., Pinto, E., Jacob-Lopes, E., and Pereira, C.M.P. Cultivation of algae in photobioreator and obtention of biodiesel. *Rev. Bras. Farmacogn.* 2011. 21: 361–364.

Hsieh, C.H., and Wu, W.T. Cultivation of microalgae for oil production with a cultivation strategy of urea limitation. *Bioresour. Technol.* 2009. 100: 3921–3926.

Hulatt, C.J., and Thomas, D.N. Productivity, carbon dioxide uptake and net energy return of microalgal bubble column photobioreactors. *Bioresour. Technol.* 2011. 102: 5775–5787.

Illman, A.M., Scragg, A.H., and Shales, S.W. Increase in *Chlorella* strains calorific values when grown in low nitrogen medium. *Enzyme Microb. Technol.* 2000. 27: 631–635.

Jimenez, C., Cossío, B.R., and Niell, F.X. Relationship between physicochemical variables and productivity in open ponds for the production of *Spirulina*: A predictive model of algal yield. *Aquaculture* 2003. 221: 331–345.

Johnson, E.J., and Russell, R.M. Beta-carotene. In: Coates, P.M., Betz, J.M., Blackman, M.R., Cragg, G.M., Levine, M., Moss, J., and White, J.D., (Eds.), *Encyclopedia of Dietary Supplements*, 2nd edition. New York, NY, 2012. p. 920.

Jorquera, O., Kiperstok, A., Sales, E.A., Embiruçu, M., and Ghirardi, M.L. Comparative energy life-cycle analyses of microalgal biomass production in open ponds and photobioreactors. *Bioresour. Technol.* 2010. 101: 1406–1413.

Juneja, A., Ceballos, R., and Murthy, G.S. Effects of environmental factors and nutrient availability on the biochemical composition of algae for biofuels production: A review. *Energies* 2013. 6: 4607–4638.

Kent, M., Welladsen, H.M., Mangott, A., and Li, Y. Nutritional evaluation of Australian microalgae as potential human health supplements. *PLOS ONE.* 2015. 10: e0118985.

.Khan, Z., Bhadouria, P., and Bisen, P.S. Nutritional and therapeutic potential of *Spirulina*. *Curr. Pharm. Biotechnol.* 2005. 6: 373–379.

Kim, T., Choi, W.S., Ye, B.R., Heo, S.J., Oh, D., Kim, S., Choi, K.S., and Kang, D.H. Cultivating *Spirulina maxima*: Innovative approaches. In: *Cyanobacteria*. Intech Open, 2018, pp. 61–83.

Klejdus, B., Kopecký, J., Benesová, L., and Vacek, J. Solid-phase/supercritical-fluid extraction for liquid chromatography of phenolic compounds in freshwater microalgae and selected cyanobacterial species. *J. Chromatogr. A.* 2009. 1216: 763–771.

Kojima, E., and Zhang, K. Growth and hydrocarbon production of microalga *Botryococcus braunii* in bubble column photobioreactors. *J. Biosci. Bioeng.* 1999. 87: 811–815.

Komarek, J., Kastovsky, J., Mares, J., and Johansen, J.R. Taxonomic classification of cyanoprokaryotes (cyanobacterial genera), using a polyphasic approach. *Preslia.* 2014. 86: 295–335.

Lee, J.B., Hayashi, T., Hayashi, K., Sankawa, U., Maeda, M., Nemoto, T., and Nakanishi, H. Further purification and structural analysis of calcium spirulan from *Spirulina platensis*. *J. Nat. Prod.* 1998. 61: 1101–1104.

Li, X., Xu, H., and Wu, Q. Large-scale biodiesel production from microalga *Chlorella protothecoides* through heterotrophic cultivation in bioreactors. *Biotechnol. Bioeng.* 2007. 98: 764–771.

Liang, Y., Sarkany, N., and Cui, Y. Biomass and lipid productivities of *Chlorella vulgaris* under autotrophic, heterotrophic, and mixotrophic growth conditions. *Biotechnol. Lett.* 2009. 31: 1043–1049.

Lisboa, C.R., Pereira, A.M., Ferreira, S.P., and Costa, J.A.V. Utilisation of *Spirulina* sp. and *Chlorella pyrenoidosa* biomass for the production of enzymatic protein hydrolysates. *Int. J. Eng. Res. Appl.* 2014. 4: 29–38.

Liu, J., and Chen, F. Biology and industrial applications of *Chlorella*: Advances and prospects. *Adv. Biochem. Eng. Biotechnol.* 2016. 153: 1–35.

Liu, J., Sun, Z., Gerken, H., Liu, Z., Jiang, Y., and Chen, F. *Chlorella zofingiensis* as an alternative microalgal producer of astaxanthin: Biology and industrial potential. *Mar. Drugs.* 2014. 12: 3487–3515.

Liu, Z.Y., Wang, G.C., and Zhou, B.C. Effect of iron on growth and lipid accumulation in *Chlorella vulgaris*. *Bioresour. Technol.* 2008. 99: 4717–4722.

Lu, Y.M., Xiang, W.Z., and Wen, Y.H. *Spirulina* (*Arthrospira*) industry in Inner Mongolia of China: Current status and prospects. *J. Appl. Phycol.* 2011. 23: 265–269.

Luo, X., Su, P., and Zhang, W. Advances in microalgae-derived phytosterols for functional food and pharmaceutical applications. *Mar. Drugs.* 2015. 13: 4231–4254.

Lv, J.M., Cheng, L.H., Xu, X.H., Zhang, L., and Chen, H.L. Enhanced lipid production of *Chlorella vulgaris* by adjustment of cultivation conditions. *Bioresour. Technol.* 2010. 101: 6797–6804.

Majdoub, H., Mansour, M.B., Chaubet, F., Roudesli, M.S., and Maaroufi, R.M. Anticoagulant activity of a sulfated polysaccharide from the green alga *Arthrospira platensis*. *Biochim. Biophys. Acta.* 2009. 1790: 1377–1381.

Manirafasha, E., Ndikubwimana, T., Zeng, X., Lu, Y., and Jing, K. Phycobiliprotein: Potential microalgae derived pharmaceutical and biological reagent. *Biochem. Eng. J.* 2016. 109: 282–296.

Mata, T.M., Martins, A.A., and Caetano, N.S. Microalgae for biodiesel production and other applications: A review. *Renew. Sustain. Energy Rev.* 2010. 14: 217–232.

Miller, M.A., Kudela, R.M., Mekebri, A., Crane, D., Oates, S.C., Tinker, M.T., Staedler, M., Miller, W.A., Toy-Choutka, S., Dominik, C., Hardin, D., Langlois, G., Murray, M., Ward, K., and Jessup, D.A. Evidence for a novel marine harmful algal bloom: Cyanotoxin (microcystin) transfer from land to sea otters. *PLOS ONE.* 2010. 5: e12576.

Milovanović, I., Mišan, A., Simeunović, J., Kovač, D., Jambrec, D., and Mandić, A. Determination of volatile organic compounds in selected strains of cyanobacteria. *J. Chem.* 2015. 2015.

Mohamed, A.G., Abo-El-Khair, B.E., and Shalaby, S.M. Quality of novel healthy processed cheese analogue enhanced with marine microalgae *chlorella vulgaris* biomass. *World Appl. Sci. J.* 2013. 23: 914–925.

Molina Grima, E., Belarbi, E.H., Acién Fernández, F.G., Robles Medina, A., and Chisti, Y. Recovery of microalgal biomass and metabolites: Process options and economics. *Biotechnol. Adv.* 2003. 20: 491–515.

Moorhead, K., Capelli, B., and Cysewski, G.R. *Spirulina Nature's Superfood*, 3rd edition. Cyanotech Corporation, Kailua-Kona, HI, 2011. ISBN 0-9637511-3-1.

Nadal, N.G.M. Sterols of *Spirulina maxima*. *Phytochemistry* 1971. 10: 2537–2538.

Nadis, S. The cells that rule the seas. *Sci. Am.* 2003. 289: 52–53.

Nakashima, Y., Ohsawa, I., Konishi, F., Hasegawa, T., Kumamoto, S., Suzuki, Y., and Ohta, S. Preventive effects of *Chlorella* on cognitive decline in age-dependent dementia model mice. *Neurosci. Lett.* 2009. 464: 193–198.

Nicoletti, M. Microalgae nutraceuticals. *Foods*. 2016. 5: 54.

Nicoletti, M. Nutraceuticals and botanicals: Overview and perspectives. *Int. J. Food Sci. Nutr.* 2012. 63: 2–6.

Ogawa, T., and Aiba, S. Bioenergetic analysis of mixotrophic growth in *Chlorella vulgaris* and *Scenedesmus acutus*. *Biotechnol. Bioeng.* 1981. 23: 1121–1132.

Oh, S.H., Kwon, M.C., Choi, W.Y., Seo, Y.C., Kim, G.B., Kang, D.H., Lee, S.Y., and Lee, H.Y. Long-term outdoor cultivation by perfusing spent medium for biodiesel production from *Chlorella minutissima*. *J. Biosci. Bioeng.* 2010. 110: 194–200.

Panahi, Y., Pishgoo, B., Jalalian, H.R., Mohammadi, E., Taghipour, H.R., Sahebkar, A., and Abolhasani, E. Investigation of the effects of *Chlorella vulgaris* as an adjunctive therapy for dyslipidemia: Results of a randomised open-label clinical trial. *Nutr. Diet.* 2012. 69: 13–19.

Parada, J.L., Zulpa de Caire, G., Zaccaro de Mulé, M.C., and Storni de Cano, M.M. Lactic acid bacteria growth promoters from *Spirulina platensis*. *Int. J. Food Microbiol.* 1998. 45: 225–228.

Patterson, G.W. Sterols of some green algae. *Comp. Biochem. Physiol. B Comp. Biochem.* 1974. 47: 453–457.

Price Waterhouse Coopers. Leveraging growth in the emerging functional foods industry: Trends and market opportunities. Functional Foods Reports, 2009. pp. 1–22.

Pugh, N., Ross, S.A., ElSohly, H.N., ElSohly, M.A., and Pasco, D.S. Isolation of three high molecular weight polysaccharide preparations with potent immune-stimulatory activity from *Spirulina platensis, Aphanizomenon flos-aquae* and *Chlorella pyrenoidosa*. *Planta Med.* 2001. 67: 737–742.

Pulz, O., and Gross, W. Valuable products from biotechnology of microalgae. *Appl. Microbiol. Biotechnol.* 2004. 65: 635–648.

Pushparaj, B., Pelosi, E., Tredici, M.R., Pinzani, E., and Materassi, R. As integrated culture system for outdoor production of microalgae and cyanobacteria. *J. Appl. Phycol.* 1997. 9: 113–119.

Qiang, H., and Richmond, A. Productivity and photosynthetic efficiency of *Spirulina platensis* as affected by light intensity, algal density and rate of mixingina flat plate-photobioreactor. *J. Appl. Phycol.* 1996. 8: 139–145.

Raja, R., Hemaiswarya, S., Ganesan, V., and Carvalho, I.S. Recent developments in therapeutic applications of Cyanobacteria. *Crit. Rev. Microbiol.* 2016. 42: 394–405.

Ramakrishnan, C.M., Haniffa, M.A., Manohar, M., Dhanaraj, M., Jesu Arockiaraj, A., Seetharaman, S., and Arunsingh, S.V. Effects of probiotics and *Spirulina* on survival and growth of juvenile common carp (*Cyprinus carpio*). *Isr. J. Aquac.* 2008. 60: 128–133.

Ranga Rao, A., Baskaran, V., Sarada, R., and Ravishankar, G.A. *In vivo* bioavailability and antioxidant activity of carotenoids from micro algal biomass – A repeated dose study. *Food Res. Int.* 2013. 54: 711–777.

Ranga Rao, A., Deepika, G., Ravishankar, G.A., Sarada, R., Narasimharao, B.P., Lei, B., and Su, Y. Industrial potential of carotenoid pigments from microalgae: Current trends and future prospects. *Crit. Rev. Food Sci. Nutr.* 2018. 25: 1–22.

Ranga Rao, A., Phang, S.M., Sarada, R., and Ravishankar, G.A. Astaxanthin: Sources, extraction, stability, biological activities and its commercial applications – A Review. *Mar. Drugs.* 2014. 12: 128–152.

Ranga Rao, A., Raghunath Reddy, R.L., Baskaran, V., Sarada, R., and Ravishankar, G.A. Characterization of microalgal carotenoids by mass spectrometry and their bioavailability and antioxidant properties elucidated in rat model. *J. Agric. Food Chem.* 2010b. 58: 8553–8559.

Ranga Rao, A., Reddy, A.H., and Aradhya, S.M. Antibacterial properties of *Spirulina platensis, Haematococcus pluvialis, Botryococcus braunii* micro algal extracts. *Curr. Trends Biotechnol. Pharm.* 2010a. 4: 809–819.

Ranga Rao, A., Yingchun, G., Hu, Z., Sommerfeld, M., and Hu, Q. Comparative study of predators in mass cultures of *Chlorella zofingiensis, Scenedesmus dimorphus* and *Nannochloropsis oceanica* in ponds and photobioreactors. Algae Biomass Summit, 24–27th Sep 2012, Algae Biomass Organization.

Regulation (EC) 1924/2006. 2007. Official Journal of the European Communities (18.1.2007 EN L 12/3) http://www.ecolex.org/ecolex/ledge/view.

Richmond, A. *Handbook of Microalgal Culture: Biotechnology and Applied Phycology*. Blackwell, Oxford, UK, 2004, p. 588. ISBN: 978-0-632-05953-9.

Richmond, A. Spirulina. In: Borowitzka, M.A., and Borowitzka, L.J., (Eds.), *Microalgal Biotechnology*. Cambridge University Press, Cambridge, London, 1988, pp. 85–121.

Richmond, A., Lichtenberg, E., Stahl, B., and Vonshak, A. Quantitative assessment of the major limitations on productivity of *Spirulina platensis* in open raceways. *J. Appl. Phycol.* 1990. 2: 195–206.

Romay, C., Armesto, J., Remirez, D., González, R., Ledon, N., and García, I. Antioxidant and anti-inflammatory properties of C-phycocyanin from blue-green algae. *Inflam. Res.* 1998. 47(1): 36–41.

Romay, CH., González, R., Ledón, N., Remirez, D., and Rimbau, V. C-phycocyanin: A biliprotein with antioxidant, anti-inflammatory and neuroprotective effects. *Curr. Protein Pept. Sci.* 2003. 4: 207–216.

Romero García, J.M., Acién Fernández, F.G., and Fernández Sevilla, J.M. Development of a process for the production of l-amino-acids concentrates from microalgae by enzymatic hydrolysis. *Bioresour. Technol.* 2012. 112: 164–170.

Ruela-de-Sousa, R.R., Fuhler, G.M., Blom, N., Ferreira, C.V., Aoyama, H., and Peppelenbosch, M.P. Cytotoxicity of apigenin on leukemia cell lines: Implications for prevention and therapy. *Cell Death Dis.* 2010. 1: e19.

Safi, C., Charton, M.P., Pignolet, O., Silvestre, F., Vaca-Garcia, C., and Pontalier, P.Y. Influence of microalgae cell wall characteristics on protein extractability and determination of nitrogen-to-protein conversion factors. *J. Appl. Phycol.* 2013. 25: 523–529.

Safi, C., Zebib, B., Merah, O., Pontalier, P.Y., and Vaca-Garcia, C. Morphology, composition, production, processing and applications of *Chlorella vulgaris*: A review. *Renew. Sustain. Energy Rev.* 2014. 35: 265–278.

Sheng, J., Yu, F., Xin, Z., Zhao, L., Zhu, X., and Hu, Q. Preparation, identification and their antitumor activities *in vitro* of polysaccharides from *Chlorella pyrenoidosa*. *Food Chem.* 2007. 105: 533–539.

Singh, N.K., and Dhar, D.W. Phylogenetic relatedness among *Spirulina* and related cyanobacterial genera. *World J. Microbiol. Biotechnol.* 2011. 27: 941–951.

Siva Kiran, R.R., Madhu, G.M., and Satyanarayana, S.V. *Spirulina* in combating Protein Energy Malnutrition (PEM) and Protein Energy Wasting (PEW) – A review. *J. Nutr. Res.* 2015. 3: 62–79.

Solomons, N.W. Vitamin A. In: Erdman, J.W., Macdonald, I.A., and Zeisel, S.H., (Eds.), *Present Knowledge in Nutrition*. Wiley-Blackwell, Oxford, UK, 2012, pp.149–184.

Sonani, R.R., Rastogi, R.P., Patel, R., and Madamwar, D. Recent advances in production, purification and applications of phycobiliproteins. *World J. Biol. Chem.* 2016. 7: 100–109.

Soni, R.A., Sudhakar, K., Rana, R.S. *Spirulina* from growth to nutritional product: A review. *Trends Food Sci. Technol.* 2017. 69: 157–171.

Spolaore, P., Joannis-Cassan, C., Duran, E., and Isambert, A. Commercial applications of microalgae. *J. Biosci. Bioeng.* 2006. 101: 87–96.

Stengel, D.B., Connan, S., and Popper, Z.A. Algal chemodiversity and bioactivity: Sources of natural variability and implications for commercial application. *Biotechnol. Adv.* 2011. 29: 483–501.

Tokuşoglu, O., and Unal, M.K. Biomass nutrient profiles of three microalgae: *Spirulina platensis*, *Chlorella vulgaris*, and *Isochrisis galbana*. *J. Food Sci.* 2003. 68: 1144–1148.

Torzillo, G., Pushparaj, B., Bocci, F., Balloni, W., Materassi, R., and Florenzano, G. Production of *Spirulina* biomass in closed photobioreactors. *Biomass.* 1986. 11: 61–74.

Vigani, M., Parisi, C., Rodríguez-Cerezo, E., Barbosa, M.J., Sijtsma, L., Ploeg, M., and Enzing, C. Food and feed products from micro-algae: Market opportunities and challenges for the EU. *Trends Food Sci. Technol.* 2015. 42: 81–92.

Volkman, J.K. Sterols in microalgae. In: Borowitzka, A.M., Beardall, J., and Raven, J.A., (Eds.), *The Physiology of Microalgae*. Springer, 2016, pp. 485–505.

Vonshak, A. Outdoor mass production of *Spirulina*: The basic concept. In: Vonshak, A., (Ed.), *Spirulina platensis (Arthrospira): Physiology, Cell-Biology and Biotechnology*. Taylor and Francis Ltd., London, 1997, pp. 79–99.

Vonshak, A., and Guy, R. Photoadaptation, photoinhibition and productivity in the blue-green alga, *Spirulina platensis* grown outdoors. *Plant Cell Environ.* 1992. 15: 613–616.

Vonshak, A., Laorawat, S., Bunnag, B., and Tanticharoen, M. The effect of light availability on the photosynthetic activity and productivity of outdoor cultures of *Arthrospira platensis (Spirulina)*. *J. Appl. Phycol.* 2014. 26: 1309–1315.

Vonshak, A., and Richmond, A. Mass production of the blue-green alga *Spirulina*: An overview. *Biomass.* 1988. 15: 233–247.

Waghmare, A.G., Salve, M.K., LeBlanc, J.G., and Arya, S.S. Concentration and characterization of microalgae proteins from *Chlorella pyrenoidosa*. *Bioresour. Bioprocess.* 2016. 3: 1–11.

Wawrik, B., and Harriman, B.H. Rapid colorimetric quantification of lipid from algal cultures. *J. Microbiol. Methods.* 2010. 80: 262–266.

Wells, M.L., Potin, P., Craigie, J.S., Raven, J.A., Merchant, S.S., Helliwell, K.E., Smith, A.G., Camire, M.E., and Brawley, S.H. Algae as nutritional and functional food sources: Revisiting our understanding. *J. Appl. Phycol.* 2017. 29: 949–982.

Widjaja, A., Chien, C.C., and Ju, Y.H. Study of increasing lipid production from fresh water microalgae-*Chlorella vulgaris*. *J. Taiwan Inst. Chem. Eng.* 2009. 40: 13–20.

Xu, H., Miao, X., and Wu, Q. High quality biodiesel production from a microalgae *Chlorella protothecoides* by heterotrophic growth in fermenters. *J. Biotechnol.* 2006. 126: 499–507.

Yasukawa, K., Akihisa, T., Kanno, H., Kaminaga, T., Izumida, M., Sakoh, T., Tamura, T., and Takido, M. Inhibitory effects of sterols isolated from *Chlorella vulgaris* on 12-O-tetradecanoylphorbol-13-acetate-Induced inflammation and tumor promotion in mouse skin. *Biol. Pharm. Bull.* 1996. 19: 573–576.

Yeh, K.L., and Chang, J.S. Effects of cultivation conditions and media composition on cell growth and lipid productivity of indigenous microalgae *Chlorella vulgaris* ESP-31. *Bioresour. Technol.* 2012. 105: 120–127.

Yim, H.E., Yoo, K.H., Seo, W.H., Won, N.H., Hong, Y.S., and Lee, J.W. Acute tubulointerstitial nephritis following ingestion of *Chlorella* tablets. *Pediatr. Nephrol.* 2007. 22: 887–888.

Zakaria, S.M., Kamal, S.M.M., Harun, M.R., Omar, R., and Siajam, S.I. Subcritical water technology for extraction of phenolic compounds from *Chlorella* sp. microalgae and assessment on its antioxidant activity. *Molecules* 2017. 22: 1105.

Zhang, K., Miyachi, S., and Kurano, N. Evaluation of a vertical flat-plate photobioreactor for outdoor biomass production and carbondioxide biofixation: Effects of reactor dimensions, irradiation and cell concentration on the biomass productivity and irradiation utilization efficiency. *Appl. Microbiol. Biotechnol.* 2001. 55: 428–433.

Zhang, Z., Huang, J.J., Sun, D., Lee, Y., and Chen, F. Two-step cultivation for production of astaxanthin in *Chlorella zofingiensis* using a patented energy-free rotating floating photobioreactor (RFP). *Bioresour. Technol.* 2017, 224: 515–522.

6 Seaweed Antimicrobials
Present Status and Future Perspectives

María José Pérez, Elena Falqué, and Herminia Domínguez

CONTENTS

Abbreviations 61
Introduction 61
Bioactive Compounds 62
Methods and Antibacterial Activity 62
Antifungal Activity 64
Antiprotozoal Activity 65
Antiviral Activity 65
Conclusions 66
Acknowledgment 66
References 66

BOX 6.1   SALIENT FEATURES

Seaweeds contain compounds from different chemical families, including polysaccharides, fatty acids, phenolics, pigments, lectins, alkaloids, terpenoids and halogenated compounds with reported antimicrobial action. Different Chlorophyta, Rhodophyta and Phaeophyta species collected in several coasts around the world were reported showing antibacterial, antifungal, antiprotozoal and/or antiviral activity among other biological properties. The most widely used methods to detect and quantify the antimicrobial (antibacterial and antifungal) activity of algae extracts and/or their purified compounds are the diffusion agar test and micro and macro dilution tests. Their results are discussed although it is difficult to unify results because different concentrations of tested algal extracts and microbial strains are used. *In vivo* assays are less numerous and their performance depends on target organisms or substrates. The search for active compounds and advances in the formulation of novel materials as nanoparticles can offer effective tools against spoilage microorganisms and especially against antibiotic resistant pathogens.

ABBREVIATIONS

AgNPs: Silver nanoparticles
HIV: Human immunodeficiency virus
HSV: Herpes simplex viruses
IC$_{50}$:   Inhibitory concentration causing 50% reduction in microbial growth
LB: Luria Bertani medium
MBC: Minimal bactericidal concentration
MBEC: Biofilm eradication concentration
MIC: Minimum inhibitory concentration

INTRODUCTION

Seaweeds represent an important source of active components to avoid microbial growth, particularly of pathogenic microorganisms, causing diseases to humans, animals and plants or responsible for food, constructions or materials spoilage. The increasing antibiotic resistance to conventional drugs is also generating an urgent need to discover new potent agents effective in controlling multiresistant strains of pathogens. In the last three years, studies made with different seaweed species collected from different areas have been published (Kailas and Nair 2016; Carbalho et al. 2017; Chingizova et al. 2017; Moubayed et al. 2017; Sasikala and Ramani 2017; Chan et al. 2018; Martins et al. 2018; Otero et al. 2018).

The advances in this area have been summarized in different reviews (Pérez et al. 2016; Catarino et al. 2017; Koirala et al. 2017; Pinteus et al. 2018). Seaweeds can also provide hydrocolloids that can be designed for drug delivery, regenerative medicine and wound healing, where the combination with antimicrobial is promising. Other actions of seaweed bioactives confirm the interest in different applications including agriculture, food, cosmetics or pharmaceuticals (Anand et al. 2016). Janarthanan and Kumar (2018) discussed the properties of seaweed compounds related to medical textiles, which are experiencing rapid growth and offer protection against harmful pathogens. The present chapter presents an update of the recent contributions to the evaluation of antimicrobial activity of macroalgae, comprising a brief introduction to the detection methods used for determining antibacterial activity and sections describing the activity against different microorganisms and viruses.

BIOACTIVE COMPOUNDS

Seaweeds are a great source of bioactive compounds due to the large number of secondary metabolites they synthesize and possess antioxidant and some phytochemical activities. According to the literature (Michalak and Chojnacka 2015) they are a rich source of carbohydrates (brown algae [*Phaeophyta*]: alginate, fucoidan, laminarin; red algae [*Rhodophyta*]: agar, carrageenan, porphyran, mannan, xylan; green algae [*Chlorophyta*]: cellulose, xylan, ulvan), proteins, minerals, vitamins, oils, fats, polyunsaturated fatty acids and other bioactive compounds (polyphenols, pigments, etc.). The different seaweed compound families are presented in Figure 6.1.

METHODS AND ANTIBACTERIAL ACTIVITY

The most widely used method to detect and quantify the antimicrobial activity of algal extracts and/or their purified compounds is the diffusion agar test, an *in vitro* test often performed as a first assay and, sometimes, the only way to detect antimicrobial activity against bacterial and fungal strains, but it is difficult to unify the results because different concentrations of extracts and strains are used (Pérez et al. 2016; Chingizova et al. 2017).

Kailas and Nair (2016) observed the different inhibition rates using 100 µg/mL of sulfated sugars of *Ulva fasciata, Enteromorpha prolifera, Chaetomorpha antennina, Gracilaria corticata,* and *Gracilaria foliifera* against *Bacillus cereus, Staphylococcus aureus, Escherichia coli* and *Salmonella abony*. Stabili et al. (2016) reported that the disks impregnated with 100 µL of lipidic extract of *Caulerpa cylindracea* showed antibacterial activity against several *Vibrio* species, suggesting its potential use in the control of mariculture diseases. Chingizova et al. (2017) evaluated the anti-*S. aureus* and anti-*E. coli*. activity of 0.1 mL of hydrophilic and lipophilic seaweed extracts deposited in 10 mm holes punched out of plates. Moubayed et al. (2017) reported the *in vitro* antimicrobial effects of the methanol and acetone extracts of *Sargassum latifolium, S. platycarpum* and *Cladophora socialis* against several human bacterial pathogens.

Parallel streak method is also an *in vitro* test used to evaluate antimicrobial activity. The bacterial strains are inoculated by parallel streaks 1 cm apart of the centre of appropriate agar plates. The test specimen is pressed transversely across the inoculums of streaks and after incubation, the interrupted streak of bacterial growth is recorded. Janarthanan and Kumar (2017) investigated the antimicrobial properties of methanol and aqueous extracts from *Ulva reticulata, U. lactuca, S. wightii, Padina tetrastomatica* and *Acanthophora spicefera* in order to apply them on cotton fabric. The best result was by using the dye extract from *U. lactuca* against *S. aureus* and *E. coli* and confirmed using the disk diffusion and parallel streak method. They prepared cotton fabric with microcapsules of brown seaweed extracts, observing a bacterial reduction of 96% (by minimum inhibitory concentration, MIC, method) and a zone of inhibition against *E. coli* and *S. aureus* about 35 mm and 40 mm. Lakshimi et al. (2017) reported the antimicrobial activity against *E. coli* cultures of films prepared with cellulose extracted from *U. fasciata* using diffusion assay.

For further in-depth study on the antimicrobial activity of seaweeds components, several dilution methods as broth (micro and macro) dilution tests are used. The lowest concentration of crude or algal extracts that inhibits the bacterial or fungal growth is represented as MIC and determined by the concentration serial broth dilution assay. A microdilution test carried out in 96-wells microplates was used at concentrations ranging from 1 to 500 µg/mL for extracts of *Cystosphaera jacquinotii, Iridaea cordata, Himantothallus grandifolius* and *Pyropia endiviifolia* against *S. aureus, Enterococcus faecalis, E. coli* and *Pseudomonas aeruginosa* (Martins et al. 2018).

Several authors have determined the antibacterial activity of algal extracts using both *in vitro* methods, the diffusion agar test and broth dilution method. Patra et al. (2017) indicated that 25 mg/disk of *Undaria pinnatifida* essential oil showed antibacterial activity against *S. aureus* and *Salmonella typhimurium*. MIC ranged from 12.5 to 25 mg/mL and MBC (minimal bactericidal concentration) was 25 mg/mL for both bacterial strains. Otero et al. (2018) found that the

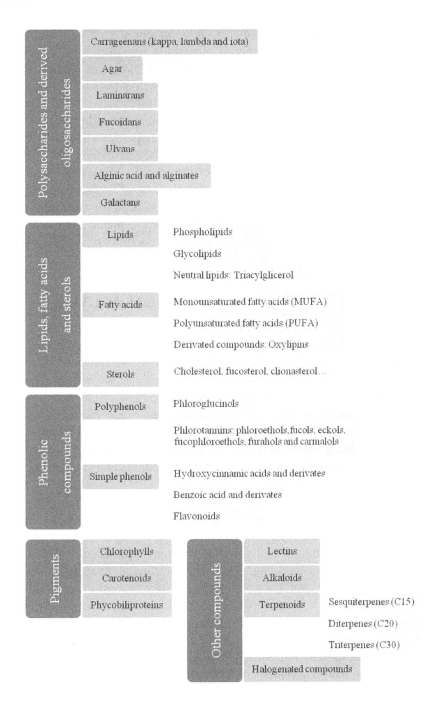

FIGURE 6.1 Classification of bioactive compounds of seaweed.

ethanolic extracts ranged from 50 to 10 mg/mL for six algal species against *S. aureus* and *E. coli* using a micro dilution method. Omar et al. (2018) screened the antibacterial activity of 100 mg/mL of dichloromethane, dichloromethane:methanol, methanol and water extracts of *Laurencia papillosa* against *S. aureus*, *S. saprophyticus* and *Streptococcus agalactiae* group B, *E. coli*, *P. aeruginosa* and *Proteus mirabilis*. They used Mueller–Hinton agar plates and calculated the degree of tolerance from MBC/MIC, considering the extract was ineffective when the ratio of MBC/MIC was 16.0.

Assays performed to determine growth inhibition by seaweed extracts were reported by different authors and, in some cases, the use of an indicator of live/dead cells was necessary (Baliano et al. 2016). Telles et al. (2017) evaluated the inhibition of *Staphylococcus epidermidis* growth and *Klebsiella pneumoniae* by heterofucans from *Sargassum filipendula* and evaluated the difference of absorbance after incubation.

In vivo assays are less numerous and their performance depends on target organisms or substrates. Esserti et al. (2017) reported the anti-*Agrobacterium* activity of

(100 µg/mL) methanolic extracts of *Cystoseira myriophylloides*, *Laminaria digitata* and *Fucus spiralis* using agar disk diffusion test. An *in vivo* assay was performed in greenhouse to evaluate the protection of tomato seedlings against the oncogenic strain C58 of *A. tumefaciens*. Plants were sprayed with aqueous seaweed extracts and then inoculated 2 or 4 days later. The diameter of tumour was recorded after 4 weeks and extracts from *C. myriophylloides* and *F. spiralis* reduced the crown gall disease. Oucif et al. (2018) employed *Cystoseira compressa* ethanol-aqueous extract for the preservation of chilled fish, in the icing storage medium of horse mackerel for 11 days and observed antimicrobial effect on the several microbial groups (aerobes, psychrotrophs, proteolytic, lipolytic and *Enterobacteriaceae* bacteria).

According to the World Health Organization, it is very important to find new potent antibiotics because of the problematic occurrence of resistance observed in recent years. The results obtained with seaweed extracts to prevent the growth of multiresistant microorganisms are encouraging. Buseti el al. (2017) reported that biofilms of *S. aureus* MRSA ATCC 33593 and *S. aureus* MRSA NCTC 10442 were susceptible to *Halidrys siliquosa* methanolic extract which achieved minimum biofilm eradication concentration (MBEC) values of 1.25 mg/mL and 5 mg/mL, respectively. Moubayed et al. (2017) observed that *Cladophora* methanolic extract had an obvious effect on methicillin resistant *S. aureus*.

The antimicrobial effectiveness of crude seaweed extracts or their purified components in nanoparticles can be used for treatment of human, animal or plant diseases, agriculture, food and textile industries, aquaculture, anti-biofouling and also for algal bloom control. The conventional antifouling paints are toxic towards non-target organisms, and some of them have been banned due to increasing evidence of their environmental risks, seaweed components can be used as antifouling agents, with advantages over other methods (Rittschof 2017). Specific problematic areas have been reviewed more recently, such as membranes (Nagaraj et al. 2018), microalgal refineries (Liao et al. 2018), water and wastewater treatments (Bagheri and Mirbagheri 2018), and the corrosive action of biofilms (Vigneron et al. 2018).

Carbalho et al. (2017) evaluated the antibacterial activity of six algal species against Gram negative marine strains as *V. aestuarianus, Pseudoalteromonas elyakovii, Polaribacter irgensii* and *Pseudomonas fluorescens*, involved in marine fouling, and *Shewanella putrefaciens* in corrosion of metal; and pathogenic *Vibrio* strains (*V. communis, V. alginolyticus* and *V. coralliilyticus*). The authors observed that *Canistrocarpus cervicornis* was the most effective one and inhibited the growth of *S. putrefaciens* and *V. aestuarianus*; *Padina* sp. and *Colpomenia sinuosa* inhibited the growth of *V. communis* and *V. alginolyticus*, respectively. *S. vulgare* inhibited the growth of the *Vibrio* species suggesting that this species possesses active compounds which protect them against fouling and pathogenic bacteria.

Silver nanoparticles (AgNPs) synthesized using different seaweed have shown antibacterial activity. Pugazhendhi et al. (2018) reported the anti-microfouling potential of AgNPs synthesized using *Gelidium amansi* against Gram negative and Gram positive biofilm forming pathogens. González-Mendoza et al. (2018) observed antibacterial activity of AgNPs from *S. vulgare* against *Bacillus cereus*.

ANTIFUNGAL ACTIVITY

Recent studies have confirmed the antifungal capacity of seaweeds, but the mechanism of action is not completely known. Martins et al. (2018) evaluated the antimicrobial activity of extracts from Antarctic seaweeds using the broth microdilution method against *Candida albicans*, and some clinical isolates from the human oral cavity and observed that the ethyl acetate extract of *Himantothallus grandifolius* was active against all strains tested, including fluconazole-resistant samples. Khan et al. (2017) found that both aqueous and methanolic extracts of *Sargasssum tenerrimum* were the most potent antifungal agents among the extracts of different seaweeds and inhibited growth of *Fusarium oxypsorum, Macrophomina phaseolina* and *Rhizoctonia solani*.

The antifungal properties were also observed for aqueous extracts. Sellimi et al. (2017) reported the antimicrobial activities of *Cystoseira barbata* glyco-conjugates, containing 50% polysaccharides with 14% sulfate, 10% proteins and 5% phenolic compounds, against filamentous fungi *Aspergillus niger, Fusarium solani, F. oxysporum, Botrytis cinerea* and *Alternaria solani*. Reis et al. (2018) reported no activity of *U. fasciata* (whole seaweed and aqueous extracts) against *Stemphylium solani*.

De Corato et al. (2017) proposed the use of supercritical carbon dioxide extracts from seaweeds (*Laminaria digitata, Undaria pinnatifida, Porphyra umbilicalis, Eucheuma denticulatum* and *Gelidium pusillum*) during fruit postharvesting against *Botrytis cinerea, Monilini alaxa* and *Penicillium digitatum* and observed that the brown seaweeds were more active inhibiting mycelia growing and conidial germination of the first two mentioned pathogens at 30 g/L. They also reported preventive and curative action on wounded fruit and ascribed the effect to fatty acids rather than to phenolics. Since fungicides are also demanded for postharvest control, seaweeds could provide efficient alternatives with lower human and environmental risks. More recently, the synthesis of nanomaterials with algal components offered a promising via for the preparation of antifungal agents (Rajeshkumar et al. 2017; Sathe et al. 2017; Mashjoor et al. 2018).

ANTIPROTOZOAL ACTIVITY

Several studies of the biocompounds and extracts produced by algae have been reported, focusing on their human and animal therapeutics (García and Monzote 2014). A fractionated lipophilic extract from the brown alga *Stypopodium zonale* revealed the presence of atomaric acid with inhibition of the growth of *Leishmania amazonensis* intracellular amastigotes in infected macrophages, with $IC_{50} = 9$ µg/mL (Soares et al. 2016). França et al. (2017) reviewed the compounds from marine organisms with potential antileishmanial activity. The sesquiterpene elatol and the tripterpene obtusol, both isolated from the red alga *Laurencia dendroidea*, and the terpenoids dolabelladienetriol and (4R,9S,14S)-4α-acetoxy-9β,14α-dihydroxydolast-1(15),7-diene, isolated from the brown seaweeds *Dictyota pfaffii* and *Canistrocarpus cervicornis*, respectively, presented antileishmanial activity against promastigotes and amastigotes of *L. amazonensis*. The review by Tchokouaha et al. (2017) showed that only few secondary metabolites presented antileishmanial activity, but more than 50% of these compounds belong to the family of diterpenes. Extracts obtained with different solvents from 151 seaweeds showed IC_{50} values as low as 0.27 µg/mL, showing a high potential to inhibit different *Leishmania* parasites. The terpenoids cited previously and fucosterol, and the species from the Dictyotaceae family and *Anadyomene*, *Laurencia* complex, *Ulva* and *Asparagopsis* genera were the most interesting compounds in macroalgae, respectively, to obtain effective antileishmanial drugs. Other compounds with interesting leishmanicidal activity are the phycocolloids, which mainly include fucoidan and laminarans in brown algae, carrageenan in red algae and ulvan in green algae. Seven species of Mediterranean macroalgae were tested against the protozoan *L. infantum*: the green algae *Chaetomorpha linum*, the red algae *Gracilaria bursa-pastoris*, *G. viridis*, *Agardhiella subulata* and *Hypnea cornuta*, and the brown algae *S. muticum* and *Undaria pinnatifida*. Dried seaweeds were extracted successively with ethanol, acetone and distilled water, and the polysaccharide fraction was precipitated with ethanol. All samples of macroalgae, mainly the brown seaweeds, showed a positive effect against the protozoan, except *Agardhiella subulata* and *Hypnea cornuta* (Armeli Minicante et al. 2016).

Fucosterol, isolated from *S. linearifolium*, showed higher antiplasmodial (*Plasmodium falciparum*) activity ($IC_{50} = 7.48$ µg/mL) (Perumal et al. 2017). Sulfated galactan extracted from the red alga *Botryocladia occidentalis* exerts their antimalarial activity *in vitro* through the inhibition of the invasion, and the *in vivo* study the survival of the infected female mice (Marqués et al. 2016). Another line of attack against this disease is through the vector control, drawing on the mosquito-larvicidal property of *Turbinaria ornata* mediated gold nanoparticles (Deepak et al. 2017).

The *in vitro* antiprotozoal activity of a new diterpene, bifurcatriol, isolated from the Irish brown alga *Bifurcaria bifurcata*, was tested against *P. falciparum*, *Trypanosoma brucei rhodesiense*, *Trypanosoma cruzi* and *Leishmania donovani*. This compound showed remarkable activity against the malaria parasite with an $IC_{50} = 0.65$ µg/mL and only moderate activity against other protozoan species, $IC_{50} = 47.8$ µg/mL (*T. cruzi*), 18.8 µg/mL (*L. donovani*) and 11.8 µg/mL (*T. brucei rhodesiense*) (Smyrniotopoulos et al. 2017).

Trichomoniasis is a sexually transmitted infection caused by the flagellated protozoan *Trichomonas vaginalis*. Telles et al. (2017) isolated five sulfated fucose-rich fractions from *S. filipendula*, and three of them showed a high inhibition capacity against trophozoites of this parasite after the 24-h treatment. Likewise, a positive correlation was found between the sugar/sulfate ratio and this anti-*T. vaginalis* activity.

Extracts from green algae (Chlorophyceae) obtained with *n*-hexane and methanol and tested *in vitro* for anti-*Toxoplasma* activities showed that the methanol extract, containing higher content of flavonoids/polyphenols, alkaloids (elaeocarpidine and auramine) and artemisic acid, had higher activity ($IC_{50} = 4.43$ µg/mL) than the *n*-hexane extract ($IC_{50} = 23.32$ µg/mL) (Powers et al. 2017).

ANTIVIRAL ACTIVITY

Some viruses cause important human, animal and plant diseases with high levels of morbidity and mortality. The prevalence in the body, the virus variants resistant to antiviral drugs, the lack of vaccines or drugs for the treatment involves the search of new drugs or bioactive compounds to circumvent these problems. Over the last number of years, more studies using fresh, dried and processed seaweeds bioactives with antiviral activity were published (Anand et al. 2016; Shi et al.

2017). Numerous studies were carried out to test the antiviral potential of various algae-derived polysaccharides and show their underlying mechanism of action. Carrageenans are sulfated polysaccharides with potent activity against several viruses, such as herpes simplex viruses (HSV), human immunodeficiency virus (HIV), human papilloma virus, flaviviruses (dengue virus), respiratory viruses (human rhinovirus, influenza virus and human metapneumovirus) and human enterovirus, among others, and some herpes viruses of veterinary interest such as bovine, suid and feline herpes viruses 1. Mainly, this antiviral activity is based on the interference in the viral entry into host cells (Shi et al. 2017). These compounds are abundant in red algae. *Solieria chordalis* in carrageenans against HSV-1 showed an IC_{50} = 3.2–54.4 µg/mL (Boulho et al. 2017). Hot water extract from red alga *Acanthophora specifira* and brown alga *Hydroclathrus clathratus* tested against Rift valley fever virus, showed great inhibition, mainly with red algae, and the log 10 reduction of viral replication (Gomaa and Elshoubaky 2016).

Fucoidan is a sulfated polysaccharide found in brown seaweeds, which usually blocks the interaction of viruses (e.g. HIV, HSV-1,2, dengue virus and cytomegalovirus) to the cells and inhibits viral-induced syncytium formation. Dinesh et al. (2016) isolated two bioactive fucoidan fractions from *S. swartzii*, and the anti-HIV-1 activity was studied. The fraction, with more sugar and sulfate content, exhibited significant higher anti-HIV-1 activity at 1.56 and 6.25 g/mL and >50% reduction in HIV-1 p24 antigen levels and reverse transcriptase activity.

Morán-Santibáñez et al. (2018) demonstrated that polyphenol-rich extracts from *Macrocystis pyrifera*, *Eisenia arborea*, *Pelvetia compressa*, *U. intestinalis* and *Solieria filiformis* prevented viral adsorption and replication against Measles virus *in vitro*, and they also observed a synergistic effect with sulfated polysaccharides from the same seaweeds. The anti-HSV-1 activity of the diterpene dolabelladienetriol isolated from *Dictyota pfaffii* was tested in rats and showed a similar efficacy to acyclovir (Garrido et al. 2017). The genus *Dictyota*, mainly *D. dichotoma*, is a rich source of other diterpenes with antiviral activity (Chen et al. 2018).

Halogenated compounds also showed antiviral activity. The new bioactive compounds chloromethyl 2-(dodecahydro-4,6,7,9-tetrahydroxy-9a-methyl-1H-cyclopenta[α]naphtalen-3-yl) acetate isolated from *Cystoseira myrica* and 2-((1E,3E)-4-chloro-2-mercaptobuta-1,3-dienyl)-1,2,3,4,4b,5,6,8a,9,10-decahydrophenanthrene-3,5,10-triol obtained from *U. lactuca*, have a noticeable activity against the hepatitis A virus, HSV-1,2 and Coxsackie B4 virus and have a growth and cytopathic effect in *Vero* cells (Zaid et al. 2016).

While the protective effects of seaweeds extracts against viruses are rather well documented, not much information is available on their role in defense response against animal and plant viruses. Lower-molecular-weight polysaccharides from *Grateloupia filicinia* could significantly inhibit avian leucosis virus subgroup J blocking virus adsorption to host cells and exhibited a stronger suppression (Sun et al. 2017). Ethanolic extracts from *U. prolifera*, *G. lemaneiformis*, *S. fusiforme*, *U. pertusa*, *Gloiopeltis furcata* and *Ishige okamurae* were tested against the white spot syndrome virus and the three firsts showed the most potent immunity *in vitro* (Li et al. 2018). A combined extract (0.2%) from *Ascophyllum nodosum*, 2-nitromethyl phenol and zinc-nanoparticles, reduces significantly disease severity and incidence of Cucumber mosaic virus in eggplants (*Solanum melongena*) inducing the activation of free and total phenols (El-Sawy et al. 2017).

CONCLUSIONS

Marine algae are among the most promising natural sources of novel antimicrobials, but the detailed mechanism of action of the different components and their mixtures has to be unveiled in order to advance their application. Advances in the formulation of novel materials, i.e. nanoparticles, can offer novel effective tools against spoilage microorganisms and resistant pathogens.

ACKNOWLEDGMENT

The authors are grateful to the Ministry of Science and Innovation of Spain (CTM2012-38095, partially funded by the FEDER Program of the European Union) for the financial support of this work.

REFERENCES

Anand, N., Rachel, D., Thangaraju, N., and Anantharaman, P. Potential of marine algae (sea weeds) as source of medicinally important compounds. *Plant Gen Res-C*. 2016. 14: 303–313.

Armeli Minicante, S., Michelet, S., Bruno, F., Castelli, G., Vitale, F., Sfriso, A., Morabito, M., and Genovese, G. Bioactivity of phycocolloids against the Mediterranean protozoan *Leishmania infantum*: An inceptive study. *Sustainability*. 2016. 8: 1131.

Bagheri, M., and Mirbagheri, S. A. Critical review of fouling mitigation strategies in membrane bioreactors treating water and wastewater. *Bioresour Technol*. 2018. 258: 318–334.

Baliano, A. P., Pimentel, E. F., Buzin, A. R., Vieira, T. Z., Romão, W., Tose, L. V., Lenz, D., de Andrade, T. U., Fronza, M., Kondratyuk, T. P., and Endringer, D. C. Brown seaweed *Padina gymnospora* is a prominent natural wound-care product. *Rev Bras Farmacog*. 2016. 26: 714–719.

Besednova, N. N., Kuznetsova, T. A., Zaporozhets, T. S., and Zvyagintseva, T. N. Brown seaweeds as a source of new pharmaceutical substances with antibacterial action. *Antibiot Khimioter*. 2015. 60: 31–41.

Boulho, R., Marty, C., Freile-Pelegrín, Y., Robledo, D., Bourgougnon, N., and Bedoux, G. Antiherpetic (HSV-1) activity of carrageenans from the red seaweed *Solieria chordalis* (Rhodophyta, Gigartinales) extracted by microwave-assisted extraction (MAE). *J Appl Phycol*. 2017. 29: 2219–2228.

Busetti, A., Maggs, C. A., and Gilmore, B. F. Marine macroalgae and their associated microbiomass as a source of antimicrobial chemical diversity. *Eur J Phycol*. 2017. 52: 452–465.

Carvalho, A. P., Batista, D., Dobretsov, S., and Coutinho, R. Extracts of seaweeds as potential inhibitors of quorum sensing and bacterial growth. *J Appl Phycol*. 2017. 29: 789–797.

Catarino, M. D., Silva, A. M. S., and Cardoso, S. M. Fucaceae: A source of bioactive phlorotannins. *Int J Mol Sci*. 2017. 18: 1327.

Chan, Y. S., Ong, C. W., Chuah, B. L., Khoo, K. S., and Sit, N. W. Antimicrobial, antiviral and cytotoxic activities of selected marine organisms collected from the coastal areas of Malaysia. *J Mar Sci Technol*. 2018. 26: 128–136.

Chen, J., Li, H., Zhao, Z., Xia, X., Li, B., Zhang, J., and Yan, X. Diterpenes from the marine algae of the genus *Dictyota*. *Mar Drugs*. 2018. 16: 159.

Chingizova, E. A., Skriptsova, A. V., Anisimov, M. M., and Aminin, D. L. Antimicrobial activity of marine algal extracts. *Int J Phytomed*. 2017. 9: 113–122.

De Corato, U., Salimbeni, R., De Pretis, A., Avella, N., and Patruno, G. Antifungal activity of crude extracts from brown and red seaweeds by a supercritical carbon dioxide technique against fruit postharvest fungal diseases. *Postharvest Biol Tec*. 2017. 131: 16–30.

Deepak, P., Sowmiya, R., Balasubramani, G., Aisearya, D., Arul, D., Josebin, M. P. D., and Perumal, P. Mosquito-larvicidal efficacy of gold nanoparticles synthesized from the seaweed, *Turbinaria ornata* (Turner) J. Agardh 1848. *Partic Sci Technol*. 2017. 36: 974–980.

Dinesh, S., Menon, T., Hanna, L. E., Suresh, V., Sathuvan, M., and Manikannan, M. *In vitro* anti-HIV-1 activity of fucoidan from *Sargassum swartzii*. *Int J Biol Macromol*. 2016. 82: 83–88.

El-Sawy, M. M., Elsharkawy, M. M., Abass, J. M., and Kasem, M. H. Antiviral activity of 2-nitromethyl phenol, zinc nanoparticles and seaweed extract against cucumber mosaic virus (CMV) in eggplant. *J Virol Antivir Res*. 2017. 6: 2.

Esserti, S., Smaili, A., Rifai, L. A., Koussa, T., Makroum, K., Belfaiza, M., Kabil, E. M., Faize, L., Burgos, L., Alburquerque, N., and Faize, M. Protective effect of three brown seaweed extracts against fungal and bacterial diseases of tomato. *J Appl Phycol*. 2017. 29: 1081–1093.

França, P. H. B., da Silva-Júnior, E. F., Santos, B. V. O., Alexandre-Moreira, M. S., Quintans-Júnior, L. J., de Aquino, T. M., and de Araújo-Júnior, J. X. Antileishmanial marine compounds: A review. *Rec Nat Prod*. 2017. 11: 92–113.

García, M., and Monzote, L. Marine products with anti-protozoal activity: A review. *Curr Clin Pharmacol*. 2014. 9: 258–270.

Garrido, V., Barros, C., Melchiades, V. A., Fonseca, R. R., Pinheiro, S., Ocampo, P., Teixeira, V. L., Cavalcanti, D. N., Giongo, V., Ratcliffe, N. A., and Teixeira, G. Subchronic toxicity and anti-HSV-1 activity in experimental animal of dolabelladienetriol from the seaweed *Dictyota pfaffii*. *Regul Toxicol Pharmacol*. 2017. 86: 193–198.

Gomaa, H. H. A., and Elshoubaky, G. A. Antiviral activity of sulfated polysaccharides carrageenan from some marine seaweeds. *Int J Curr Pharm Rev Res*. 2016. 7: 34–42.

González-Mendoza, D., Valdez-Salas, B., Carrillo-Beltrán, M., Castro-López, S., Méndez-Trujillo, V., Gutiérrez-Miceli, F., Rodríguez-Hernández, L., Durán-Hernández, D., and Arce-Vázquez, N. Antimicrobial effects of silver-phyconanoparticles from *Sargassum vulgare* against spoilage of fresh vegetables caused by *Bacillus cereus*, *Fusarium solani* and *Alternaria alternata*. *Int J Agric Biol*. 2018. 20: 1230–1234.

Janarthanan, M., and Kumar, M. S. New bioactive non-implantable textile material using green seaweed for medical applications. *Int J Cloth Sci Tech*. 2017. 29: 69–83.

Janarthanan, M., and Kumar, M. S. The properties of bioactive substances obtained from seaweeds and their applications in textile industries. *J Ind Tex*. 2018. 48: 361–401.

Kailas, A. P., and Nair, S. M. HPLC profiling of antimicrobial and antioxidant phycosugars isolated from the South West Coast of India. *Carbohydr Polym*. 2016. 151: 584–592.

Khan, S. A., Abid, M., and Hussain, F. Antifungal activity of aqueous and methanolic extracts of some seaweeds against common soil-borne plant pathogenic fungi. *Pak J Bot*. 2017. 49: 1211–1216.

Koirala, P., Jung, H. A., and Choi, J. S. Recent advances in pharmacological research on *Ecklonia* species: A review. *Arch Pharm Res*. 2017. 40: 981–1005.

Lakshmi, D. S., Trivedi, N., and Reddy, C. R. K. Synthesis and characterization of seaweed cellulose derived carboxymethylcellulose. *Carbohydr Polym*. 2017. 157: 1604–1610.

Li, Y., Sun, S., Pu, X., Yang, Y., Zhu, F., Zhang, S., and Xu, N. Evaluation of antimicrobial activities of seaweed resources from Zhejiang Coast, China. *Sustainability*. 2018. 10: 2158.

Liao, Y., Bokhary, A., Maleki, E., and Liao, B. A review of membrane fouling and its control in algal-related membrane processes. *Bioresour Technol*. 2018. 264: 343–358.

Marques, J., Vilanova, E., Mourão, P. A. S., and Fernàndez-Busquets, X. Marine organism sulfated polysaccharides exhibiting significant antimalarial activity and inhibition of red blood cell invasion by *Plasmodium*. *Sci Rep*. 2016. 6: 24368.

Martins, R. M., Nedel, F., Guimarães, V. B. S., da Silva, A. F., Colepicolo, P., de Pereira, C. M. P., and Lund, R. G. Macroalgae extracts from Antarctica have antimicrobial and anticancer potential. *Front Microbiol*. 2018. 9: 412.

Mashjoor, S., Yousefzadi, M., Zolgharnain, H., Kamrani, E., and Alishahi, M. Organic and inorganic nano-Fe3O4: Alga *Ulva flexuosa*-based synthesis, antimicrobial effects and acute toxicity to briny water rotifer *Brachionus rotundiformis*. *Environ Pollut*. 2018. 237: 50–64.

Michalak, I., and Chojnacka, K. Review: Algae as production systems of bioactive compounds. *Eng Life Sci*. 2015. 15: 160–176.

Morán-Santibañez, K., Peña-Hernández, M. A., Cruz-Suárez, L. E., Ricque-Marie, D., Skouta, R., Vasquez, A. H., Rodríguez-Padilla, C., and Trejo-Avila, L. M. Virucidal and synergistic activity of polyphenol-rich extracts of seaweeds against Measles virus. *Viruses*. 2018. 10: 465.

Moubayed, N. M., Al Houri, H. J., Al Khulaifi, M. M., and Al Farraj, D. A. Antimicrobial, antioxidant properties and chemical composition of seaweeds collected from Saudi Arabia (Red Sea and Arabian Gulf). *Saudi J Biol Sci*. 2017. 24: 162–169.

Nagaraj, V., Skillman, L., Li, D., and Ho, G. Review – Bacteria and their extracellular polymeric substances causing biofouling on seawater reverse osmosis desalination membranes. *J Environ Manage*. 2018. 223: 586–599.

Omar, H., Al-Judaibiand, A., and El-Gendy, A. Antimicrobial, antioxidant, anticancer activity and phytochemical analysis of the red alga, *Laurencia papillosa*. *Int J Pharm*. 2018. 14: 572–583.

Otero, P., Quintana, S. E., Reglero, G., Fornari, T., and García-Risco, M. R. Pressurized liquid extraction (PLE) as an innovative green technology for the effective enrichment of Galician algae extracts with high quality fatty acids and antimicrobial and antioxidant properties. *Mar Drugs*. 2018. 16: 156.

Oucif, H., Miranda, J. M., Mehidi, S. A., Abi-Ayad, S. M., Barros-Velázquez, J., and Aubourg, S. P. Effectiveness of a combined ethanol–aqueous extract of alga *Cystoseira compressa* for the quality enhancement of a chilled fatty fish species. *Eur Food Res Technol*. 2018. 244: 291–299.

Patra, J. K., Lee, S.-W., Park, J. G., and Baek, K. H. Antioxidant and antibacterial properties of essential oil extracted from an edible seaweed *Undaria pinnatifida*. *J Food Biochem*. 2017. 41: e12278.

Pérez, M. J., Falqué, E., and Domínguez, H. Antimicrobial action of compounds from marine seaweed. *Mar Drugs*. 2016. 14: 52.

Perumal, P., Sowmiya, R., Prasanna Kumar, S., Ravikumar, S., Deepak, P., and Balasubramani, G. Isolation, structural elucidation and antiplasmodial activity of fucosterol compound from brown seaweed, *Sargassum linearifolium* against malarial parasite *Plasmodium falciparum*. *Nat Prod Res*. 2017. 32: 1316–1319.

Pinteus, S., Lemos, M. F. L., Alves, C., Neugebauer, A., Silva, J., Thomas, O. P., Botana, L. M., Gaspar, H., and Pedrosa, R. Marine invasive macroalgae: Turning a real threat into a major opportunity – The biotechnological potential of *Sargassum muticum* and *Asparagopsis armata*. *Algal Res*. 2018. 34: 217–234.

Powers, J. L., Zhang, X., Kim, C. Y., Abugri, D. A., and Witola, W. H. Activity of green algae extracts against *Toxoplasma gondii*. *Med Aromat Plants*. 2017. 6: 293.

Pugazhendhi, A., Prabakar, D., Jacob, J. M., Karuppusamy, I., and Saratale, R. G. Synthesis and characterization of silver nanoparticles using *Gelidium amansii* and its antimicrobial property against various pathogenic bacteria. *Microb Pathog*. 2018. 114: 41–45.

Rajeshkumar, S., Malarkodi, C., and Venkat Kumar, S. Synthesis and characterization of silver nanoparticles from marine brown seaweed and its antifungal efficiency against clinical fungal pathogens. *Asian J Pharm Cli Res*. 2017. 10: 190–193.

Reis, R. P., de Carvalho Junior, A. A., Facchinei, A. P., dos Santos Calheiros, A. C., and Castelar, B. Direct effects of ulvan and a flour produced from the green alga *Ulva fasciata* Delile on the fungus *Stemphylium solani* Weber. *Algal Res*. 2018. 30: 23–27.

Rittschof, D. Off the shelf fouling management. *Mar Drugs*. 2017. 15: 176.

Sasikala, C., and Ramani, G. D. Comparative study on antimicrobial activities of seaweeds. *Asian J Pharm Clin Res*. 2017. 10: 384–386.

Sathe, P., Laxman, K., Myint, M. T. Z., Dobretsov, S., Richter, J., and Dutta, J. Bioinspired nanocoatings for biofouling prevention by photocatalytic redox reactions. *Sci Rep*. 2017. 7: 3624.

Sellimi, S., Benslima, A., Barragán-Montero, V., Hajji, M., and Nasri, M. Polyphenolic-protein-polysaccharide ternary conjugates from *Cystoseira barbata* Tunisian seaweed as potential biopreservatives: Chemical, antioxidant and antimicrobial properties. *Int J Biol Macromol*. 2017. 105: 1375–1383.

Shi, Q., Wang, A., Lu, Z., Qin, C., Hu, J., and Yin, J. Overview on the antiviral activities and mechanisms of marine polysaccharides from seaweeds. *Carbohydr Res*. 2017. 453–454: 1–9.

Smyrniotopoulos, V., Merten, C., Kaiser, M., and Tasdemir, D. Bifurcatriol, a new antiprotozoal acyclic diterpene from the brown alga *Bifurcaria bifurcata*. *Mar Drugs*. 2017. 15: 245.

Soares, D., Szlachta, M., Teixeira, V., Soares, A., and Saraiva, E. The brown alga *Stypopodium zonale* (Dictyotaceae): A potential source of anti-Leishmania drugs. *Mar Drugs*. 2016. 14: 163.

Stabili, L., Fraschetti, S., Acquaviva, M. I., Cavallo, R. A., De Pascali, S. A., Fanizzi, F. P., Gerardi, C., Narracci, M., and Rizzo, L. The potential exploitation of the Mediterranean invasive alga *Caulerpa cylindracea*: Can the invasion be transformed into a gain? *Mar Drugs*. 2016. 14: 210.

Sun, Y., Chen, X., Cheng, Z., Liu, S., Yu, H., Wang, X., and Li, P. Degradation of polysaccharides from *Grateloupia filicina* and their antiviral activity to avian leucosis virus subgroup J. *Mar Drugs*. 2017. 15: 345.

Tchokouaha Yamthe, L. R., Appiah-Opong, R., Tsouh Fokou, P. V., Tsabang, N., Fekam Boyom, F., Nyarko, A. K., and Wilson, M. D. Marine algae as source of novel antileishmanial drugs: A review. *Mar Drugs*. 2017. 15: 323.

Telles, C. B. S., Mendes-Aguiar, C., Fidelis, G. P., Frasson, A. P., Pereira, W. O., Scortecci, K. C., Camara, R. B. G., Nobre, L. T. D. B., Costa, L. S., Tasca, T., and Rocha, H.A.O. Immunomodulatory effects and antimicrobial activity of heterofucans from *Sargassum filipendula*. *J Appl Phycol*. 2017. 30: 569–578.

Vigneron, A., Head, I. M., and Tsesmetzis, N. Damage to offshore production facilities by corrosive microbial biofilms. *Appl Microbiol Biotechnol*. 2018. 102: 2525–2533.

Zaid, S. A. A. L., Hamed, N. N. E. D., Abdel-Wahab, K. S. E. D., Abo El-Magd, E. K., and Salah El-Din, R. A. L. Antiviral activities and phytochemical constituents of Egyptian marine seaweeds (*Cystoseira myrica* (S.G. Gmelin) C. Agardh and *Ulva lactuca* Linnaeus) aqueous extract. *Egypt J Hosp Med*. 2016. 64: 422–429.

7 Marine-Algal Bioactive Compounds
A Comprehensive Appraisal

Iahtisham-Ul-Haq, Masood Sadiq Butt, Natasha Amjad, Iqra Yasmin, and Hafiz Ansar Rasul Suleria

CONTENTS

Introduction 71
Algal Bioactives 72
Health Claims of Algal Bioactives 73
Anti-Cancer 73
Anti-Viral 73
Antioxidant 75
Anti-Obesity 75
Anti-Angiogenic 75
Neuroprotective 76
Photoprotection 76
Conclusion 77
References 77

BOX 7.1 SALIENT FEATURES

Bioactive compounds play a significant role in disease prevention and maintenance of normal physiological functions. Due to increased demand of functional foods and nutraceuticals, numerous food sources are being exploited for novel bioactive compounds fit for their subsequent use in food, cosmetic and pharmaceutical products. Among various sources, marine algal bioactives are gaining popularity because of their potential health benefits. Such health effects are delivered by phlorotannins, natural pigments, diterpenes and sulfated polysaccharides present in different algal species. Numerous studies have identified antioxidant, anti-obesity, anti-cancer, anti-tumor, anti-viral, anti-angiogenic and neuroprotective activities of algal bioactive agents. In this respect, the health claims associated with marine algal bioactives are comprehensively discussed and possible mechanisms involved in delivering such effects are summarized in this chapter.

INTRODUCTION

Functional foods and the nutraceuticals market is one of the fastest growing segments in the health and fitness industries with an estimated share of 5.5–6 billion US$ annually from seaweed products (Chen et al. 2017). Marine algae are important sources of such valuable compounds capable of providing a plethora of biological activities (Li et al. 2011a). Algae belong to heterogeneous group of autotrophic organisms with smaller and less complex structure than plants (Barrow and Shahidi 2008; El Gamal 2010). Marine algae are considered a good source of food because they have less lipids, more polysaccharides and vitamins, polyunsaturated fatty acids and minerals along with other active compounds (Gupta and Abu-Ghannam 2011). Besides, algae are capturing significant importance in the functional food and nutraceutic ingredients market around the globe due to a plethora of bioactive compounds they contain thus providing sufficient nutritional benefits (Wells et al. 2017).

Edible marine algae also known as seaweeds are grouped as brown algae (phaeophyte), green algae (chlorophyte) and red algae (rhodophyte) based on their chemical composition of pigments. Seaweeds possess color because of pigments they contain such as chlorophyll a & b providing a green color while beta carotene and xanthophyll exhibit yellowish or brownish colors. In particular, fucoxanthin gives a brown color to seaweeds. In red seaweeds, fucoxanthin covers the chroma of phycoerythrin while beta-carotene and chlorophyll are masked by phycocyanin that ultimately give rise to a red colored appearance to the seaweed. Seaweeds produce numerous bioactive compounds and are a potential contributor of vitamins such as A, B1, B3, B12, C, D, E and folic acid alongside rich contents of Ca, K, Na and P (Gupta and Abu-Ghannam 2011).

Algae have been traditionally used in several Asian countries viz. Korea, China and Japan for food and health purposes. In western countries, their utilization is limited to the cosmetics and pharmaceutical industries. However, some application in the food industry is observed as a source of thickening agent (Imeson 2012). Nutritionally, algae contain a rich profile of polysaccharides, fatty acids, vitamins, polyphenols and minerals (Cardoso et al. 2015; Wells et al. 2017). Marine algal compounds are promising in alleviating various maladies owing to their rich chemistry and diverse beneficial activities (Cardoso et al. 2015). Consequently, these compounds provide significant antioxidant, anticancer, antidiabetic, antimicrobial, antiviral, anti-HIV, hepatoprotective, antihistaminic, anti-osteoporotic, cardioprotective and anticholinergic activities (Artan et al. 2008; Cardoso et al. 2015; Cho et al. 2011; Das et al. 2010; El Gamal 2010; Martínez–Hernández et al. 2018; Pangestuti and Kim 2011a). Particularly, macro algae have grabbed sufficient attention due to the presence of natural bioactives capable of delivering broad biological activities such as anti-tumor, antioxidant, antibiotic, antiviral and anti-inflammatory. Among algal bioactives, steroids, terpenoids, amino acids, phenolic halogenated ketones, phlorotannins, cyclic polysulphides and alkenes have been reported in various studies. Pharmacological effects of these phytochemicals have been studied particularly in relation to steroids and terpenoids (Bhagavathy et al. 2011).

However, there is a continued quest for investigations into the algal constituents and evaluating their possible health and/or adverse effects. Also, less availability of nutritional information and the effect of geographical regions and seasons on algae are limiting to fully understand the beneficial/adverse effects of algae (Wells et al. 2017). This chapter deals with chemical composition and possible health effects of algal bioactives that can potentially be used in functional foods and the nutraceutics market.

ALGAL BIOACTIVES

Algae are considered an important source of biologically active substances due to their abundance and the advantage of environmentally friendly cultivation methods (Michalak and Chojnacka 2015). Optimal extraction of bioactive compounds from food matrix is of crucial importance for obtaining acceptable biological activity and subsequent use in product manufacture. Algae are rich in vital nutrients and bioactive compounds such as terpenoids, tocopherols, carotenoids, polyphenols, phlorotannins, diterpenes and sulfated polysaccharides. These bioactive compounds have antiviral, antimicrobial, antioxidant, tumor suppressing and anti-inflammatory properties (Martínez–Hernández et al. 2018; Ponce et al. 2003; Wells et al. 2017). Hence, proper extraction must be carried out retaining their natural beneficial effects (Roohinejad et al. 2018).

Among bioactive compounds of algae, its natural pigments, phlorotannins, diterpenes and sulfated polysaccharides have gained significant importance. Algae being a photosynthetic organism contains chlorophylls as a major pigment (Chen et al. 2017). This pigment is converted to pheophorbide, pheophytin and pyropheophytin upon processing which possess plausible cancer preventive perspectives (Chen and Roca 2018a; Chen and Roca 2018b). Among other things, pheophytin has been found to acquire more cytotoxic effects than pheophorbide (Lin et al. 2014). The content of pigments is season dependent as are most of the fucoxanthin, violaxanthin, chlorophyll A and carotene in brown algae (*Ascophyllum*). Similar seasonal variation is also present in *Fucus serratus* where lower levels are seen during autumn. Moreover, chlorophyll contents are also influenced with the light exposure as harbored algae showed more pigment as compared to openly grown algae (Chen and Roca 2018b).

Polysaccharides have prominent immunomodulatory and anticancer roles in pharmacological perspectives. Algae are promising in providing a range of polysaccharides including fucoidans, alginates and laminarans. Fucoidans and laminarans are mainly water soluble, whereas alkali soluble polysaccharides include high molecular alginic acids. Laminarans are storage glucans of brown algae. Two types of extra-cellular acid polysaccharides exist in algae namely alginic acid and sulfated fucans. Further classification of fucans can be made as glycorunogalactofucans, xylofucoglycuronans and fucoidans (Gupta and Abu-Ghannam 2011). Sulfated polysaccharides from marine algae have claimed to possess anticoagulant, antiviral, antioxidant and antitumor properties (Gupta and Abu-Ghannam 2011; Wijesekara et al. 2011).

Another class of significant bioactive moieties from algal sources include diterpenes. Various types of diterpenoids including dolabellanes, xenicanes, hydroazulenoids and sesquiterpenoids are present in algae. These are actually produced by species of the *Dictyotaceae* family as main secondary metabolites and possess high cytotoxic, algicidal and anti-viral properties (Li et al. 2011a).

Phlorotannins are vital bioactives present in marine algae. Biologically, these are synthesized through polyketide or acetate-malonate pathway. Chemically, phlorotannins are derived from polymerization of phloroglucinol (1,3,5-trihydroxybenzene) monomer units. The compounds vary in molecular size from 126 Da to 650 kDa with a highly hydrophilic nature (Ragan and Glombitza 1986). These varying phloroglucinol-based polyphenols contain both phenoxy and phenyl units. Furthermore, phlorotannins can be classified based on their linkage into four groups as fuhalols and phlorethols (having ether linkage), eckols (with a dibenzodioxin linkage), fucols (containing phenyl linkage) and fucophloroethols (having both ether and phenyl linkage). Among marine brown algae, *E. cava* is the richest source of phlorotannins compared to others, including eckol, dieckol, phloroglucinol, phlorofucofuroeckol A, 7-phloroeckol and dioxinodehydroeckol as potent biologic agents (Michalak and Chojnacka 2015). Some studies have shown up to 70% bioavailability of plant polyphenolics, however, extensive research is needed to assess the biological availability of phlorotannins from marine algae. Nevertheless, studies have proven the biologic worth of these compounds as anti-cancer, anti-HIV, bactericidal, antioxidant and with enzyme inhibitory functions (El Gamal 2010; Li et al. 2011a).

HEALTH CLAIMS OF ALGAL BIOACTIVES

Algal bioactive have been studied extensively for potential health benefits. Bioactive ingredients from algae and their accompanying health claims are summarized in Table 7.1. A detailed summary of functions of various compounds with special reference to their health benefits is discussed under the following headings.

ANTI-CANCER

Cancer is uncontrolled cell growth that can be directly induced by free radicals in the human body. Carcinogenesis is manifested due to an imbalance in free radicals in biological systems. Hence, antioxidants or chemoprotective agents can play a vital role in etiology of cancer insurgence. Natural anticancer bioactives are a vital tool to control the progression of cancer as a number of compounds are capable of preventing or curing oncogenic events (Li et al. 2011a). In this regard, sulfated polysaccharides from algae have been endorsed exhibiting a prophylactic role against cancer insurgence. These polysaccharides control the free-radical generation by scavenging their radical activities hence lowering the incidence of cancer formation in the human body (Wijesekara et al. 2011).

Sulfated polysaccharides impart anti-metastasis activity by blocking interaction between basement membrane and cancer cells. However, mechanisms are yet unknown involving a blockage of tumor cells propagation and adhesion to various substances. Nevertheless, studies have suggested that oral intake of seaweeds significantly reduces in vivo carcinogenesis. It is worth mentioning that the sulfated polysaccharides extracted from *Ecklonia cava* have shown a vital anti-proliferative effect and stimulated apoptosis in human leukemic monocyte lymphoma cell line (U-937) (Athukorala et al. 2009).

Fucoidans from brown algae have been reported as potential anticancer agents attributed to their molecular weight and sulfate content. Broadly, depolymerization and lowered molecular weight of fucoidans enhanced their anticancer prospects (Gupta and Abu-Ghannam 2011; Malyarenko and Ermakova 2017). The seaweeds derived sulfated polysaccharides possess protective effects against oxidative damage which may also be involved in cancer prevention mechanisms endorsing their use as promising chemo-preventing agents (Wijesekara et al. 2011).

Phlorotannins from brown algae are also useful to ameliorate cancer insurgence. Animal studies have shown that laminaria lowers the risk of intestinal and mammary cancer development based on antioxidant and cell proliferation reduction mechanisms (Yuan and Walsh 2006). Similarly, dioxinodehydroeckol, a derivative of phlorotannin, induces apoptosis in MCF-7 cell lines and reduces the risk of breast cancer (Kong et al. 2009b). Furthermore, eckol, dieckol, phlorofucofuroeckol and fucodiphloroethol G extracted form *E. cava* potentially exhibit cytotoxic effects on HT-29, HT-1080, A549 and HeLa human cancer cell lines (Li et al. 2011b).

ANTI-VIRAL

Anti-viral polysaccharides from algae possess very low cytotoxic effects on mammalian cells. This attribute is of

TABLE 7.1
Algal Bioactives with Special Reference to Health Claims

Bioactive Compounds	Algae Source	Health Claims	References
Dieckol	*Ecklonia cava*	Whitening agent, photoprotection	Brunt and Burgess (2018); Pallela (2014)
Eckol	*Corallina pilulifera*		
Dioxinodehydroeckol	*Ecklonia cava*	Anticancer	Pangestuti and Kim (2015)
Diphlorethohydroxycarmalol	*Ishige okamurae*	Antidiabetic, whitening agent	Murray et al. (2018), Senevirathne and Kim (2011)
Fucan	*Fucus vesiculosus*	Antithrombin, avian RT inhibitor	Lahrsen et al. (2018)
Fucoidan	*Fucus evanescens*	Anticancer, anti-metastatic, antitumor, HSV-1 & 2 and HCMV inhibitors	Abrantes et al. (2010), Gupta and Abu-Ghannam (2011), Heo et al. 2010; Kwak (2014), Malyarenko and Ermakova (2017)
Galactofuran	*Adenocystis*		
Galactofucan sulfate	*utrlicularis*		
	Undria pinnatifida		
Fucosterol	*Pelvetia siliquosa*	Antidiabetic	Abdul et al. (2016), Gupta and Abu-Ghannam (2011)
Fucoxanthins	*Hizikia fusiformis*	Neuroprotective	Garcia-Vaquero and Hayes (2016), Zhang et al. (2017)
Laminarin	*Laminaria japonica*	Anti-apoptotic	Kadam et al. (2015)
Pheophytin a	*Sargassum fulvellum*	Neuroprotective	Alghazwi et al. (2016), Ikeda et al. (2003), Khodosevich and Monyer (2010)
Phlorofucofuroeckol-A	*Ecklonia kurome*	Algicidal	Chowdhury et al. (2014)
Phlorofucofuroeckol-B	*Ecklonia kurome*	Antiallergy	Sugiura et al. (2015)
Polysaccharide fractions	*Sargassum* species	Anti-HIV, anti-HSV-1 & 2 cytotoxic	Dinesh et al. (2016), Vaseghi et al. (2018), Yende et al. (2014)
Siphonoxanthin	*Codium fragile*	Antiangiogenic	Yue et al. (2017)
Sulfated polysaccharides	*Undaria pinnatifida*	Anti-viral	Ahmadi et al. (2015)
4,18-dihydroxy dictyolactone	*Dictyota* sp.	Cytotoxic	Cheng et al. (2014)
8,10,18-trihydroxy-2,6-dolabelladiene	*Dictyota pfaffii*	HSV-1 inhibitor and early protein modulator	Abrantes et al. (2010)
8,8′-bieckol	*Ecklonia cava*	HIV-1 RT inhibitors	Ahn et al. (2004), Artan et al. (2008))
8,4‴-dieckol			
6,6′-bieckol			
Alginic acid	*Sargassum vulgare*	Antitumor	Gutiérrez-Rodríguez et al. (2018)
Diterpenes	*Dictyota* sp.	Anti-retroviral	Chen et al. (2018)

vital significance when using drugs as antiviral agents. Among algal polysaccharides, fucoidan possesses significant anti-viral activities against human cytomegalovirus (HCMV) and human immunodeficiency virus (HIV) (Gupta and Abu-Ghannam 2011). *Adenocystis utricularis* contains galactofuran that possesses strong inhibitory properties against the herpes simplex virus (HSV) 1 & 2 and showed no cytotoxic effects (Ponce et al. 2003). Similarly, aqueous polysaccharide fraction from *Sargassum patens* also exhibited strong activity against HSV-1 & 2 with low cytotoxic effects (Zhu et al. 2003). Likewise, anti-viral activities of *U. pinnatifida* derived galactofucan sulfate extract against HCMV and HSV-1 & 2 have also been reported (Hemmingson et al. 2006).

Human immunodeficiency virus (HIV) is one of the biggest threats to human life around the globe. The people suffering from the infection of HIV receive a combination of antiretroviral therapy (ART) for its treatment/management (Langebeek et al. 2014). Scientists have been seeking novel bioactive compounds to prepare drugs for treating the HIV infection with minor side effects comparative to already existing drugs (Artan et al. 2008). In this respect, marine brown algae possess a significant amount of innovative anti-HIV constituents for the production of drugs (Cardoso et al. 2014; Li et al. 2011a).

The brown algae especially *Ecklonia cava* is potentially important due to phlorotannin compounds (Li et al. 2011a). *In vitro* studies have shown an inhibitory effect of phlorotannins 8,8′-bieckol and 8,4‴-dieckol on HIV-1 protease and reverse transcriptase. Both of these compounds are dimmers of eckol but their relative inhibitory effect is different. For instance, 8,8′-bieckol exhibited a ten-fold higher inhibitory effect than dieckol (Ahn et al. 2004). This bioactive showed a selective inhibitory effect on reverse transcriptase which is comparable to nevirapine (Artan et al. 2008).

Among major phloroglucinol derivatives, 6,6′-bieckol also acquires strong inhibitory effects against syncytia formation induced by HIV-1, lytic responses and *in vitro* and *in vivo* viral p24 antigen production (Artan et al. 2008). It also blocked entry of HIV-1 and its reverse transcriptase enzyme activity when exposed to a dose rate of 1.07 μM. Besides, it did not show any cytotoxic effects at such concentrations where HIV-1 replication is completely inhibited. Hence, phlorotannins from marine algae have remarkable potential to produce novel therapeutics against HIV with minor side effects (Artan et al. 2008).

ANTIOXIDANT

The inequity between production and neutralization of free radicals causes the onset of oxidative stress which takes part in etiology of several degenerative diseases. To reduce the radical load in the human body, bioactive agents from dietary sources hold a cardinal significance (Iahtisham-Ul-Haq et al. 2018). Purposely, phlorotannins from marine brown algae are promising in alleviating oxidative stress by controlling free-radical species production and propagation. These bioactives exhibit strong antioxidant activity against oxidative cell damage by oxygen derived free radicals (Li et al. 2011a).

Earlier, eckstolonol, a novel phlorotannin extracted from *E. stolonifera*, exhibited potential *in vitro* antioxidant activities (Kang et al. 2003). Additionally, phlorotannins from *H. fusiform* are proven as potential radical scavengers (Siriwardhana et al. 2005). Kang et al. (2005b) reported that the bioactive eckol compound from *E. cava* effectively attenuates oxidative stress-mediated lung cell damage. Likewise, bioactives from *E. cava*, *E. bicyclis*, *E. kurome* and *H. fusiformis* have been endorsed for their antioxidant properties in various researches (Li et al. 2011a). Shibata et al. (2008) endorsed eckol, dieckol, 8,8′-dieckol and phlorofucofuroeckol A as promising antioxidants to control phospholipid peroxidation. Particularly, phloroglucinol and triphlorethol-A are among the promising algal compounds capable of ameliorating oxidative cell damages (Kang et al. 2005a; Kang et al. 2006).

ANTI-OBESITY

A simple definition of obesity may involve the accumulation of excessive fat causing increase in weight of an individual (Kong et al. 2009a). Critically, obesity is described as the accumulation of fat in adipose tissue which is due to hyperplasia and hypertrophy of fat cells (Pozza and Isidori 2018). In this context, hindering the adipogenesis process can treat obesity. Purposely, algal bioactives especially fucoxanthin and fucoxanthinol have been seen to inhibit 3T3-L1 preadipocytes differentiation to adipocytes. Mechanistically, these compounds down-regulate adipogenic transcription factors (Maeda et al. 2006). Another mechanism involving structure specific suppression in adipocyte differentiation has been suggested in a study directed by Okada et al. (2008) where neoxanthin and fucoxanthin exhibited significant suppressive responses on adipocyte differentiation. Woo et al. (2009) stated that induction of uncoupling protein 1 (UCP1) mediated by fucoxanthin administration showed anti-obesity effects. This protein can bring about uncoupled oxidative phosphorylation causing production of heat instead of ATP especially in white adipose tissues (WAT). Hence, most of the heat producing mechanisms are held in brown adipose tissues (BAT) so the UCP1 is fully expressed in these tissues. Whereas, BAT is not primarily involved in human weight reduction as adults contain less BAT than WAT which contributes to weight gain (Trayhurn and Wood 2005). Also, WAT is currently known to secrete adipokines which is an important biological mediator (Curat et al. 2006). As it is now discovered that fucoxanthin controls the expression of UCP1, hence this bioactive molecule can serve as a potential anti-obesity agent in clinical trials (Pangestuti and Kim 2011a). Similarly, other researchers have also depicted the anti-obesity effects of Xanthigen™ administering pomegranate oil and fucoxanthin to non-diabetic obese women which offered potential liver protecting and weight reducing effects (Abidov et al. 2010).

ANTI-ANGIOGENIC

Angiogenesis denotes development of new blood vessels from already present vasculature in response to physiological or pathological circumstances. Unregulated spread of blood vessels is noted in pathological events including rheumatoid arthritis, inflammatory diseases and tumor metastasis that further worsens the pathological state (Elshabrawy et al. 2015).

The control of angiogenesis is an effective strategy to hinder cancer growth and other angiogenic diseases alongside lessening their severity (Pangestuti and Kim

2011a). In this context, algal bioactives exhibited promising anti-angiogenic effects. For instance, siphonoxanthin obtained from green algae (*Codium fragile*) momentously blocked the human umbilical vein endothelial cells (HUVEC) propagation at dose rates of 2.5 µM and above. HUVEC tube formation was sufficiently inhibited by siphonoxanthin treatment and no tube formation was identified when the concentrations of the compound were raised to 25 µM suggesting the suppression of angiogenic mediators (Ganesan et al. 2010).

In an alternative investigation, Sugawara et al. (2006) demonstrated that fucoxanthin significantly hinder tube development and spread of HUVEC at a dose rate of 10 µM. It efficiently reduces the differentiation of endothelial cells from endothelial progenitor cells. Alongside, *in vivo* and *ex vivo* microvessel outgrowth was also suppressed by fucoxanthin and fucoxanthinol. From the experiments of Sugawara et al. (2006) and Ganesan et al. (2010) it is evident that algal bioactives i.e. fucoxanthin and siphonoxanthin have strong anti-angiogenic properties that may be due to hydroxyl groups as both of these compounds are structurally similar having hydroxyl groups at 3 and 3′ positions, respectively.

NEUROPROTECTIVE

Neurodegenerative ailments are predicted as the second most prevalent reason for mortality among the elderly by the 2040s, surpassing death due to cancer (Ansari et al. 2010). Neuroprotection can be achieved by avoiding damage to the neuronal cells, degeneration or dysfunction in the central nervous system (CNS), apoptosis and limiting post injury death or dysfunction in CNS (Tucci and Bagetta 2008). Protection of CNS dysfunctions and neural damage is currently being attained by numerous drugs from natural and synthetic sources. However, synthetic drugs carry various side effects causing serious health concerns. In this respect, drugs or neuroprotective agents from natural sources are being investigated due to their purportedly safer nature (Pangestuti and Kim 2011a). Therefore, an extensive research is being conducted to identify neuroprotective agents from marine algae (Pangestuti and Kim 2011b).

In an earlier study, fucoxanthin from *H. fusiformis* have been observed to inhibit the progression of GOTO cells and expression of N-myc in human neuroblastoma cell line. Though, the complete mechanism is yet to be identified but a dose of 10 µg/mL showed a 38% reduction in growth rate of these cells (Okuzumi et al. 1990). In stroke-prone spontaneously hypertensive rats (SHRSP), Ikeda et al. (2003) found attenuating effect of "wakame" against hypertension. Lately, fucoxanthins were confirmed to possess significant neuroprotective activity by protecting cortical neurons cell damage during oxygen reperfusion and hypoxia (Khodosevich and Monyer 2010). This protective effect can be assumed due to radical scavenging abilities of fucoxanthins as reoxygenation and hypoxia liberate free radicals which are causative of degenerative neural damages (Pangestuti and Kim 2011a). Furthermore, pheophytin a isolated from *Sargassum fulvellum* was shown to protect neurite outgrowths which are cardinal in neuronal development in embryogenesis (Ina et al. 2007). Such protective effects of pheophytin a are attributed to its low molecular weight due to which it acquires easy penetration in cells and effectively promote the growth of neurite (Pangestuti and Kim 2011a). Considering the scientific evidences, algal bioactives can be endorsed for their neuroprotective abilities and potential utilization in clinical therapies.

PHOTOPROTECTION

Marine organisms are vital in the production of cosmeceutical and pharmaceutical compounds. Certain compounds from marine organisms are of significant importance to protect against the harmful effects of ultraviolet (UV) radiations and photoaging complications (Pallela et al. 2010). Photo-oxidation of living fragments particularly in skin cells is a prominent phenomenon involved in oxidative stress that primarily involves interaction of ultraviolet radiations which have strong oxidative properties (Cardozo et al. 2011). In this context, phlorotannins including dieckol and eckol from *E. cava* have exhibited strong protective effects to reduce the damages caused by exposure to a spectrum of light. The compounds can lower the intracellular ROS arising from γ-rays exposure. Furthermore, DNA damage and lipid peroxidation in membranes due to exposure to radiations is curtailed by eckol administration. Based on scientific evidences, it can be stated that phlorotannins possess consequential competencies to reduce skin damage arising from oxidative stress induced by radiations. Further, these compounds can profoundly be utilized in pharmaceutical, functional foods and cosmetic industries to alleviate skin damages (Li et al. 2011a).

It has further been studied that methanolic extract of red seaweed (*Corallina pilulifera*) possesses significant photoprotective agents which protects DNA from UV-A induced oxidative stress alongside inhibiting matrix metalloproteinases (Pallela et al. 2010). Similarly, naturally occurring photoprotective ingredients from algae (*E. cava*, *C. pelulifera* and *P. rosengurttii*) have also been reported by Saewan and Jimtaisong (2015). These compounds include plastoquinones, sargachromenol, fucoxanthins, astaxanthins, dieckol and phlorotannins (Saewan and Jimtaisong 2015).

CONCLUSION

Algal bioactives provide sufficient diversity for functional foods and cosmeceutical applications. Although the functional food market is a dynamically developing segment in the food industry, there are always challenges in designing and developing novel food and cosmetic products due to complexity, expensive technologies and market risks involved in acceptance of the products. In this milieu, technical requirements, legislation and consumer demands must be considered to attain a sustainable product market. Furthermore, carefully forecasted models should precisely be utilized to predict customer behavior. There exists a need for commercial exploitation of seaweeds in food products for routine utilization. As the consumer is a key contributor in sustaining the functional food market, so consumer-led products should be launched alongside introducing novel and less expensive technologies considering the demands of industry.

REFERENCES

Abdul, Q. A., Choi, R. J., Jung, H. A. and Choi, J. S. Health benefit of fucosterol from marine algae: A review. *J Sci Food Agric*. 2016. 96: 1856–1866.

Abidov, M., Ramazanov, Z., Seifulla, R. and Grachev, S. The effects of Xanthigen™ in the weight management of obese premenopausal women with non-alcoholic fatty liver disease and normal liver fat. *Diab Obes Metab*. 2010. 12: 72–81.

Abrantes, J. L., Barbosa, J., Cavalcanti, D., Pereira, R. C., Fontes, C. L. F., Teixeira, V. L., Souza, T. L. M. and Paixão, I. C. The effects of the diterpenes isolated from the Brazilian brown algae *Dictyota pfaffii* and *Dictyota menstrualis* against the herpes simplex type-1 replicative cycle. *Planta Med*. 2010. 76: 339–344.

Ahmadi, A., Zorofchian Moghadamtousi, S., Abubakar, S. and Zandi, K. Antiviral potential of algae polysaccharides isolated from marine sources: A review. *BioMed Res Int*. 2015.

Ahn, M. J., Yoon, K. D., Min, S. Y., Lee, J. S., Kim, J. H., Kim, T. G., Kim, S. H., Kim, N. G., Huh, H. and Kim, J. Inhibition of HIV-1 reverse transcriptase and protease by phlorotannins from the brown alga *Ecklonia cava*. *Biol Pharm Bull*. 2004. 27: 544–547.

Alghazwi, M., Kan, Y. Q., Zhang, W., Gai, W. P., Garson, M. J. and Smid, S. Neuroprotective activities of natural products from marine macroalgae during 1999–2015. *J Appl Phycol*. 2016. 28: 3599–3616.

Ansari, J., Siraj, A. and Inamdar, N. Pharmacotherapeutic approaches of Parkinson's disease. *Int J Pharmacol*. 2010. 6: 584–590.

Artan, M., Li, Y., Karadeniz, F., Lee, S. H., Kim, M. M. and Kim, S. K. Anti-HIV-1 activity of phloroglucinol derivative, 6,6'-bieckol, from *Ecklonia cava*. *Bioorg Med Chem*. 2008. 16: 7921–7926.

Athukorala, Y., Ahn, G. N., Jee, Y. H., Kim, G. Y., Kim, S. H., Ha, J. H., Kang, J. S., Lee, K. W. and Jeon, Y. J. Antiproliferative activity of sulfated polysaccharide isolated from an enzymatic digest of *Ecklonia cava* on the U-937 cell line. *J Appl Phycol*. 2009. 21: 307–314.

Barrow, C. and Shahidi, F. *Marine Nutraceuticals and Functional Foods*. 2008. CRC Press: Boca Raton, FL.

Bhagavathy, S., Sumathi, P. and Bell, I. J. S. Green algae *Chlorococcum humicola* – A new source of bioactive compounds with antimicrobial activity. *Asian Pac J Trop Biomed*. 2011. 1: 1–7.

Brunt, E. G. and Burgess, J. G. The promise of marine molecules as cosmetic active ingredients. *Int J Cosmet Sci*. 2018. 40: 1–15.

Cardoso, S., Carvalho, G. L., Silva, J. P., Rodrigues, S. M., Pereira, R. O. and Pereira, L. Bioproducts from seaweeds: A review with special focus on the *Iberian Peninsula*. *Curr Org Chem*. 2014. 18: 896–917.

Cardoso, S. M., Pereira, O. R., Seca, A. M., Pinto, D. C. and Silva, A. Seaweeds as preventive agents for cardiovascular diseases: from nutrients to functional foods. *Mar Drugs*. 2015. 13: 6838–6865.

Cardozo, K. H. M., Marques, L. G., Carvalho, V. M., Carignan, M. O., Pinto, E., Marinho-Soriano, E. and Colepicolo, P. Analyses of photoprotective compounds in red algae from the Brazilian coast. *Rev Bras Farmacogn*. 2011. 21: 202–208.

Chen, J., Li, H., Zhao, Z., Xia, X., Li, B., Zhang, J. and Yan, X. Diterpenes from the marine algae of the genus *Dictyota*. *Mar Drugs*. 2018. 16: 159.

Chen, K., Ríos, J. J., Pérez-Gálvez, A. and Roca, M. Comprehensive chlorophyll composition in the main edible seaweeds. *Food Chem*. 2017. 228: 625–633.

Chen, K. and Roca, M. Cooking effects on chlorophyll profile of the main edible seaweeds. *Food Chem*. 2018a. 266: 368–374.

Chen, K. and Roca, M. *In vitro* bioavailability of chlorophyll pigments from edible seaweeds. *J Func Foods*. 2018b. 41: 25–33.

Cheng, S., Zhao, M., Sun, Z., Yuan, W., Zhang, S., Xiang, Z., Cai, Y., Dong, J., Huang, K. and Yan, P. Diterpenes from a Chinese collection of the brown alga *Dictyota plectens*. *J Nat Prod*. 2014. 77: 2685–2693.

Cho, M., Lee, H. S., Kang, I. J., Won, M. H. and You, S. Antioxidant properties of extract and fractions from *Enteromorpha prolifera*, a type of green seaweed. *Food Chem*. 2011. 127: 999–1006.

Chowdhury, M., Sukhan, Z., Kang, J., Ehsan, M., Hannan, M., Shahrin, T., Gatachow, P., Far, M., Kim, C. and Hong, Y. Algicidal activity of the brown seaweed, *Ecklonia cava* against red tide microalgae. In: Proceedings of the 5th International Conference on Environment 2014. 101–103.

Curat, C., Wegner, V., Sengenes, C., Miranville, A., Tonus, C., Busse, R. and Bouloumie, A. Macrophages in human visceral adipose tissue: Increased accumulation in obesity and a source of resistin and visfatin. *Diabetologia*. 2006. 49: 744.

Das, S. K., Ren, R., Hashimoto, T. and Kanazawa, K. Fucoxanthin induces apoptosis in osteoclast-like cells differentiated from RAW264. 7 cells. *J Agric Food Chem*. 2010. 58: 6090–6095.

Dinesh, S., Menon, T., Hanna, L. E., Suresh, V., Sathuvan, M. and Manikannan, M. In vitro anti-HIV-1 activity of fucoidan from *Sargassum swartzii*. *Int J Biol Macromol*. 2016. 82: 83–88.

El Gamal, A. A. Biological importance of marine algae. *Saudi Pharm J*. 2010. 18: 1–25.

Elshabrawy, H. A., Chen, Z., Volin, M. V., Ravella, S., Virupannavar, S. and Shahrara, S. The pathogenic role of angiogenesis in rheumatoid arthritis. *Angiogenesis*. 2015. 18: 433–448.

Ganesan, P., Matsubara, K., Ohkubo, T., Tanaka, Y., Noda, K., Sugawara, T. and Hirata, T. Anti-angiogenic effect of siphonaxanthin from green alga, *Codium fragile*. *Phytomedicine*. 2010. 17: 1140–1144.

Garcia-Vaquero, M. and Hayes, M. Red and green macroalgae for fish and animal feed and human functional food development. *Food Rev Int*. 2016. 32: 15–45.

Gupta, S. and Abu-Ghannam, N. Bioactive potential and possible health effects of edible brown seaweeds. *Trends Food Sci Technol*. 2011. 22: 315–326.

Gutiérrez-Rodríguez, A. G., Juárez-Portilla, C., Olivares-Bañuelos, T. and Zepeda, R. C. Anticancer activity of seaweeds. *Drug Discov Today*. 2018. 23: 434–447.

Hemmingson, J. A., Falshaw, R., Furneaux, R. H. and Thompson, K. Structure and antiviral activity of the galactofucan sulfates extracted from *Undaria pinnatifida* (Phaeophyta). *J Appl Phycol*. 2006. 18: 185–193.

Heo, S. J., Ko, S. C., Kang, S. M., Cha, S. H., Lee, S. H., Kang, D. H., Jung, W. K., Affan, A., Oh, C. and Jeon, Y. J. Inhibitory effect of diphlorethohydroxycarmalol on melanogenesis and its protective effect against UV-B radiation-induced cell damage. *Food Chem Toxicol*. 2010. 48: 1355–1361.

Iahtisham-Ul-Haq, Shamshad, A., Butt, M. S. and Suleria, H. A. R. Heath benefits of anthocyanins in black carrot (*Daucus carota*). In: Goyal, M. R. and Suleria, H. A. R. (eds.) *Human Health Benefits of Plant Bioactive Compounds; Potentials and Prospects*. 2018. Apple Academic Press, Inc: Waretown, NJ.

Ikeda, K., Kitamura, A., Machida, H., Watanabe, M., Negishi, H., Hiraoka, J. and Nakano, T. Effect of *Undaria pinnatifida* (Wakame) on the development of cerebrovascular diseases in stroke-prone spontaneously hypertensive rats. *Clin Exp Pharmacol Physiol*. 2003. 30: 44–48.

Imeson, A. P. *Thickening and Gelling Agents for Food*. 2012. Springer Science & Business Media.

Ina, A., Hayashi, K.-I., Nozaki, H. and Kamei, Y. Pheophytin a, a low molecular weight compound found in the marine brown alga *Sargassum fulvellum*, promotes the differentiation of PC12 cells. *Int J Dev Neurosci*. 2007. 25: 63–68.

Kadam, S. U., Tiwari, B. K. and O'Donnell, C. P. Extraction, structure and biofunctional activities of laminarin from brown algae. *Int J Food Sci Technol*. 2015. 50: 24–31.

Kang, H. S., Chung, H. Y., Jung, J. H., Son, B. W. and Choi, J. S. A new phlorotannin from the brown alga Ecklonia stolonifera. *Chem Pharm Bull*. 2003. 51: 1012–1014.

Kang, K. A., Lee, K. H., Chae, S., Koh, Y. S., Yoo, B. S., Kim, J. H., Ham, Y. M., Baik, J. S., Lee, N. H. and Hyun, J. W. Triphlorethol-A from *Ecklonia cava* protects V79-4 lung fibroblast against hydrogen peroxide induced cell damage. *Free Radic Res*. 2005a. 39: 883–892.

Kang, K. A., Lee, K. H., Chae, S., Zhang, R., Jung, M. S., Ham, Y. M., Baik, J. S., Lee, N. H. and Hyun, J. W. Cytoprotective effect of phloroglucinol on oxidative stress induced cell damage via catalase activation. *J Cell Biochem*. 2006. 97: 609–620.

Kang, K. A., Lee, K. H., Chae, S., Zhang, R., Jung, M. S., Lee, Y., Kim, S. Y., Kim, H. S., Joo, H. G., Park, J. W., Ham, Y. M., Lee, N. H. and Hyun, J. W. Eckol isolated from *Ecklonia cava* attenuates oxidative stress induced cell damage in lung fibroblast cells. *FEBS Lett*. 2005b. 579: 6295–6304.

Khodosevich, K. and Monyer, H. Signaling involved in neurite outgrowth of postnatally born subventricular zone neurons in vitro. *BMC Neurosci*. 2010. 11: 18.

Kong, C. S., Kim, J. A. and Kim, S. K. Anti-obesity effect of sulfated glucosamine by AMPK signal pathway in 3T3-L1 adipocytes. *Food Chem Toxicol*. 2009a. 47: 2401–2406.

Kong, C. S., Kim, J. A., Yoon, N. Y. and Kim, S. K. Induction of apoptosis by phloroglucinol derivative from *Ecklonia cava* in MCF-7 human breast cancer cells. *Food Chem Toxicol*. 2009b. 47: 1653–1658.

Kwak, J. Y. Fucoidan as a marine anticancer agent in preclinical development. *Mar Drugs*. 2014. 12: 851–870.

Lahrsen, E., Schoenfeld, A. K. and Alban, S. Size-dependent pharmacological activities of differently degraded fucoidan fractions from *Fucus vesiculosus*. *Carbohydr Polym*. 2018. 189: 162–168.

Langebeek, N., Gisolf, E. H., Reiss, P., Vervoort, S. C., Hafsteinsdóttir, T. B., Richter, C., Sprangers, M. A. and Nieuwkerk, P. T. Predictors and correlates of adherence to combination antiretroviral therapy (ART) for chronic HIV infection: A meta-analysis. *BMC Med*. 2014. 12: 142.

Li, Y., Qian, Z. J., Kim, M. M. and Kim, S. K. Cytotoxic activities of phlorethol and fucophlorethol derivatives isolated from Laminariaceae Ecklonia cava. *J Food Biochem*. 2011b. 35: 357–369.

Li, Y.-X., Wijesekara, I., Li, Y. and Kim, S.-K. Phlorotannins as bioactive agents from brown algae. *Proc Biochem*. 2011a. 46: 2219–2224.

Lin, C. Y., Lee, C. H., Chang, Y. W., Wang, H. M., Chen, C. Y. and Chen, Y. H. Pheophytin a inhibits inflammation via suppression of LPS-induced nitric oxide synthase-2, prostaglandin E2, and interleukin-1β of macrophages. *Int J Mol Sci.* 2014. 15: 22819–22834.

Maeda, H., Hosokawa, M., Sashima, T., Takahashi, N., Kawada, T. and Miyashita, K. Fucoxanthin and its metabolite, fucoxanthinol, suppress adipocyte differentiation in 3T3-L1 cells. *Int J Mol Med.* 2006. 18: 147–152.

Malyarenko, O. S. and Ermakova, S. P. Fucoidans: Anticancer activity and molecular mechanisms of action. In: *Seaweed Polysaccharides.* 2017.

Martínez-Hernández, G. B., Castillejo, N., Carrión-Monteagudo, M. D. M., Artés, F. and Artés-Hernández, F. Nutritional and bioactive compounds of commercialized algae powders used as food supplements. *Food Sci Technol Int.* 2018. 24: 172–182.

Michalak, I. and Chojnacka, K. Algae as production systems of bioactive compounds. *Eng Life Sci.* 2015. 15: 160–176.

Murray, M., Dordevic, A. L., Bonham, M. P. and Ryan, L. Do marine algal polyphenols have antidiabetic, antihyperlipidemic or anti-inflammatory effects in humans? A systematic review. *Crit Rev Food Sci Nutr.* 2018. 58: 2039–2054.

Okada, T., Nakai, M., Maeda, H., Hosokawa, M., Sashima, T. and Miyashita, K. Suppressive effect of neoxanthin on the differentiation of 3T3-L1 adipose cells. *J Oleo Sci.* 2008. 57: 345–351.

Okuzumi, J., Nishino, H., Murakoshi, M., Iwashima, A., Tanaka, Y., Yamane, T., Fujita, Y. and Takahashi, T. Inhibitory effects of fucoxanthin, a natural carotenoid, on N-myc expression and cell cycle progression in human malignant tumor cells. *Cancer Lett.* 1990. 55: 75–81.

Pallela, R. Antioxidants from marine organisms and skin care. In: *Systems Biology of Free Radicals and Antioxidants.* 2014. Springer.

Pallela, R., Na-Young, Y. and Kim, S. K. Anti-photoaging and photoprotective compounds derived from marine organisms. *Mar Drugs.* 2010. 8: 1189–1202.

Pangestuti, R. and Kim, S. K. Biological activities and health benefit effects of natural pigments derived from marine algae. *J Func Foods.* 2011a. 3: 255–266.

Pangestuti, R. and Kim, S. K. Neuroprotective effects of marine algae. *Mar Drugs.* 2011b. 9: 803–818.

Pangestuti, R. and Kim, S.-K. Seaweeds-derived bioactive materials for the prevention and treatment of female's cancer. In: *Handbook of Anticancer Drugs from Marine Origin.* 2015. Springer.

Ponce, N. M., Pujol, C. A., Damonte, E. B., Flores, M. L. and Stortz, C. A. Fucoidans from the brown seaweed Adenocystis utricularis: Extraction methods, antiviral activity and structural studies. *Carbohydr Res.* 2003. 338: 153–165.

Pozza, C. and Isidori, A. M. What's behind the obesity epidemic. In: *Imaging in Bariatric Surgery.* 2018. Springer.

Ragan, M. and Glombitza, K. *Handbook of Physiological Methods.* 1986. Cambridge University Press: Cambridge, UK.

Roohinejad, S., Nikmaram, N., Brahim, M., Koubaa, M., Khelfa, A. and Greiner, R. Potential of novel technologies for aqueous extraction of plant bioactives. In: *Water Extraction of Bioactive Compounds.* 2018. Elsevier.

Saewan, N. and Jimtaisong, A. Natural products as photoprotection. *J Cosm Dermatol.* 2015. 14: 47–63.

Senevirathne, M. and Kim, S. K. 14 Brown algae-derived compounds as potential cosmeceuticals. *Mar Cosmeceuticals: Trends and Prospects.* 2011. 179.

Shibata, T., Ishimaru, K., Kawaguchi, S., Yoshikawa, H. and Hama, Y. Antioxidant activities of phlorotannins isolated from Japanese Laminariaceae. *J Appl Phycol.* 2008. 20: 705–711.

Siriwardhana, N., Lee, K.-W. and Jeon, Y.-J. Radical scavenging potential of hydrophilic phlorotannins of *Hizikia fusiformis. Algae.* 2005. 20: 69–75.

Sugawara, T., Matsubara, K., Akagi, R., Mori, M. and Hirata, T. Antiangiogenic activity of brown algae fucoxanthin and its deacetylated product, fucoxanthinol. *J Agric Food Chem.* 2006. 54: 9805–9810.

Sugiura, Y., Nagayama, K., Kinoshita, Y., Tanaka, R. and Matsushita, T. The anti-allergic effect of the ethyl acetate fraction from an Ecklonia kurome extract. *Food Agric Immunol.* 2015. 26: 181–193.

Trayhurn, P. and Wood, I. *Signalling Role of Adipose Tissue: Adipokines and Inflammation in Obesity.* 2005. Portland Press Limited.

Tucci, P. and Bagetta, G. *How to Study Neuroprotection?* 2008. Nature Publishing Group.

Vaseghi, G., Sharifi, M., Dana, N., Ghasemi, A. and Yegdaneh, A. Cytotoxicity of *Sargassum angustifolium* partitions against breast and cervical cancer cell lines. *Adv Biomed Res.* 2018. 7: 43.

Wells, M. L., Potin, P., Craigie, J. S., Raven, J. A., Merchant, S. S., Helliwell, K. E., Smith, A. G., Camire, M. E. and Brawley, S. H. Algae as nutritional and functional food sources: Revisiting our understanding. *J Appl Phycol.* 2017. 29: 949–982.

Wijesekara, I., Pangestuti, R. and Kim, S.-K. Biological activities and potential health benefits of sulfated polysaccharides derived from marine algae. *Carb Polym.* 2011. 84: 14–21.

Woo, M. N., Jeon, S. M., Shin, Y. C., Lee, M. K., Kang, M. A. and Choi, M. S. Anti-obese property of fucoxanthin is partly mediated by altering lipid-regulating enzymes and uncoupling proteins of visceral adipose tissue in mice. *Mol Nutri Food Res.* 2009. 53: 1603–1611.

Yende, S. R., Harle, U. N. and Chaugule, B. B. Therapeutic potential and health benefits of Sargassum species. *Pharmacogn Rev.* 2014. 8: 1–7.

Yuan, Y. V. and Walsh, N. A. Antioxidant and antiproliferative activities of extracts from a variety of edible seaweeds. *Food Chem Toxicol.* 2006. 44: 1144–1150.

Yue, P. Y., Leung, H. M., Li, A. J., Chan, T. N., Lum, T. S., Chung, Y. L., Sung, Y. H., Wong, M. H., Leung, K. S. and Zeng, E. Y. Angiosuppressive properties of marine-derived compounds – A mini review. *Environ Sci Pollut Res Int.* 2017. 24: 8990–9001.

Zhang, L., Wang, H., Fan, Y., Gao, Y., Li, X., Hu, Z., Ding, K., Wang, Y. and Wang, X. Fucoxanthin provides neuroprotection in models of traumatic brain injury via the Nrf2-ARE and Nrf2-autophagy pathways. *Sci Rep.* 2017. 7: 46763.

Zhu, W., Ooi, V., Chan, P. and Ang Jr, P. Inhibitory effect of extracts of marine algae from Hong Kong against Herpes simplex viruses. In: Proceedings of the 17th International Seaweed Symposium. Oxford University Press: Oxford. 2003. 159–164.

8 Ultrasound-Assisted Extraction of Bioactive Compounds from Microalgae

Mona Ahmed J. Alzahrani and Conrad O. Perera

CONTENTS

Introduction 81
Microalgae 82
Ultrasound-Assisted Extraction (UAE) 82
Principle of UAE 82
Physical Parameters That Influence UAE 82
Solvent Parameters That Influence UAE 84
Applications of Ultrasound in the Extraction of Bioactive Compounds 84
UAE Application on Fresh Water and Marine Microalgae 86
Bioactive Compounds from Fresh Water and Marine Microalgae 86
Bioactive Proteins, Peptides and Amino Acids 86
Conclusion 86
Acknowledgments 86
References 86

BOX 8.1 SALIENT FEATURES

Ultrasound-assisted extraction (UAE) is an emerging extraction technique that has been widely used on different matrices by enhancing heat and mass transfer. It can break cell walls of microalgae to facilitate the diffusion of intracellular bioactive compounds. UAE has advantages over the classical methods in terms of reduction of solvent use, reduction of unit operations, reduction of extraction time, reduction in energy use, lower environmental impact and rapid return on investment. Sound waves that are at human hearing frequencies range from 16 Hz to under 20 kHz, while ultrasound waves have frequencies ranging from 20 kHz to 10 MHz, well above the human hearing range but below microwave frequencies. The mechanism of the extraction process of the ultrasound is due to the rarefaction and compression phenomena, as well as the critical molecular distance of the medium above which it starts to break down to generate bubbles in the liquid medium. The mechanisms of action assisting the extraction by ultrasound are: fragmentation by cavitation, erosion, sonocapillary effect, sonoporation, local shear stress and destruction and de-texturization of the materials used for extraction. The extraction of various bioactive compounds using UAE is discussed in this chapter.

INTRODUCTION

Numerous studies have focused on techniques that are able to break the cell walls to facilitate the diffusion of intracellular bioactive compounds of interest (Paniagua-Michel 2015; Chemat et al. 2017). These techniques have been applied on different microalgae to extract biologically active compounds (Pico 2013; Al-Zuhair et al. 2017). In order to comply with the economics of the process, reduce waste and reduce energy consumption that are usually associated with the conventional extraction technologies, a considerable number of "environmentally friendly" extraction techniques have been used on algae, most of which are only applicable at laboratory scale (Ibañez et al. 2012; Michalak & Chojnacka 2015). Scaling up of an optimised extraction process is quite challenging, especially when dealing with microalgal species (Torres et al. 2013; Ghasemi Naghdi et al. 2016). Among the technologies used, ultrasound-assisted extraction (UAE) has received considerable attention due to its positive influence on heat and mass transfer and since it is considered a green technology (Legay et al. 2011). In this chapter, the current status and applications of ultrasound-assisted extraction of biologically active compounds from microalgae is presented.

MICROALGAE

Microalgae belong to the subgroup algae and are a heterogeneous group of organisms that possess characteristics that distinguish them, for example, cell size, colour, habitation of aquatic environments, and above all, they are unicellular organisms that are naturally photoautotrophic (Singh & Saxena 2015). Microalgae can be prokaryotic or eukaryotic, and they comprise several thousand species (Heimann & Huerlimann 2015). They are classified as: 1) cyanobacteria (blue-green algae), 2) rhodophyta (red microalgae), 3) chlorophyta (green microalgae) and 4) chromophyta (all others). This diversity makes microalgae a valuable source of biologically active compounds that could support human and animal nutrition, cosmetics, pharmaceuticals and fuel industries (Singh et al. 2005; Mercer & Armenta 2011; Mimouni et al. 2012; Paniagua-Michel 2015).

ULTRASOUND-ASSISTED EXTRACTION (UAE)

UAE is an emerging extraction technique that has been widely used on different matrices by enhancing heat and mass transfer. UAE has the following advantages over the classical methods: reduction of solvent use, reduction of unit operations, reduction of extraction time, reduction in energy use, use of renewable plant resources, better security and safety, lower environmental impact and rapid return on investment (Chemat et al. 2017). Several classes of vital compounds, such as antioxidants, pigments, aromas and other organic and mineral compounds have been successfully extracted by UAE (Chemat et al. 2011; Chemat et al 2017). UAE has been used for: 1) extraction of vital, thermo-labile food compounds such as phenolic compounds, betacyanin and betaxanthin; 2) extraction of lipids and proteins from different matrices; 3) improving oil extraction from oil seeds; 4) improved permeabilization of cell's membrane; 5) extraction of fruit juices, and processing of sauces, purees and dairy products; 6) improving the stability of dispersions by reducing settling of dispersed particles; 7) improving emulsifying properties of proteins; 8) extraction of bioactive compounds from microalgae (Knorr et al. 2004; Baysal & Demirdoven 2012; Pico 2013; Chemat et al. 2017; Yang et al. 2018).

PRINCIPLE OF UAE

Ultrasound waves are high frequency sound waves, which are above the human hearing capacity to perceive (Kadam et al. 2013). The main difference between ultrasound, sound and infrasound is the wave's frequency. Sound waves that are at human hearing frequencies range from 16 Hz to under 20 kHz, while ultrasound waves have frequencies ranging from 20 kHz to 10 MHz, over the human hearing range but below microwave frequencies, and the infrasound wave frequencies are below 16 Hz (Chemat et al. 2011; Pico 2013). Ultrasound waves are characterized by their frequency and wavelength, and the mathematical product of these two parameters results in the wave speed through the medium (Figure 8.1). In addition, ultrasound wave intensity or amplitude is also an important parameter and is used to classify the industrial application. The intensity of an ultrasound wave is defined as the amount of energy passing through the unit cross-sectional area perpendicularly to the beam per unit time at that point, and it is measured in watts per square centimetre (W/cm^2) (Hendee & Ritenour 2003). The frequency of ultrasound exerts significant influences on the extraction yield and kinetics.

PHYSICAL PARAMETERS THAT INFLUENCE UAE

The power in a beam of ultrasound is defined as the total energy passing over the whole cross-sectional area of the beam per unit time (rate). If the intensity is uniform, then:

$$\text{Power} = \text{intensity}\left(W/cm^2\right) \times \text{area}\left(cm^2\right)$$

When the intensity is not uniform, then the spatial variations within the beam must be taken into account. The industrial application of ultrasound can be classified as: low intensity ultrasound (LIU) with less than 1 W/cm^2 (high frequency 100 kHZ–1 MHz and low power typically less than 1 W/cm^2) and high intensity ultrasound (HIU) with 10–1000 W/cm^2. HIU is applied at low frequencies (16–100 kHz) but high power (typically 10–1000 W/cm^2) to modify processes or products by physical disruption of tissue, which is mainly used for extraction processes (Pico 2013). Waves propagate through the solid–liquid media, moving in the longitudinal and perpendicular (as shear waves) directions of particles or close to the surface of the particle. However, waves can propagate in gases and liquids only in the longitudinal direction.

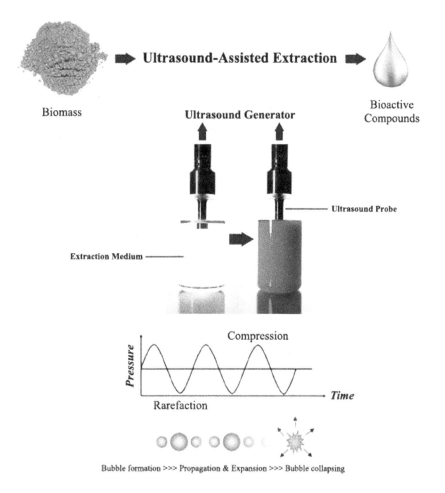

FIGURE 8.1 Ultrasound-assisted extraction.

Ultrasound intensity is defined as the energy transmitted per second per square meter of the emitting surface (Tiwari 2015). Knowledge of the intensity of ultrasound is required for two reasons: first, the output intensity of an ultrasound instrument affects its sensitivity, and hence the signal size. Secondly, knowing the amounts of energy dissipated in biological tissues on the exposure to ultrasound can help with the assessment of potential biological consequences (Hendee & Ritenour 2003). Intensity is correlated to the amplitude of the transduce and the amplitude of the sound wave created (Chemat et al. 2017).

The mechanism of the extraction process of the ultrasound can be addressed as follows: besides the rarefaction and compression phenomena, every medium has a critical molecular distance, where below this critical distance the medium remains intact, but above which the medium starts breaking down and bubbles can be generated in the liquid (Picó 2013). A recent review by Chemat et al. (2017) identified several mechanisms of action assisting the extraction using ultrasound. They are: fragmentation by cavitation, erosion, sonocapillary effect, sonoporation, local shear stress and destruction and de-texturization of plant structures. The effect of ultrasound wave and the mechanism of cavitation are shown in Figure 8.2.

In case of ultrasound processing, the molecules of the medium can reach or even exceed the critical molecular distance and the cavitation bubbles are created as a response to the ultrasound application. Cavitation causes high shear forces in the medium (Chemat et al. 2017). In addition, the size of cavitation bubbles can grow during rarefaction phases and decrease in size during the compression phases. The collapse of these cavitation bubbles occurs when they reach a certain critical point during the compression phase, and consequently releasing large amounts of energy (Chemat et al. 2017).

Ultrasound extraction efficiency can be explained by the fact that sonication simultaneously facilitates mass transfer of solutes to the solvent used for extraction while enhancing the hydration and fragmentation process (Soria & Villamiel 2010; Baysal & Demirdoven 2012). As in other solvent extraction processes, temperature and polarity of the solvent used influence the extraction procedure. The other important physical factors governing

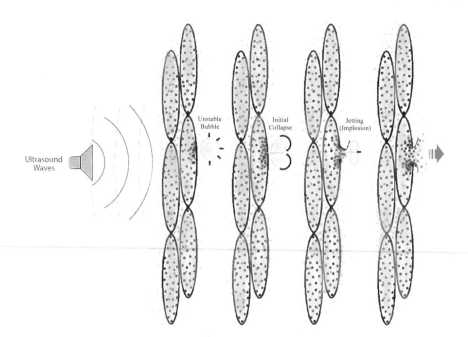

FIGURE 8.2 Effect of ultrasound at the cellular level.

the ultrasound-assisted extraction are frequency and sonication time (Takeuchi et al. 2009) and electrical acoustic intensity (Dey & Rathod 2013).

SOLVENT PARAMETERS THAT INFLUENCE UAE

The choice of solvent depends on the targeted bioactive compounds and also on the solvent properties such as viscosity, surface tension and vapour pressure (Chemat et al. 2017). These physical properties of the solvent could affect the cavitation phenomenon and the cavitation threshold. For example, an increase in viscosity or an increase in surface tension creates an increase in the molecular interactions and the cavitation threshold.

Another factor that influences the solvent properties is the temperature (Picó 2013; Chemat et al. 2017). An increase in temperature decreases the viscosity and surface tension and causes the vapour pressure to increase. An increase in vapour pressure causes the solvent vapour to enter the bubble cavity and causes a reduction in the sonication effect.

Extraction kinetics and yield of bioactive compounds are dependent on the structure of the raw material used, and on the compounds to be extracted as well as the polarity of the solvent used for extraction of certain compounds. The acceleration of the kinetics of extraction is probably obtained by the increase of the intra-particular diffusion of the solute that results from the disruption of the cell walls. However, in some cases, lower frequencies are required in the processes to avoid degradation of bioactive compounds (Takeuchi et al. 2009).

There are two main types of ultrasound units used for extraction purposes: 1) ultrasound cleaning bath; and 2) ultrasound probe system fitted with horn transducers (probe system). Extraction of small volumes can be successfully made using an ultrasonic horn with the tip submerged in the liquid. However, for large extraction volumes, ultrasonic baths, continuous or recycled-flow sono-reactors may be used (Ibañez et al. 2012).

APPLICATIONS OF ULTRASOUND IN THE EXTRACTION OF BIOACTIVE COMPOUNDS

As has been mentioned earlier, the significant mechanism that UAE delivers is the ability to enhance the extraction process by increasing heat and mass transfer between solvent and the matrix of choice. The formation and collapse of cavitation bubbles due to ultrasonication leads to better cell poration/disruption, which improves the penetration of solvent into the matrix selected (Toma et al. 2001; Chemat et al. 2017).

Despite the possibilities that the UAE technique offers, few have investigated ultrasound extraction of bioactive compounds from marine species and aquaculture products, especially the extraction/isolation from these species. The existing studies concerning UAE on algae were mostly concentrated on the extraction of pigments and phenolic compounds from different sea and fresh water microalgae (Table 8.1).

TABLE 8.1
UAE from Fresh Water and Marine Microalgae and the Processing Conditions (2005–2018)

Algae	UAE Conditions	Solvent	Extract	References
Porphyra, Palmaria (red marine algae), *Undariapinnatifida, Himanthaliaelongata* and *Laminaria ochroleuca* (brown marine algae)	• Frequency 17 kHz. • Treatment temperature 65°C.	N/A	Minerals	Domínguez-González et al. (2005)
Dunaliellaspsalina	• Treatment time 3 min.	Methanol and *N,N'*-dimethylformamide (DMF)	Carotenoids and Chlorophylls	Macías-Sánchez et al. (2009)
Sargassum Hypneaspinella, Porphyra sp., *Undariapinnatifida, Chondruscrispus, Halopytisincurvus*	• Treatment time 30 min.	MeOH:H$_2$O 1:9, v/v	Isoflavones	Klejdus et al. (2010)
*Chaetocerosgracilis, Chaetocerosmulleri, Dunaliella*sp.*, Isochrysis*sp.*, Nannochloropsisoculata, Tetraselmis*sp.*, Tetraselmischui, Tetraselmistetrathele* and *Thalassiosiraweissflogii*.	• Frequency 40 kHz. • 80 W ultrasound power.	N/AT	Lipids	Araujo et al. (2011)
Chlorella sp., *Nostoc* sp., *Tolypothrix* sp.	• Frequency 50 kHz. • Treatment time 15 min.	ddH$_2$O	Lipids	Prabakaran and Ravindran (2011)
Chlorella vulgaris	• Frequency 40 kHz. • Ultrasonic intensity of 2.68 W/cm^2.	Bligh and Dyer (1959): methanol + chloroform + water Chen et al. (1981): methanol + Dichlorometane Folch et al. (1957): methanol + chloroform Hara and Radin (1978): isopropanol + hexane	Lipids	Araujo et al. (2013)
Spirulina platensis	• Frequency 20 kHz. • Ultrasonic intensity of 167 W/cm^2. • Treatment time 8 min.	Methanol (pre-soaked for 2 min) *n*-heptane	β-carotene	Dey and Rathod (2013)
Chlorella vulgaris	• Power = 200 W • Treatment time 78.7 min.	Ethanol (80%)	Chlorophylls	Kong et al. (2014)
Nannochloropsis spp.	• Frequency 24 kHz. • Power = 100–400 W. • Treatment time 30 min.	Ethanol and DMSO	Phenolics and chlorophylls	Parniakov et al. (2015)
Spirulina platensis	• Frequency 42 kHz. • Power = 200 W. • Treatment time 42 min.	Ethanol	Phycocyanin	Hadiyanto and Suttrisnorhadi (2016)
Nitzschialaevis, Spirulina platensis, Chlorella vulgaris	• Frequency 25 kHz. • Power = 750 W. • Ultrasonic amplitude of 25%. • Treatment time 5 min.	ddH$_2$O	Bioactive proteins	Alzahrani et al. (2018)
Spirulina platensis	• Frequency 20 kHz. • Ultrasonic amplitude of 50%. • Treatment time 1.5 min.	Dry biomass	C-phycocyanin	Tavanandi et al. (2018)

UAE APPLICATION ON FRESH WATER AND MARINE MICROALGAE

The recent study by Araujo et al. (2011, 2013) examined oil recovery extracted by UAE from different algal species (*Chaetocerosgracilis, Chaetocerosmulleri, Chlorella vulgaris, Dunaliella* sp., *Isochrysis* sp., *Nannochloropsisoculata, Tetraselmis* sp., *Tetraselmischui, Tetraselmistetrathele* and *Thalassiosiraweissflogii*). It has been demonstrated that the salinity of the culture media is an important parameter for oil production, and both *C. vulgaris* and *C. gracilis* microalgae were the most suitable microalgae for industrial-scale oil production (Ibañez et al. 2012). Additionally, UAE was found to be a more suitable oil extraction technique from diatoms than supercritical fluid extraction (Mendiola et al. 2007), which can be due to better penetration of solvent through cell walls because of cavitation phenomenon (Ibañez et al. 2012). Another recent study by Prabakaran and Ravindran (2011) showed that UAE was found to be the most efficient and reproducible technique for oil extraction from microalgae *Chlorella* sp., *Nostoc* sp. and *Tolypothrix* sp.

BIOACTIVE COMPOUNDS FROM FRESH WATER AND MARINE MICROALGAE

Recent studies have proven that marine and fresh water algae possess functional compounds that can be used as food ingredients, nutraceuticals and pharmaceuticals with a wide variety of biological activities (Smit 2004; Kadam et al. 2013; Alzahrani 2018). Microalgae, including diatoms, produce secondary metabolites, otherwise known as bioactive compounds, which may have varying degrees of activity against other microorganisms as a defence mechanism, or acting against certain physiological disorders. These bioactive compounds are profoundly used as antibiotics and may be effective against infectious diseases such as some bacterial infections, neural tube defects and neuropsychiatric sequelae, and HIV-1 (most common and pathogenic strain of HIV) (Bhatnagar & Kim 2010). Moreover, some algal-bioactive compounds, specifically peptides, were also found to have potent antitumor, antimalarial and multidrug reversing activities. Two of the cyanobactins known to have drug reversing activity are Ulithiacyclamide and Patellamide A (Sivonen et al. 2010).

Other recently discovered bioactive peptides were also isolated from *Cyanophyceae* microalgal species, including Viridamides A and B, which are linear lipopeptides with a terminal acetylene group and novel terminal proline methyl ester and a 5-methoxydec-9-ynoic acid moiety. These peptides were found to have anti-trypanosomal and anti-leishmanial activities (Allmendinger et al. 2010; Veerabadhran et al. 2014).

BIOACTIVE PROTEINS, PEPTIDES AND AMINO ACIDS

Bioactive peptides are those with a particular sequence of amino acids that have additional physiological health benefits beyond their basic nutritional value. The protein content of algae varies from one species to another. Proteins and peptides from algae were found to have pharmacological and nutritional values, namely antioxidative, antihypertensive, anticoagulant, antiproliferative and neuroprotective activities (Samarakoon & Jeon 2012; Alzahrani et al. 2018).

CONCLUSION

Ultrasound can be applied on the microalgal biomass to enhance extracts recovery efficiency through breaking or weakening the microalgal cell walls, which in turn facilitates easier cellular extraction of biologically active constituents. However, UAE at frequencies over 20 kHz may affect the bioactive compounds through the formation of free-radicals (Azwanida 2015).

ACKNOWLEDGMENTS

The first author acknowledges funding from the King Abdullah Scholarships Program and support from the Saudi Arabian Cultural Mission in New Zealand and School of Chemical Sciences – Faculty of Science at the University of Auckland for her PhD studies.

REFERENCES

Allmendinger, A., Spavieri, J., Kaiser, M., Casey, R., Hingley-Wilson, S., Lalvani, A., Guiry, M., Blunden, G., & Tasdemir, D. (2010). Antiprotozoal, antimycobacterial and cytotoxic potential of twenty-three British and Irish red algae. *Phytotherapy Research*, 24(7), 1099–1103.

Alzahrani, M. A. J. (2018). Proteins and their enzymatic hydrolysates from the marine diatom *Nitzschialaevis* and screening for their in vitro antioxidant, antihypertension, anti-inflammatory and antimicrobial activities. Auckland, New Zealand: University of Auckland. Retrieved from http://hdl.handle.net/2292/37581.

Alzahrani, M. A. J., Perera, C. O., & Hemar, Y. (2018). Production of bioactive proteins and peptides from the diatom *Nitzschialaevis* and comparison of their *in vitro* antioxidant activities with those from *Spirulina platensis* and *Chlorella vulgaris*. *International Journal of Food Science and Technology*, 53(3), 676–682.

Al-Zuhair, S., Ashraf, S., Hisaindee, S., Al-Darmaki, N. A., Battah, S., Svistunenko, D., Reeder, B., Stanway, G., & Chaudhary, A. (2017). Enzymatic pre-treatment of microalgae cells for enhanced extraction of proteins. *Engineering in Life Sciences*, 17(2), 175–185.

Araujo, G. S., Matos, L. J., Fernandes, J. O., Cartaxo, S. J., Gonçalves, L. R., Fernandes, F. A., & Farias, W. R. (2013). Extraction of lipids from microalgae by ultrasound application: Prospection of the optimal extraction method. *Ultrasonics Sonochemistry*, 20(1), 95–98.

Araujo, G. S., Matos, L. J., Gonçalves, L. R., Fernandes, F. A., & Farias, W. R. (2011). Bioprospecting for oil producing microalgal strains: Evaluation of oil and biomass production for ten microalgal strains. *Bioresource Technology*, 102(8), 5248–5250.

Azwanida, N. N. (2015). A review on the extraction methods use in medicinal plants, principle, strength and limitation. *Medicinal and Aromatic Plants*, 04(3), 196. doi:10.4172/2167-0412.1000196.

Baysal, T., & Demirdoven, A. (2012). Ultrasound in food technology. In: *Handbook on Applications of Ultrasound* (p. 163). CRC Press, Boca Raton, FL.

Bhatnagar, I., & Kim, S. K. (2010). Immense essence of excellence: Marine microbial bioactive compounds. *Marine Drugs*, 8(10), 2673–2701.

Bligh, E. G., & Dyer, W. J. (1959). A rapid method of total lipid extraction and purification. *Canadian journal of biochemistry and physiology*, 37(8), 911–917.

Chemat, F., Rombaut, N., Sicaire, A. G., Meullemiestre, A., Fabiano-Tixier, A. S., & Abert-Vian, M. (2017). Ultrasound assisted extraction of food and natural products. Mechanisms, techniques, combinations, protocols and applications. A review. *Ultrasonics Sonochemistry*, 34, 540–560.

Chemat, F., Zill-e-Huma, & Khan, M. K. (2011). Applications of ultrasound in food technology: Processing, preservation and extraction. *Ultrasonics Sonochemistry*, 18(4), 813–835.

Chen, I., Shen, C., & Sheppard, A. (1981). Comparison of methylene chloride and chloroform for the extraction of fats from food products. *Journal of the American Oil Chemists'*, 58(5), 599–601.

Dey, S., & Rathod, V. K. (2013). Ultrasound assisted extraction of β-carotene from *Spirulina platensis*. *Ultrasonics Sonochemistry*, 20(1), 271–276.

Domínguez-González, R., Moreda-Piñeiro, A., Bermejo-Barrera, A., & Bermejo-Barrera, P. (2005). Application of ultrasound-assisted acid leaching procedures for major and trace elements determination in edible seaweed by inductively coupled plasma-optical emission spectrometry. *Talanta*, 66(4), 937–942.

Folch, J., Lees, M., & Sloane Stanley, G. (1957). A simple method for the isolation and purification of total lipids from animal tissues. *Journal of Biological Chemistry*, 226, 497–509.

Ghasemi Naghdi, F., González González, L. M., Chan, W., & Schenk, P. M. (2016). Progress on lipid extraction from wet algal biomass for biodiesel production. *Microbial Biotechnology*, 9(6), 718–726.

Hadiyanto, H., & Suttrisnorhadi, S. (2016). Response surface optimization of ultrasound assisted extraction (UAE) of phycocyanin from microalgae *Spirulina platensis*. *Emirates Journal of Food and Agriculture*, 28(4), 227–234.

Hara, A., & Radin, N. (1978). Lipid extraction of tissues with a low-toxicity solvent. *Analytical Biochemistry*, 90(1), 420–426.

Heimann, K., & Huerlimann, R. (2015). Chapter 3 – Microalgal classification: Major classes and genera of commercial microalgal species. In: S.-K. Kim (Ed.), *Handbook of Marine Microalgae* (pp. 25–41). Boston: Academic Press.

Hendee, W. R., & Ritenour, E. R. (2003). Medical imaging physics. In: W. R. Hendee, & E. R. Ritenour (Eds.), *Medical Imaging Physics* (4th Ed., pp. 303–316). New York, NY: John Wiley & Sons, Inc.

Ibañez, E., Herrero, M., Mendiola, J. A., & Castro-Puyana, M. (2012). Extraction and characterization of bioactive compounds with health benefits from marine resources: macro and micro algae, cyanobacteria, and invertebrates. In: *Marine Bioactive Compounds* (pp. 55–98). Springer.

Kadam, S. U., Tiwari, B. K., & O'Donnell, C. P. (2013). Application of novel extraction technologies for bioactives from marine algae. *Journal of Agricultural and Food Chemistry*, 61(20), 4667–4675.

Klejdus, B., Lojková, L., Plaza, M., Šnóblová, M., & Štěrbová, D. (2010). Hyphenated technique for the extraction and determination of isoflavones in algae: Ultrasound-assisted supercritical fluid extraction followed by fast chromatography with tandem mass spectrometry. *Journal of Chromatography A*, 1217(51), 7956–7965.

Knorr, D., Zenker, M., Heinz, V., & Lee, D.-U. (2004). Applications and potential of ultrasonics in food processing. *Trends in Food Science and Technology*, 15(5), 261–266.

Kong, W., Liu, N., Zhang, J., Yang, Q., Hua, S., Song, H., & Xia, C. (2014). Optimization of ultrasound-assisted extraction parameters of chlorophyll from *Chlorella vulgaris* residue after lipid separation using response surface methodology. *Journal of Food Science and Technology*, 51(9), 2006–2013.

Legay, M., Gondrexon, N., Le Person, S., Boldo, P., & Bontemps, A. (2011). Enhancement of heat transfer by ultrasound: Review and recent advances. *International Journal of Chemical Engineering*, 2011.

Macías-Sánchez, M. D., Mantell, C., Rodríguez, M., Martínez de la Ossa, E., Lubián, L. M., & Montero, O. (2009). Comparison of supercritical fluid and ultrasound-assisted extraction of carotenoids and chlorophyll a from *Dunaliella salina*. *Talanta*, 77(3), 948–952.

Mendiola, J. A., Torres, C. F., Toré, A., Martín-Álvarez, P. J., Santoyo, S., Arredondo, B. O., Señoráns, F. J., Cifuentes, A., & Ibáñez, E. (2007). Use of supercritical CO_2 to obtain extracts with antimicrobial activity from *Chaetocerosmuelleri* microalga. A correlation with their lipidic content. *European Food Research and Technology*, 224(4), 505–510.

Mercer, P., & Armenta, R. E. (2011). Developments in oil extraction from microalgae. *European Journal of Lipid Science and Technology*, 113(5), 539–547.

Michalak, I., & Chojnacka, K. (2015). Algae as production systems of bioactive compounds. *Engineering in Life Sciences*, 15(2), 160–176.

Mimouni, V., Ulmann, L., Pasquet, V., Mathieu, M., Picot, L., Bougaran, G., Cadoret, J. P., Morant-Manceau, A., & Schoefs, B. (2012). The potential of microalgae for the production of bioactive molecules of pharmaceutical interest. *Current Pharmaceutical Biotechnology*, 13(15), 2733–2750.

Paniagua-Michel, J. (2015). Chapter 16: Microalgal Nutraceuticals. In Se-Kwon Kim (Ed.). *Handbook of Marine Microalgae – Biotechnology Advances*, 255–267. Academic Press, South Korea.

Pico, Y. (2013). Ultrasound-assisted extraction for food and environmental samples. *TrAC Trends in Analytical Chemistry*, 43, 84–99.

Prabakaran, P. & Ravindran, A. D. (2011). A comparative study on effective cell disruption methods for lipid extraction from microalgae. *Letters in applied microbiology*, 53(2), 150–154.

Samarakoon, K., & Jeon, Y.-J. (2012). Bio-functionalities of proteins derived from marine algae – A review. *Food Research International*, 48(2), 948–960.

Singh, J., & Saxena, R. C. (2015). Chapter 2. An introduction to microalgae: Diversity and significance. In: S.-K. Kim (Ed.), *Handbook of Microalgae: Biotechnology Advances* (pp. 11–24). Academic Press, Elsevier.

Singh, S., Kate, B. N., & Banerjee, U. C. (2005). Bioactive compounds from cyanobacteria and microalgae: An overview. *Critical Reviews in Biotechnology*, 25(3), 73–95.

Sivonen, K., Leikoski, N., Fewer, D. P., & Jokela, J. (2010). Cyanobactins—Ribosomal cyclic peptides produced by cyanobacteria. *Applied Microbiology and Biotechnology*, 86(5), 1213–1225.

Smit, A. J. (2004). Medicinal and pharmaceutical uses of seaweed natural products: A review. *Journal of Applied Phycology*, 16(4), 245–262.

Soria, A. C., & Villamiel, M. (2010). Effect of ultrasound on the technological properties and bioactivity of food: A review. *Trends in Food Science and Technology*, 21(7), 323–331.

Takeuchi, T., Pereira, C., Braga, M., Maróstica, M., Leal, P., & Meireles, M. (2009). Low-pressure solvent extraction (solid-liquid extraction, microwave assisted, and ultrasound assisted) from condimentary plants. *Extracting Bioactive Compounds for Food Products*, 137–218.

Tavanandi, H. A., Mittal, R., Chandrasekhar, J., & Raghavarao, K. S. M. S. (2018). Simple and efficient method for extraction of C-Phycocyanin from dry biomass of *Arthospira platensis*. *Algal Research*, 31, 239–251.

Tiwari, B. K. (2015). Ultrasound: A clean, green extraction technology. *TrAC Trends in Analytical Chemistry*, 71, 100–109.

Toma, M., Vinatoru, M., Paniwnyk, L., & Mason, T. J. (2001). Investigation of the effects of ultrasound on vegetal tissues during solvent extraction. *Ultrasonics Sonochemistry*, 8(2), 137–142.

Torres, C. M., Ríos, S. D., Torras, C., Salvadó, J., Mateo-Sanz, J. M., & Jiménez, L. (2013). Microalgae-based biodiesel: A multicriteria analysis of the production process using realistic scenarios. *Bioresource Technology*, 147, 7–16.

Veerabadhran, M., Manivel, N., Mohanakrishnan, D., Sahal, D., & Muthuraman, S. (2014). Antiplasmodial activity of extracts of 25 cyanobacterial species from coastal regions of Tamil Nadu. *Pharmaceutical Biology*, 52(10), 1291–1301.

Yang, X., Li, Y., Li, S., Oladejo, A., Wang, Y., Huang, S., Zhou, C., Ye, X., Ma, H., & Duan, Y. (2018). Effects of ultrasound-assisted α-amylase degradation treatment with multiple modes on the extraction of rice protein. *Ultrasound Sonochemistry*, 40(A), 890–899.

9 Biogeneration of Volatile Organic Compounds in Microalgae-Based Systems

Pricila Nass Pinheiro, Karem Rodrigues Vieira, Andriéli Borges Santos, Eduardo Jacob-Lopes, and Leila Queiroz Zepka

CONTENTS

Abbreviations 89
Introduction 89
Factors Affecting the Microalgae Volatile Organic Compounds 90
Biosynthesis of Volatile Organic Compounds in Microalgae-Based System 91
Techniques for Volatile Organic Compounds Recovery 95
Commercial Perspective of Volatile Compounds Produced by Microalgae 95
Conclusions and Future Perspectives 95
Acknowledgments 96
References 96

BOX 9.1 SALIENT FEATURES

Microalgae are considered a potentially new and valuable source of biologically active metabolites for industrial applications. The volatile organic compounds (VOCs) are secondary metabolites obtained from microalgae that could be used as an important alternative source of biobased chemicals. Depending on species, culture medium, and environmental conditions, microalgae are capable of produce a wide variety of the VOCs of different chemical classes such as alcohol, aldehydes, ketones, hydrocarbons, esters, terpenes, and sulfurized compounds. Thus, this chapter covers topics that refer to the factors affecting the production of volatile organic compounds, metabolic pathways, techniques for volatile organic compound recovery, and commercial perspective of volatile compounds produced by microalgae.

ABBREVIATIONS

AAD: aldehyde decarbonylase
AAR: acyl-acyl reductase
ACCoA: acetyl-Coenzyme A
BPG: 1,3-bisphosphoglycerate
CO_2: carbon dioxide
DMADP: dimethylallyl diphosphate
DMDS: dimethyldisulfide
DMS: dimethylsulfide
DMSP: dimethylsulphoniopropionate
DMTS: dimethyltrisulfide
FPP: farnesyl diphosphate
G3P: glyceraldehyde-3 phosphate
O_3: ozone
OAA: oxaloacetate
PYR: pyruvate
VOCs: volatile organic compounds

INTRODUCTION

Microalgae are a group of photosynthetic microorganisms typically unicellular and eukaryotic. Although cyanobacteria belong to the domain of bacteria, and are photosynthetic prokaryotes, often they are considered microalgae (Buono et al. 2014). Microalgae-based systems are considered a potentially new and valuable source of biologically active compounds for applications in several biotechnology sectors (Lauritano et al. 2018).

Depending on species, culture, and environmental conditions, microalgae are capable of producing a variety of volatile organic compounds (VOCs) (Hosoglu 2018). The biosynthesis of these compounds will depend on the availability of building blocks, such as carbon, nitrogen, and energy supply from the primary metabolism. Therefore, the availability of these compounds has great impact on the concentration of VOCs (Santos et al. 2016a).

The identification of these organic molecules could therefore be a source of useful chemical products, based on a nonconventional technological route. Although several commercial uses have been found for microalgae, little is known about the integration of microalgae-based systems applied to utilization of volatile organic compounds generated in this process. Thus, the aim of this chapter is to present the biogeneration of volatile organic compounds in microalgae-based systems, focusing on the culture conditions, biosynthesis, and industrial applications of these compounds.

FACTORS AFFECTING THE MICROALGAE VOLATILE ORGANIC COMPOUNDS

Odors compounds desirable and undesirable are ubiquitous in the environment, arising from natural and artificial processes, by the generation of volatile organic compounds (VOCs). Jacob-Lopes and Franco (2013) reported that the VOCs are the main bioproducts formed during microalgae cultivation. The emission of volatile compounds in microalgae is influenced by several biotic and abiotic factors such as growth phase, species type, stresses (temperature, light intensity, pH, salinity), nutrients, gases (H_2O, CO_2, O_3), aeration (mixing/turbulence), or static culture (Aychuchan et al. 2017).

To achieve the largest possible microalgal variety VOCs in a cost-effective way, selection of a microalgae mode of cultivation is of vital importance. Three major modes of microalgae cultivation can be adopted, namely photoautotrophic, heterotrophic, and mixotrophic (Figure 9.1).

In general, microalgae are commonly grown by converting dissolved, inorganic carbon (CO_2) and absorbing solar energy. They have pigments such as chlorophyll and carotenoids, and in some cases phycobiliproteins which are involved in capturing luminous energy to perform photosynthesis. For the CO_2 converted into carbohydrates, catalyzed by the enzyme ribulose 1,5-bisphosphate carboxylase/oxygenase (Rubisco), this process is referred to as the Calvin cycle. The Calvin cycle is the metabolic mechanism for fixing CO_2 in microalgae. This process comprises three stages; carboxylation, reduction, and regeneration. The end of the cycle forms one molecule of glyceraldehyde-3-phosphate that through the action of enzymes forms phosphoenolpyruvate, and finally pyruvate (Santos et al. 2016a).

Additionally, some species of these microorganisms have the versatility to maintain their structures in the absence of light, being able to grow heterotrophically. Heterotrophic growth is an aerobic process where assimilation of organic substrates generates energy through oxidative phosphorylation accompanied by oxygen consumption as the final electron acceptor (Perez-Garcia and Bashan 2015). To use these organic compounds, transport occurs through the membrane. This substrate will be converted into glucose 6-phosphate, so can start the route oxidative pentose phosphate pathway. During metabolism, there is the formation of two molecules of ATP (adenosine triphosphate). The final product is also pyruvate (Fay 1983).

The heterotrophic metabolic route serves as the exclusive source of energy for maintenance and biosynthesis, besides providing the carbon required as building blocks for biosynthesis (Francisco et al. 2014). The biosynthesis of volatile compounds depends mainly of the availability of carbon and nitrogen as well as energy provided by primary metabolism. Therefore, the availability of these building blocks has a major impact on the concentration of any secondary metabolites, including VOCs (Jacob-Lopes et al. 2010; Santos et al. 2016a).

Some microalgae are mixotrophic and can simultaneously drive phototrophy and heterotrophy to utilize both inorganic (CO_2) and organic carbon substrates, thus leading to an additive or synergistic effect of the two processes that enhance the productivity of biomass and consequently a production of volatile compounds (Bhatnagar et al. 2011). CO_2 is fixed through photosynthesis, which is influenced by illumination, while organic compounds are assimilated through aerobic respiration, which is affected by the availability of organic carbon. Several species are able to switch between photoautotrophic and heterotrophic growth (Perez-Garcia and Bashan 2015).

Regardless of metabolism, the biosynthesis of volatile organic compounds occurs through the formation of pyruvate molecules and with the growth conditions of controlled and appropriate microalgae, these microorganisms have the capacity to produce volatile compounds desirable with a pleasant perception threshold (Santos et al. 2016a). Therefore, understanding the microalgae culture conditions can provide a better structure for the production of volatile organic compounds with industrial potential.

Biogeneration of Volatile Organic Compounds

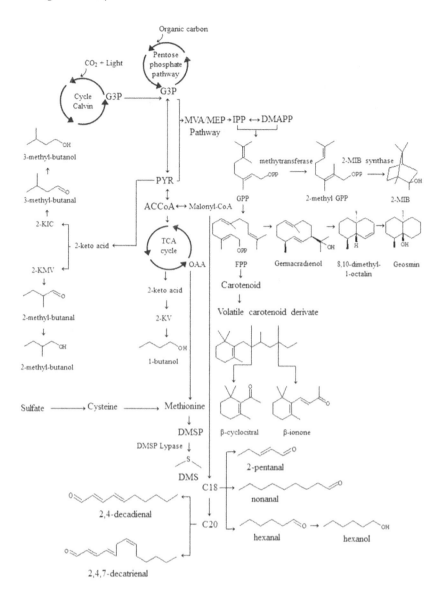

FIGURE 9.1 Overview of biosynthetic pathways leading to the emission of microalgae volatile organic compounds (VOCs). Compound abbreviations are following specified. *ACCoA* acetyl-Coenzyme A, C18 linoleic or linolenic acid, C20 arachidonic or eicosapentaenoic acid, CO2 carbon dioxide, DMAPP dimethylallyl diphosphate, DMS dimethylsulphide, DMSP dimethylsulphoniopropionate, G3P glyceraldehyde-3-phosphate, FPP farnesyl diphosphate, GPP geranyl diphosphate, IPP isopentenyl pyrophosphate, 2-KIC 2-ketoisocaproate, 2-KMV 2-keto-3-methylvalerate, 2-KV 2-ketovalerate, 2-MIB 2-Methylisoborneol, MVA mevalonate pathway, MEP methylerythritol phosphate pathway, OAA oxaloacetate, PYR pyruvate.

BIOSYNTHESIS OF VOLATILE ORGANIC COMPOUNDS IN MICROALGAE-BASED SYSTEM

The different VOCs produced through the metabolism of microalgae may belong to organic classes such as aldehydes, alcohols, terpenes, sulfurized compounds, esters, ketones, and hydrocarbons (Table 9.1).

Several biosynthetic pathways are involved in the synthesis of volatile organic compounds released by microalgae. Figure 9.1 shows the overview of biosynthetic pathways to the emission of volatile organic compounds from microalgae. The main routes of formation of these compounds may be enzymatically or degradation reaction.

Microalgae fatty acid derivatives as aldehydes are the most prevalent volatile organic compounds, due to their low odor threshold values (Santos et al. 2016b). The fatty acid synthesis starts from the acetyl-CoA and the extender malonyl-CoA through a cyclic series of reactions catalyzed by fatty acid synthases, aldehydes are derivatives by the conversion of fatty acids by a lipoxygenase. Compounds such as 2,4-decadienal and 2,4,7-decatrienal are derived from the lipoxygenase/

TABLE 9.1
Major VOCs Found in Microalgae-Based Systems, as Reported in Previous Literatures

Chemical Name	Odor Descriptor	OTC (µg L⁻¹)	Species That Produce the Odorants	References
Alcohols				
1-butanol	Medicine/fruit	0.2	*Phormidium autumnale*.	Santos et al. (2016b).
1-hexanol	Flower/green	2.5	*Chlorella vulgaris, Nannochloropsis oculata, Phormidium autumnale, Tetraselmis* sp.	Van Durme et al. (2013); Santos et al. (2016b).
2-methylbutanol	Whiskey	0.045	*Tetraselmis* sp.	Van Durme et al. (2013).
3-methylbutanol	Whiskey/malt	0.04	*Chlorella vulgaris, Microcystis aeruginosa, Nannochloropsis oculata, Phormidium autumnale, Tetraselmis* sp.	Hasegawa et al. (2012); Van Durme et al. (2013); Santos et al. (2016b).
1-penten-3-ol	Green	0.4	*Botryococcus braunii, Chlorella vulgaris, Nannochloropsis oculata, Nitzschia closterium, Rhodomonas* sp.	Van Durme et al. (2013); Zhou et al. (2017).
Aldehydes				
2,4,7-decatrienal	Rancid/fishy	19.8	*Microcystis papillosa, Microcystis varians*.	Lee et al. (2017).
2,4-decadienal	Rancid/fishy	19.5	*Botryococcus braunii*.	Watson (2003); Van Durme et al. (2013); Lee et al. (2017).
hexanal	Grass/tallow	0.0024	*Botryococcus braunii, Chlorella vulgaris, Nannochloropsis oculata, Phormidium autumnale, Rhodomonas* sp., *Schizochytrium limacinum, Tetraselmis* sp.	Van Durme et al. (2013); Santos et al. (2016b); Hosoglu (2018).
2-methylbutanal	Cocoa/almond	0.1	*Phormidium autumnale*.	Santos et al. (2016b).
3-methylbutanal	Malt	0.002	*Chlorella vulgaris, Nannochloropsis oculata, Phormidium autumnale, Rhodomonas* sp., *Tetraselmis* sp.	Van Durme et al. (2013); Santos et al. (2016b).
nonanal	Citrus/green	0.001	*Botryococcus braunii, Chaetoceros calcitrans, Chlorella prothecoides, Chlorella vulgaris, Crypthecodinium cohnii, Microcystis* sp., *Nannochloropsis oculata, Nitzschia closterium, Platymonas helgolandica, Rhodomonas* sp., *Schizochytrium limacinum, Thalassiosira weissflogii*.	Van Durme et al. (2013); Zhou et al. (2017); Hosoglu (2018).
2-pentenal	Strawberry/fruit	1.5	*Botryococcus braunii, Chlorella vulgaris, Nannochloropsis oculata, Nitzschia closterium, Rhodomonas* sp., *Tetraselmis* sp.	Van Durme et al. (2013); Zhou et al. (2017).
Esters				
methyl phenylacetate	Honey	0.65	*Botryococcus braunii, Chlorella prothecoides, Crypthecodinium cohnii, Rhodomonas* sp., *Schizochytrium limacinum, Tetraselmis chuii*.	Van Durme et al. (2013); Hosoglu (2018).
methyl octanoate	Fruit/orange	0.2	*Botryococcus braunii, Chlorella prothecoides, Crypthecodinium cohnii, Rhodomonas* sp., *Schizochytrium limacinum, Tetraselmis chuii*.	Van Durme et al. (2013); Hosoglu (2018).
Hydrocarbons				
hexadecane	–	–	*Chaetoceros calcitrans, Dicrateria inornata, Nannochloropsis* sp., *Nitzschia closterium, Spirulina platensis, Platymonas helgolandica, Thalassiosira weissflogii*.	Milovanovic et al. (2015); Zhou et al. (2017).

(Continued)

TABLE 9.1 (CONTINUED)
Major VOCs Found in Microalgae-Based Systems, as Reported in Previous Literatures

Chemical Name	Odor Descriptor	OTC (µg L^{-1})	Species That Produce the Odorants	References
8-heptadecene	–	–	*Chaetoceros calcitrans, Dicrateria inornata, Nannochloropsis* sp., *Nitzschia closterium, Spirulina platensis, Platymonas helgolandica, Thalassiosira weissflogii.*	Milovanovic et al. (2015); Zhou et al. (2017).
heptadecane	–	–	*Chaetoceros calcitrans, Dicrateria inornata, Nannochloropsis* sp., *Nitzschia closterium, Spirulina platensis, Platymonas helgolandica, Thalassiosira weissflogii.*	Milovanovic et al. (2015); Zhou et al. (2017).
Ketones				
2,3-pentenedione	Toasted/caramellic	5.50	*Botryococcus braunii, Chlorella vulgaris, Nannochloropsis oculata, Nitzschia closterium, Rhodomonas* sp., *Tetraselmis* sp.	Van Durme et al. (2013); Zhou et al. (2017).
1-Penten-3-one	Fish	0.013	*Botryococcus braunii, Chaetoceros calcitrans, Chlorella vulgaris, Dicrateria inornata, Nannochloropsis oculata, Nitzschia closterium, Platymonas helgolandica, Rhodomonas* sp., *Tetraselmis* sp.	Van Durme et al. (2013); Zhou et al. (2017).
Terpenes				
α-ionone	Tropical fruity	0.003	*Botryococcus braunii, Nannochlopsis, Rhodomonas* sp., *Tetraselmis* sp.	Van Durme et al. (2013).
β-cyclocitral	Mint, tobacco	0.005	*Anabaena* sp., *Botryococcus braunii, Chlorella vulgaris, Microcystis aeruginosa, Microcystis botrys, Microcystis viridis, Microcystis wesenbergii, Nannochloropsis oculata, Nostoc* sp., *Phormidium autumnale, Rhodomonas* sp., *Spirulina platensis, Tetraselmis* sp.	Van Durme et al. (2013); Milovanovic et al. (2015); Santos et al. (2016b); Lee et al. (2017).
β-ionone	Violet/flower	0.035	*Anabaena* sp., *Botryococcus braunii, Nannochlopsis oculata, Nostoc* sp., *Rhodomonas* sp., *Spirulina platensis, Tetraselmis* sp.	Van Durme et al. (2013); Milovanovic et al. (2015).
geosmin	Earthy/musty	0.007	*Anabaena circinalis, Anabaena crassa, Anabaena lemmermannii, Anabaena macrospora, Anabaena planctonica, Anabaena solitaria, Anabaena viguieri, Anabaena millerii, Aphanizomenon gracile, Geitlerinema splendidum, Leibleinia subtilis, Microcoleus* sp., *Phormidium allorgei, Phormidium amoenum, Phormidium breve, Phormidium cortianum, Phormidium formosum, Phormidium simplicissimum, Phormidium uncinatum, Phormidium viscosum, Phormidium* sp.	Watson (2003); Liato and Aïder (2017); Lee et al. (2017).
2-methylisoborneol (2-MIB)	Earthy/musty	0.005	*Oscillatoria curviceps, Oscillatoria limosa, Oscillatoria tenuis, Oscillatoria variabilis, Phormidium autumnale, Phormidium breve, Phormidium calcícola, Phormidium favosum, Phormidium tenue, Phormidium* sp.	Watson et al. (2016); Lee et al. (2017).
Sulfurized compounds				
dimethyl sulfide (DMS)	Cabbage/sulfurous	0.00084	*Anacystis nidulans, Chaetoceros calcitrans, Chlorella protothecoides, Chlorella vulgaris, Crypthecodinium cohnii, Nannochloropsis* sp., *Oscillatoria chalybea, Oscillatoria tenuis, Phormidium autumnale, Platymonas helgolandica, Plectonema boryanum, Schizochytrium limacinum, Synechococcus cedrorum, Tetraselmis chuii, Tetraselmis* sp., *Thalassiosira weissflogii.*	Watson (2003); Van Durme et al. (2013); Zhou et al. (2017); Lee et al. (2017); Hosoglu (2018).

(Continued)

TABLE 9.1 (CONTINUED)
Major VOCs Found in Microalgae-Based Systems, as Reported in Previous Literatures

Chemical Name	Odor Descriptor	OTC (µg L^{-1})	Species That Produce the Odorants	References
dimethyl disulfide (DMDS)	Septic/garlic/putrid	<4.0	*Microcystis aeruginosa, Microcystis wesenbergii, Rhodomonas* sp., *Tetraselmis* sp.	Van Durme et al. (2013); Lee et al. (2017).
dimethyl trisulfide (DMTS)	Septic/garlic/putrid/swampy	0.01	*Microcystis aeruginosa, Microcystis wesenbergii, Rhodomonas* sp., *Tetraselmis* sp.	Van Durme et al. (2013); Lee et al. (2017).

hydroperoxid lyase dependent degradation of arachidonic or eicosapentaenoic acid, however, the fatty acids linoleic or linolenic acid are a precursor of aldehydes such as nonanal, hexanal, and 2-pentanal, as well as alcohol hexanol (Adolph et al. 2003; Yu et al. 2014; Santos et al. 2016b; Jerković et al. 2018).

Hydrocarbons and ketones can also be formed from the lipid degradation (Santos et al. 2016a). Hydrocarbon production is mainly achieved by two enzymes, an acyl-acyl carrier protein reductase (AAR) and an aldehyde decarbonylase (AAD), which play a crucial role in converting fatty acid intermediates to alkanes and alkenes (Milovanovic et al. 2015). Ketones can be formed in many ways; aliphatic ketones can be lipid oxidation products or ketone and methyl degradation (C3-C17) could be formed from the oxidative cleavage of carotenoids (Santos et al. 2016a, b).

Alcohols are another significant chemical class detected in microalgae cultures (Hosoglu 2018). They can be produced through the 2-ketoacid pathway, where the corresponding aldehydes are converted using a 2-ketoacid decarboxylase and then reduced to the alcohols with an alcohol dehydrogenase. The precursors, 2-ketoisocaproate (2-KIC) and 2-keto-3-methylvalerate (2-KMV), can be converted to 3-methyl-butanal and 2-methyl-butanal subsequently reduced to 3-methyl-butanol and 2-methyl-butanol, respectively (Tashiro et al. 2015). These 2-keto acids can be further subjected to decarboxylation, followed by reduction, oxidation and/or esterification, and additionally forming acids and esters (Santos et al. 2016a).

Many of the compounds detected in microalgae originate from the terpenoid pathways. β-ionone is produced by double-bond cleavage enzymes between carbons 9 and 10 of β-carotene. In addition, β-cyclocitral can be formed from the enzymatic cleavage of the double bond between carbons 7 and 8 of the same carotenoid, catalyzed by β-carotene-oxygenases bound to the cell membrane (Chang et al. 2011; Santos et al. 2016b).

Geosmin and 2-methylisoborneol (2-MIB) are synthesized through the isoprenoid pathways, isopentenyl diphosphate (IPP) and dimethylallyl diphosphate (DMADP), and are the central intermediates in the isoprenoid biosynthesis. They can be produced in the mevalonate pathway (MVA) or methylerythritol phosphate (MEP) pathway. Subsequently these starter molecules are converted to immediate prenyl diphosphate precursors, such as geranyl diphosphate (GDP) and farnesyl diphosphate (FDP) (Liato and Aïder 2017; Meena et al. 2017). The cyclization of farnesyl diphosphate (FDP) to geosmin is catalyzed by geosmin synthase via three steps (farnesyl diphosphate to germacradienol, germacradienol to 8,10-dimethyl-1-octalin, and 8,10-dimethyl-1-octalin to geosmin) in cyanobacteria (Giglio et al. 2008). The 2-MIB synthase mechanism is based 2-C-methyltransferase catalyzed methylation of geranyl diphosphate, C10 monoterpene precursor, into 2-methylgeranyl diphosphate. Then, 2-MIB synthase catalyzes cyclization of the 2-methylgeranyl diphosphate to 2-MIB (Lee et al. 2017).

Sulfur compounds are another group of potent odorous products that are liberated by many microalgae, compounds with human odor threshold concentrations sufficiently low to cause malodors (Achyuthan et al. 2017; Watson and Jüttner 2017). Dimethylsulfide (DMS), dimethyldisulfide (DMDS), and dimethyltrisulfide (DMTS) are the major components responsible for the strong offensive odor, generated by a diversity of biota, biochemical pathways, enzymes, and precursors (Graham et al. 2010; Watson and Jüttner 2017; Huang et al. 2018).

In microalgae-based systems, the most important biogenic volatile sulfide produced is dimethylsulphide (DMS) (Watson et al. 2016). Microbial catabolism of the dimethylsulphoniopropionate (DMSP) is thought to be the major biological process generating the volatile compound (Carrión et al. 2015). The DMSP arises from the sulfur-containing amino acid methionine, initially from the enzymatic action of methionine decarboxylase and subsequently undergoes decarboxylation, oxidation, and methylation reactions to yield the final product. The demethiolation of DMSP leads to methanethiol which can be converted to DMS by methylation (Achyuthan et al. 2017; Curson et al. 2017).

The establishment of biochemical pathways can target specific biomolecules production of microalgae metabolism to better knowledge of their ecological function and also to compounds of interest for application in several industries.

TECHNIQUES FOR VOLATILE ORGANIC COMPOUNDS RECOVERY

Commercial production of volatile organic compounds by biotechnology often requires economic profitability. The microalgae biosynthesis is generally limited by low productivity or low concentrations of main compounds in the bioreactor. In order to gain high yields and productivity, it is important to choose the reactor design and the convenient system for the recovery of volatile compounds (Akachaa and Gargouri 2015). There are several possible ways for the simultaneous recovery of volatile products from the bioreactor, such as: adsorption; condensation; and the membrane-based techniques (Saffarionpour and Ottens 2018; Try et al. 2018).

In the condensation-based recovery system, the air from the bioreactor passes through the vertical trap column placed in a cryogenic bath containing liquid nitrogen that allows VOC vapor to condense. Another technique is the adsorption, widely used in the recovery of VOCs from the bioreactor, being a process based on the ability of a solid (e.g. adsorbent) to connect a gaseous component (e.g. adsorbate) to its surface (Saffarionpour and Ottens 2018). Membrane-based techniques have been used for more than two decades to recover VOCs. Two membrane-based techniques are pervaporation and pertraction. The principle of pervaporation is the separation of liquid mixtures by partial vaporization through a dense membrane with a gas flow. The same principle is applied in pertraction but the downstream state is a liquid phase (Feron et al. 1999; Try et al. 2018).

The techniques proposed for the recovery of volatile organic compounds aim to minimize their losses and recover the major components which are valuable in producing a high-quality final product for industrial application.

COMMERCIAL PERSPECTIVE OF VOLATILE COMPOUNDS PRODUCED BY MICROALGAE

Chemicals obtained from microalgae-based systems are sold at prices 1000 times higher than those synthetic chemicals, which show great potential for the exploitation of these processes. Typical applications of microalgae correspond to a variety of metabolites with potential application in products such as cosmetics, food ingredients, and bioenergy. They can also be used as environmental indicators (Jacob-Lopes et al. 2008; Abdehl-Raouf et al. 2012).

Volatile organic compounds generated by microalgae with commercial appeal include 3-methyl-butanol, hexanol, hexanal, β-cyclocitral, and β-ionone (Smith et al. 2010; Santos et al. 2016b). Due to their low odor thresholds aldehydes are important VOCs generated by microalgae because they contribute with desirable aromas as well as rancid odors and flavors. Saturated aldehydes have a green-like, hay-like, paper-like odor, whereas unsaturated aldehydes have a fatty, oily, frying odor (Santos et al. 2016a; Hosoglu 2018).

There is a growing interest in the production of biofuels from renewable sources, offering sustainable solutions for the energy sector as a promising alternative to the traditional petrochemical industry (Si et al. 2014; Severo et al. 2018). The production of hydrocarbons is of particular interest due to their potential for use as advanced biofuels (Choi and Lee 2013). Aliphatic alcohols with higher carbon chain length or equal to five are attractive targets for biofuels in order to have a high energy density and low water solubility. Other alcohol having substantial energy interest is the 1-butanol in order to have a comparable gasoline energy (Zhang et al. 2008).

Advances in microalgal biotechnology are the beginning of microbial production as a viable means of biochemical synthesis, emerging as an alternative means for flexible, efficient, and low impact production to be able to produce a variety of industrially relevant volatile compounds.

CONCLUSIONS AND FUTURE PERSPECTIVES

Microalgae can produce a variety of volatile compounds with important odor characteristics, and the knowledge about the biosynthesis of these structures from microalgae might prove useful to help elucidated ways. Nonetheless, there are many bottlenecks related to microalgae technology for it to become an industrial reality, so strategies should be developed for controlling the many factors that contribute to changes in metabolic behaviors that influence VOCs production and emission which is challenging even under laboratory conditions where small cultures can be grown under carefully defined and known/measurable parameters. This becomes increasingly difficult during commercial, large-scale production. Despite these challenges, advancements in culture techniques and careful design of experiments, including statistically meaningful replicates and the use of pattern-analysis techniques can provide critical information that will contribute towards the understanding of the microalgae volatile profile. Thus, the information supplied through this and previous studies presents a consistent argument for the "fingerprinting" of volatile organic compounds from microalgae species.

ACKNOWLEDGMENTS

Inc and National Council for Scientific and Technological Development of Brazil (CNPq) and National Council for the Improvement of Higher Education (CAPES).

REFERENCES

Abdel-Raouf, N., Al-Homaidan, A. A., Ibraheem, I. B. M. Microalgae and wastewater treatment. *Saudi J Biol Sci.* 2012. 19: 257–275.

Achyuthan, K. E., Harper, J. C., Manginell, R. P., Moorman, M. W. Volatile metabolites emission by in vivo microalgae – An overlooked opportunity? *Metabolites.* 2017. 7: 39–85.

Adolph, S., Poulet, S. A., Pohnert, G. Synthesis and biological activity of α, β, γ, δ-unsaturated aldehydes from diatoms. *Tetrahedron.* 2003. 59(17): 3003–3008.

Akacha, N. B., MohamedGargourib, M. Microbial and enzymatic technologies used for the production of natural aroma compounds: Synthesis, recovery modeling, and bioprocesses. *Food and Bioprod Process.* 2015. 94: 675–706.

Bhatnagar, A., Chinnasamy, S., Singh, M., Das, K. C. Renewable biomass production by mixotrophic algae in the presence of various carbon sources and wastewaters. *Appl Energy.* 2011. 88: 3425–3431.

Buono, S., Langellotti, A. L., Martello, A., Rinna, F., Fogliano, V. Functional ingredients from microalgae. *Food Funct.* 2014. 5: 1669–1685.

Carrión, O., Curson, A. R. J., Kumaresan, D., Fu, Y., Lang, A. S., Mercadé, E., Todd, J. D. A novel pathway producing dimethylsulphide in bacteria is widespread in soil environments. *Nat Commun.* 2015. 6: 6579.

Chang, D. W., Hsieh, M. L., Chen, Y. M., Lin, T. F., Chang, J. S. Kinetics of cell lysis for *Microcystis aeruginosa* and *Nitzschia palea* in the exposure to β-cyclocitral. *J Hazard Mater.* 2011. 185: 1214–1220.

Choi, J. Y., Lee, S. Y. Microbial production as short-chain alkane. *Nature.* 2013: 1–6.

Curson, A. R., Liu, J., Bermejo Martínez, A., Green, R. T., Chan, Y., Carrión, O., Williams, B. T., Zhang, S. H., Yang, G. P., Bulman Page, P. C., Zhang, X. H., Todd, J. D. Dimethylsulfoniopropionate biosynthesis in marine bacteria and identification of the key gene in this process. *Nat Microbiol.* 2017. 2: 17009.

Van Durme, J., Goiris, K., De Winne, A., De Cooman, L., Muylaert, K. Evaluation of the volatile composition and sensory properties of five species of microalgae. *J Agric Food Chem.* 2013. 61: 10881–10890.

Fay, P. *The Blue-Greens (Cyanophyta-Cyanobacteria)*, 5th edition, 1983. Great Britain.

Feron, G., Dufossé, L., Souchon, I., Voilley, A., Spinnler, H.-E. The production of lactone by micro-organisms: A review with particular interest to the model: Ricinoleic. *Rec Res Dev Microbiol.* 1999. 3: 23–40.

Francisco, E. C., Franco, T. T., Wagner, R., Jacob-Lopes, E. Assessment of different carbohydrates as exogenous carbon source in cultivation of cyanobacteria. *Bioprocess Biosyst Eng.* 2014. 1: 2–11.

Giglio, S., Jiang, J., Saint, C. P., Cane, D. E., Monis, P. T. Isolation and characterization of the gene associated with geosmin production in cyanobacteria. *Environ Sci Technol.* 2008. 42: 8027–8032.

Graham, J. L., Loftin, K. A., Meyer, M. T., Ziegler, A. C. Cyanotoxin mixtures and taste-and-odor compounds in cyanobacterial blooms from the Midwestern United States. *Environ Sci Technol.* 2010. 44: 7361–7368.

Hasegawa, M., Nishizawa, A., Tsuji, K., Kimura, S., Harada, K.-I. Volatile organic compounds derived from 2-keto-acid decarboxylase in *Microcystis aeruginosa*. *Microbes Environ.* 2012. 27(4): 525–528.

Hosoglu, M. I. Aroma characterization of five microalgae species using solid-phase microextraction and gas chromatography-mass spectrometry/olfactometric. *Food Chem.* 2018. 240: 1210–1218.

Huang, H., Xu, X., Liu, X., Han, R., Liu, J., Wang, G. Distributions of four taste and odor compounds in the sediment and overlying water at different ecology environment in Taihu Lake. *Sci Rep.* 2018. 8: 6179.

Jacob-Lopes, E., Cacia Ferreira Lacerda, L. M., Franco, T. T. Biomass production and carbono dioxide fixation by *Aphanothece microscopica Nageli* in a bubble column photobioreactor. *Biochem Eng J.* 2008. 40: 27–34.

Jacob-Lopes, E., Franco, T. T. From oil refinery to microalgal biorefinery. *J CO2 Utilization.* 2013. 2: 1–7.

Jacob-Lopes, E., Gimenes Scoparo, C. H., Queiroz, M. I., Franco, T. T. Biotransformations of carbon dioxide in photobiorreactors. *Energ Convers Manage.* 2010. 51: 894–900.

Jerković, I., Marijanović, Z., Roje, M., Kús, P. M., Jokić, S., Čož-Rakovac, R. Phytochemical study of the headspace volatile organic compounds of fresh algae and seagrass from the Adriatic Sea (single point collection). *PLOS ONE.* 2018. 13(5): e0196462.

Lauritano, C., Martin, J., Cruz, M., Reyes, F., Romano, G., Ianora, A. First identification of marine diatoms with anti-tuberculosis activity. *Sci Rep.* 2018. 8: 1–10.

Lee, J., Rai, P. K., Jeon, Y. J., Kim, K. H., Kwon, E. E. The role of algae and cyanobacteria in the production and release of odorants in water. *Environ Pollut.* 2017. 227: 252–262.

Liato, V., Aïder, M. Geosmin as a source of the earthy-musty smell in fruits, vegetables and water: Origins, impact on foods and water, and review of the removing techniques. *Chemosphere.* 2017. 181: 9–18.

Meena, S., Rajeev Kumar, S., Dwivedi, V., Kumar Singh, A., Chanotiya, C. S., Akhtar, M. Q., Kumar, K., Kumar Shasany, A., Nagegowda, D. A. Transcriptomic insight into terpenoid and carbazole alkaloid biosynthesis, and functional characterization of two terpene synthases in curry tree (*Murraya koenigii*). *Sci Rep.* 2017. 7: 44126.

Milovanovic, I., Mišan, A., Simeunovic, J., Kova, D., Dubravka Jambrec, D., Anamarija Mandi, A. Determination of volatile organic compounds in selected strains of syanobacteria. *J Chem.* 2015. 1: 1–6.

Perez-Garcia, O., Bashan, Y. Microalgal heterotrophic and mixotrophic culturing for bio-refining: From metabolic routes to techno-economics. In: A. Proko, et al. (Ed.), *Algal Biorefineries: Algal Growth, Products and Optimization*, 1st edition. Springer International Publishing, Switzerland. 2015. 2: 61–131.

Saffarionpour, S., Ottens, M. Recent advances in techniques for flavor recovery in liquid food processing. *Food Eng Rev.* 2018. 10: 81–94.

Santos, A. B., Fernandes, A. F., Wagner, R., Jacob-Lopes, E., Zepka, L. Q. Biogeneration of volatile organic compounds produced by *Phormidium autumnale* in heterotrophic bioreactor. *J Appl Phycol.* 2016b. 1: 1–10.

Santos, A. B., Vieira, K. R., Nogara, G. P., Wagner, R., Jacob-Lopes, E., Zepka, L. Q. Biogeneration of volatile organic compounds by microalgae: Occurrence, be-havior, ecological implications and industrial applications. In: J. P. Moore. (Ed.). *Volatile Organic Compounds: Occurrence, Behavior and Ecological Implications*, 1st edition. Nova Science Publishers. 2016a. 1: 1–18.

Severo, I. A., Deprá, M. C., Barin, J. S., Wagner, R., De Menezes, C. R., Zepka, L. Q., Jacob-Lopes, E. Bio-combustion of petroleum coke: The process integration with photobioreactors. *Chem Eng Sci.* 2018. 177: 422–430.

Si, T., Luo, Y., Xiao, H., Zhao, H. Utilizing an endogenous pathway for 1-butanol production in *Saccharomyces cerevisiae*. *Metab Eng.* 2014. 22: 60–68.

Smith, K. M., Cho, K. M., Liao, J. C. Engineering Corynebacterium glutamicum fr isobutanol production. *Appl Microbiol Biotechnol.* 2010. 87: 1045–1055.

Tashiro, Y., Rodriguez, G. M., Atsumi, S. 2-Keto acids based biosynthesis pathways for renewable fuels and chemicals. *J Ind Microbiol Biotechnol.* 2015. 42(3): 361–373.

Try, S., Voilley, A., Chunhieng, T., De-Coninck, J., Waché, Y. Aroma compounds production by solid-state fermentation, importance of in situ gas-phase recovery systems. *Appl Microbiol Biotechnol.* 2018. 102: 7239–7255.

Watson, S. B. Cyanobacterial and eukaryotic algal odour compounds: Signals or by-products? A review of their biological activity. *Phycologia.* 2003. 42: 332–350.

Watson, S. B., Jüttner, F. Malodorous volatile organic sulfur compounds: Sources, sinks and significance in inland waters. *Crit Rev Microbiol.* 2017. 43: 210–237.

Watson, S. B., Monis, P., Baker, P., Giglio, S. Biochemistry and genetics of taste- and odor-producing cyanobacteria. *Harmful Algae.* 2016. 54: 112–127.

Yu, A.-Q., Juwono, N., Leong, K. P. S. S. J., Chang, M. W., Production of fatty acid-derived valuable chemicals in synthetic microbes. *Front Bioeng Biotechnol.* 2014. 2(78): 1–12.

Zhan, J., Rong, J., Wang, Q. Mixotrophic cultivation, a preferable microalgae cultivation mode for biomass/bioenergy production, and bioremediation, advances and prospect. *Int J Hydr Energy.* 2017. 42: 8505–8517.

Zhang, K., Sawaya, M. R., Eisenberg, D. S., Liao, J. C. Expanding metabolism for biosynthesis of non-natural alcohols. *Proc Natl Acad Sci USA.* 2008. 105: 20653–20658.

Zhou, L. V., Chen, J., Xu, J., Li, Y., Zhou, C., Yan, X. Change of volatile components in six microalgae with different growth phases. *J Sci Food Agric.* 2017. 97(3): 761–769.

10 Antidiabetic Properties of Brown Seaweeds (*Sargassum polycystum* C.Ag)

Suhaila Mohamed and Mahsa Motshakeri

CONTENTS

Introduction 99
Mechanisms of Action of Algal Extracts 100
The Effects of Brown Seaweed on Diabetic Pancreas, Liver, and Kidneys 102
Antidiabetic Effects of Other Seaweeds 103
Conclusion 104
References 104

BOX 10.1 SALIENT FEATURES

Seaweeds contain many nutrients and bioactive compounds and are beneficial for many human diseases. Certain seaweeds such as the brown *Sargassum polycystum* improved insulin sensitivity, blood sugar levels and blood lipid levels in experimental Type 2 diabetes mammals. Both the *Sargassum polycystum* alcoholic and water extracts, effectively reduced blood glucose and glycosylated hemoglobin (HbA1C), serum total cholesterol, triglyceride levels and plasma atherogenic index. In the experimental mammals, unlike Metformin, the brown seaweed extract did not change plasma insulin levels, but increased the responses to insulin. The brown seaweed ethanolic and aqueous extracts also prevented pathological lesions of the livers and kidneys in experimental diabetic mammals. Low doses of brown seaweed (equivalent 30mg/kg for humans) experimentally protected or restored the pancreas islets, reduced the liver and kidney damages in the diabetic rodents and may produce beneficial homeostatic effects. The brown seaweed extract consumption produced liver and kidney toxicity at high doses. The brown seaweed extract reduced dyslipidemia in Type 2 diabetic mammals, by being an insulin sensitizer, and at low doses may be organ protective against Type 2 diabetes complications, besides helping reduce atherogenic risk.

INTRODUCTION

Diabetes mellitus is a chronic metabolic disorder of more than 221 million people worldwide and is expected to reach 11% of the adult global population (Cheng, 2005). The majority (about 90%) of diabetes is of Type 2 (T2DM) or non-insulin-dependent diabetes mellitus (NIDDM), which is associated with uncontrolled hyperglycemia, attributed to high calorie malnutrition together with slightly malfunctioning insulin production due to changes in pancreatic β-cells functions, insulin secretions, and insulin insensitivity. The economic burden of diabetes is over $132 billion (medical costs, disability, work loss, and mortality) (Cheng, 2005). Diabetic individuals have a three times higher heart disease risk and five times higher stroke risk compared to normal (Peters et al., 2014).

Seaweeds are important to the marine environments as food, habitat, shelters, and for their global oxygen generation contribution. Seaweeds are sustainable sources of beneficial nutrients and bioactive compounds. Several seaweeds such as *Petalonia binghamiae*, *Sargassum cystoseria* (Phaeophyceae), *Padina gymnospora*, and *Spyridia fusiformis* (red) have hypoglycemic effects in experimental diabetic mammals (Mohamed et al., 2012). Marine algae and algal polysaccharides also showed hypocholesterolaemic effects in mammals (Mohamed et al., 2011). Among the very common brown seaweeds, *Sargassum polycystum* possess antioxidant and healing properties. *S. polycystum* is traditionally used to alleviate eczema, scabies and psoriasis, ulcers, lung diseases, renal dysfunction, viral hepatitis, heart ailments, and to promote bile secretion

(Motshakeri et al., 2013). *Sargassum polycystum* has anti-lipidemic, antioxidant, membrane stabilizing, drug metabolizing enzymes (protective), TNF-α suppressive, lipid peroxidation (preventive), hepatic enzymes (protective), and non-enzymatic antioxidant defense (enhancing) properties *in vivo* (Mohamed et al., 2012). The brown seaweed exclusive polyphenols, Phlorotannins, help mitigate T2DM via various molecular mechanisms and targets, involving enzymes (pancreatic, hepatic, and intestinal), glucose metabolism and transport, glucose-induced toxicity, and β-cell cytoprotection (Lopes et al., 2016).

Diabetes mellitus is an endocrine disorder characterized by defects in (i) carbohydrate, lipid, and protein metabolism, (ii) body antioxidant defence systems, (iii) cellular membranes, (iv) subcellular organelles, (v) DNA, and (vi) cell biochemistry. The resulting diabetic complications include coronary heart disease, stroke, retinopathy, nephropathy, liver disease, and peripheral neuropathy (Cade, 2008). Hyperglycemia increases the production of free radicals, induces oxidative injuries to the liver and carbohydrate metabolism disorders, that are evidenced by hepatocytes degenerations, pyknotic nuclei, cellular necrosis, and increased lipid accumulation (Giacco and Brownlee, 2010). However, livers have regenerative abilities after initial injuries. Plant-derived antioxidants are effective, safe, and economical therapeutics for diabetes management and organ protection against these damages. Brown seaweed was hepatoprotective under a high-fat/ high-cholesterol diet (Motshakeri et al., 2014). The consumption of a brown seaweed ethanolic or water extracts dose dependently reduced blood glucose, glycosylated haemoglobin (HbA1C) levels, and dyslipidemia in Type 2 diabetic animals. Brown seaweed apparently functioned as an insulin sensitizer in T2DM mammals, besides helping reduce atherogenic risk (Motshakeri et al., 2014).

MECHANISMS OF ACTION OF ALGAL EXTRACTS

The defects caused by Type 2 diabetes mellitus on fat, carbohydrate, and protein metabolism, can lead to microvascular or macrovascular complications. The hypoglycemic effects of brown seaweed were caused by enhanced insulin sensitivity for glucose uptake in insulin-target tissues. Insulin sensitizing agents together with lifestyle modification are effective for both delaying the onset of Type 2 diabetes and preserving β-cell. There is also a possibility that the brown seaweed may help regulate postprandial glucose by retarding carbohydrate digestion or absorption. Both the brown seaweed water (high dose) and ethanolic extracts (both doses) helped reduce fasting blood glucose and the HbA1C levels in experimental Type 2 diabetic mammals (Figure 10.1) (Motshakeri et al., 2013).

The prevention of enhancement of HbA1C levels in seaweed extract treated animals could be due to the antioxidative compounds, such as the photosynthetic pigments, various polyphenols, amino acids, proteins, minerals, and other organic complexes present in the seaweed. Brown seaweed is known to contain chlorophyll a, chlorophyll b, zeaxanthin, β-carotene, fucoxanthin and dinoxanthin (Matanjun et al., 2008). Dietary fucoxanthin (carotenoid), which is present in some brown seaweed, reduced blood glucose and plasma insulin levels in diabetic/obese KK-A^y mice, and lowered blood glucose and HbA1C in non-diabetic, high-fat diet rodents. Brown seaweed alcoholic extract possesses flavonoids, triterpenoids, and phenolic compounds that have a radical-scavenging, oxidation-reducing ability and anti-diabetic properties (Mohamed et al., 2012).

HbA1C is a very stable but reversible glycation reaction product between glucose and the erythrocytes haemoglobin protein which correlated well with blood glucose levels. The HbA1C level was significantly increased in non-treated diabetic mammals, and the brown seaweed extracts help decrease the HbA1C levels (Figure 10.1). Antioxidants and free-radical scavengers in brown seaweed help prevent or reverse oxidative protein glycation. Chronic hyperglycaemia induces the glycosylation of many proteins, including haemoglobin and the eye lens β-crystalline protein.

The gluco-toxicity and lipotoxicity caused by prolonged hyperglycemia, dyslipidemia, and hyperinsulinemia damages the malfunctioning β-cells further. Unlike metformin, the brown seaweed *Sargassum polycystum* supplementation showed no insulinotropic effects in experimental diabetic mammals (Figure 10.2). Metformin has insulinotropic affects on human islets at high glucose concentration, which may put additional stress to the Beta cells, but helped reduce advanced-glycated end products generation, hepatic glucose output, fatty acid synthesis, dyslipidemia and increase the antioxidative enzymes activities in the red blood cells to retard HbA1C formation (Nowotny et al., 2015).

Various brown seaweed water extracts contain sulphated polysaccharides, such as alginate, fucoidan, and laminaran, that possess antioxidant and free-radical quenching properties, which are related to their sulphates, anion groups, and molecular weights (Mohamed et al., 2012). Their very viscous soluble fibres passively

Antidiabetic Properties of Brown Seaweeds

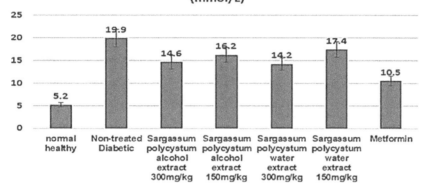

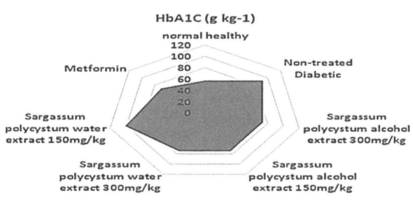

FIGURE 10.1 Sargassum polycystum, blood glucose and HbA1C levels of diabetic rats supplemented with brown seaweeds. (Adapted from Motshakeri et al., 2013.)

influence carbohydrate intestinal absorption and insulin response. Brown seaweed contains about 6 % soluble dietary fibers, 42 % minerals on a dry weight (DW) basis (macrominerals Na, K, Ca, and Mg; and trace minerals Fe, Zn, Cu, Se, and I), 0.029 % lipids, 5.4 crude protein, ~35 mg % vitamin C and 11 mg % α-tocopherol (Matanjun et al., 2009).

Increased blood glucose levels would subsequently increase blood TC (total cholesterol) and TG (triglyceride) levels, due to increased free fatty acids mobilization from adipose tissue and hormone-induced lipolysis. The brown seaweed extracts help reduce serum TG, and TC levels in experimental diabetic mammals (Figure 10.2). The atherogenic index of plasma (AIP) were decreased in all experimental diabetic mammals treated with brown seaweed extract. The AIP indicates the equilibrium between the atherogenic and protective lipoproteins which are related to the size of the pro- and anti-atherogenic lipoproteins. AIP values below 0.11 is low risk, AIP values between 0.11–0.21 is of medium risk and AIP values over the 0.21 point is high risk. The decrease in AIP values in experimental diabetic

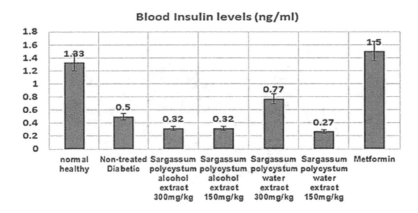

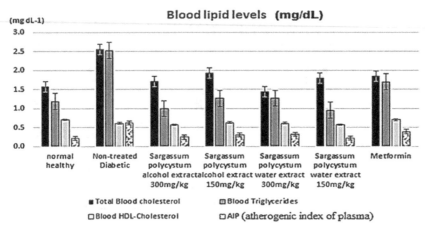

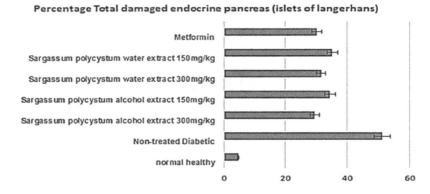

FIGURE 10.2 Blood insulin, blood lipid levels and percentage total damaged endocrine pancreas (islets of langerhans) of diabetic rats supplemented with brown seaweeds. (Adapted from Motshakeri et al., 2013 and Motshakeri et al., 2014.)

mammals by the brown seaweed extract (to near normal control rats values) indicated that the brown seaweed extracts were protective against atherosclerosis and CVD risk, which was reportedly better than metformin (Matanjun et al., 2010).

The brown seaweed extracts help reduce cholesterol and are beneficial for diabetics with high blood cholesterol and an atherosclerosis risk. Seaweeds water-soluble fraction or sulphated polysaccharides (e.g. alginate and fucoidan) were repeatedly reported to have hypocholesterolemic effects in mammals. The soluble polysaccharides probably reduce intestinal cholesterol absorption and bile acid reabsorption, while enhancing bile acid and cholesterol excretion (Matanjun et al., 2010).

THE EFFECTS OF BROWN SEAWEED ON DIABETIC PANCREAS, LIVER, AND KIDNEYS

Diabetic mammals can eventually develop pancreatic atrophy (islets shrinkage) and pancreatic injuries evidenced by pyknotic nuclei, acidophilic cytoplasm in the necrotic cells, and vacuolar changes in degenerative cells. Low doses of brown seaweed or metformin may reduce the severity of these injuries in diabetic individuals, and may even help the pancreas to recover towards normal conditions (Motshakeri et al., 2014).

Some seaweeds (e.g. *Sargassum fusiforme*) may have arsenic toxicity, hence it is essential to check their safety or efficacy of the seaweeds by observing any changes. The kidney and liver are crucial for the excretion and elimination of toxins from the body. The pancreas regulates micronutrient metabolism, and progressive degenerations of the pancreatic β-cells during diabetes development are often not easily detected. Any changes in the pancreatic islet structure, function, or size usually alters systemic metabolic insulin sensitivity, insulin secretion, and blood glycemic control (Motshakeri et al., 2014).

Final stage diabetes in humans frequently causes β-cell mass reduction, intra-islet deposition of fat and amyloid, pancreatic islets shrinkage (atrophy), irregular islets, cellular swelling, β-cell vacuolation, cell apoptosis, and necrosis. Habitual hyper-calorie diets often cause insulin resistance, with the eventual islets of langerhans degeneration and β-cell loss. These cells remain as thick layers of non-β peripheral cells in Type 2 diabetes development. Brown seaweed such as *Sargassum polycystum* may potentially reduce these injuries, by suppressing endocrine cells damage and necrotic cells (Figure 10.2). Beta-cell regeneration is one known mechanism by which plants and seaweeds (e.g. *Ulva rigida*) demonstrate anti-hyperglycaemic activity. Brown seaweed such as *Sargassum polycystum* may ameliorate β-cells death and recovery of the non-critically injured β-cells, via antioxidant-related mechanisms (Motshakeri et al., 2014).

The liver cellular structures of diabetic mammals are usually disoriented with deteriorations characterised by glycogen deposition, fatty changes (nucleus located at the peripheral cell membrane), pyknotic nuclei with acidophilic cytoplasm, hydropic swelling, hepatocytes disarrangement, micro-vesicular vacuolization, granular degeneration, and necrotic cells. These hepatocytes fatty degenerations are linked to insulin deficiency and the dysregulation of mitochondrial β-oxidation of fatty acids, causing the esterification of fatty acids to triglyceride in the cytoplasm, characterized by the multiple, triglyceride droplets within the hepatocytes. These diabetic injuries were attenuated by low doses of brown seaweed such as *Sargassum polycystum* or metformin in animal studies (Motshakeri et al., 2014).

Diabetes, over-nutrition, high calorie diet, obesity, and atherosclerosis are inflammatory conditions, which activate mobile macrophages like kupffer cells, that adhere to the periportal sinusoid endothelial lining. Kupffer cells can either mediate damage or protect regeneration and repair. The *Sargassum polycystum* ethanolic extract (200 mg/kg, body weight) improved the hepatic mitochondrial antioxidant defence system against free radicals, which could partly account for the enhancement in insulin sensitivity (Motshakeri et al., 2014). The hepato-protective properties of the brown seaweed *Sargassum polycystum* in non-diabetic conditions were also reported in rats induced with hepatitis, acetaminophen-related lipid peroxidation, or hypercalorie diet (Mohamed et al., 2011). Diabetes damages renal tissues by hyperglycemia and hyperlipidemia that result in degenerations in convoluted tubules in the cortex, and inflammation. Diabetic nephropathy, a common diabetes complication, is characterized by glomerular basement membrane thickening, hypertrophy, atrophy of the glomerular and tubular cells, glomerular hyperfiltration, extra-cellular matrix components accumulation in the glomerular mesangium and tubular interstitium, with the ultimate renal tissue and functions loss. These degenerations and necrosis were decreased by low doses of *Sargassum polycystum* extract in the diabetic animal study, but were increased at high doses indicating possible toxicity (Motshakeri et al., 2014)

ANTIDIABETIC EFFECTS OF OTHER SEAWEEDS

Seaweed consumption was shown to decrease the diabetes risk in men (a Korean national survey). Seaweed extracts from the brown *Petalonia binghamiae* or the red *Hypnea musciformis* show insulin-like actions, while the brown *Sargassum polycystum* extract demonstrated insulin sensitizing action in the cells of diabetic mammals (Motshakeri et al., 2013). *Petalonia binghamiae* and many other seaweeds stimulated the 3T3-L1 preadipocytes differentiation, that increased peroxisome proliferator-activated receptor gamma (PPARγ) transcriptional activities and influenced glucose uptake in mature adipocytes (Kang et al., 2012). The PPARγ improves insulin sensitivity for glucose and lipid metabolism. *Eklonia stolonifera* and sea tangle were also anti-antioxidative and anti-diabetic in mammals. Various seaweeds including *Ascophyllum nodosum*, *Pelvetia babingtonii,* and *Grateloupian elliptica,* contain phlorotannins and polyphenols that strongly inhibit α-glucosidase that help reduce postprandial hyperglycemia, noteworthy for Type 2 diabetics (Mohamed et al., 2012). The polyphenols in seaweeds stimulated cells glucose uptake (insulin sensitizer). Most seaweeds contain high levels of soluble dietary fibres (carrageenan, agar, alginates) that passively retard digestion and glucose absorption (Mohamed et al., 2012). Dietary brown seaweed did not prevent obesity in mice, but reduced insulin resistance (IR) and systemic inflammation, as well as in the adipose tissues and bone marrow-derived immune cells (Oh et al., 2016).

CONCLUSION

Based on the 'conversion of animal doses to human equivalent doses guidelines' from the Center for Drug Evaluation and Research (CDER) of the United States Food and Drug Administration (USFDA), the brown seaweed *Sargassum polycystum* extract equates to not more than 2 g /day extract for a 60 kg human (probably as functional food ingredients or as a health supplement). At the given dose, brown seaweed showed slightly less anti-hyperglycemic and anti-lipidemic activities in rats compared to metformin. Hence it functions more as a complementary therapy or health food supplement for insulin resistant or mildly-diabetic individuals, controlled on exercise and diet. Low doses of brown seaweed extracts were beneficial in alleviating histological injuries in experimental diabetic mammals' tissues and organs. At high doses it may be beneficial to the pancreas but may be toxic to the kidneys and liver of diabetic mammals.

REFERENCES

Cade, W.T., 2008. Diabetes-related microvascular and macrovascular diseases in the physical therapy setting. *Phys. Ther.* 88, 1322–1335. doi:10.2522/ptj.20080008

Cheng, D., 2005. Prevalence, predisposition and prevention of type II diabetes. *Nutr. Metab. (Lond).* 2, 29. doi:10.1186/1743-7075-2-29

Giacco, F., Brownlee, M., 2010. Oxidative stress and diabetic complications. *Circ. Res.* 107, 1058–1070. doi:10.1161/CIRCRESAHA.110.223545

Kang, S.I., Shin, H.S., Kim, H.M., Yoon, S.A., Kang, S.W., Kim, J.H., Ko, H.C., Kim, S.J., 2012. Petalonia binghamiae extract and its constituent fucoxanthin ameliorate high-fat diet-induced obesity by activating AMP-activated protein kinase. *J. Agric. Food Chem.* 60, 3389–3395. doi:10.1021/jf2047652

Lopes, G., Andrade, P.B., Valentão, P., 2016. Phlorotannins: Towards new pharmacological interventions for diabetes mellitus type 2. *Molecules* 22, 56. doi:10.3390/molecules22010056

Matanjun, P., Mohamed, S., Muhammad, K., Mustapha, N.M., 2010. Comparison of cardiovascular protective effects of tropical seaweeds, *Kappaphycus alvarezii*, *Caulerpa lentillifera*, and *Sargassum polycystum*, on high-cholesterol/high-fat diet in rats. *J. Med. Food* 13, 792–800. doi:10.1089/jmf.2008.1212

Matanjun, P., Mohamed, S., Mustapha, N.M., Muhammad, K., 2009. Nutrient content of tropical edible seaweeds, *Eucheuma cottonii*, *Caulerpa lentillifera* and *Sargassum polycystum*. *J. Appl. Phycol.* 21, 75–80. doi:10.1007/s10811-008-9326-4

Matanjun, P., Mohamed, S., Mustapha, N.M., Muhammad, K., Ming, C.H., 2008. Antioxidant activities and phenolics content of eight species of seaweeds from north Borneo. *J. Appl. Phycol.* 20, 367–373. doi:10.1007/s10811-007-9264-6

Mohamed, S., Hashim, S.N., Rahman, H.A., 2012. Seaweeds: A sustainable functional food for complementary and alternative therapy. *Trends Food Sci. Technol.* 23, 83–96.

Mohamed, S., Matanjun, P., Hashim, S.N., Rahman, H.A., Mustapha, N.M., 2011. Edible seaweeds: A functional food with organ protective and other therapeutic applications. In: *Seaweed: Ecology, Nutrient Composition and Medicinal Uses*. Nova Science Publishers, Inc., pp. 67–98.

Motshakeri, M., Ebrahimi, M., Goh, Y.M., Matanjun, P., Mohamed, S., 2013. *Sargassum polycystum* reduces hyperglycaemia, dyslipidaemia and oxidative stress via increasing insulin sensitivity in a rat model of type 2 diabetes. *J. Sci. Food Agric.* 93, 1772–1778. doi:10.1002/jsfa.5971

Motshakeri, M., Ebrahimi, M., Goh, Y.M., Othman, H.H., Hair-Bejo, M., Mohamed, S., 2014. Effects of brown seaweed (*Sargassum polycystum*) extracts on kidney, liver, and pancreas of type 2 diabetic rat model. *Evid.Based Complement. Alternat. Med.* 1–11. doi:10.1155/2014/379407

Nowotny, K., Jung, T., Höhn, A., Weber, D., Grune, T., 2015. Advanced glycation end products and oxidative stress in type 2 diabetes mellitus. *Biomolecules* 5, 194–222. doi:10.3390/biom5010194

Oh, J.H., Kim, J., Lee, Y., 2016. Anti-inflammatory and anti-diabetic effects of brown seaweeds in high-fat diet-induced obese mice. *Nutr. Res. Pract.* 10, 42–48. doi:10.4162/nrp.2016.10.1.42

Peters, S.A.E., Huxley, R.R., Woodward, M., 2014. Diabetes as risk factor for incident coronary heart disease in women compared with men: A systematic review and meta-analysis of 64 cohorts including 858,507 individuals and 28,203 coronary events. *Diabetologia* 57, 1542–1551. doi:10.1007/s00125-014-3260-6

11 Biologically Active Vitamin B_{12} from Edible Seaweeds

Tomohiro Bito, Fei Teng and Fumio Watanabe

CONTENTS

Abbreviations 105
Introduction 105
Vitamin B_{12} Content of Edible Seaweeds 106
Characterization of Vitamin B_{12} Compounds from Edible Seaweeds 107
Bioavailability of Vitamin B_{12} from Edible Seaweeds 108
Conclusion 108
Author Contributions 108
Notes 109
References 109

BOX 11.1 SALIENT FEATURES

Vitamin B_{12} has the largest molecular mass and the most complex structure among all vitamins. Vitamin B_{12} functions as coenzymes of methionine synthase and methylmalonyl-CoA mutase involved in methionine biosynthesis and in amino acid and fatty acid metabolism, respectively, in humans. Vitamin B_{12} is synthesized by certain bacteria and archaea only and not by the majority of plants. Animal-based, but not plant-based, foods are considered to be a major dietary B_{12} source. Thus, strict vegetarians or vegans who do not consume any animal-based foods are reportedly at a greater risk of developing vitamin B_{12} deficiency. The major symptoms of vitamin B_{12} deficiency include megaloblastic anemia and neuropathy. Although vitamin B_{12} is generally absent in plant-derived foods, we have identified edible seaweeds that naturally contain large vitamin B_{12} amounts. *Porphyra* spp. contain substantial vitamin B_{12} amounts, whereas other edible seaweeds contain none or only trace amounts. This chapter describes the characterization of vitamin B_{12} from edible seaweeds.

ABBREVIATIONS

LC/MS-MS: Liquid chromatography/electrospray ionization-tandem mass spectrometry
OH-B_{12}: Hydroxocobalamin

INTRODUCTION

Vitamin B_{12} (B_{12}) has the largest molecular mass (1355.4) and the most complex structure among all vitamins (Watanabe and Bito 2016). Scientific use of the term "vitamin B_{12}" is restricted to cyanocobalamin (CN-B_{12}), which has a lower axial ligand containing a cobalt-coordinated nucleotide (5,6-dimethylbenzimidazole) as the base (Figure 11.1).

In this chapter, B_{12} refers to all potentially biologically active B_{12} compounds. CN-B_{12} is found in dietary supplements and is readily converted to coenzyme forms, namely methylcobalamin (MeB$_{12}$) and 5'-deoxyadenosylcobalamin (AdoB$_{12}$), in the body (Watanabe and Bito 2016). Since most B_{12} compounds [e.g., hydroxocobalamin (OH-B_{12}), AdoB$_{12}$, and MeB$_{12}$] are chemically more labile than CN-B_{12} (Watanabe and Miyamoto 2003, Juzeniene and Nizauskaite 2013), food and biological samples are treated with KCN prior to analytical testing to convert their B_{12} compounds into a single and stable form (CN-B_{12}), which can then be analyzed using microbiological assay methods (Angyal 1996), thin-layer chromatography (Tanioka et al. 2008), high-performance liquid chromatography (Ueta et al. 2011), liquid chromatography/electrospray ionization-tandem mass spectrometry (LC/MS-MS), and so on (Bito et al. 2016).

MeB$_{12}$ is a coenzyme of methionine synthase (EC 2.1.1.13), which is involved in methionine biosynthesis, while AdoB$_{12}$ is a coenzyme of methylmalonyl-CoA mutase (EC 5.4.99.2), which is involved in amino acid and odd-chain fatty acid metabolism in animal

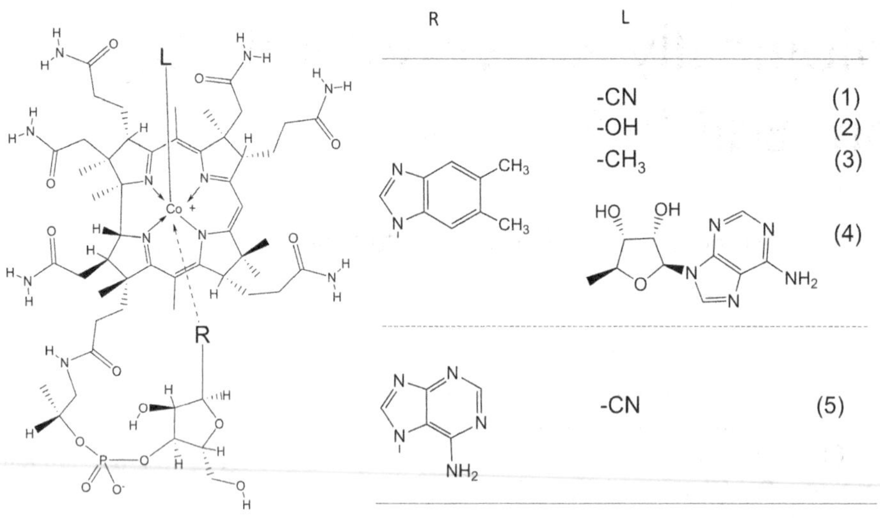

FIGURE 11.1 Structure of vitamin B_{12} and partial structures of vitamin B_{12}-related compounds (1) Cyanocobalamin (vitamin B_{12}), (2) hydroxocobalamin, (3) methylcobalamin, (4) 5'-deoxyadenosylcobalamin, and (5) pseudovitamin B_{12}.

cells (Bito and Watanabe 2016). Notably, there is a significant accumulation of methylmalonic acid and homocysteine during B_{12} deficiency, which is usually considered indicative of the vitamin deficiency (Bito and Watanabe 2016).

B_{12} is synthesized by certain bacteria and archaea only and not by the majority of plants. B_{12} accumulates in animal tissues from the food chain (Watanabe and Bito 2018). Thus, animal-based, but not plant-based, foods are considered to be a major dietary B_{12} source (Watanabe 2007). Thus, strict vegetarians or vegans who do not consume any animal-based foods are reportedly at a greater risk of developing B_{12} deficiency (Pawlak et al. 2013). We have identified edible seaweeds that naturally contain large B_{12} amounts (Watanabe et al. 2002, 2014). However, some algae have been shown to contain inactive corrinoid compounds, such as pseudovitamin B_{12} (PseudoB_{12}) (Figure 11.1) (Watanabe et al. 1999a, 2006, Miyamoto et al. 2006). This chapter summarizes the characterization and bioavailability of B_{12} compounds from edible seaweeds.

VITAMIN B_{12} CONTENT OF EDIBLE SEAWEEDS

Various types of edible seaweeds are available worldwide, and these are known to be rich in vitamins, minerals, and dietary fibers (Wells et al. 2017). The B_{12} content of foods is generally determined by bioassays using a B_{12}-requiring lactic acid bacterium, *Lactobacillus delbrueckii* subsp. *lactis* ATCC7830 (Watanabe 2007). The edible brown algae *Laminaria angustata* (Kombu), *Undaria pinnatifida* (Wakame), *Eisenia bicyclis* (Arame), and *Sargassum fusiforme* (Hijiki) contain no or trace B_{12} amounts (Watanabe et al. 2002) (Figure 11.2). Although only trace B_{12} amounts have been detected in some green and red algae [*Monostroma nitidum* (Aosa-nori), *Gelidium elegans* (Tengusa), and so on], *Enteromorpha prolifera* (Ao-nori) (Watanabe et al. 1999b) and *Porphyra yezoensis* (Susabi-nori) (Watanabe et al. 2000) have been reported to contain substantial B_{12} amounts.

Various species of *Porphyra*, which contain substantial B_{12} amounts (approximately 78 µg/100 g dry weight), are most commonly consumed in the form of dried nori products (Bito et al. 2017). In addition, various types of toasted nori products are commercially available in the local retail markets in Japan. The B_{12} content of these products has been reported to be approximately 58 µg per 100 g weight (Bito et al. 2017). These results clearly indicate that dried and toasted nori products contain substantial B_{12} amounts (Table 11.1).

In aquatic environments, algae appear to acquire B_{12} through a symbiotic relationship with B_{12}-synthesizing bacteria because half of all the algae require B_{12} (Croft et al. 2005). *Porphyra* spp. reportedly have the ability to take up and accumulate exogenous B_{12} (Yamada et al. 1996), which is derived through such microbial interactions. Even algae that do not require B_{12} for growth can accumulate substantial B_{12} amounts and can use it as a cofactor for B_{12}-dependent methionine synthase (Helliwell et al. 2016).

Members of the genus *Porphyra*, specifically *P. tenera* and *P. yesoensis*, are widely cultivated and consumed in Japan, Korea, and China (Levine and Sahoo 2010). In addition, *P. dentata* and *P. haitanesis* are cultivated in

FIGURE 11.2 Major edible seaweeds that are commercially available at local retail markets in Japan. Odd numbers represent dried seaweeds products, even numbers represent seaweed products soaked in water.

Korea and China, respectively (Niu et al. 2010), while wild *Porphyra* spp. are harvested and consumed in all these counties.

Dried products of *Porphyra* spp. (mainly nori) are packaged and commercially available in Japan and Korea. However, Chinese *Porphyra* preparations constitute a dried and flattened nori cake (zicai), which contains approximately 60.2 µg of B_{12} per 100 g dry weight (Fei Teng, Northeast Agricultural University and China, unpublished data).

CHARACTERIZATION OF VITAMIN B_{12} COMPOUNDS FROM EDIBLE SEAWEEDS

B_{12} compounds purified from various *Porphyra* (red algae) species and *E. prolifera* (a green alga) using thin-layer chromatography and high-performance liquid chromatography have been shown to be true B_{12} and not a corrinoid, which is inactive in humans (Watanabe et al. 1999b, 2000, Miyamoto et al. 2009). Although B_{12} has been identified as a major corrinoid compound

TABLE 11.1
Vitamin B_{12} Contents of Various *Porphyra* Products

Porphyra Products	Vitamin B_{12} Content (μg/100 g Dry Weight)	References
Dried (Japan)[a]	51.5	Watanabe et al. (2000)
Dried (Korea)	133.8	Miyamoto et al. (2009)
Dried (China)	60.2	Bito et al. (2017)
Dried (New Zealand)	28.5	Bito et al. (2017)
Toasted (Japan)	57.6	Bito et al. (2017)
Seasoned and toasted (Korea)	51.7	Miyamoto et al. (2009)
Laverbread (Welsh, UK)	28.5[b]	Bito et al. (2017)

[a] *Porphyra* has various names, being called nori (Japan), kim (Korea), zicai (China), and karengo (New Zealand). Vitamin B_{12} content was assayed in *Porphyra* products obtained from these countries using a microbiological method.

[b] (μg/100 g wet weight).

in Chinese dried products produced from *Porphyra* spp. using *Escherichia coli* 215 bioautography following thin-layer chromatography PseudoB_{12} has also been detected as a minor peak using LC-MS/MS (Fei Teng, Northeast Agricultural University, China, unpublished data).

BIOAVAILABILITY OF VITAMIN B_{12} FROM EDIBLE SEAWEEDS

The effects of feeding nori produced from *P. yezonensis* to B_{12}-deficient rats have been investigated to clarify the bioavailability of B_{12} in such dried products (Takenaka et al. 2001). When B_{12}-deficient rats began to excrete large amounts of methylmalonic acid in their urine, they were fed a diet supplemented with nori (10 μg/kg diet) for 20 days. Urinary methylmalonic acid excretion fell to undetectable levels, and hepatic B_{12} and, in particular, AdoB_{12} levels significantly increased. Moreover, B_{12} from the dried product of *P. tenera* was shown to be significantly absorbed by B_{12}-deficient rats (van den Berg et al. 1991). These results indicate that B_{12} from *Porphyra* spp. is bioavailable to rats.

In addition, a nutritional analysis of vegans who consumed diets, including brown rice and dried nori, for 4–10 years has suggested that the consumption of dried nori prevented B_{12} deficiency in this group (Suzuki 1995). Vegans who consume nori products have also been shown to exhibit higher serum or plasma B_{12} concentrations than those who do not consume these (Dagnelie et al. 1991, Rauma et al. 1995); however, mean corpuscular volume values deteriorate further in those consuming nori products (Dagnelie et al. 1991). On the contrary, when vegans consumed dried nori for 8 months, their total serum B_{12} and mean corpuscular volume were normal and serum holotranscobalamin (a B_{12}-transport protein), and homocysteine levels were within tolerable levels, but their methylmalonic acid values were elevated (Schwarz et al. 2014). Thus, the bioavailability of *Porphyra* B_{12} in humans remains to be determined in detail.

CONCLUSION

This chapter presented evidence that some edible seaweeds (specifically *Porphyra* spp.) contain substantial amounts of biologically available B_{12}, which is lacking in other plant-derived foods. Consumption of dried nori products of *Porphyra* spp. should be beneficial for human health because they contain various biologically active compounds and essential nutrients (Bito et al. 2017). Although many analytical studies have indicated that dried and toasted nori products contain substantial amounts of intact B_{12} (Watanabe et al. 1999b, 2000, Miyamoto et al. 2009), B_{12} may still be destroyed and/or converted into inactive B_{12} compounds during the drying process and subsequent storage (Yamada et al. 1999). Thus, we would like to emphasize that B_{12} compounds in nori products be precisely identified and quantified (using LC-MS/MS) if nori products are to be consumed as a sole B_{12} source.

AUTHOR CONTRIBUTIONS

All authors contributed equally to the preparation of this manuscript and have approved the final version.

NOTES

The authors declare that they have no competing financial interests. This work was supported by JSPS KAKENHI Grant number 25450168 (FW).

REFERENCES

Angyal, G. *Methods for the Microbiological Analysis of Selected Nutrients*. 1996. 63–65. Rockville, MD: AOAC International.

Bito, T., Bito, M., Asai, Y., Takenaka, S., Yabuta, Y., Tago, K., Ohnishi, M., Mizoguchi, T., and Watanabe, F. Characterization and quantitation of vitamin B12 compounds in various *Chlorella* supplements. *J Agric Food Chem*. 2016. 64: 8516–8524.

Bito, T., Teng, F., and Watanabe, F. Bioactive compounds of edible purple laver *Porphyra* sp. (nori). *J Agric Food Chem*. 2017. 65: 10685–10692.

Bito, T., and Watanabe, F. Biochemistry, function, and deficiency of vitamin B12 in *Caenorhabditis elegans*. *Exp Biol Med*. 2016. 241: 1663–1668.

Croft, M. T., Lawrence, A. D., Raux-Deery, E., Warren, M. J., and Smith, A. G. Algae acquire vitamin B12 through a symbiotic relationship with bacteria. *Nature*. 2005. 438: 90–93.

Dagnelie, P. C., van Staveren, W. A., and van den Berg, H. Vitamin B12 from algae appears not to be bioavailable. *Am J Clin Nutr*. 1991. 53: 695–697.

Helliwell, K. E., Lawrence, A. D., Holzer, A., Kudahl, U. J., Sasso, S., Kräutler, B., Scanlan, D. J., Warren, M. J., and Smith, A. G. Cyanobacteria and eukaryotic algae use different chemical variants of vitamin B12. *Curr Biol*. 2016. 26: 999–1008.

Juzeniene, A., and Nizauskaite, Z. Photodegradation of cobalamins in aqueous solutions and in human blood. *J Photochem Photobiol B* 2013. 122: 7–14.

Levine, I. A., and Sahoo, D. *Porphyra: Harvesting Gold from the Sea*. 2010. Delhi, India: I.K. International Publishing House Pvt. Ltd.

Miyamoto, E., Tanioka, Y., Nakao, T., Barla, F., Inui, H., Fujita, T., Watanabe, F., and Nakano, Y. Purification and characterization of a corrinoidcompound in an edible cyanobacterium *Aphanizomenon flos-aquae* as a nutritional supplementary food. *J Agric Food Chem*. 2006. 54: 9604–9607.

Miyamoto, E., Yabuta, Y., Kwak, C. S., Enomoto, T., and Watanabe, F. Characterization of vitamin B12 compounds from Korean purple laver (*Porphyra* sp.) products. *J Agric Food Chem*. 2009. 57: 2793–2796.

Niu, J. F., Chen, Z. F., Wang, G. C., and Zhou, B. C. Purification of phycoerythrin from *Porphyra yezoensis* Ueda (Bangiales, Rhodophyta) using expanded bed absorption. *J Appl Phycol*. 2010. 22: 25–31.

Pawlak, R., Parrott, S. J., Raj, S., Cullum-Dugan, D., and Lucus, D. How prevalent is vitamin B12 deficiency among vegetarians? *Nutr Rev*. 2013. 71: 110–117.

Rauma, A. L., Törrönen, R., Hänninen, O., and Mykkänen, H. Vitamin B12 status of long-term adherents of a strict uncooked vegan diet ("living food diet") is compromised. *J Nutr*. 1995. 125: 2511–2515.

Schwarz, J., Dschietzig, T., Schwarz, J., Dura, A., Nelle, E., Watanabe, F., Wintgens, K. F., Reich, M., and Armbruster, F. P. The influence of a whole food vegan diet with nori algae and wild mushrooms on selected blood parameters. *Clin Lab*. 2014. 60: 2039–2050.

Suzuki, H. Serum vitamin B12 levels in young vegans who eat brown rice. *J Nutr Sci Vitaminol*. 1995. 41: 587–594.

Takenaka, S., Sugiyama, S., Ebara, S., Miyamoto, E., Abe, K., Tamura, Y., Watanabe, F., Tsuyama, S., and Nakano, Y. Feeding dried purple laver (nori) to vitamin B12-deficient rats significantly improves vitamin B12 status. *Br J Nutr*. 2001. 85: 699–703.

Tanioka, Y., Yabuta, Y., Miyamoto, E., Inui, H., and Watanabe, F. Analysis of vitamin B12 in food by silica gel 60 TLC and bioautography with vitamin B12-dependent *Escherichia coli* 215. *J Liq Chrom Rel Technol*. 2008. 31: 1977–1985.

Ueta, K., Takenaka, S., Yabuta, Y., and Watanabe, F. Broth from canned clams is suitable for use as an excellent source of free vitamin B12. *J Agric Food Chem*. 2011. 59: 12054–12058.

Van den Berg, H., Brandsen, L., and Sinkeldam, B. J. Vitamin B12 content and bioavailability of spirulina and nori in rats. *J Nutr Biochem*. 1991. 2: 314–318.

Watanabe, F. Vitamin B12 sources and bioavailability. *Exp Biol Med*. 2007. 232: 1266–1274.

Watanabe, F., and Bito, T. Corrinoids in food and biological samples. *Front Nat Prod Chem*. 2016. 2: 229–244.

Watanabe, F., and Bito, T. Vitamin B12 sources and microbial interaction. *Exp Biol Med*. 2018. 243: 148–158.

Watanabe, F., Katsura, H., Takenaka, S., Fujita, T., Abe, K., Tamura, Y., Nakatsuka, T., and Nakano, Y. Pseudovitamin B12 is the predominate cobamide of an algal health food, spirulina tablets. *J Agric Food Chem*. 1999b. 47: 4736–4741.

Watanabe, F., and Miyamoto, E. Hydrophilic vitamins. In: *Handbook of Thin-Layer Chromatography*, 3rd ed. revised and expanded. 2003. pp. 589–605. New York, NY: Marcel Dekker.

Watanabe, F., Miyamoto, E., Fujita, T., Tanioka, Y., and Nakano, Y. Characterization of a corrinod compound in the edible (blue-green) algae, suizenji-nori. *Biosci Biotechnol Biochem*. 2006. 70: 3066–3068.

Watanabe, F., Takenaka, S., Katsura, H., Masumder, S. A., Abe, K., Tamura, Y., and Nakano, Y. Dried green and purple lavers (nori) contain substantial amounts of biologically active vitamin B12 but less of dietary iodine relative to other edible seaweeds. *J Agric Food Chem*. 1999a. 47: 2341–2343.

Watanabe, F., Takenaka, S., Katsura, H., Miyamoto, E., Abe, K., Tamura, Y., Nakatsuka, T., and Nakano, Y. Characterization of a vitamin B12 compound in the edible purple laver, *Porphyra yezoensis*. *Biosci Biotechnol Biochem*. 2000. 64: 2712–2715.

Watanabe, F., Takenaka, S., Kittaka-Katsura, H., Ebara, S., and Miyamoto, E. Characterization and bioavailability of vitamin B12-compounds from edible algae. *J Nutr Sci Vitaminol*. 2002. 48: 325–331.

Watanabe, F., Yabuta, Y., Bito, T., and Teng, F. Vitamin B12-containing plant food sources for vegetarians. *Nutrients*. 2014. 6: 1861–1873.

Wells, M. L., Potin, P., Craigie, J. S., Raven, J. A., Merchant, S. S., Helliwell, K. E., Smith, A. G., Camire, M. E., and Brawley, S. H. Algae as nutritional and functional food sources: Revisiting our understanding. *J Appl Phycol*. 2017. 29: 949–982.

Yamada, K., Yamada, Y., Fukuda, M., and Yamada, S. Bioavailability of dried asakusanori (*Porphyra tenera*) as a source of cobalamin (vitamin B12). *Int J Vitam Nutr Res*. 1999. 69: 412–418.

Yamada, S., Sasa, M., Yamada, K., and Fukuda, M. Release and uptake of vitamin B12 by Asakusanori (*Porphyra tenera*) seaweed. *J Nutr Sci Vitaminol*. 1996. 42: 507–515.

12 Potentials and Challenges in the Production of Microalgal Pigments with Reference to Carotenoids, Chlorophylls, and Phycobiliproteins

Delia B. Rodriguez-Amaya and Iriani R. Maldonade

CONTENTS

Introduction 111
Carotenoids 112
Astaxanthin 113
β-carotene 114
Lutein 114
Chlorophylls 115
Phycobiliproteins 116
Conclusion and Future Perspectives 116
References 116

BOX 12.1 SALIENT FEATURES

Stimulated by the current effort to substitute artificial food colorants with natural pigments, accentuated with the potential health benefits of the latter, production of pigments from microalgae has drawn considerable attention in recent years, especially in relation to carotenoids (e.g., astaxanthin, β-carotene, lutein), chlorophylls, and phycobiliproteins (phycocyanin, phycoerythrin). In spite of the perceived advantages, the large scale microalgal production of pigments is still limited. Only two microalgal carotenoids are produced industrially: β-carotene of *Dunaliella salina* and astaxanthin of *Haematococcus pluvialis*. Potential for commercial expansion is high, but more work is needed to make the process cost-competitive and sustainable. The entire process consists of cell cultivation, biomass harvesting, cell disruption, pigment extraction, purification, and storage; all these operations are being optimized. Selection of species with appropriate production time and yield, improving the design of the culture system, and optimizing the culture conditions to maximize biomass and pigment productions at low cost are being widely investigated. Making downstream processing more efficient and affordable is also being pursued. The scientific and technology advancements, however, have been mostly achieved at the laboratory stage, needing scaling up to the industrial level. Research findings indicate that microalgal pigment production can be commercially viable and advantageous.

INTRODUCTION

Microalgal pigments have drawn considerable research interest, resulting in a voluminous literature. This short chapter focuses on publications of the last decade.

Microalgae contain carotenoids (yellow, orange, and red), chlorophylls (green), and phycobiliproteins (blue, red and orange) (Figure 12.1), with the first two as the major pigments. Because of food safety concerns, consumer preference for natural products, and health benefits, microalgal pigments are gaining greater commercial importance.

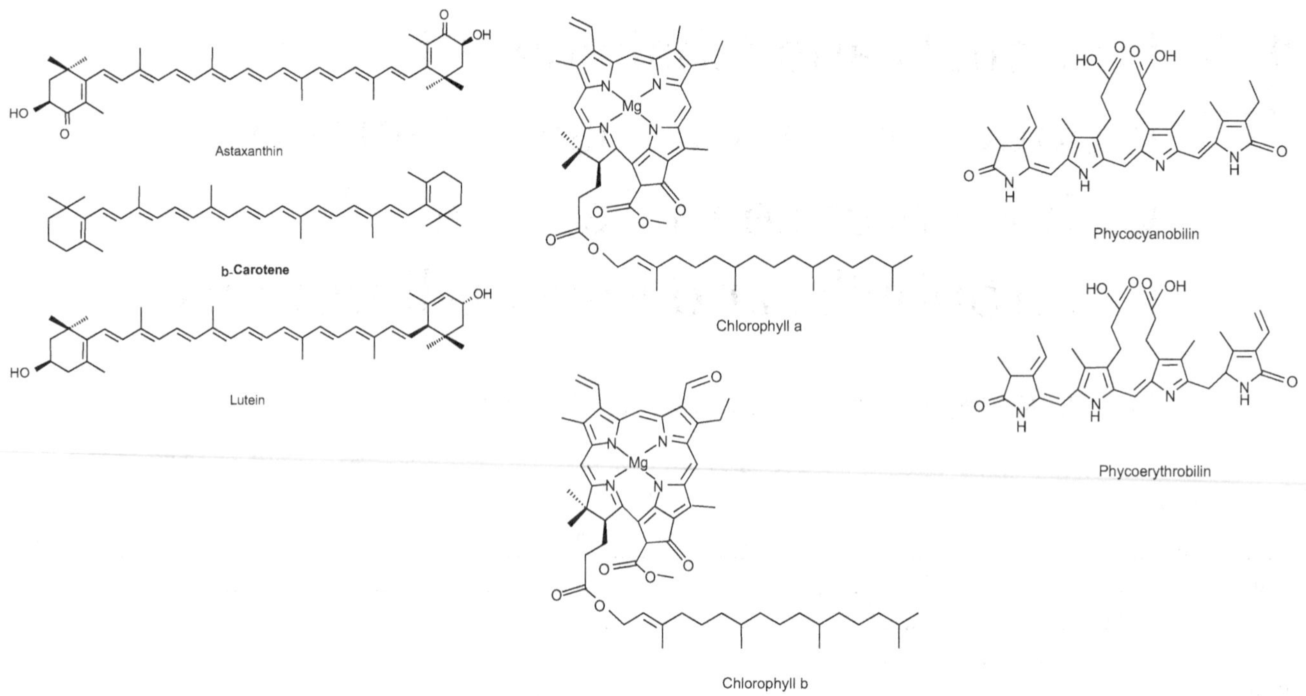

FIGURE 12.1 Structures of important microalgal pigments.

CAROTENOIDS

The carotenoid market is expected to grow from approximately $1.5 billion in 2017 to $2.0 billion by 2022 (BCC Research, 2018). Presently, the major carotenoids of market interest are capsanthin, astaxanthin, β-carotene, lutein, lycopene, canthaxanthin, and zeaxanthin. Commercial carotenoids are mostly products of chemical synthesis, but some are also produced by extraction from plant and microbial sources.

Two microalgal carotenoids have been produced industrially: β-carotene of *Dunaliella salina* (since the 1980s), and astaxanthin of *Haematococcus pluvialis* (since the 1990s). The main carotenoids of microalgae are astaxanthin, β-carotene, lutein, lycopene, zeaxanthin, violaxanthin, and fucoxanthin. The first three are the most studied and are highlighted in this chapter.

Microalgal production of carotenoids have many advantages (Fernandez-Sevilla et al., 2010; Gupta et al., 2015; Lin et al., 2015; Perez-Lopez, 2014; Ambati et al., 2018b). The growth rate is 5–10 times that of higher plants. Only a small area of non-arable land is needed, avoiding competition for resources with conventional agriculture. Wastewater can be used as growing medium, eliminating dependency on costly culture media. Microalgal cultivation may clean the environment through CO_2 sequestration and wastewater treatment. It can be done year-round and can adapt to a wide range of conditions and climates.

In spite of the perceived advantages, the large scale production of carotenoids from microalgae is still limited. It is not yet considered sufficiently cost-effective to compete with chemical synthesis and extraction from plant-based sources.

The entire process consists of cell cultivation, biomass harvesting, cell disruption, pigment extraction, purification, and storage. To be cost-effective, all these operations should be optimized. The selection of species with appropriate production time and yield of biomass and pigment is the first hurdle.

Microalgae can be cultivated in open or closed systems under autotrophic, heterotrophic, or mixotrophic conditions. The cost of open ponds is much lower than that of closed photobioreactors. Cheapest to construct and maintain, consuming less energy, the raceway ponds are the most commercially employed. Open ponds have the following drawbacks: uneven light intensity, higher evaporation losses, greater requirement for water, reduced temperature control, poor mass transfer rates, diffusion of CO_2 to the atmosphere, and high risk of contamination (Singh and Sharma, 2012). Though much more expensive, closed photobioreactors are subject to minimal risk of contamination, have better control of culture conditions, require less light and area, and can result in higher biomass and pigment yield (Gupta et al., 2015). Flat or tubular photobioreactors are the basic design structures.

Downstream processing, especially cell harvesting and disruption, is challenging. Centrifugation is fast, efficient, suitable for most strains, and most widely applied for harvesting microalgae (Gong and Bassi, 2016). However, it demands high capital cost and continuous energy investment. Filtration is time and energy consuming for small size microalgae. Gravity sedimentation is inexpensive, but requires a long time for small, uniformly suspended cells when no additional flocculants are present. Chemical harvesting methods require lower capital investment and consume much less energy but are not as efficient as mechanical methods. Flocculation has received much attention because of the possibility of treating large-scale microalgal suspensions at a lower cost.

The existence of rigid cell walls in many microalgal species poses a barrier to full recovery of the pigments. The efficiency of cell disruption depends on the microalgae species, particularly the cell membrane composition and morphology (Günerken et al., 2015).

Methods for breaking cells are mechanical or non-mechanical. Mechanical methods are best for industry, but energy consumption is high (D'Alessandro and Antoniosi Filho, 2016). Non-mechanical methods consume less energy and disrupt cell membrane uniformly, but take more time, may affect product quality, and are more difficult to control.

Conventional solvent extraction is a simple approach but suffers from inherent limitations: use of large volumes of often toxic solvents, long extraction times, low efficiency and selectivity, and disposal of potentially hazardous solvents to the environment (Michalak and Chojnacka, 2015; Poojary et al., 2016). Recent trends in the extraction of carotenoids from microalgae consist of innovative techniques, such as supercritical fluid, microwave-assisted, ultrasound-assisted, enzyme-assisted, and pressurized liquid extractions. Advantages of these techniques are: extraction of biologically active compounds without degradation or loss of activity (Michalak and Chojnacka, 2015), environmentally friendly, higher extraction yield, and shorter process time. Poojary et al. (2016) reviewed the various methods, denoting differences in yield, selectivity, and economic and environmental sustainability.

Aside from carotenoids, microalgae are sources of other valuable compounds, including vitamins, lipids, proteins, and polysaccharides. Several compounds can be extracted from the same biomass and used as commodities in biotechnology industries. The biorefinery approach (production of biofuels along with other high value co-products) for sustainable and economically feasible production of marketable microalgal products is gaining wide support (e.g., Chew et al., 2017; Koller et al., 2014; Shah et al., 2016).

ASTAXANTHIN

Astaxanthin is a xanthophyll responsible for the reddish color of salmon, trout, and crustaceans. Synthetic astaxanthin is a mixture of three isomers, (3S, 3′S), (3R, 3′S), and (3R, 3′R). Astaxanthin from *Haematococcus* is exclusively a (3S, 3′S) isomer and is mostly esterified with fatty acids (Higuera-Ciapara et al., 2006).

Astaxanthin is widely used in the nutraceutical, pharmaceutical, cosmetic, feed, and food industries. As a feed additive, it is essential for salmon and trout farming for pigmentation and adequate growth and reproduction (Higuera-Ciapara et al., 2006). Potential toxicity has been raised for synthetic astaxanthin (Li et al., 2011; Milledge, 2011), due to its different stereochemistry and potential carryover of synthesis intermediates (Shah et al., 2016). Synthetic astaxanthin can only be used as a fish feed additive (Li et al., 2011). *Haematococcus* astaxanthin now has a significant, well-established, and growing market in the nutraceutical area (Borowitzka, 2013).

Biological activities attributed to astaxanthin include antioxidant, anti-lipid peroxidation, anti-inflammation, immunomodulation, suppression of LDL-cholesterol oxidation, raising HDL-cholesterol, improving blood flow, UV-light protection, anti-insulin resistance, and neuroprotective properties (Grimmig et al., 2017; Kidd, 2011; Ambati et al., 2014; Yang et al., 2013). Astaxanthin therefore has potential action against various diseases including cancer, diabetes, cardiovascular, gastrointestinal, liver, neurodegenerative, and skin diseases.

The astaxanthin market reached $288.7 million in 2017 and is projected to grow to $426.9 million by 2022 (BCC Research, 2018). The market value of astaxanthin usually varies from $2500 to $7000/kg (Koller et al., 2014; Milledge, 2011; Perez-Lopez et al., 2014). More than 95% consists of synthetic astaxanthin because of its lower production cost (about $1000/kg). The microalgal alternative accounts for <1% of commercialized astaxanthin. Li et al. (2011) surveyed the costs and affirmed that China can sell astaxanthin at a lower price (US$718/kg).

H. pluvialis, the major source of microalgal astaxanthin, accumulates up to 3.8% astaxanthin on the dry weight basis. Some companies established along this production line are Cyanotech (Hawaii, USA), Algatechnologies (Israel), Astareal (Japan), and Algacan (Canada) (Gong and Bassi, 2016).

The production of astaxanthin by *H. pluvialis* is more problematic than the production of β-carotene by

D. salina. Since *Haematococcus* is a freshwater alga, it is susceptible to contamination with other organisms, making open-air culture extremely difficult. The optimal conditions for cell growth differ from those of astaxanthin biosynthesis. Thus, a two-stage process is usually adopted for industrial production (Wan et al., 2014; Wichuk et al., 2014). The first stage is done photoautotrophically under controlled culture conditions suitable for microalgal growth, in either tubular, bubble column, or airlift photobioreactors. The following stage, which is less prone to contamination, is done in open cultivation ponds, subject to environmental and nutrient stress to stimulate carotenoid accumulation.

Numerous studies have been undertaken to boost astaxanthin production, modifying the design of the culture system, and optimizing the culture conditions (e.g., Poonkum et al., 2015; Yang et al., 2016; Yoo et al., 2012; Zhang et al., 2014, 2017). Besides improving technology, stress must be induced to increase carotenoid concentration (Li et al., 2011), such as nitrogen deprivation, strong light intensity, salt stress, and phosphate deficiency (e.g., Kang et al., 2007; Sarada et al., 2012).

C. zofingiensis is considered the best alternative source of astaxanthin (Liu et al., 2014). The astaxanthin-rich *H. pluvialis* is already used commercially, but slow growth, low biomass yield, high light requirement, and vulnerability to contamination limit its industrial application. *C. zofingiensis* has a high growth rate and high cell density that can be achieved through heterotrophic glucose-fed cultivation.

Investigating large-scale astaxanthin production by *H. pluvialis* in two European cities, Panis and Carreon (2016) concluded that for Europe, natural astaxanthin is not a competitive alternative to the synthetic form for aquaculture. However, astaxanthin production by *H. pluvialis* cultivated in sites characterized by high solar radiation and high temperatures was considered an attractive venture. Haque et al. (2016) reported intensified astaxanthin production using bioethanol wastewater streams as potential 'green' media to culture *H. pluvialis*.

The thick cell wall of *H. pluvialis* is made up of sporopollenin-like material, which hinders astaxanthin extraction. Methods applied to cell-wall breakage and astaxanthin extraction from *Chlorella* and *Haematococcus* were discussed by Kim et al. (2016), comparing efficiency, energy consumption, type and dosage of solvent, biomass concentration, toxicity, scalability, and synergistic combinations.

HCl pretreatment facilitated astaxanthin extraction (Dong et al., 2014). Compared to solvent and ultrasound-assisted extractions, microwave-assisted extraction gave the highest astaxanthin recovery (Ruen-ngam et al., 2010). Astaxanthin was also efficiently extracted using supercritical fluid extraction with ethanol or sunflower oil as co-solvent (Pan et al., 2012; Wang et al., 2012).

β-CAROTENE

β-carotene is ubiquitous in foods, conferring a yellow-to-orange color. As a food color additive, it finds application in dairy products, cakes, soups, margarine, and confectionary. The market value of β-carotene, which was $259.4 million in 2017, should reach $335.5 million in 2022 (BCC, 2018).

β-carotene is one of the most investigated food bioactives. Aside from its well-established provitamin A activity, it has been associated with reduced risk for certain types of cancer (e.g., breast, esophageal, gastric, pancreatic cancers) (Aune et al., 2012; Ge et al., 2013; Zhou et al., 2016). However, high-dose, long-term supplementation with β-carotene is harmful (higher incidence of lung cancer) to smokers and asbestos workers (ATBC, 1994; Omenn et al., 1996).

Industrial production of *D. salina* β-carotene is a well-established technology. It is in operation in Australia, China, India, Israel, Japan, and the United States (Borowitzka, 2013; Ambati et al., 2018a). *D. salina* has the advantage of not having a cell wall and producing high levels of β-carotene (up to 14%). Its halotolerant nature allows it to be cultivated in open saline mass culture (e.g., coastal sea water), relatively free of competing microorganisms.

Accumulation of β-carotene in *D. salina* can be enhanced by high light intensity, extreme temperatures, high salinity, and nutrient limitation (Çelekli et al., 2014; Lamers et al., 2010, 2012; Michalak and Chojnacka, 2015).

LUTEIN

Lutein, a yellow xanthophyll, is the main carotenoid of green vegetables, and together with zeaxanthin, the predominant pigments of egg yolk and corn. The worldwide market for lutein was worth about US$235 million in 2017; it is expected to reach US$293 million in 2018 (BCC-Research, 2018).

Marketed lutein is mainly extracted from marigold flowers. It is used as food and feed colorant, the latter to improve pigmentation of the bird's skin and egg yolk. It is also further processed into health foods.

Lutein and zeaxanthin, which accumulate in the macula of the human retina, are credited with reduced risk of age-related macular degeneration (Ma et al., 2012) and cataracts (Ma et al., 2014). Preferentially taken up into neural tissue, these carotenoids are also linked with better cognitive performance (Feeney et al., 2017).

One of the most studied microalgae for lutein production is *Muriellopsis* sp. with a lutein content of 0.4% to 0.6% per dry biomass (Blanco et al., 2007). The growth rate and the capacity to endure harsh environmental conditions make *S. almeriensis* a promising lutein source (Sánchez et al., 2008a,b). *Coccomyxa onubensis* is also a potential lutein producer. It survives at low pH and high concentrations of heavy metals, preventing outdoor cultivation from being contaminated with undesired microorganisms while producing high concentrations of lutein (Vaquero et al., 2012).

Fernandez-Sevilla et al. (2010) affirmed that microalgae could compete with marigold even without counting on any of the improvements in microalgal technology in the near future. Lin et al. (2015) compared the different stages of lutein production from marigold flowers and microalgae. Microalgae had faster growth rates and 3–4 times higher lutein yield. Marigold needed more land and water, but required less nutrients (N, P, K) and less energy.

In spite of the advantages of lutein production by microalgae, no microalgal lutein product has reached the market. Technical obstacles cited are: lutein values not high enough to be economically feasible on an industrial scale, high harvesting cost, and high energy demand for cell disruption and extraction. Rapid cultivation of algal strains with high lutein content and efficient downstream processing at affordable costs are deemed necessary. Lutein productivity can be achieved by selecting adequate species, obtaining high lutein yielding mutants, and optimizing culture conditions.

Optimization of culture systems and conditions for enhanced lutein production has been widely performed (e.g., Chen et al., 2018; Dineshkumar et al., 2016; Ho et al., 2015; Jeon et al., 2014). Augmenting lutein production by stress conditions is difficult (Cordero et al., 2011; Mulders et al., 2014).

Unlike astaxanthin and β-carotene, which are secondary carotenoids, lutein is a primary carotenoid required for the structure and function of the light-harvesting complexes in photosynthesis. As such, stress conditions do not necessarily promote lutein accumulation (Ho et al., 2015; Xie et al., 2013). In *C. zofingiensis*, while strong light and nitrogen starvation enhanced significantly the synthesis of astaxanthin, lutein decreased under these conditions (Cordero et al., 2011).

In *Desmodesmus* sp. F2 and *Coelastrella* sp. F50, cultivated under outdoor tropical conditions, lutein content did not change significantly in microalgae grown in different carbon sources or different seasons (Chiu et al., 2016). The major factor influencing productivity was duration of effective irradiance.

Coagulation employed to harvest cells enhanced separation efficiency, but lowered the lutein yield from *Chlorella* sp. ESP-6 cells, attributed to hydroxyl-amine interactions between lutein and chitosan molecules (Utomo et al., 2013).

For *Scenedesmus obliquus* CNW-N, cell disruption was more efficient with a bead-beater than with an autoclave or sonicator (Chan et al., 2013). Enzymatic pretreatment increased the extraction of lutein from *C. vulgaris* (Deenu et al., 2013). High-pressure cell disruption increased recovery of lutein from *C. sorokiniana* MB-1 (Chen et al., 2016). Supercritical CO_2 with ethanol as entrainer was efficient in extracting lutein from *C. vulgaris* (Kitada et al., 2009).

CHLOROPHYLLS

Chlorophylls a and b, the typical green pigments of higher plants, are the main chlorophylls of microalgae. Microalgal chlorophyll and carotenoid contents are generally higher than those of plant sources (Villarruel-López et al., 2017).

Chlorophyll is used as a food coloring agent and as an additive in pharmaceutical and cosmetic products (Hosikian et al., 2010). In Japan this colorant is commonly added to chewing gums, shakes, beverages, desserts, ice creams (Begum et al., 2016). Chlorophyll has antioxidant, anti-mutagenic, anticancer, anti-inflammatory, and wound-healing properties (Balder et al., 2006; Ferruzi and Blakeslee, 2007).

Da Silva Ferreira and Sant'Anna (2017) reviewed the strategies to maximize chlorophyll production in microalgae, including variation in light intensity, culture agitation, and changes in temperature and nutrient availability. These factors affect chlorophyll concentration in a species-specific manner. Phytohormones stimulated *S. quadricauda* cell growth, biomass production, as well as accumulation of chlorophyll a and carotenoids (Kozlova et al., 2017).

Hosikian et al. (2010) reviewed chlorophyll production from microalgae cultivation to chlorophyll fractionation and purification. Extraction begins with dewatering and desalting of the highly dilute culture, followed by extraction of the chlorophyll from the dried biomass, then fractionation to separate the chlorophyll pigments and derivatives. Supercritical fluid extraction appeared superior to solvent extraction.

PHYCOBILIPROTEINS

Phycobiliproteins (phycobilins bonded to protein) are deep-colored, fluorescent proteins that can be found in microalgae, especially cyanobacteria (e.g., *Spirulina (Arthrospira) platensis*) and rhodophytes (e.g., *Galdieria sulphuraria)*. Generally, they are classified into phycoerythrin (red), phycocyanin (blue), allophycocyanin (bluish green), and phycoerythrocyanin (orange).

Phycocyanin derived from *S. platensis* is used as a colorant in foods, such as chewing gum, ice sherberts, popsicles, candies, soft drinks, dairy products, and jellies (Begum et al., 2016). Phycoerythrin derived from *Phorphyridium aerugineum* and *S. platensis* is also used in color confectionary, gelatin, fermented milk products, ice creams, cake decoration, and milk shakes. Because of their intense fluorescence, phycobilins are employed as indicators in clinical and immunological analysis. They can also have applications in the cosmetic industry. Dietary supplements of *Chlorella* and *Spirulina* have been used in powder form, after drying and sieving the biomass (Borowitzka, 2013).

Phycobiliproteins are credited with antioxidant, anticancer, anti-inflammatory, immunomodulatory, hepatoprotective, nephroprotective, and neuroprotective effects (Eriksen, 2008; Mysliwa-Kurdziel and Solymosi, 2017).

The production of microalgal phycobiliproteins, although highlighted in recent years, still faces technological obstacles. The various aspects of C-phycoyanin production were reviewed by Eriksen (2008). Kuddus et al. (2013) discussed the sources and production of C-phycocyanin, techniques for extraction and purification, and potential industrial applications. Begum et al. (2016) reviewed the different factors that affect pigment production. Cuellar-Bermudez et al. (2015) summarized the existing methods to extract and purify microalgal metabolites, including phycoerythrin and phycocyanin.

To Manirafasha et al. (2016), microalgal phycobilin production still confronts two major obstacles: (1) the upstream and downstream processing is hindered by selection of suitable strains, bioreactor design, culture conditions, etc.; (2) purification of phycobiliproteins from microalgae is still low.

Statistical optimization of four variables (nitrate, phosphate, pH, and light intensity) increased the maximum phycobiliprotein contents in *Synechocystis* sp. PCC 6701 over 400% (Hong and Lee, 2008). Low light intensity and high initial biomass concentration led to increased C-phycocyanin accumulation in *Spirulina* (Xie et al., 2015). Fed-batch cultivation proved to be an effective strategy to further enhance C-phycocyanin production, which also required nitrogen-sufficient condition and other nutrients. Biomass production and phycocyanin accumulation were enhanced in *Spirulina* in fed-batch cultivation by adding sodium glutamate and succinic acid (Manirafasha et al., 2018).

Methods for C-phycocyanin isolation were reported by various authors and summarized in a review by Sekar and Chandramohan (2008). These involve various steps: cell disruption, primary isolation, purification, drying, and characterization of the end products. Ammonium sulfate is utilized to precipitate phycocyanin from extracts of microalgae biomass (Kuddus et al., 2013). Kumar et al. (2014) and Sonani et al. (2017) purified phycocyanins from *S. platensis* CCC540 and *Synechococcus* sp., respectively, in this manner.

Because of the difficulty and high cost of extraction and purification, the dried cells, without the extraction of the pigments, can be used in commercial formulations. Spray-dried *Porphyridium cruentum* biomass, for example, was used in chewing gum as a natural colorant (Toker, 2018). Dry biomass of *Spirulina* has been used in food and beverages for blue colouring.

CONCLUSION AND FUTURE PERSPECTIVES

Uncertainty about the safety of synthetic colorants and increasing evidence for the health-promoting effects of natural pigments provide an excellent opportunity for microalgal pigments to expand in the global market. Intense research has been carried out on improving the design of the culture system and optimizing the culture conditions to maximize biomass and pigment productions at low cost. Making downstream processing more efficient and affordable has also been pursued. The scientific and technology advancements, however, have been mostly achieved at the laboratory stage, needing scaling up to the industrial level. Current findings indicate that microalgal pigment production can be cost-effective and sustainable, rendering this process commercially viable.

REFERENCES

Ambati, R. R., Gogisetty, D., Aswathanarayana, R. G., Ravi, S., Bikkina, P. N., Bo, L., and Yuepeng, S. Industrial potential of carotenoid pigments from microalgae: Current trends and future prospects. *Crit. Rev. Food Sci. Nutr.* 2018a: 1–22. doi: 10.1080/10408398.2018.1432561.

Ambati, R. R., Gogisetty, D., Aswathnarayana Gokare, R., Ravi, S., Bikkina, P. N., Su, Y., and Lei, B. *Botryococcus* as an alternative source of carotenoids and its possible applications – An overview. *Crit. Rev. Biotechnol.* 2018b. 38 (4): 541–558.

Ambati, R. R., Phang, S. M., Ravi, S., and Aswathanarayana, R. G. Astaxanthin: Sources, extraction, stability, biological activities and its commercial applications – A review. *Mar. Drugs.* 2014. 12: 128–152.

ATBC (Alpha-Tocopherol, Beta-Carotene Cancer Prevention Group). The effect of vitamin E and beta-carotene on the incidence of lung cancer and other cancers in male smokers. *N. Engl. J. Med.* 1994. 330: 1029–1035.

Aune, D., Chan, D. S., Vieira, A. R., Navarro Rosenblatt, D. A., Vieira, R., Greenwood, D. C., and Norat, T. Dietary compared with blood concentrations of carotenoids and breast cancer risk: A systematic review and meta-analysis of prospective studies. *Am. J. Clin. Nutr.* 2012. 96: 356–373.

Balder, H. F., Vogel, J., Jansen, M. C., Weijenberg, M. P., van den Brandt, P. A., Westenbrink, S., van der Meer, R., and Goldbohm, R. A. Heme and chlorophyll intake and risk of colorectal cancer in the Netherlands cohort study. *Cancer Epidemiol. Biomarkers Prev.* 2006. 15 (4): 717–725.

BCC Research. *The Global Market for Carotenoids*. 2018.

Begum, H., Yusoff, F. M. D., Banerjee, S., Khatoon, H., and Shariff, M. Availability and utilization of pigments from microalgae. *Crit. Rev. Food Sci. Nutr.* 2016. 56: 2209–2222.

Blanco, A. M., Moreno, J., Del Campo, J. A., Rivas, J., and Guerrero, M. G. Outdoor cultivation of lutein-rich cells of *Muriellopsis* sp. in open ponds. *Appl. Microbiol. Biotechnol.* 2007. 73: 1259–1266.

Borowitzka, M. A. High-value products from microalgae – Their development and commercialisation. *J. Appl. Phycol.* 2013. 25: 743–756.

Çelekli, A., Bozkurt, H., and Dönmez, G. Predictive modeling of β-carotene accumulation by *Dunaliella salina* as a function of pH, NaCl, and irradiance. *Russ. J. Plant Physiol.* 2014. 61: 215–223.

Chan, M.-C., Ho, S.-H., Lee, D.-J., Chen, C.-Y., Huang, C.-C., and Chang, J.-S. Characterization, extraction and purification of lutein produced by an indigenous microalga *Scenedesmus obliquus* CNW-N. *Biochem. Eng. J.* 2013. 78: 24–31.

Chen, C. Y., Jesisca, Hsieh, C., Lee, D. J., Chang, C. H., and Chang, J. S. Production, extraction and stabilization of lutein from microalga *Chlorella sorokiniana* MB-1. *Bioresour. Technol.* 2016. 200: 500–505.

Chen, C. Y., Lu, I. C., Nagarajan, D., Chang, C. H., Ng, I. S., Lee, D. J., and Chang, J. S. A highly efficient two-stage cultivation strategy for lutein production using heterotrophic culture of *Chlorella sorokiniana* MB-1-M12. *Bioresour. Technol.* 2018. 253: 141–147.

Chew, K. W., Yap, J. Y., Show, P. L., Suan, N. H., Juan, J. C., Ling, T. C., Lee, D. J., and Chang, J. S. Microalgae biorefinery: High value products perspectives. *Bioresour. Technol.* 2017. 229: 53–62.

Chiu, P. H., Soong, K., and Chen, C.-N. N. Cultivation of two thermotolerant microalgae under tropical conditions: Influences of carbon sources and light duration on biomass and lutein productivity in four seasons. *Bioresour. Technol.* 2016. 212: 190–198.

Cordero, B. F., Obraztsova, I., Couso, I., Leon, R., Vargas, M. A., and Rodriguez, H. Enhancement of lutein production in *Chlorella sorokiniana* (Chorophyta) by improvement of culture conditions and random mutagenesis. *Mar. Drugs.* 2011. 9: 1607–1624.

Cuellar-Bermudez, S. P., Aguilar-Hernandez, I., Cardenas-Chavez, D. L., Ornelas-Soto, N., Romero-Ogawa, M. A., and Parra-Saldivar, R. Extraction and purification of high-value metabolites from microalgae: Essential lipids, astaxanthin and phycobiliproteins. *Microb. Biotechnol.* 2015. 8: 190–209.

D'Alessandro, E. B., and Antoniosi Filho, N. R. Concepts and studies on lipid and pigments of microalgae: A review. *Renew. Sust. Energ. Rev.* 2016. 58: 832–841.

Da Silva Ferreira, V., and Sant'Anna, C. Impact of culture conditions on the chlorophyll content of microalgae for biotechnological applications. *World J. Microbiol. Biotechnol.* 2017. 33: 20. doi: 10.1007/s11274-016-2181-6.

Deenu, A., Naruenartwongsakul, S., and Kim, S. M. Optimization and economic evaluation of ultrasound extraction of lutein from *Chlorella vulgaris*. *Biotechnol. Bioprocess Eng.* 2013. 18: 1151–1162.

Dineshkumar, R., Subramanian, G., Dash, S. K., and Sen, R. Development of an optimal light-feeding strategy coupled with semi-continuous reactor operation for simultaneous improvement of microalgal photosynthetic efficiency, lutein production and CO_2 sequestration. *Biochem. Eng. J.* 2016. 113: 47–56.

Dong, S., Huang, Y., Zhang, R., Wang, S., and Liu, Y. Four different methods comparison for extraction of astaxanthin from green alga *Haematococcus pluvialis*. *Sci. World J.* 2014. 2014. doi: 10.1155/2014/694305.

Eriksen, N. T. Production of phycocyanin-a pigment with applications in biology, biotechnology, foods and medicine. *Appl. Microbiol. Biotechnol.* 2008. 80: 1–14.

Feeney, J., O'Leary, N., Moran, R., O'Halloran, A. M., Nolan, J. M., Beatty, S., Young, I. S., and Kenny, R. A. Plasma lutein and zeaxanthin are associated with better cognitive function across multiple domains in a large population-based sample of older adults: Findings from the Irish Longitudinal Study on Aging. *J. Gerontol.* 2017. 72: 1431–1436.

Fernández-Sevilla, J. M., Acién Fernández, F. G., and Molina Grima, E. Biotechnological production of lutein and its applications. *Appl. Microbiol. Biotechnol.* 2010. 86: 27–40.

Ferruzzi, M. G., and Blakeslee, J. Digestion, absorption, and cancer preventative activity of dietary chlorophyll derivatives. *Nutr. Res.* 2007. 27: 1–12.

Ge, X. X., Xing, M. Y., Yu, L. F., and Shen, P. Carotenoid intake and esophageal cancer risk: A meta-analysis. *Asian Pac. J. Cancer Prev.* 2013. 14: 1911–1918.

Gong, M., and Bassi, A. Carotenoids from microalgae: A review of recent developments. *Biotechnol. Adv.* 2016. 34: 1396–1412.

Grimmig, B., Kim, S. H., Nash, K., Bickford, P. C., and Douglas Shytle, R. Neuroprotective mechanisms of astaxanthin: A potential therapeutic role in preserving cognitive function in age and neurodegeneration. *GeroScience*. 2017. 39: 19–32.

Günerken, E., D'Hondt, E., Eppink, M. H. M., Garcia-Gonzalez, L., Elst, K., and Wijffels, R. H. Cell disruption for microalgae biorefineries. *Biotechnol. Adv*. 2015. 33: 243–260.

Gupta, P. L., Lee, S. M., and Choi, H. J. A mini review: Photobioreactors for large scale algal cultivation. *World J. Microbiol. Biotechnol*. 2015. 31: 1409–1417.

Haque, F., Dutta, A., Thimmanagarib, M., and Chiang, Y. W. Intensified green production of astaxanthin from *Haematococcus pluvialis*. *Food Bioprod. Process*. 2016. 99: 1–11.

Higuera-Ciapara, I., Félix-Valenzuela, L., and Goycoolea, F. M. Astaxanthin: A review of its chemistry and applications. *Crit. Rev. Food Sci. Nutr*. 2006. 46: 185–196.

Ho, S. H., Xie, Y., Chan, M. C., Liu, C. C., Chen, C. Y., Lee, D. J., Huang, C. C., and Chang, J. S. Effects of nitrogen source availability and bioreactor operating strategies on lutein production with *Scenedesmus obliquus* FSP-3. *Bioresour. Technol*. 2015. 184: 131–138.

Hong, S.-J., and Lee, C.-G. Statistical optimization of culture media for production of phycobiliprotein by *Synechocystis* sp. PCC 6701. *Biotechnol. Bioprocess. Eng*. 2008. 13: 491–498.

Hosikian, A., Lim, S., Halim, R., and Danquah, M. K. Chlorophyll extraction from microalgae: A review on the process engineering aspects. *Int. J. Chem. Eng*. 2010. 2010: 1–11. doi: 10.1155/2010/391632.

Jeon, J. Y., Kwon, J. S., Kang, S. T., Kim, B. R., Jung, Y., Han, J. G., Park, J. H., and Hwang, J. K. Optimization of culture media for large-scale lutein production by heterotrophic *Chlorella vulgaris*. *Biotechnol. Prog*. 2014. 30: 736–743.

Kang, C. D., Lee, J. S., Park, T. H., and Sim, S. J. Complementary limiting factors of astaxanthin synthesis during photoautotrophic induction of *Haematococcus pluvialis*: C/N ratio and light intensity. *Appl. Microbiol. Biotechnol*. 2007. 74: 987–994.

Kidd, P. Astaxanthin, cell membrane nutrient with diverse clinical benefits and anti-aging potential. *Altern. Med. Rev*. 2011. 16: 355–364.

Kim, D. Y., Vijayan, D., Praveenkumar, R., Han, J. I., Lee, K., Park, J. Y., Chang, W. S., Lee, J. S., and Oh, Y. K. Cell-wall disruption and lipid/astaxanthin extraction from microalgae: *Chlorella* and *Haematococcus*. *Bioresour. Technol*. 2016. 199: 300–310.

Kitada, K., Machmudah, S., Sasaki, M., Goto, M., Nakashima, Y., Kumamoto, S., and Hasegawa, T. Supercritical CO_2 extraction of pigment components with pharmaceutical importance from *Chlorella vulgaris*. *J. Chem. Technol. Biotechnol*. 2009. 84: 657–661.

Koller, M., Muhr, A., and Braunegg, G. Microalgae as versatile cellular factories for valued products. *Algal Res*. 2014. 6: 52–63.

Kozlova, T. A., Hardy, B. P., Krishna, P., and Levin, D. B. Effect of phytohormones on growth and accumulation of pigments and fatty acids in the microalgae *Scenedesmus quadricauda*. *Algal Res*. 2017. 27: 325–334.

Kuddus, M., Singh, P., Thomas, G., and Al-Hazimi, A. Recent developments in production and biotechnological applications of C-phycocyanin. *Biomed. Res. Int*. 2013. 2013: 742859. doi: 10.1155/2013/742859.

Kumar, D., Dhar, D. W., Pabbi, S., Kumar, N., and Walia, S. Extraction and purification of C-phycocyanin from *Spirulina platensis* (CCC540). *Indian J. Plant Physiol*. 2014. 19: 184–188.

Lamers, P. P., Janssen, M., de Vos, R. C., Bino, R. J., and Wijffels, R. H. Carotenoid and fatty acid metabolism in nitrogen–starved *Dunaliella salina*, a unicellular green microalga. *J. Biotechnol*. 2012. 162: 21–27.

Lamers, P. P., van de Laak, C. C., Kaasenbrood, P. S., Lorier, J., Janssen, M., de Vos, R. C., Bino, R. J., and Wijffels, R. H. Carotenoid and fatty acid metabolism in light-stressed *Dunaliella salina*. *Biotechnol. Bioeng*. 2010. 106: 638–648.

Li, J., Zhu, D., Niu, J., Shen, S., and Wang, G. An economic assessment of astaxanthin production by large scale cultivation of *Haematococcus pluvialis*. *Biotechnol. Adv*. 2011. 29 (6): 568–574.

Lin, J. H., Lee, D. J., and Chang, J. S. Lutein production from biomass: Marigold flowers versus microalgae. *Bioresour. Technol*. 2015. 184: 421–428.

Liu, J., Sun, Z., Gerken, H., Liu, Z., Jiang, Y., and Chen, F. *Chlorella zofingiensis* as an alternative microalgal producer of astaxanthin: Biology and industrial potential. *Mar. Drugs*. 2014. 12: 3487–3515.

Ma, L., Dou, H. L., Wu, Y. Q., Huang, Y. M., Huang, Y. B., Xu, X. R., Zou, Z. Y., and Lin, X. M. Lutein and zeaxanthin intake and the risk of age-related macular degeneration: A systematic review and meta-analysis. *Br. J. Nutr*. 2012. 107: 350–359.

Ma, L., Hao, Z. X., Liu, R. R., Yu, R. B., Shi, Q., and Pan, J. P. A dose–response meta-analysis of dietary lutein and zeaxanthin intake in relation to risk of age-related cataract. *Graefes Arch. Clin. Exp. Ophthalmol*. 2014. 252: 63–70.

Manirafasha, E., Murwanashyaka, T., Ndikubwimana, T., Rashid Ahmed, N., Liu, J., Lu, Y., Zeng, X., Ling, X., and Jing, K. Enhancement of cell growth and phycocyanin production in *Arthrospira* (*Spirulina*) *platensis* by metabolic stress and nitrate fed-batch. *Bioresour. Technol*. 2018. 255: 293–301.

Manirafasha, E., Ndikubwimana, T., Zeng, X., Lu, Y., and Jing, K. Phycobiliprotein: Potential microalgae derived pharmaceutical and biological reagent. *Biochem. Eng. J*. 2016. 109: 282–296.

Michalak, I., and Chojnacka, K. Algae as production systems of bioactive compounds. *Eng. Life Sci*. 2015. 15: 160–176.

Milledge, J. J. Commercial application of microalgae other than as biofuels: A brief review. *Rev. Environ. Sci. Biotechnol*. 2011. 10: 31–41.

Mulders, K. J., Lamers, P. P., Martens, D. E., and Wijffels, R. H. Phototrophic pigment production with microalgae: Biological constraints and opportunities. *J. Phycol.* 2014. 50: 229–242.

Mysliwa-Kurdziel, B., and Solymosi, K. Phycobilins and phycobiliproteins used in food industry and medicine. *Mini Rev. Med. Chem.* 2017. 17: 1173–1193.

Omenn, G. S., Goodman, G. E., Thornquist, M. D., Balmes, J., Cullen, M. R., Glass, A., Keogh, J. P., Meyskens, F. L., Valanis, B., Williams, J. H., Barnhart, S., and Hammar, S. Effects of a combination of beta carotene and vitamin A on lung cancer and cardiovascular disease. *N. Engl. J. Med.* 1996. 334: 1150–1155.

Pan, J.-L., Wang, H.-M., Chen, C.-Y., and Chang, J.-S. Extraction of astaxanthin from *Haematococcus pluvialis* by supercritical carbon dioxide fluid with ethanol modifier. *Eng. Life Sci.* 2012. 12: 638–647.

Panis, G., and Carreon, J. R. Commercial astaxanthin production derived by green alga *Haematococcus pluvialis*: A microalgae process model and a techno-economic assessment all through production line. *Algal Res.* 2016. 18: 175–190.

Pérez-López, P., González-García, S., Jeffryes, C., Agathos, S. N., McHugh, E., Walsh, D. J., Murray, P. M., Moane, S., Feijoo, G., and Moreira, M. T. Life cycle assessment of the production of the red antioxidant carotenoid astaxanthin by microalgae: From lab to pilot scale. *J. Clean. Prod.* 2014. 64: 332–344.

Poojary, M. M., Barba, F. J., Aliakbarian, B., Donsì, F., Pataro, G., Daniel, A., Dias, D. A., and Juliano, P. Innovative alternative technologies to extract carotenoids from microalgae and seaweeds. *Mar. Drugs.* 2016. 14: 214. doi: 10.3390/md14110214.

Poonkum, W., Powtongsook, S., and Pavasant, P. Astaxanthin induction in microalga *H. pluvialis* with flat panel airlift photobioreactors under indoor and outdoor conditions. *Prep. Biochem. Biotechnol.* 2015. 45: 1–17.

Ruen-ngam, D., Shotipruk, A., and Pavasant, P. Comparison of extraction methods for recovery of astaxanthin from *Haematococcus pluvialis*. *Sep. Sci. Technol.* 2010. 46: 64–70.

Sánchez, J. F., Fernández, J. M., Acién, F. G., Rueda, A., Pérez-Parra, J., and Molina, E. Influence of culture conditions on the productivity and lutein content of the new strain *Scenedesmus almeriensis*. *Proc. Biochem.* 2008b. 43: 398–405.

Sánchez, J. F., Fernández-Sevilla, J. M., Acién, F. G., Cerón, M. C., Pérez-Parra, J., and Molina-Grima, E. Biomass and lutein productivity of *Scenedesmus almeriensis*: Influence of irradiance, dilution rate and temperature. *Appl. Microbiol. Biotechnol.* 2008a. 79: 719–729.

Sarada, R., Ranga Rao, A., Sandesh, B. K., Dayananda, C., Anila, N., Chauhan, V. S., and Ravishankar, G. A. Influence of different culture conditions on yield of biomass and value added products in microalgae. *Dyn. Biochem. Proc. Biotechnol. Mol. Biol.* 2012. 6: 77–85.

Sekar, S., and Chandramohan, M. Phycobiliproteins as a commodity: Trends in applied research, patents and commercialization. *J. Appl. Phycol.* 2008. 20: 113–136.

Shah, M. M. R., Liang, Y., Cheng, J. J., and Daroch, M. Astaxanthin-producing green microalga *Haematococcus pluvialis*: From single cell to high value commercial products. *Front. Plant Sci.* 2016. 7: 531. doi: 10.3389/fpls.2016.00531.

Singh, R. N., and Sharma, S. Development of suitable photobioreactor for algae production – A review. *Renew. Sust. Energ. Rev.* 2012. 16: 2347–2353.

Sonani, R. R., Patel, S., Bhastana, B., Jakharia, K., Chaubey, M. G., Singh, N. K., and Madamwar, D. Purification and antioxidant activity of phycocyanin from *Synechococcus* sp. R42DM isolated from industrially polluted site. *Bioresour. Technol.* 2017. 245: 325–331.

Toker, O. S. *Porphyridumcruentum* as a natural colorant in chewing gum. *Food Sci. Technol.* 2018. doi: 10.1590/fst.41817.

Utomo, R. P., Chang, Y. R., Lee, D. J., and Chang, J. S. Lutein recovery from *Chlorella* sp. ESP-6 with coagulants. *Bioresour. Technol.* 2013. 139: 176–180.

Vaquero, I., Ruiz-Domínguez, M. C., Márquez, M., and Vílchez, C. Cu mediated biomass productivity enhancement and lutein enrichment of the novel microalga *Coccomyxa onubensis*. *Process Biochem.* 2012. 47: 694–700.

Villarruel-López, A., Ascencio, F., and Nuño, K. Microalgae, a potential natural functional food source – A review. *Pol. J. Food Nutr. Sci.* 2017. 67: 251–264.

Wan, M., Zhang, J., Hou, D., Fan, J., Li, Y., Huang, J., and Wang, J. The effect of temperature on cell growth and astaxanthin accumulation of *Haematococcus pluvialis* during a light–dark cyclic cultivation. *Bioresour. Technol.* 2014. 167: 276–283.

Wang, L., Yang, B., Yan, B., and Yao, X. Supercritical fluid extraction of astaxanthin from *Haematococcus pluvialis* and its antioxidant potential in sunflower oil. *Innov. Food Sci. Emerg. Technol.* 2012. 13: 120–127.

Wichuk, K., Brynjólfsson, S., and Fu, W. Biotechnological production of value-added carotenoids from microalgae: Emerging technology and prospects. *Bioengineered.* 2014. 5: 204–208.

Xie, Y., Ho, S. H., Chen, C.-N. N., Chen, C. Y., Ng, I. S., Jing, K. J., Chang, J. S., and Lu, Y. Phototrophic cultivation of a thermo-tolerant *Desmodesmus* sp. for lutein production: Effects of nitrate concentration, light intensity and fed-batch operation. *Bioresour. Technol.* 2013. 144: 435–444.

Xie, Y., Jin, Y., Zeng, X., Chen, J., Lu, Y., and Jing, K. Fed-batch strategy for enhancing cell growth and C-phycocyanin production of *Arthrospira* (*Spirulina*) *platensis* under phototrophic cultivation. *Bioresour. Technol.* 2015. 180: 281–287.

Yang, Y., Kim, B., and Lee, J.-Y. Astaxanthin structure, metabolism, and health benefits. *J. Hum. Nutr. Food Sci.* 2013. 1: 1003.

Yang, Z., Cheng, J., Li, K., Zhou, J., and Cen, K. Optimizing gas transfer to improve growth rate of *Haematococcus pluvialis* in a raceway pond with chute and oscillating baffles. *Bioresour. Technol.* 2016. 214: 276–283.

Yoo, J. J., Choi, S. P., Kim, B. W., and Sim, S. J. Optimal design of scalable photo-bioreactor for phototropic culturing of *Haematococcus pluvialis*. *Bioprocess. Biosyst. Eng.* 2012. 35: 309–315.

Zhang, W., Wang, J., Wang, J., and Liu, T. Attached cultivation of *Haematococcus pluvialis* for astaxanthin production. *Bioresour. Technol.* 2014. 158: 329–335.

Zhang, Z., Huang, J. J., Sun, D., Lee, Y., and Chen, F. Two-step cultivation for production of astaxanthin in *Chlorella zofingiensis* using a patented energy-free rotating floating photobioreactor (RFP). *Bioresour. Technol.* 2017. 224: 515–522.

Zhou, Y., Wang, T., Meng, Q., and Zhai, S. Association of carotenoids with risk of gastric cancer: A meta-analysis. *Clin. Nutr.* 2016. 35: 109–116.

13 Potential Health and Nutraceutical Applications of Astaxanthin and Astaxanthin Esters from Microalgae

Ambati Ranga Rao and Gokare A. Ravishankar

CONTENTS

Introduction 122
Origin and Occurrence of Astaxanthin 122
Life Cycle of *H. Pluvialis* and Carotenoid Accumulation 123
Biomass and Astaxanthin Production through Photobioreactor Cultivation 123
Technology of Breaking Aplanospores of *H. Pluvialis* or Effluxing of Pigment for Downstream Processing 123
Carotenoid Composition in *H. Pluvialis* 124
Commercial Production of Astaxanthin...124
Astaxanthin for Salmon and Trout Feeds...124
Safety of Astaxanthin..126
Bioavailability of Astaxanthin 127
Human Clinical Studies 129
Astaxanthin as Source of Health and Nutraceutical Applications 129
Astaxanthin Offers Protection from Ultraviolet Radiation 129
Astaxanthin Supports Immune System 130
Astaxanthin in Prevention of Inflammatory Disorders 130
Astaxanthin Prevents Cardiovascular Disease 130
Astaxanthin Prevents Ulcers and Gastric Injury 130
Astaxanthin for Diabetics 130
Astaxanthin Inhibits Cancer 131
Astaxanthin Prevents Neurodegenerative Diseases 131
Conclusion 131
Acknowledgment 133
References 133

BOX 13.1 SALIENT FEATURES

Microalgae are a potential source of bioactive molecules such as lipids, fatty acids, carotenoids, proteins, minerals, amino acids, hydrocarbons, carbohydrates and so on for commercial applications such as the food, feed, nutraceutical, pharmaceutical and cosmeceutical industries. Among bioactive molecules from algae, carotenoids are gaining recognition in the global market for food applications. As per the literature, several algal species such as *Haematococcus pluvialis, Dunaliella salina, Chlorococcum sps., Botryococcus braunii* and *Spirulina platensis* have been used to produce carotenoids, and more work is required to make the process cost effective. The global market for astaxanthin for use in various commercial applications is estimated to reach $1.1 billion by 2020. In this context, the current chapter is focused on the production of astaxanthin and its esters from *H. pluvialis*. Astaxanthin and its esters are accumulated in large amounts in *H. pluvialis*. Astaxanthin acts as a strong colorant agent and also has antioxidant properties. Further, it is used in medical, nutraceutical, cosmeceutical, food and feed applications. This chapter covers the occurrence of astaxanthin in nature, its accumulation, biomass production, downstream processing, chemistry and its possible role in various biological activities. Further available scientific literatures in the above studies are presented to provide a understanding of the utility of algal astaxanthin and its esters for varied applications.

BOX 13.2   SALIENT FEATURES

Astaxanthin is a keto-carotenoid found in algae, fungi, bacteria, seafood, quail, flamingo and complex plants. It is a one of the renowned molecules renowned for use in food, feed, cosmetics, nutraceutical and pharmaceutical applications. Astaxanthin is approved as food colorant in animal feed by the US Food and Drug Administration (USFDA) and also it is approved for use as natural food dye by the European Commission. It is commercially used for pigmentation in the aquaculture industry to increase astaxanthin content in farmed salmonids. Astaxanthin and its esters have shown potential antioxidant properties in *in vitro* and *in vivo* models. These studies reported that astaxanthin showed potential health benefits for both animals and humans. Recently, our research team reported that astaxanthin and its esters from *Haematococcus pluvialis* showed potential biological activities such as antioxidant, heptoprotective, and anti-skin-cancer activity in rat model. Astaxanthin mono-diesters showed better biological activity than astaxanthin. The current literature shows that astaxanthin can be used as a nutritional supplement which inhibits oxidation and free-radical formation in the metabolism, stimulates immunization and prevents ulcers, gastric injury, cardiovascular disease, diabetes and neurodegenerative disorders as well as cancer. Astaxanthin products such as whole cell biomass, capsules, soft gels, oils, syrups, tablets, creams and powder are available on the market. They are marketed under different brand names such as AlgaBerry™, Neoalgae, Bicosome®, KEYnatura, Regenurex, Supreme Asta Oil®, Supreme Asta Powder®, AstaZine™, NaturAsta™, AstaBio®, AstaFirst™, Astalif™, Zanthin®, AstaReal®, Astaxanthin Premia-Ex, Bioastin®, AstaPure™, Zanthin®, AstaZine® and Stazen®. This chapter also provides up-to-date information on the biological activities, health benefits and industrial applications of astaxanthin and its esters. Further research needs to be carried out on the metabolic pathways, and molecular studies are also required of mono and di- esters of astaxanthin for their use in nutraceutical and pharmaceutical applications.

INTRODUCTION

Astaxanthin is a keto-carotenoid found in various microorganisms (Ambati et al. 2014). It is a pink-colored and fat-soluble pigment which does not have provitamin-A activity in the human body (Davinelli et al. 2018). Astaxathin exhibits more potent antioxidant activity than other carotenoids (Nakagawa et al. 2011; Patel et al. 2018). It is approved as food colorant in animal feed by The United States Food and Drug Administration (USFDA) (Pashkow et al. 2008), and several European countries have approved its marketing as dietary supplement for human consumption (Roche 1987). *Haematococcus pluvialis* is a unicellular green microalga that is commonly found in freshwater bodies (Shah et al. 2016). The green cells of *H. pluvialis* experience environmental stress such as high light intensity and phosphate deprivation; they differentiate from a vegetative stage to form aplanospores in a resting stage. At this stage, the cell volume increases, producing a hard cellular wall that accumulates astaxanthin and its derivatives (Shah et al. 2016). *H. pluvialis* can grow in autotrophic and heterotrophic culture conditions (Sarada et al. 2012). Astaxanthin content in this organism varies depending upon the culture conditions such as salt, nitrogen, temperature and light levels (Sarada et al. 2012). Astaxanthin is used in food, feed, pharmaceutical, nutraceutical and cosmeceutical applications (Xiaofei 2015; Ranga Rao et al. 2018, 2010b). Consumption of astaxanthin and its esters can prevent various disorders in humans and animals (Fakhri et al. 2018). In view of this, this chapter provides a detailed view of the potential health benefits of astaxanthin and its esters from various algal forms.

ORIGIN AND OCCURRENCE OF ASTAXANTHIN

Astaxanthin is a naturally occurring carotenoid and it is found in various organisms such as salmon, trout, lobsters, shrimps, algae, bacteria and animals (Ambati et al. 2014). Astaxanthin is responsible for the pink coloration in the flesh of fish, and it is not synthesized *de novo* in salmonoids but is entirely obtained from their diet.

Astaxanthin is also found in yeast (*Phaffia rhodozyma*) (Yuan et al. 2011). Astaxanthin is also obtained from different sources such as *Haematococcus pluvialis* (3.8%), *Chlorococcum* (0.2%), *Chlorella zofingiensis* (0.001%), *Neochloris wimmeri* (0.6%), *Enteromorpha intestinalis* (0.02%), *Ulva lactuca* (0.01%), *Agrobacterium aurantiacum* (0.01%), *Paracoccus carotinifaciens* (2.2%), *Xanthophyllomyces* (0.5%), and *Thraustochytrium sp.* (0.2%) (Ambati et al. 2014).

LIFE CYCLE OF *H. PLUVIALIS* AND CAROTENOID ACCUMULATION

The life cycle of *H. pluvialis* comprises four stages: zoospores, microzooids, palmella and hematocysts (Elliot 1934). The zoospore, microzooid and palmella stages are called *vegetative* phases. Aplanospores are referred to as the red nonmotile astaxanthin accumulated encysted phase of the *H. pluvialis* life cycle. The zoospores are spherical ellipsoids which contain two flagella of equal length emerging from the anterior end and a cup-shaped chloroplast with numerous scattered pyrenoids. The zoospore cells are between 8 and 20 μm long with a distinct gelatinous extra-cellular matrix of variable thickness. The flagellated cells are major ones in the vegetative growth phase under favorable culture conditions. The macrozooids are divided into daughter cells by mitosis, and their cell size is stretched in unfavorable culture conditions (Wayama et al. 2013). The nonmotile palmella cells become the resting vegetative cells (Hagen et al. 2002). Palmella cells will turn into aplanospores under stress conditions such as high light intensity, high salinity and nutrient deficiency (Imamoglu et al. 2009). At this stage, cells contain a thick and rigid trilaminar sheath and a secondary cell wall of acetolysis-resistant material. Astaxanthin accumulates in the mature aplanospores and the lipid droplets deposited into the cytoplasm (Hagen et al. 2002). In some exceptional cases, it is reported that astaxanthin accumulates in *H. pluvilias* without forming aplanospores (Brinda et al. 2004). Under the optimal conditions, the red aplanospores turn into flagellated zoospores to initiate a new vegetative growth cycle.

BIOMASS AND ASTAXANTHIN PRODUCTION THROUGH PHOTOBIOREACTOR CULTIVATION

H. pluvialis is a unicellular freshwater alga and it undergoes a morphological change from green motile cells to red aplanospores, with the accumulation of astaxanthin taking place in the aplanospore phase (Boussiba 2000). Various photobioreactors are used for the production of astaxanthin to avoid contamination risks, as the outdoor cultivation of extremeophyllic forms like *Dunaliella* or *Spirulina* is challenging. Production of high-density biomass is also difficult, creating a huge bottleneck in commercialization (Lorenz and Cysewski 2000). Cell growth depends on pH light intensity, temperature, dissolved oxygen and nutrients (Xi et al. 2016; Oncel et al. 2011). The biomass concentration of *H. pluvialis* was reported to have reached 10 g/L in a photo-bioreactor cultivation (Suh et al. 2006). A two-stage cultivation was introduced for *H. pluvialis* wherein maximal biomass was achieved under growth-supporting favorable conditions, and the carotenogenesis accumulation was induced under light and nutrient stress conditions (Suh et al. 2006; Aflalo et al. 2007, 2009; Zhang et al. 2014). In order to produce greater amounts of green vegetative cells, an approach involving heterotrophic cultivation of this alga was also explored (Kobayashi et al. 1997; Tripathi et al. 1999). The optimized light and nitrogen regimes enabled dark-grown *H. pluvialis* cells to rebuild competent photosynthetic machinery to better utilize strong light for astaxanthin production with minimum ill-effect. There are several reports of cultivation of *H. pluvialis* in raceway ponds and photobioreactors under the photoautotrophic, heterotrophic, mixotrophic conditions for astaxanthin production (Olaizola 2000; Olaizola and Huntley, 2003; Fábregas et al. 2001; Garcia-Malea et al. 2009; Issarapayup et al. 2009; Zhang et al. 2009; Wang et al. 2013a, b).

TECHNOLOGY OF BREAKING APLANOSPORES OF *H. PLUVIALIS* OR EFFLUXING OF PIGMENT FOR DOWNSTREAM PROCESSING

Several techniques are applied in order to break open the aplanospore cells and recover the astaxanthin content (Mercer and Armenta 2011). Mechanical processes, expeller pressing and bead milling are used for commercial-scale production. Microalgae cells are squeezed under different pressure conditions in order to rupture the thick aplanospore cells. Bead milling is the most effective method when biomass concentration in the algal cake after harvesting is between 100 and 200 g/l (Greenwell et al. 2010). Other extraction methods such as the use of edible oils, acids, solvents and supercritical CO_2 are often coupled with microwave-assisted and pretreatment y enzyme for maximizing the recovery of astaxanthin from *H. pluvialis*. The supercritical carbon-dioxide method is applied in industrial applications for astaxanthin extraction from *H. pluvialis* due to its many processing advantages; for example, it uses optimal temperatures, prevents the degradation of valuable substances, involves a shorter extraction time and obviates the use of toxic solvents (Machmudah et al. 2006; Guedes et al. 2011). Moreover, the astaxanthin extraction efficiency was increased to 80%90% by using the supercritical carbon-dioxide extraction method. However, the astaxanthin extraction from *H. pluvialis* has been reported by various other extraction methods such as the use of pressurized liquids, enzymes, acids and edible oils and solvents (Sarada et al. 2006; Kang and Sim 2008; Jaime et al. 2010; Zou et al. 2013; Dong et al. 2014).

CAROTENOID COMPOSITION IN *H. PLUVIALIS*

A closer look at the typical ranges for the amounts of individual carotenoid pigments found in different sources of *Haematococcus pluvialis* have indicated the following percentage (%) variation in astaxanthin (total) 81–99, free astaxanthin 1–5, astaxanthin mono-esters 46–79, astaxanthin di-esters 10–39, β–carotene 0–5, lutein 1–11, canthaxanthin 0–5.5 and other carotenoids 1–9 (Ranga et al. 2009; Ranga Rao et al. 2010b; Ranga Rao et al. 2013a). A biosynthetic pathway for astaxanthin is shown in Figure 13.1. Some examples of astaxanthin ester found in *H. pluvialis* are shown in Figure 13.2.

COMMERCIAL PRODUCTION OF ASTAXANTHIN

The commercial production of *Haematococcus pluvialis* is undertaken by several companies around the world. Both Cyanotech and Mera Pharmaceuticals in North America are cultivating *H. pluvialis* in open-pond systems (Figure 13.3).

The companies involved in astaxanthin and algal biomass production are shown in Table 13.1 and 13.2, and the products made worldwide based on astaxanthin are shown in Table 13.3.

ASTAXANTHIN FOR SALMON AND TROUT FEEDS

The predominant source of carotenoids for salmonids has been synthetic astaxanthin, which has been used for pigmentation for over 20 years, more so after

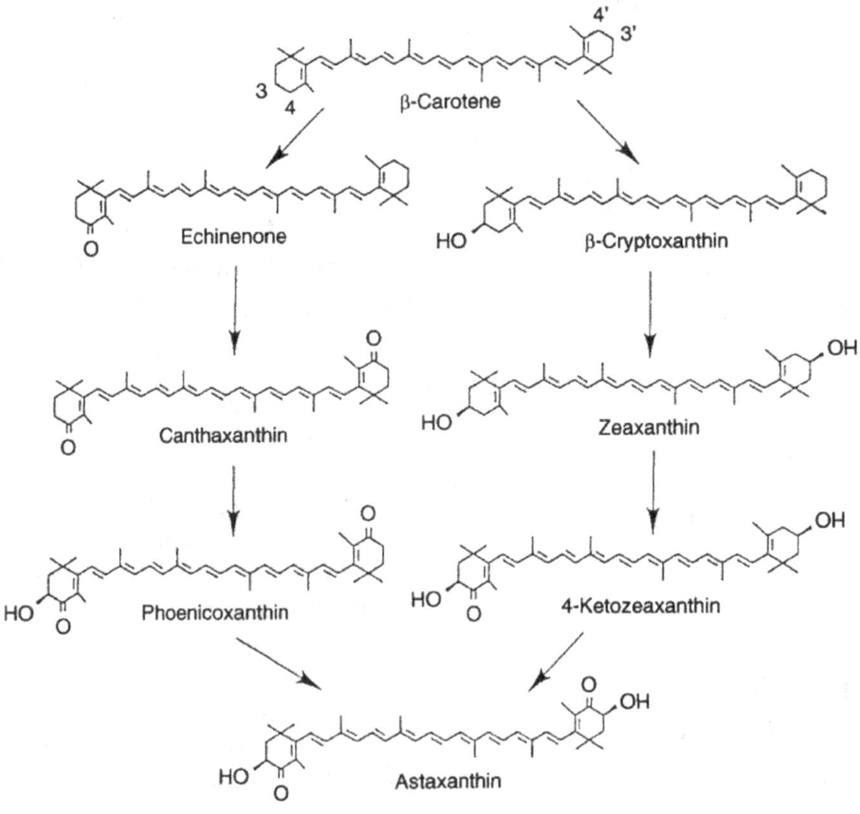

FIGURE 13.1 Biosynthesis pathway of astaxanthin from *H. pluvialis* (Fraser et al. 1997).

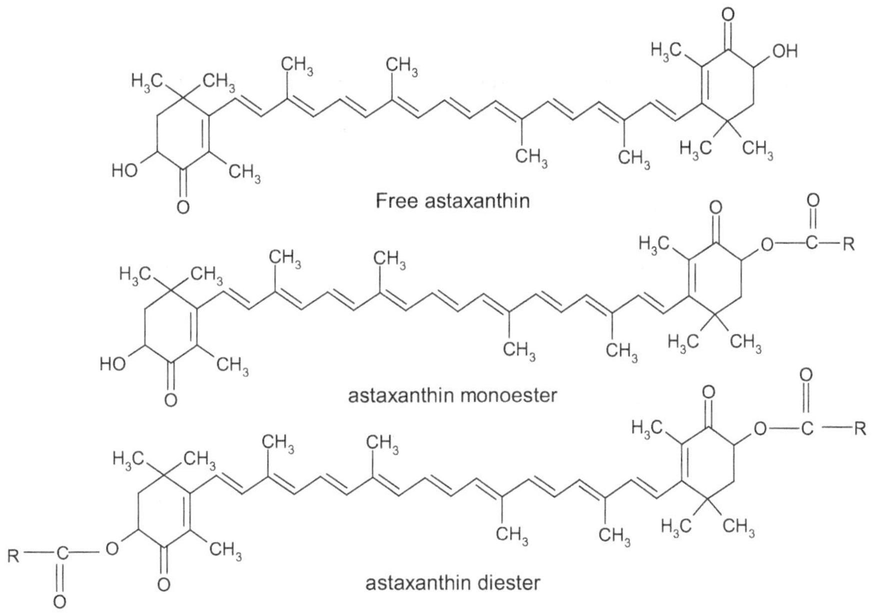

Potential Health and Nutraceutical Applications of Astaxanthin and Astaxanthin Esters from Microalgae

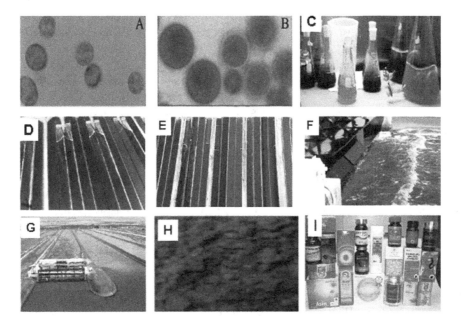

FIGURE 13.2 Some examples of astaxanthin and its esters found in *H. pluvialis*.

FIGURE 13.3 (**See color insert.**) Commercial production of astaxanthin: A) green motile vegetative phase, B) carotenoid rich encysted phase, C) *Haematococcus* culture in the experimental stage, D) *Haematococcus* in raceway ponds, E) astaxanthin accumulated in raceway ponds, F) *Haematococcus* before harvesting stage, G) *Haematococcus* culture in a pond ready for harvest, F) astaxanthin powder, I.) astaxanthin products on the market.

the approval by the USFDA in 1996 (Serdoz 2017). Processed crustacean waste from krill, shrimp, crab and crawfish is used as natural source of the pigment containing astaxanthin in the ingredients in salmonids and trout feeds (Lim et al. 2018; Sato et al. 2018; Koppe et al. 2012). Astaxanthin is stable at higher temperature due to its unsaturated molecular structure (Ranga Rao et al. 2007; De Bruijn et al. 2016). However, good storage conditions are important for the final animal feed products to promote stability and control the degradation of astaxanthin (Martínez-Delgado et al. 2017). Choubert and Heinrich (1993) showed that feeding rainbow trout with algae up to 6% of the diet had no major effect on growth or mortality. Moreover, it was found to support the growth and health of the rainbow trout. Thus, it was concluded that the astaxanthin from algae were a safe

TABLE 13.1
Various Companies Involved in Production of Astaxanthin for Health Benefits

Company Name	Country	Product Name	Description
Algaen Corporation	USA	AlgaBerry™	Human health, nutrition,
Neoalgae MicroSeaweeds Products	Spain	Neoalgae	Health foods
Cape Carotene	South Africa	—	Fish feed
NVI Algae Products	USA	—	Dietary supplements, nutraceuticals, pharmaceuticals and health products
Bicosome SL	Spain	Bicosome®	Antioxidant, UV-protection
Algae Health Ltd.	Ireland	—	Nutritional supplements
Keynatura	Iceland	KEYnatura	Nutritional products
Regenurex Health Corporation	Canada	Regenurex	Soft gel, capsules, nutraceutical
Atacama Bio Natural	Chile	Supreme Asta Oil® Supreme Asta Powder®	Oleoresin, powder, aquaculture and nutraceutical
Beijing Gingko Group Biological Technology Co. Ltd.	China	AstaZine™	Algae meal, oleoresin; aquaculture, nutraceutical and pharmaceutical
Jingzhou Natural Astaxanthin Inc.	China	NaturAsta™	Algae meal, oleoresin; aquaculture, nutraceutical and pharmaceutical
Kunming Biogenic Co. Ltd.	China	AstaBio®	Algae meal, oleoresin; aquaculture, nutraceutical and pharmaceutical
Wefirst Biotechnology Co. Ltd.	China	AstaFirst™	Algae meal, oleoresin; aquaculture, and pharmaceutical
Algalif	Iceland	Astalif™	Algae meal, oleoresin; aquaculture, nutraceutical and pharmaceutical
Parry Nutraceuticals Ltd	India	Zanthin®	Beadlets, softgel, capsules, nutraceutical
Evergen Resources	Indonesia	Evergen Astaxanthin	Oleoresin, aquaculture and nutraceutical
Alga Technologies Ltd.,	Israel	AstaPure™	Algae meal, oleoresin, beadlets, aquaculture, nutraceutical and pharmaceutical
AstaReal Co. Ltd.	Japan	AstaReal®	Oil, powder, nutraceutical and pharmaceutical
Algaetech International Sdn. Bhd.	Malaysia	Astaxanthin Premia-Ex	Algae meal, oleoresin, soft gels, aquaculture, nutraceutical and pharmaceutical
FEBICO (Far East Bio-tech. Co. Ltd.,)	Taiwan	ORG-ASTA	Capsules, nutraceuticals
Zestlife	United Kingdom	Zestlife Astaxanthin,	Softgel capsules; nutraceutical
Cyanotech Corporation	USA	Bioastin®, Naturose®	Softgel capsules; nutraceutical
Mera Pharamceuticals Inc.	USA	AstaFactor®	Dietary supplements
Stazen Inc.	USA	Stazen®	Softgel capsules; health benefits
Valensa International	USA	Zanthin®	Softgel capsules, nutraceutical
Algae Health Sciences	USA	AstaZine®	Nutraceuticals, health benefits

(Source from Lim et al. 2018).

and effective source of pigment. The largest market for astaxanthin is aquaculture, constituting 24% of total global fisheries production and currently valued at $35 billion per annum (Ravishankar et al. 2005).

SAFETY OF ASTAXANTHIN

Safety and toxicity studies were conducted on astaxanthin derived from *H. pluvialis*. (Yuan et al. 2011; Ranga Rao et al. 2015) in animal models without negative effect. Interestingly, several antioxidant enzymes levels in rats were enhanced after feeding of astaxanthin (Ranga Rao et al. 2013b). Astaxanthin provided significant protection against naproxen-induced gastric ulcers and also CCl_4-induced rat models (Kamath et al. 2008; Ranga Rao et al. 2015). In a significant study on humans, no side effects were observed when astaxanthin was administered at 2, 4 and 6 mg/day (Spiller 2003). These results support future investigations and clinical studies on astaxanthin.

TABLE 13.2
Companies Involved in Producing Algal Biomass

Company Name	Country	Focus Area of Research
Varicon Aqua Solutions Ltd.	Hallow, United Kingdom	Manufacture of PBRs and the Cell-Hi line of algal nutrients
Algaepower, Inc.	Vermont, USA	Renewable algoil and algae by-products
Aqualgae	Viana do Castelo, Portugal	Scale-up solutions for starter culture
Subitec	Stuttgart, Germany	Culture of algae for nutraceuticals, aquaculture, cosmetics and health care
MassReaction, Inc.	Florida, USA	Renewable fuels, transport fuels
Commercial Algae Management	North Carolina, USA	Cultivation of algae in bio-ponds, PBRs, supercritical CO2 extraction, downstream encapsulation of high-value products
Culturing Solutions	Florida, USA	Carbon capture from Ethanol production, Power Generation, Concrete manufacturing, and other industrial flue gasses
Cellana LLC	Hawaii, USA	Producing algal-based bioproducts-omega-3 EPA and DHA oils; animal feed and bio-fuel feed stocks
QDS Environmental Technology Inc.	New Jersey, USA	CO2 mitigation
Aquaculture Partners	Sant Gregori, Spain	Designing, manufacturing, developing, and marketing aquaculture equipment
Phenometrics	Michigan, USA	Enhancing the production capabilities of algae and algal products
Valoya	Helsinki, Finland	Provides LED lights for algal growth
ECO2Capture, Inc.	Ohio, USA	Enhances the capture of CO2 for use in the production of algae products for multiple end-use applications
Algenol Biotech	Florida, USA	Algal product development and manufacturing
Aquaculture Freelance Expertise	Challans, France	Algal production, nutrients and algal pasts
Redono Oy	Lohja, Finland	Recycling and treating industrial wastewater
LGem b.v.	Voorhout, Netherlands	Commercial-scale production; algae
Algae Dynamics	Ontario, Canada	Engaged in the development of health products and pharmaceuticals from algal oils
Algosource	Saint-Nazaire, France	Industrial development of algae
Industrial Plankton	British Columbia, Canada	Creates highly productive systems for algal cultivation
Necton SA	Portugal	Cultivation of algae for commercialization for cosmetics and food
EnAlgae	Swansea, United Kingdom	Developing sustainable technologies for algal biomass production for energy and other products
Reed Mariculture Inc.	California, USA	Production of algal concentrates
Algae Energy Inc.	Atlanta, Georgia, USA	Developing technology to commercially produce the highest quality algae products for use in both human and animal applications
Symbiotic EnviroTek Inc.	Alberta, Canada	Engaged in commercializing its advanced modular algae cultivation system
Advanced Biofuel Center	Jaipur, India	Large-scale replacement of petroleum and other fossil fuels
Algae Lab Systems	North Carolina, USA	Provides monitoring and control systems for the commercial algae growing industry
Solarvest Bioenergy Inc.	Nova Scotia, Canada	Development of natural and genetically modified microalgae strains for high-value markets

BIOAVAILABILITY OF ASTAXANTHIN

Astaxanthin absorption is found to increase when consumed with dietary oils. Bioavailability studies have been conducted on astaxanthin and its esters in rat models and also in broiler chicks by several researchers (Ranga Rao et al. 2013b; Madhavi et al. 2018; Sun et al. 2018). The bioavailability of astaxanthin in tissues and liver was enhanced after feeding of *H. pluvialis* biomass as source of astaxanthin (Rang Rao et al. 2010b, 2013b). Astaxanthin esters from *H. pluvialis* were more effectively absorbed into the body than the free form in rat studies (Ranga Rao et al. 2013a; Zhou et al. 2015). Østerlie et al. (2000) confirmed that bioavailability of astaxanthin in human subjects with a single dosage (100

TABLE 13.3
Products Worldwide Based on Astaxanthin for Health Benefits

Brand Name	Dosage Form	Ingredients	Company Name	Purpose
Source Naturals	Softgels	Astaxanthin	Source Naturals Inc.	Antioxidant, dietary supplement
AstaPure	Softgels	Astaxanthin	Jarrow Formulas	Supports skin and eye health
Now	Softgels	Astaxanthin	Now Foods	Potent free-radical neutralizer, supports balanced immune system responses
Nutrigold	Softgels	Astaxanthin	NutriGold Inc.	Dietary supplements; antioxidant
LifeExtension	Softgels	Astaxanthin; phospholipids	Life Extension Inc.	Dietary supplements
Futurebiotics	Softgels	Astaxanthin	Futurebiotics	Protects against free radicals
Viva Naturals	Softgels	Astaxanthin	Viva Naturals Inc.	Supports joint health, flights the signs of aging; promotes cellular health
Bioastin	Gel cap	Astaxanthin	Nutrex Hawaii Inc.	Fights cellular; supports joint, skin & eye health
Nutricost	Softgels/capsules	Astaxanthin; Edible oils, tocopherol	Nutricost LLC.	Anti-inflammatory agent
Solgar	Softgels	Astaxanthin	Solgar Inc.	Dietary supplement; antioxidant, supports healthy skin
Science Based Nutrition	Softgels	Astaxanthin; Olive oil	Doctor's Best Inc.	Natural antioxidant; dietary supplements
Skin Apeel	Cream	Astaxanthin, Vitamin-E	Serum	Antiaging, replenishing and rejuvenation
Vegan	Capsules	Astaxanthin	Deva	Antioxidant
AstaRich	Soft gel capsules	Astaxanthin	Acme ProGen Biotech Pvt. Ltd.	IDiabetes; brain health; antiaging; skin, eye and immune function
Physician Formulas	Soft gels/tablets	Astaxanthin	Physician Formulas Vitamin Company	Antioxidant
Eyesight Rx	Tablets	Astaxanthin, vitamin-C, plant extracts	Physician Formulas Vitamin Company	Vision function
KriaXanthin	Softgels	Astaxanthin, EPA, DHA	Physician Formulas Vitamin Company	Antioxidant
Astaxanthin Ultra	Softgels	Astaxanthin,	AOR	Cardiovascular health/gastrointestinal
Astaxanthin GoldTM	Softgels	Astaxanthin,	Nutrigold	Eye/joint/skin/immune health
Best Astaxanthin	Softgels	Astaxanthin, Canthaxanthin	Bioastin	Cell membrane/blood flow
Dr. Mercola	Capsules	Astaxanthin, Omega-3 ALA	Dr. Mercola Premium Supplements	Aging/muscle
Astaxanthin	Cream	Astaxanthin, herbal extracts	True Botanica	Face moisturizing
Astavita ex	Capsules	Astaxanthin, total carotenoid	Fuji Chemical Industry	Aging care
Astavita SPORT	Capsules	Astaxanthin, total carotenoid and zinc	Fuji Chemical Industry	Sports nutrition
AstaREAL	Oil/powder/biomass	Astaxanthin; astaxanthin esters	Fuji Chemical Industry	Soft gel, tablet, beverages, animal feed
AstaTROL	Extracts	Astaxanthin	Fuji Chemical Industry	Cosmetics
AstaFX	Capsules	Astaxanthin	Purity and Products Evidence Based Nutritional Supplements	Skin/cardiovascular function
Pure Encapsulations	Capsules	Astaxanthin	Synergistic nutrition	Antioxidant

(Continued)

TABLE 13.3 (CONTINUED)
Products Worldwide Based on Astaxanthin for Health Benefits

Brand Name	Dosage Form	Ingredients	Company Name	Purpose
Zanthin Xp-3	Softgels/capsules	Astaxanthin	Valensa	Human body
Micro Algae Super Food	Softgel	Astaxanthin	Anumed Intel Biomed company	Heart/eye/joint

mg) was enhanced with lipid-based formulations due to solubilization of a high amount of carotenes.

HUMAN CLINICAL STUDIES

Human clinical trials have been carried out on astaxanthin without any side effects on the subjects (Davinelli et al. 2018). Moreover, immune function tests showed no deleterious effects with astaxanthin supplementation in human subjects. The majority of these studies were conducted in healthy participants who volunteered to assess astaxanthin dose, bioavailability, safety and oxidative stress (Miyawaki et al. 2008; Park et al. 2010; Choi et al. 2011).

ASTAXANTHIN AS SOURCE OF HEALTH AND NUTRACEUTICAL APPLICATIONS

Medical researchers have shown that astaxanthin has significant nutraceutical and pharmaceutical applications. *In vitro* experiments, *in vivo* pre-clinical studies and early-stage clinical trials have clearly indicated the possibility that astaxanthin itself, in conjunction with other components, behaves like a prophylactic and curing agent against various diseases and health conditions (Table 13.4).

ASTAXANTHIN OFFERS PROTECTION FROM ULTRAVIOLET RADIATION

Carotenoids have an important role in nature in protecting tissues against ultraviolet (UV) light and sunburn (Savouré et al. 1995; Nakajima et al. 2016). Astaxanthin significantly protected tissues against UV-light induced photo-oxidation (Ito et al. 2018). Preclinical research showed that astaxanthin performed better than β-carotene either by itself or in combination with retinol. Moreover, it was extremely proficient at preventing photo-aging of the skin (Savouré et al. 1995; Toomey and Schulick 2017). Astaxanthin successfully protected UV-induced inflammation by decreasing inducible nitric oxide and the Cox-2 enzyme *in vitro*. This property helps to reduce cell death of skin cells in murine models (Yoshihisa et al. 2016). Astaxanthin and its esters

TABLE 13.4
Clinical Studies Using Natural Astaxanthin from *H. pluvialis*

Experiment Duration	Human Subjects	Dosage (mg)	Benefits of Astaxanthin
12	Females, males	6 and 12	Antiaging/cardiovascular/neurological
2	Females, males	6	Antiaging/eye health
4	Females, males	12	Cardiovascular/antiaging
12	Females, males	6 and 12	Neurological
12	N/A	1 and 3	Bioavailability, aging
12	Females, males	6,12 and 18	Cardiovascular
12	Males	4	Muscle performance, eye health
12	Males	4	Muscle performance
3	Females, males	5 and 20	Oxidative stress
3	Females, males	5,20, and 40	Oxidative stress
8	Females	2 and 8	Immunity/oxidative stress
8	Females, males	6	Skin health
8	Females, males	4	Skin/antiaging

(Source from www.algalif.com); NA-not applicable.

performed well against UV-DMBA induced ski- cancer model in rat studies (Ranga Rao et al. 2013a) with tumor suppression effects.

ASTAXANTHIN SUPPORTS IMMUNE SYSTEM

Oxidants have been directly linked to the stimulation of inflammation genes in endothelial cells. Similarly, reactive oxygen species (ROS) have been attributed an aggravating role in the inflammation that accompanies asthma and exercise-induced muscle damage. These immune-modulating properties are not related to provitamin-A activity, because astaxanthin, unlike β-carotene, does not have such activity (Sayahi and Shirali 2017; Toti et al. 2018). Immunoglobulin production in response to T-dependent stimuli in human blood cells was enhanced by astaxanthin (Biswal 2014). Immune response in humans was enhanced by astaxanthin during inflammation (Park et al. 2010). Enhanced antibody production and decreased humoral immune response in older animals was observed after dietary supplementation with astxanthin (Jyonouchi et al. 1994, 1995).

ASTAXANTHIN IN PREVENTION OF INFLAMMATORY DISORDERS

Astaxanthin showed potent antioxidant and immune responses, which terminate the induction of inflammation in biological systems (El-Agamy et al. 2018; Park et al. 2018). The structure of astaxanthin is very close to that of lutein and zeaxanthin but has a stronger antioxidant activity and UV-light protection effect (Ranga Rao et al. 2013a; Levy et al. 2018). Astaxanthin di-esters appear to exert influence as synergistic anti-inflammatory agents, increasing the effectiveness of aspirin when the two are administered together (Yamashita 1995). Feeding astaxanthin, *Ginkgo biloba* extract and vitamin C to asthmatic animals resulted significantly lowering bronchoalveolar lavage fluid cells and enhanced lung tissues when compared with the anti-inflammatory drug ibuprofen (Haines et al. 2010). Potential anti-inflammatory activity was observed in UV-induced photokeratitis in mice model after oral administration of nanoastaxanthin (Harada et al. 2017). Astaxanthin has also been shown to be beneficial for the treatment of ocular inflammation (Suzuki et al. 2006).

ASTAXANTHIN PREVENTS CARDIOVASCULAR DISEASE

Astaxanthin is a potent antioxidant with anti-inflammatory activity and its effect has been examined in both experimental animals and human subjects. It has also shown potential therapeutic effect against atherosclerotic cardiovascular disease (Fassett and Coombes 2011; Fakhri et al. 2018). Astaxanthin prevented the oxidation of low-density lipoprotein in *in vitro* experiments and also in a study with human subjects. It enhanced the production of low-density lipoproteins in the bloodstream of rats and inhibited LDL oxidation, possibly leading to the prevention of atherosclerosis (Iwamoto et al. 2001). Astaxanthin can exert beneficial effects in protecting against hypertension and stroke and improving memory in vascular dementia (Hussein, et al. 2005). A significant improvement was found in blood flow in the group treated with astaxanthin (Miyawaki 2008). Astaxanthin can decrease exercise-induced damage in the heart as well as in the skeletal muscle as reported by Aoi et al. (2003). The disodium disuccinate astaxanthin, a water-dispersible synthetic astaxanthin derivative, reduced myocardial damage in a rabbit model of ischemia/reperfusion (Lauver et al. 2005).

ASTAXANTHIN PREVENTS ULCERS AND GASTRIC INJURY

Helicobacter pylori is the initial manifestation of chronic gastritis and stomach ulcers. Left untreated, it can lead to more serious consequences including stomach cancer and lymphoma. It can be caused by eating a diet deficient in nutrients such as carotenoids. A low dietary intake of antioxidants such as carotenoids and vitamin C is known to be an important factor for the acquisition of *H. pylori* by humans (Bennedsen et al. 1999). A combination of astaxanthin with vitamin C and β-carotene was tested on ulcer properties in stressed rats (Kamath et al. 2008). In this experiment, the rats that were fed astaxanthin as well as β-carotene were protected from the formation of gastric ulcerations. Both astaxanthin and its esters showed potential anti-ulcer properties in rates with ethanol-induced gastric ulcers; this may be due to inhibition of H1, K1 ATPase, up-regulation of mucin content and an increase in antioxidant activities (Kamath et al. 2008).

ASTAXANTHIN FOR DIABETICS

Oxidative stress levels are very high in diabetes mellitus patients. It is induced by hyperglycemia, due to the dysfunction of pancreatic β-cells and tissue damage in patients. Astaxanthin exerted beneficial effects in diabetes with the preservation of β-cell function (Uchiyama et al. 2002). The astaxanthin-treated group showed lower blood glucose levels compared with the non-treated group treatment with astaxanthin ameliorating the progression and acceleration of diabetic nephropathy in the rodent model of diabetes. Astaxanthin also ameliorates insulin resistance via mechanisms involving the increase of glucose uptake and by modulating the level of circulating lipid metabolites and adiponectin (Preuss et al. 2011). Astaxanthin improves glucose metabolism and reduces blood pressure in patients with type 2 diabetes mellitus (Mashhadi et al. 2018). Astaxanthin reduced the oxidative stress in kidneys and liver of rats and prevented renal cell damage (Otton et al. 2010).

ASTAXANTHIN INHIBITS CANCER

The specific antioxidant dose may be helpful for the early detection of various degenerative disorders. These oxidants contribute to cancer and degenerative disorders though oxidation of proteins, lipids and DNA. Astaxanthin prevents oxidative stress in both rats and mice (Alam et al. 2018; Anuradha 2018). Astaxanthin from algae showed potential anti-skin-cancer activity against a UV-DMBA-induced skin cancer rat model (Ranga Rao et al. 2013a; Martinez Andrade et al. 2018). Dietary administration of astaxanthin has proved to significantly inhibit carcinogenesis in the mouse urinary bladder and also the rat oral cavity (Tanaka et al. 1994, 1995). Dietary astaxanthin is also effective in fighting mammary cancer by reducing the growth of induced mammary tumors (Chew et al. 1999). Astaxanthin supplementation in rats was found to inhibit the stress-induced suppression of tumor-fighting natural killer cells (Kurihara et al. 2002). Chew and Park (2004) reported that astaxanthin, canthaxanthin and β-carotene inhibited tumor growth, with astaxanthin showing the highest antitumor activity. Astaxanthin could attenuate oxidative stress, DNA damage and cell death as well as induction of early hepatocarcinogenesis in rats induced by cyclophosphamide (Tripathi and Jena 2010). Prabhu et al. (2009) discovered that astaxanthin had an inhibitory effect against chemically induced colonic preneoplastic progression in a rat model.

ASTAXANTHIN PREVENTS NEURODEGENERATIVE DISEASES

Astaxanthin is considered to be one of the strongest antioxidants in nature, due to its high potential for scavenging free radicals in the human body. Astaxanthin had protective effects on various neurodegenerative disorders such as Alzheimer's and Parkinson's diseases (Wu et al. 2015; Galasso et al. 2018). Dietary astaxanthin accumulates in the brain after a single-dose administration which may be helpful in protecting against neurodegenerative disease (Manabe et al. 2018). Astaxanthin may offer beneficial effects in improving memory in vascular dementia (Hussein et al. 2005). Astaxanthin and its isomers 9-cis showed a higher antioxidant activity than all-trans isomer in inhibiting the generation of reactive oxygen species induced by 6-hydroxydopamine in human neuroblastoma SH-SY5Y cells (Liu and Osawa 2009). Astaxanthin attenuates cognitive disorders in *in vivo* and *in vitro* models for neurodegenerative diseases (Che et al. 2018).

CONCLUSION

The research data available on astaxanthin from green microalgae is interesting and has been well researched in various biological studies in both experimental animals and humans. Astaxanthin showed potential to affect UV radiation, carcinogenesis, immune response, inflammatory disease, cardiovascular disease, ulcers, gastric injury, diabetes and neurodegenerative disorders. Various companies are producing biomass of *Haematococcus* and astaxanthin from algae. In addition, most of the companies make astaxanthin products, such as *AlgaBerry*™, *Neoalgae*, *Bicosome*®, *KEYnatura*, *Regenurex*, *Supreme Asta Oil*®, *Supreme Asta Powder*®, *AstaZine*™, *NaturAsta*™, *AstaBio*®, *AstaFirst*™, *Astalif*™, *Zanthin*®, *Evergen Astaxanthin AstaPure*™, *AstaReal*®, *Astaxanthin Premia-Ex*, *ORG-ASTA Zestlife Astaxanthin Bioastin*®, *Naturose*®, *AstaFactor*®, *Stazen*®, *Zanthin*® and *AstaZine*®, for nutraceutical and pharmaceutical applications. Some of these new products contain other ingredients such as DHA, EPA, vitamins E and D and minerals. A large number of patents have been filed on astaxanthin and for its biological activities of commercial value (Table 13.5).

Based on the current scientific literature, we conclude that astaxanthin supplementation from *Haematococcus* might be a practical and beneficial strategy in health applications. In future, the market for natural astaxanthin will increase for making multiproducts for food, feed, cosmetic, nutraceutical and pharmaceutical applications. The present commercial producers of natural astaxanthin cannot compete with synthetic ones on price alone. However, the astaxanthin and its esters from *Haematococcus* could command a premium price over synthetic astaxanthin due to the superiority of their biological activities.

TABLE 13.5
Some Patents of Commercial Significance of Astaxanthin

Patent No.	Title	Purpose
EP1404187	The use of di-esters of astaxanthin for enhancing the growth of farmed fish	Animal feed
US20180147159	Astaxanthin anti-inflammatory synergistic combinations	Anti-inflammatory properties
US20180146698	Formula feed for poultry	Animal feed
US20170216383	Composition and method to improve blood lipid profiles and reduce low-density lipoprotein in humans using algae	Improve human blood and lipid profiles
US20180071191	Astaxanthin support to healthy hair, skin and nails	Health supplements
US 20170007555	Composition and method to alleviate joint pain using phospholipids and astaxanthin	Therapeutic use
WO2018055009	Algae comprising therapeutic or nutritional agents	Nutritional agents
CN104856055	Food supplement with combination of astaxanthin and omega-3	Food supplements
WO2017017649	Self emulsifying system with astaxanthin and its use as feed additive for livestock use	Feed additive
US20150182475	Therapeutic astaxanthin and phospholipid composition and associated method	Dietary supplements
WO2015006646	Compositions comprising hydroxytyrosol, curcumin and astaxanthin and use thereof	Prevention of inflammatory disorders
WO2013043366	Methods and compositions for improving visual function and eye health	Improve eye health
US20130004582	Composition and method to alleviate joint pain	Reduced joint pain
US20120238522	Carotenoid containing compositions and methods	Preventing bacterial infections
US20120253078	Agent for improving carcass performance in finishing hogs	Food supplements
US20120114823	Feed additive for improved pigment retention	Fish feed
US20120004297	Agent for alleviating vascular failure	Preventing vascular failure
US20100291053	Inflammatory disease treatment	Preventing inflammatory
US20100267838	Pulverulent carotenoid preparation for coloring drinks	Preparation of drinks
US20090297492	Method for improving cognitive performance	Improving brain function
US20100158984	Encapsulates	Capsules
US20090142431	Algal and algal extract dietary supplement composition	Dietary supplement
US20090069417	Carotenoid oxidation products as chemopreventive and chemotherapeutic agents	Cancer prevention
US20090118227	Formulation for oral administration with beneficial effects on the cardiovascular system	Cardiovascular protection
US20090047304	Composition for body-fat reduction	Inhibits body fat
US20080293679	Use of carotenoids and carotenoid derivatives analogs for reduction/ inhibition of certain negative effects of COX inhibitors	Inhibit of lipid peroxidation
US20080234521	Crystal forms of astaxanthin	Nutritional dosage
US20070293568	Neurocyte protective agent	Neuroprotection
US20060217445	Natural astaxanthin extract reduces DNA oxidation	Reduces oxidative damage
CA2203836	Use of astaxanthin for retarding and ameliorating central nervous system and eye damage	Prevents eye and central nervous system diseases or injuries

ACKNOWLEDGMENT

The first author acknowledges Vignan`s Foundation for Science, Technology and Research University for providing financial support and research facility for this work, and also the World Academy of Science for the award of young affiliate for the year 2014–2018.

REFERENCES

Aflalo, C., Meshulam, Y., Zarka, A., and Boussiba, S. On the relative efficiency of two-vs. one-stage production of astaxanthin by the green alga *Haematococcus pluvialis*. *Biotechnol Bioeng*. 2007. 98: 300–305.

Alam, M. N., Hossain, M. M., Rahman, M. M., Subhan, N., Mamun, M. A. A., Ulla, A., Reza, H. M., and Alam, M. A. Astaxanthin prevented oxidative stress in heart and kidneys of isoproterenol-administered aged rats. *J Diet Suppl*. 2018. 15: 42–54.

Ambati, R. R., Phang, S. M., Ravi, S., and Aswathanarayana, R. G. Astaxanthin: Sources, extraction, stability, biological activities and its commercial applications—A review. *Mar Drugs*. 2014. 12: 128–152.

Anuradha, C. V. Astaxanthin, a marine carotenoid against hepatic oxidative stress: A systematic review. In: Patel, V. B., Rajendram, R., and Preedy, V. R., (Eds.) *The Liver International (Oxidative Stress and Dietary Antioxidants)*, 2018. Academic Press, Elsevier.

Aoi, W., Naito, Y., Sakuma, K., Kuchide, M., Tokuda, H., Maoka, T., Toyokuni, S., Oka, S., Yasuhara, M., and Yoshikawa, T. Astaxanthin limits exercise-induced skeletal and cardiac muscle damage in mice. *Antioxid Redox Signal*. 2003. 5(1): 139–144.

Bennedsen, M., Wang, X., Willen, R., Wadstrom, R., and Andersen, L.P. Treatment of *H. pylori* infected mice with antioxidant astaxanthin reduces gastric inflammation, bacterial load and modulates cytokine release by splenocytes. *Immunol Lett*. 1999. 70: 185–189.

Biswal, S. Oxidative stress and astaxanthin: The novel super nutrient carotenoid. *Int J Health Allied Sci*. 2014. 3: 147–153.

Boussiba, S. Carotenogenesis in the green alga *Haematococcus pluvialis*: Cellular physiology and stress response. *Physiol Plant*. 2000. 108: 111–117.

Brinda, B. R., Sarada, R., Kamath, B. S., and Ravishankar, G. A. Accumulation of astaxanthin in flagellated cells of *Haematococcus pluvialis*—cultural and regulatory aspects. *Curr Sci*. 2004. 87: 1290–1294.

Che, H., Li, Q., Zhang, T., Wang, D., Yang, L., Xu, J., Yanagita, T., Xue, C., Chang, Y., and Wang, Y. The effects of astaxanthin and docosahexaenoic acid-acylated astaxanthin on Alzheimer's disease in APP/PS1 double transgenic mice. *J Agric Food Chem*. 2018. 66: 4948–4957.

Chew, B. P., and Park, J. S. Carotenoid action on the immune response. *J Nutr*. 2004. 34: 257S–261S.

Chew, B. P., Park, J. S., Wong, M. W., and Wong, T. S. A comparison of the anticancer activities of dietary α-carotene, canthaxanthin and astaxanthin in mice *in vivo*. *Anticancer Res*. 1999. 19(3A): 1849–1853.

Choi, H. D., Kim, J. H., Chang, M. J., Kyu-Youn, Y., and Shin, W. G. Effects of astaxanthin on oxidative stress in overweight and obese adults. *Phytother Res*. 2011. 25: 1813–1818.

Choubert, G., and Heinrich, O. Carotenoid pigments of green alga *Haematococcus pluvialis*: Assay on rainbow trout *Oncorhynchus mykiss*, pigmentation in comparison with synthetic astaxanthin and canthaxanthin. *Aquaculture*. 1993. 112: 217–226.

Davinelli, S., Nielsen, M. E., and Scapagnini, G. Astaxanthin in skin health, repair and disease: A comprehensive review. *Nutrients*. 2018. 10: 522.

De Bruijn, W. J. C., Weesepoel, Y., Vincken, J. P., and Gruppen, H. Fatty acids attached to all-*trans*-astaxanthin alter its *cis-trans* equilibrium, and consequently its stability, upon light-accelerated autoxidation. *Food Chem*. 2016. 194: 1108–1115.

Dong, S., Huang, Y., Zhang, R., Wang, S., and Liu, Y. Four different methods comparison for extraction of astaxanthin from green alga *Haematococcus pluvialis*. *Sci World J*. 2014. 7: 694–705.

El-Agamy, S. E., Abdel-Aziz, A. K., Wahdan, S., Esmat, A., and Azab, S. S. Astaxanthin ameliorates doxorubicin-induced cognitive impairment (Chemobrain) in experimental rat model: Impact on oxidative, inflammatory, and apoptotic machineries. *Mol Neurobiol*. 2018. 55: 5727–5740.

Elliot, A. M. Morphology and life history of *Haematococcus pluvialis*. *Arch Protistenk*. 1934. 82: 250–272.

Fábregas, J., Otero, A., Maseda, A., and Domínguez, A. Two-stage cultures for the production of astaxanthin from *Haematococcus pluvialis*. *J Biotechnol*. 2001. 89: 65–71.

Fakhri, S., Abbaszadeh, F., Dargahi, L., and Jorjani, M. Astaxanthin: A mechanistic review on its biological activities and health benefits. *Pharmacol Res*. 2018. 136: 1–20.

Fassett, R. G., and Coombes, J. S. Astaxanthin: A potential therapeutic agent in cardiovascular disease. *Mar Drugs*. 2011. 9: 447–465.

Fraser, P. D., Miura, Y., and Misawa, N. *In vitro* characterization of astaxanthin biosynthetic enzymes. *J Biol Chem*. 1997. 272: 6128–6135.

Galasso, C., Orefice, I., Pellone, P., Cirino, P., Miele, R., Ianora, A., Brunet, C., and Sansone, C. On the neuroprotective role of astaxanthin: New perspectives? *Mar Drugs*. 2018. 16: 247.

García-Malea, M. C., Acién, F. G., Del Río, E., Fernández, J. M., Cerón, M. C., Guerrero, M. G., and Molina-Grima, E. Production of astaxanthin by *Haematococcus pluvialis*: Taking the one-step system outdoors. *Biotechnol Bioeng*. 2009. 102: 651–657.

Greenwell, H. C., Laurens, L. M. L., Shields, R. J., Lovitt, R. W., and Flynn, K. J. Placing microalgae on the biofuels priority list: A review of the technological challenges. *J Royal Soc Interface*. 2010. 7: 703–726.

Guedes, A. C., Amaro, H. M., and Malcata, F. X. Microalgae as sources of carotenoids. *Mar Drugs*. 2011. 9: 625–644.

Hagen, C., Siegmund, S., and Braune, W. Ultrastructural and chemical changes in the cell wall of *Haematococcus pluvialis* (Volvocales, Chlorophyta) during aplanospore formation. *Euro J Phycol*. 2002. 37: 217–226.

Haines, D. D., Varga, B., Bak, I., Juhasz, B., Mahmoud, F. F., Kalantari, H., Gesztelyi, R., Lekli, I., Czompa, A., and Tosaki, A. Summative interaction between astaxanthin, *Ginkgo biloba* extract (EGb761) and vitamin C in suppression of respiratory inflammation: A comparison with ibuprofen. *Phytother Res*. 2010. 25: 128–136.

Harada, F., Morikawa, T., Lennikov, A., Mukwaya, A., Schaupper, M., Uehara, O., Takai, R., Yoshida, K., Sato, J., Horie, Y., Sakaguchi, H., Wu, C. Z., Abiko, Y., Lagali, N., and Kitaichi, N. Protective effects of oral astaxanthin nanopowder against ultraviolet induced photokeratitis in mice. *Oxid Med Cell Longev*. 2017.

Hussein, G., Nakamura, M., Zhao, Q., Iguchi, T., Goto, H., Sankawa, U., and Watanabe, H. Antihypertensive and neuroprotective effects of astaxanthin in experimental animals. *Biol Pharm Bull*. 2005. 28: 47–52.

Imamoglu, E., Dalay, M. C., and Sukan, F. V. Influences of different stress media and high light intensities on accumulation of astaxanthin in the green alga *Haematococcus pluvialis*. *New Biotechnol*. 2009. 26: 199–204.

Issarapayup, K., Powtongsook, S., and Pavasant, P. Flat panel airlift photobioreactors for cultivation of vegetative cells of microalga *Haematococcus pluvialis*. *J Biotechnol*. 2009. 142: 227–232.

Ito, N., Seki, S., and Ueda, F. The protective role of astaxanthin for UV-induced skin deterioration in healthy people—a randomized double blind, placebo-controlled trial. *Nutrients*. 2018. 10: 817.

Iwamoto, T., Hoosoda, K., Hirano, R., Kurata, H., Matsumoto, A., Miki, W., Kamiyama, M., Itakaru, H., Yamamoto, S., and Kondo, K. Inhibition of low-density lipoprotein oxidation by astaxanthin. *J Atheroscler Thromb*. 2001. 7: 216–222.

Jaime, L., Rodríguez-Meizoso, I., Cifuentes, A., Santoyo, S., Suarez, S., Ibáñez, E., and Señorans, F. J. Pressurized liquids as an alternative process to antioxidant carotenoids extraction from *Haematococcus pluvialis* microalgae. *LWT Food Sci Tech*. 2010. 43: 105–112.

Jyonouchi, H., Sun, S., and Gross, M. Effect of carotenoids on *in vitro* immunoglobulin production by human peripheral blood mononuclear cells: Astaxanthin, a carotenoid without vitamin A activity, enhances *in vitro* immunoglobulin production in response to a T-dependent stimulant and antigen. *Nutr Cancer*. 1995. 23: 171–183.

Jyonouchi, H., Zhang, L., Gross, M., and Tomita, Y. Immunomodulating actions of carotenoids: Enhancement of *in vivo* and *in vitro* antibody production to T-dependent antigens. *Nutr Cancer*. 1994. 21: 47–58.

Kamath, B. S., Srikanta, B. M., Dharmesh, S. M., Sarada, R., and Ravishankar, G. A. Ulcer preventive and antioxidative properties of astaxanthin from *Haematococcus pluvialis*. *Eur J Pharmacol*. 2008. 590: 387–395.

Kang, C. D., Lee, J. S., Park, T. H., and Sim, S. J. Comparison of heterotrophic and photoautotrophic induction on astaxanthin production by *Haematococcus pluvialis*. *Appl Microbiol Biotechnol*. 2005. 68: 237–241.

Kang, C. D., and Sim, S. J. Direct extraction of astaxanthin from *Haematococcus* culture using vegetable oils. *Biotechnol Lett*. 2008. 30: 441–444.

Kobayashi, M., Kurimura, Y., and Tsuji, Y. Light independent, astaxanthin production by the green microalga *Haematococcus pluvialis* under salt stress. *Biotechnol Lett*. 1997. 19: 507–509.

Koppe, W. M., Moeller, N. P., and Baardsen, G. K. L. Feed additive for improved pigment retention. 2012. Patent no US20120114823.

Kurihara, H., Koda, H., Asami, S., Kiso, Y., and Tanaka, T. Contribution of the anti-oxidative property of astaxanthin to its protective effect on the promotion of cancer metastasis in mice treated with resistant stress. *Life Sci*. 2002. 70: 2509–2520.

Lauver, D. A., Lockwood, S. F., and Lucchesi, B. R. Disodium disuccinate astaxanthin (Cardax) attenuates complement activation and reduces myocardial injury following ischemia/reperfusion. *J Pharmacol Exp Ther*. 2005. 314: 686–692.

Levy, R., Hadad, N., Sedlov, T., and Zelkha, M. Astaxanthin anti-inflammatory synergistic combinations. 2018. Patent no. US20180147159.

Lim, K. C., Yusoff, F. M., Shariff, M., and Kamarudin, M. S. Astaxanthin as feed supplement in aquatic animals. *Rev Aquacult*. 2018. 10: 738–773.

Lorenz, R. T., and Cysewski, G. R. Commercial potential for *Haematococcus* microalgae as a natural source of astaxanthin. *Trends Biotechnol*. 2000. 18: 160–167.

Machmudah, S., Shotipruk, A., Goto, M., Sasaki, M., and Hirose, T. Extraction of astaxanthin from *Haematococcus pluvialis* using supercritical CO_2 and ethanol as entrainer. *Ind Eng Chem Res*. 2006. 45: 3652–3657.

Madhavi, D., Kagan, D., and Seshadri, S. A study on the bioavailability of a proprietary, sustained-release formulation of astaxanthin. *Integr Med*. 2018. 17.

Manabe, Y., Komatsu, T., Seki, S., and Sugawara, T. Dietary astaxanthin can accumulate in the brain of rats. *Biosci Biotechnol Biochem*. 2018. 82(8): 1433–1436.

Martínez Andrade, K. A., Lauritano, C., Romano, G., and Ianora, A. Marine microalgae with anti-cancer properties. *Mar Drugs*. 2018. 16: 165.

Martínez-Delgado, A. A., Khandual, S., and Villanueva-Rodríguez, S. J. Chemical stability of astaxanthin integrated into a food matrix: Effects of food processing and methods for preservation. *Food Chem*. 2017. 225: 23–30.

Mashhadi, N. S., Zakerkish, M., Mohammadiasl, J., Zarei, M., Mohammadshahi, M., and Haghighizadeh, M. H. Astaxanthin improves glucose metabolism and reduces blood pressure in patients with type-2 diabetes mellitus. *Asia Pac J Clin Nutr.* 2018. 27(2): 341–346.

Mercer, P., and Armenta, R. E. Developments in oil extraction from microalgae. *Eur J Lipid Sci Technol.* 2011. 113: 539–547.

Miyawaki, H., Takahashi, J., Tsukahara, H., and Takehara, I. Effects of astaxanthin on human blood rheology. *J Clin Biochem Nutr.* 2008. 43: 69–74.

Nakagawa, K., Kiko, T., Miyazawa, T., Carpentero Burdeos, G., Kimura, F., Satoh, A., and Miyazawa, T. Antioxidant effect of astaxanthin on phospholipid peroxidation in human erythrocytes. *Br J Nutr.* 2011. 105: 1563–1571.

Nakajima, H., Terazawa, S., Niwano, T., Yamamoto, Y., and Imokawa, G. The inhibitory effects of antioxidants on ultraviolet-induced up-regulation of the wrinkling-inducing enzyme neutral endopeptidase in human fibroblasts. *PLOS ONE.* 2016. 11(9): e0161580.

Olaizola, M. Commercial production of astaxanthin from *Haematococcus pluvialis* using 25,000-liter out door photobioreactors. *J Appl Phycol.* 2000. 12: 499–506.

Olaizola, M., and Huntley, M. E. Recent advances in commercial production of astaxanthin from micro algae. In: Fingerman, M., and Nagabhushanam, R., (Eds.) *Biomaterials and Bioprocessing,* 2003. Science Publishers.

Oncel, S. S., Imamoglu, E., Gunerken, E., and Sukan, F. V. Comparison of different cultivation modes and light intensities using mono-cultures and co-cultures of *Haematococcus pluvialis* and *Chlorella zofingiensis. J Chem Technol Biotechnol.* 2011. 86: 414–420.

Østerlie, M., Bjerkeng, B., and Liaaen-Jensen, S. Plasma appearance and distribution of astaxanthin E/Z and R/S isomers in plasma lipoproteins of men after single dose administration of astaxanthin. *J Nutr Biochem.* 2000. 11: 482–490.

Otton, R., Marin, D. P., Bolin, A. P., Santos Rde, C., Polotow, T. G., Sampaio, S. C., and de Barros, M. P. Astaxanthin ameliorates the redox imbalance in lymphocytes of experimental diabetic rats. *Chem Biol Interact.* 2010. 186: 306–315.

Park, J. H., Yeo, I. J., Han, J. H., Suh, J. W., Lee, H. P., and Hong, J. T. Anti-inflammatory effect of astaxanthin in phthalic anhydride-induced atopic dermatitis animal model. *Exp Dermatol.* 2018. 27: 378–385.

Park, J. S., Chyun, J. H., Kim, Y. K., and Chew, B. Astaxanthin decreased oxidative stress and inflammation and enhanced immune response in humans. *Nutr Metab.* 2010. 71: 18.

Pashkow, F. J., Watumull, D. G., and Campbell, C. L. Astaxanthin: A novel potential treatment for oxidative stress and inflammation in cardiovascular disease. *Am J Cardiol.* 2008. 101: 58D–68D.

Patel, V. B., Rajendram, R., Preedy, V. R., Dewi, I. C., Falaise, C., Hellio, C., Bourgougnon, N., and Mouget, J. L. Anticancer, antiviral, antibacterial, and antifungal properties in microalgae. In: Levine, I. A., and Fleurence, J., (Eds.) *Microalgae in Health and Disease Prevention,* 2018, pp. 235–261. Academic Press.

Prabhu, P. N., Ashokkumar, P., and Sudhandiran, G. Antioxidative and antiproliferative effects of astaxanthin during the initiation stages of 1, 2-dimethyl hydrazine–induced experimental colon carcinogenesis. *Fund Clin Pharmacol.* 2009. 23: 225–234.

Preuss, H. G., Echard, B., Yamashita, E., and Perricone, N. V. High dose astaxanthin lowers blood pressure and increases insulin sensitivity in rats: Are these effects interdependent? *Int J Med Sci.* 2011. 8: 126–138.

Ranga Rao, A.R., Sarada, A. R., Baskaran, V., and Ravishankar, G. A. Identification of carotenoids from green alga *Haematococcus pluvialis* by HPLC and LC-MS (APCI) and their antioxidant properties. *J Microbiol Biotechnol.* 2009. 19: 1333–1341.

Ranga Rao, A.R., Baskaran, V., Sarada, R., and Ravishankar, G. A. In vivo bioavailability and antioxidant activity of carotenoids from microalgal biomass—A repeated dose study. *Food Res Int.* 2013b. 54: 711–717.

Ranga Rao, A..R., Deepika, G., Ravishankar, G. A., Sarada, R., Narasimharao, B. P., Lei, B., and Su, Y. Industrial potential of carotenoid pigments from microalgae: Current trends and future prospects. *Crit Rev Food Sci Nutr.* 2018. 25: 1–22.

Ranga Rao, A.R., Harshvardhan Reddy, A., and Aradhya, S. M. Antibacterial properties of *Spirulina platensis, Haematococcus pluvialis* and *Botryococcus braunii* micro algal extracts. *Curr Trends Biotechol Pharm.* 2010a. 4: 807–817.

Ranga Rao, A., Raghunath Reddy, R. L., Baskaran, V., Sarada, R., and Ravishankar, G. A. Characterization of microalgal carotenoids by mass spectrometry and their bioavailability and antioxidant properties elucidated in rat model. *J Agric Food Chem.* 2010b. 58: 8553–8559.

Ranga Rao, A. R., Sarada, R., and Ravishankar, G. A. Stabilization of astaxanthin in edible oils and its use as an antioxidant. *J Sci Food Agric.* 2007. 87: 957–965.

Ranga Rao, A. R., Sarada, R., Shylaja, M. D., and Ravishankar, G. A. Evaluation of hepatoprotective and antioxidant activity of astaxanthin and astaxanthin esters from microalga *Haematococcus pluvialis. J Food Sci Technol.* 2015. 52: 6703–6710.

Ranga Rao, A. R., Sindhuja, H. N., Dharmesh, S. M., Sankar, K. U., Sarada, R., and Ravishankar, G. A. Effective inhibition of skin cancer, tyrosinase, and antioxidative properties by astaxanthin and astaxanthin esters from the green alga *Haematococcus pluvialis. J Agric Food Chem.* 2013a. 61: 3842–3851.

Ravishankar, G. A., Sarada, R., Sandesh Kamath, B., and Namitha, K. K. Food applications of algae. In: Kalidas, S., Gopinadhan, P., Anthony, P., and Robert, E. L., (Eds.) *Food Biotechnology,* 2nd Edition, 2005, pp. 491–524. CRC Press: Boca Raton, FL.

Roche, F. Astaxanthin: Human food safety summary. In: *Astaxanthin As a Pigmenter in Salmon Feed, Color Additive Petition 7CO2 1 1, United States Food and Drug Administration,* 1987, p. 43. Hoffman-La Roche Ltd.: Basel, Switzerland.

Sarada, R., Ranga Rao, A., Sandesh, B. K., Dayananda, C., Anila, N., Chauhan, V. S., and Ravishankar, G. A. Influence of different culture conditions on yield of biomass and value added products in microalgae. *Dyn Biochem Process Biotechnol Mol Biol.* 2012. 6: 77–85.

Sarada, R., Vidhyavathi, R., Usha, D., and Ravishankar, G. A. An efficient method for extraction of astaxanthin from green alga *Haematococcus pluvialis*. *J Agric Food Chem.* 2006. 54: 7585–7588.

Sato, W., Nagal, H., Kawashima, Y., Ikarashi, M., and Sakai, Y. Formula feed for poultry. 2018. Patent no. US20180146698

Savouré, N., Briand, G., Amory-Touz, M. C., Combre, A., Maudet, M., and Nicol, M. Vitamin A status and metabolism of cutaneous polyamines in the hairless mouse after UV irradiation: Action of β-carotene and astaxanthin. *Int J Vitam Nutr Res.* 1995. 65: 79–86.

Sayahi, M., and Shirali, S. The anti-diabetic and anti-oxidant effects of carotenoids: A review. *Asian J Pharm Res Health Care.* 2017. 9: 186–191.

Serdoz, F. Self emulsifying system with astaxanthin and its use as feed additive for livestock use. 2017. Patent no. WO2017017649.

Shah, M. M., Liang, Y., Cheng, J. J., and Daroch, M. Astaxanthin-producing green microalga *Haematococcus pluvialis*: From single cell to high value commercial products. *Front Plant Sci.* 2016. 7: 531.

Spiller, G. A., and Dewell, A. Safety of astaxanthin-rich *Haematococcus pluvialis* algal extract: A randomized clinical trial. *J Med Food.* 2003. 6: 51–56.

Suh, I. S., Joo, H. N., and Lee, C. G. A novel double-layered photobioreactor for simultaneous *Haematococcus pluvialis* cell growth and astaxanthin accumulation. *J Biotechnol.* 2006. 125: 540–546.

Sun, T., Yin, R., Magnuson, A. D., Tolba, S. A., Liu, G., and Lei, X. G. Dietary supplemental microalgal astaxanthin produced dose-dependent enrichments and improved redox status in tissues of broiler chicks. *J Agric Food Chem.* 2018. 66: 5521–5530.

Suzuki, Y., Ohgami, K., Shiratori, K., Jin, X. H., Ilieva, I., Koyama, Y., Yazawa, K., Yoshida, K., Kase, S., and Ohno, S. Suppressive effects of astaxanthin against rat endotoxin induced uveitis by inhibiting the NF-kB signaling pathway. *Exp. Eye Res.* 2006. 82: 275–281.

Tanaka, T., Makita, H., Ohnishi, M., Mori, H., Satoh, K., and Hara, A. Chemoprevention of rat oral carcinogenesis by naturally occurring xanthophylls, astaxanthin and canthaxanthin. *Cancer Res.* 1995. 55: 4059–4064.

Tanaka, T., Morishita, Y., Suzui, M., Kojima, T., Okumura, A., and Mori, H. Chemoprevention of mouse urinary bladder carcinogenesis by the naturally occurring carotenoid astaxanthin. *Carcinogenesis.* 1994. 15: 15–19.

Toomey, J. M., and Schulick, P. Supplement to support healthy hair, skin and nails. 2017. Patent no. US20180071191.

Toti, E., Chen, O., Palmery, M., Villano Valencia, D., and Peluso, I. Non pro-vitamin A and pro-vitamin-A carotenoids as immune-modulators: Recommended dietary allowance, therapeutic index, or personalized nutrition? *Oxid Med Cell Longev.* 2018. 2018: 1–20.

Tripathi, D. N., and Jena, G. B. Astaxanthin intervention ameliorates cyclophosphamide-induced oxidative stress, DNA damage and early hepatocarcinogenesis in rat: Role of Nrf2, p53, p38 and phase-II enzymes. *Mutat Res.* 2010. 696: 69–80.

Tripathi, U., Sarada, R., Rao, S. R., and Ravishankar, G. A. Production of astaxanthin in *Haematococcus pluvialis* cultured in various media. *Bioresour Technol.* 1999. 68: 197–199.

Uchiyama, K., Naito, Y., Hasegawa, G., Nakamura, N., Takahashi, J., and Yoshikawa, T. Astaxanthin protects β-cells against glucose toxicity in diabetic db/db mice. *Redox Rep.* 2002. 7: 290–293.

Wang, J. F., Han, D. X., Sommerfeld, M. R., Lu, C. M., and Hu, Q. Effect of initial biomass density on growth and astaxanthin production of *Haematococcus pluvialis* in an outdoor photobioreactor. *J Appl Phycol.* 2013a. 25: 253–260.

Wang, J. F., Sommerfeld, M. R., Lu, C. M., and Hu, Q. Combined effect of initial biomass density and nitrogen concentration on growth and astaxanthin production of *Haematococcus pluvialis* (Chlorophyta) in outdoor cultivation. *Algae.* 2013b. 28: 193–202.

Wayama, M., Ota, S., Matsuura, H., Nango, N., Hirata, A., and Kawano, S. Three-dimensional ultrastructural study of oil and astaxanthin accumulation during encystment in the green alga *Haematococcus pluvialis*. *PLOS ONE.* 2013. 8: e53618.

Wu, H., Niu, H., Shao, A., Wu, C., Dixon, B. J., Zhang, J., Yang, S., and Wang, Y. Astaxanthin as a potential neuroprotective agent for neurological diseases. *Mar Drugs.* 2015. 13: 5750–5766.

Xi, T., Kim, D. G., Roh, S. W., Choi, J. S., and Choi, Y. E. Enhancement of astaxanthin production using *Haematococcus pluvialis* with novel LED wavelength shift strategy. *Appl Microbiol Biotechnol.* 2016. 100: 6231–6238.

Xiaofei., Z. Food supplement with combination of astaxanthin and omega. 2015. Patent no. CN104856055.

Yamashita, E. Anti-inflammatory agent. 1995. Japanese Patent No. 07300421.

Yoshihisa, Y., Andoh, T., Matsunaga, K., Rehman, M. U., Maoka, T., and Shimizu, T. Efficacy of astaxanthin for the treatment of atopic dermatitis in a murine model. *PLOS ONE.* 2016. 11(3): e0152288.

Yuan, J. P., Peng, J., Yin, K., and Wang, J. H. Potential health-promoting effects of astaxanthin: A high-value carotenoid mostly from microalgae. *Mol Nutr Food Res.* 2011. 55: 150–165.

Zhang, B. Y., Geng, Y. H., Li, Z. K., Hu, H. J., and Li, Y. G. Production of astaxanthin from *Haematococcus* in open pond by two-stage growth one-step process. *Aquaculture.* 2009. 295: 275–281.

Zhang, W., Wang, J., Wang, J., and Liu, T. Attached cultivation of *Haematococcus pluvialis* for astaxanthin production. *Bioresour Technol.* 2014. 158: 329–335.

Zhou, Q., Xu, J., Yang, S., Xue, Y., Zhang, T., Wang, J., and Xue, C. The effect of various antioxidants on the degradation of O/W microemulsions containing esterified astaxanthins from *Haematococcus pluvialis*. *J Oleo Sci.* 2015. 64: 515–525.

Zou, T. B., Jia, Q., Li, H. W., Wang, C. X., and Wu, H. F. Response surface methodology for ultrasound-assisted extraction of astaxanthin from *Haematococcus pluvialis*. *Mar Drugs*. 2013. 11: 1644–1655.

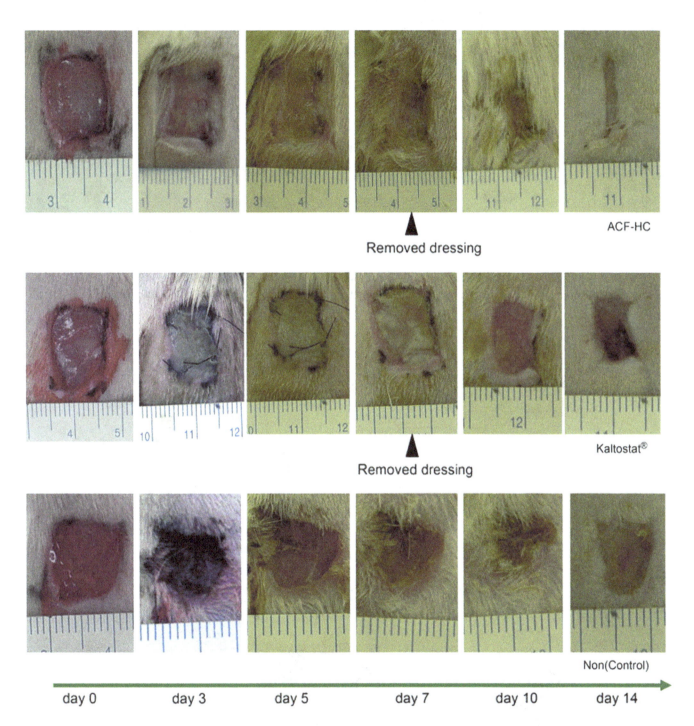

FIGURE 2.6 Photographic findings of wounds covered with chitin/chitosan, alginate, fucoidan hydrogel sheet or Kaltostat, and controls. Each wound on the indicated day is representative of eight wounds (four rats) covered. with ACF-HS or Kaltostat, or not covered (control). Figure adopted with permission (K. Murakami et al., 2010).

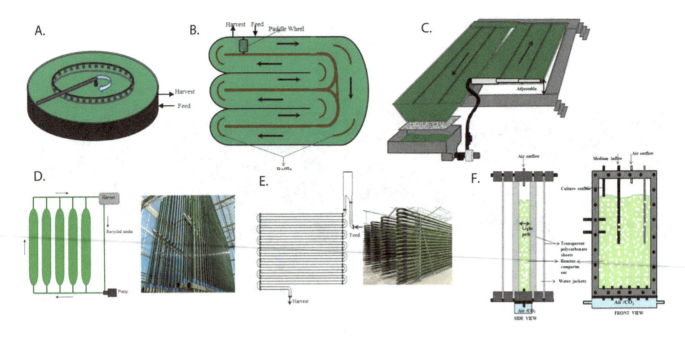

FIGURE 3.1 A Circular pond, B Raceway pond, C Sloped open bioreactor, D Vertical tubular reactor, E Horizontal tubular reactor, F Flat panel laboratory reactor (Chisti 2007; Olivieri et al. 2014; Parmar et al. 2011; Posten 2009; Sforza et al. 2014).

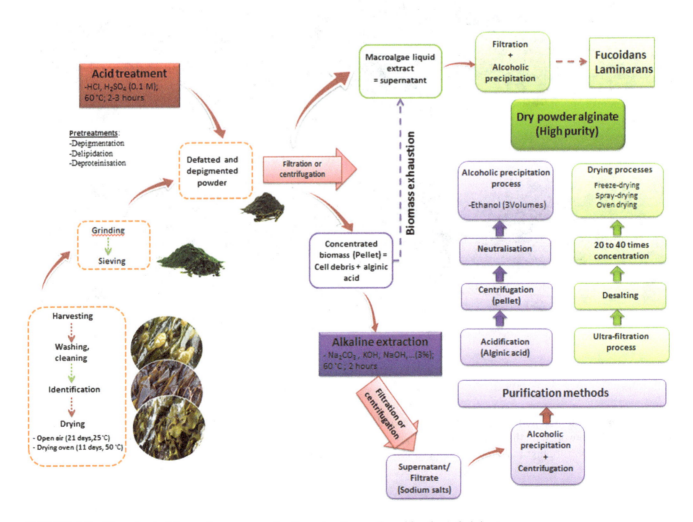

FIGURE 4.2 Flowchart of the main processes for the extraction and purification of alginates.

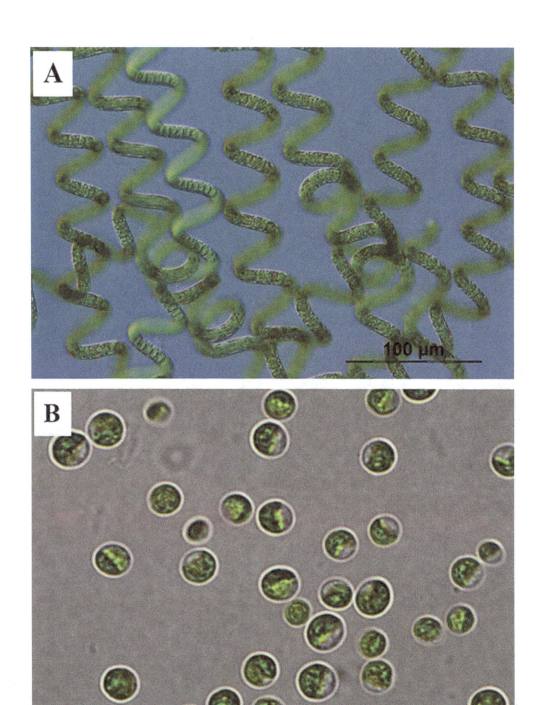

FIGURE 5.1 (**A**). Spirulina (https://sagdb.uni-goettingen.de/detailedList.php?str_number=85.79), (**B**). Chlorella

FIGURE 5.2 Production of biomass from *Spirulina* by various companies (Sources: http://www.algaeindustrymagazine.com). Earthrise-California (A), Cyanotech-Hawaii(B), Boonsom, Thailand (C), Parry-India(D), Hainan-China(E), and Yaeyama-Japan (F).

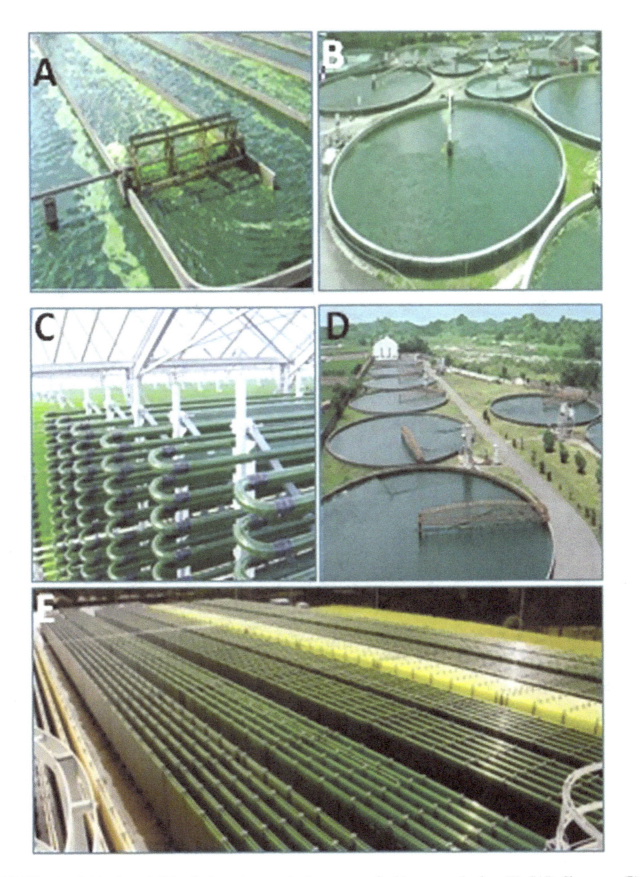

FIGURE 5.3 Cultivation of *Chlorella* in various production systems for biomass production; (A). SAB, Singapore, (B). Nutriphys, Belgium; (C). Chlorella-echlorial (https://chlorella-echlorial.com), (D). Sun Chlorella, USA and (E). Allmicroalgae-natural products, Lisboa, Portugal.

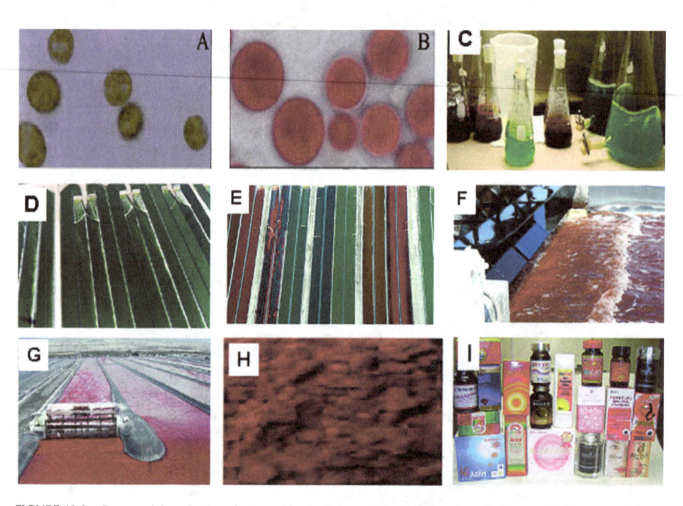

FIGURE 13.3 Commercial production of astaxanthin; **A.** Green motile vegetative phase **B.** Carotenoid-rich encysted phase, **C.** *Haematococcus* culture in the experimental stage, **D.** *Haematococcus* in raceway ponds, **E.** Astaxanthin accumulated in raceway ponds, **F.** *Haematococcus* before harvesting stage, **G.** *Haematococcus* culture a pond ready for harvest, **F.** Astaxanthin powder, **I.** Astaxanthin products in the market.

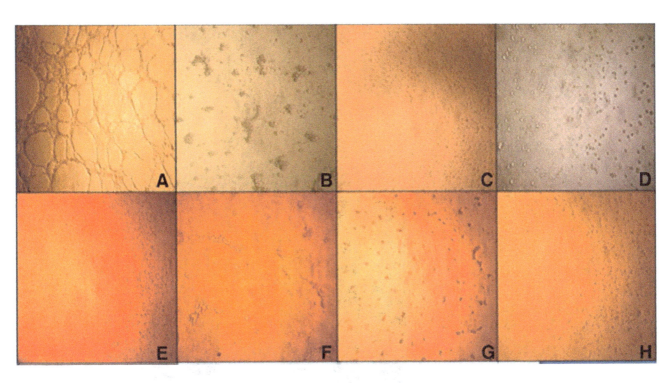

FIGURE 20.2 Effects of SV1 and PSV1 on capillary tube formation of rabbit aortic endothelial cells (RAEC). RAECs (1 × 10^5 per well) in medium containing 10% serumwere seeded into Matrigel pre-coated 24-well plates and treated with different concentrations of SV1 and PSV1 (A—Control; B—Heparin (10 µg/mL); C—SV1 25; D—SV1 50; E—SV1 100; F—PSV1 25; G—PSV1 50; H—PSV1 100). Representative phase contrast photomicrographs (100× magnification). (Adapted from Dore et al. *Microvas Res.* 2013. 88:12–18.)

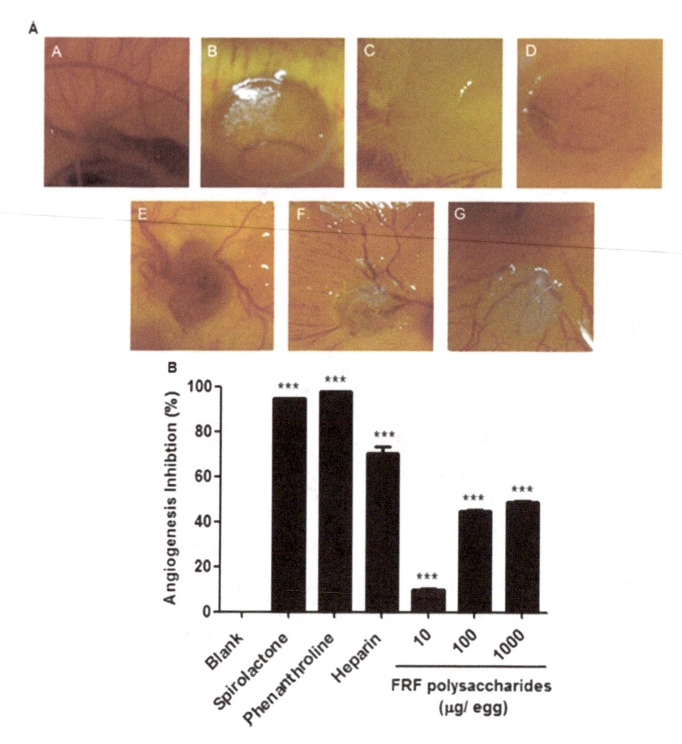

FIGURE 20.3 CAM assay to determine the antiangiogenic potential of FRF. **A:** Chorioallantoic membrane assay (CAM) of embryonated chicken eggs: blank A agarose; negative controls: B phenantroline 10 μg egg^{-1} and C spironolactone 10 μg egg^{-1}. D Heparin 10 μg egg^{-1} (positive control). FRF polysaccharides 10 μg egg^{-1}, E 100 μg egg^{-1}, and F 1000 μg egg^{-1} G. **B:** Antiangiogenic activity was obtained by score of quantitative vessels in the region of application of polysaccharides, with 0=none-antiagiogenic effect; 0.5–1=very low to medium antiagiogenic effect (no capillary-free area); 1=medium antiagiogenic effect (reduction in the capillary density: action equivalent to double that of disk area); and 2=hard antiangiogenic effect (more than double that of disk area). ***p<0.05. (Adapted from Castro et al. *J Appl Phycol.* 2015. 27:1315–1325.)

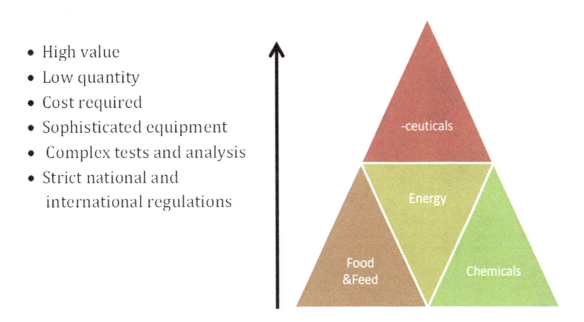

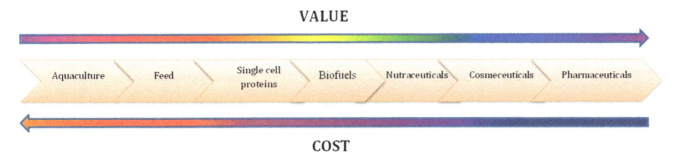

FIGURE 24.1 Value and cost relation of fundamental microalgal compounds.

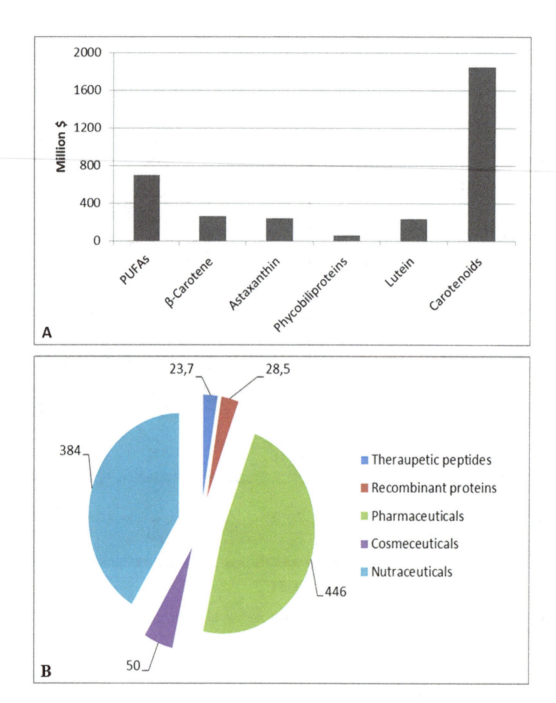

FIGURE 24.2 Market share of microalgal bioactive compounds A. fine chemicals B. % main sector share.

14 *Dunaliella salina*
Sustainable Source of β-Carotene

J. Paniagua-Michel

CONTENTS

Abbreviations 139
Introduction 139
Isoprenoid-Carotenoids: The Molecular Source of β-carotene 140
Why Carotenoids, Particularly β-carotene 140
Natural β-carotene and Human Health 141
β-carotene in Marine Organisms 142
Dunaliella salina: Among the Richest Natural Sources of β-carotene 142
Carotenoids and β-carotene Market Price 143
Attributes of *Dunaliella salina*: Health and Nutraceutical Products 143
β-carotene Processing and Manufacturing Requirements in *D. salina* 143
Conclusions 144
Acknowledgments 145
References 145

BOX 14.1 SALIENT FEATURES

The halophilic microalgae *Dunaliella salina* is among the richest and most abundant natural source of β-carotene. This isoprenoid carotenoid is the most abundant precursor of vitamin A in human and animal systems. *Dunaliella* and other carotenoids have gained scientific, biotechnological and commercial recognition in world markets, mainly for their uses in food, animal feed, pharmaceutical, nutraceutical, and cosmeceutical formulations. The growing preference for natural sources of carotenoids, as opposed to their synthetic substitutes, has encouraged demand for algae and specifically *D. salina* and *D. bardawil*. The recent approval of carotenoids from these algae as safe natural color by food and drugs authorities have positioned the alga and its constituent- β-carotene as an important and sought-after ingredient in health food formulations for humans and also in animal feed.

ABBREVIATIONS

USFDA: United States Food and Drug Administration
C40: forty carbon atoms
USDA: United States Department of Agriculture
ROS: Reactive Oxygen Species
GRAS: Generally recognized as safe
INCI: International Nomenclature of Cosmetic Ingredients

INTRODUCTION

Global demand for microalgae-based food ingredients is growing rapidly. Moreover, because of their nutritional value, algae are being marketed as functional foods or nutraceuticals. Among microalgal forms, *Dunaliella salina* is the richest natural source of beta-carotene, (~14 % on dry weight basis), and β-carotene is the most abundant provitamin A carotenoid in human beings (Paniagua-Michel et al. 2012; Shete and Quadro 2013). Among microorganisms, the industrial scale production of β-carotene is mainly supplied by molds— *Blakeslea trispora* and the algae *D. salina* and *D. bardawil* (Bogacz-Radomska and Harasym 2018). The conventional natural sources include carrot, papaya and a number of fruits and vegetables.

D. salina is an eukaryotic microalgae which naturally accumulates glycerol and β-carotene as strategic mechanisms to survive under various stress conditions such as high light intensities, high salt concentrations, and osmotic stress (Ben-Amotz and Avron 1982; Ramos et al. 2011). *Dunaliella* lacks a rigid polysaccharide cell wall, a unique characteristic feature among chlorophytes, making it easily digestible by animals and humans as well as providing a good source of protein (57% of dry matter) (Raja et al. 2007). Moreover, this chlorophyte is a well-known microalga with high pharmaceutical and nutraceutical value in industrial applications because of its rich blend of fatty acids, protein and carotenoid, a valuable source of nutrients (Spolaore et al. 2006; Larkum et al. 2012). *Dunaliella* powder for human use and dried *Dunaliella* for aquaculture feed are currently marketed worldwide, specifically in the form of products for health and nutraceutical applications (Guedes et al. 2011). Recently, carotenoids from *D. salina* and *D. bardawil* have been approved by USFDA as food coloring and are recognized as safe. In this chapter, we provide a current update on the uses and potentialities of *Dunaliella salina* and its importance for carotenoids and for use in health foods and as nutraceutical products.

ISOPRENOID-CAROTENOIDS: THE MOLECULAR SOURCE OF β-CAROTENE

Carotenoids are tetraterpenoids exhibiting a fat-soluble property. These C40 isoprenoids are considered to be the backbone of the carotenoid molecule (Paniagua Michel and Subramanian 2016; Ambati et al. 2018; Ranga Rao et al. 2018). Carotenoids are also characterized by a conjugated double-bond polyene chain. Due to this arrangement, they absorb light from the visible region of the electromagnetic spectrum, mainly in the 400–500 nm, and hence are colored. The carotenoids in *Dunaliella* exibit a characteristic yellow, orange, and red color (Paniagua-Michel et al. 2012). Carotenoids play an important role in providing light-harvesting pigments that are involved in the collection and transfer of light energy to chlorophyll for photosynthesis. The photoprotection of chlorophyll by carotenoids is achieved by dissipating the surplus energy in photosynthesis, leading to the inhibition of the formation of reactive oxygen species (Cogdell 1978). These isoprenoid carotenoids have oxygenated derivatives, or *xanthophylls*. Their chemical characteristics, as well as their solubility and polarity, are a consequence of the hydroxy, keto, epoxy, or aldehyde oxygen-containing groups it contains. In a similar way, the antioxidant function and dietary properties of carotenoids are also involved in their modulation and release from the food matrix, their uptake and accumulation by intestinal mucosal cells and tissue, as well as in metabolism (Shete and Quadro 2013; Carvalho and Carrujo 2017). The biosynthesis of β-carotene proceeds from geranylgeranyl pyrophosphate, and from eight isoprene units, thus having 40 carbons (Schwender et al. 2001). The β-carotene molecule is characterized by beta-rings at both ends of its structure (Capa-Robles et al. 2009; Paniagua-Michel et al. 2009). β-carotene is lipo soluble and composed of two retinal groups (Paniagua-Michel et al. 2015). This isoprenoid absorbs in the visible region of the spectrum, between 400 and 500 nm. β-carotene converts to vitamin A in one of two ways: either the end of the β-carotene molecule undergoes rupture, or central cleavage may occur by the enzyme beta-carotene-15,15'-dioxygenase (Figure 14.1) which has been characterized in humans. Hence, theoretically, two equivalent retinal molecules may be obtained, and the interaction of each retinal structure produces retinol (vitamin A) and retinoic acid (Shete and Quadro 2013; Al-Muhteseb and Emeish 2015; Paniagua-Michel et al. 2015).

WHY CAROTENOIDS, PARTICULARLY β-CAROTENE

Recently, carotenoid pigments from algae have received an increased attention in health-food applications (Guedes et al. 2011a; Zhang et al. 2014). These isoprenoid C40 pigments involve more than 700 different naturally occurring chemical structures; of these, astaxanthin, β-carotene, and lutein are considered to be important carotenoids for their potential biotechnological applications in foods (Zhang et al. 2014; Cardoso et al. 2017). These lipophilic molecules (Christaki et al. 2013) are characterized by their double bonds (Figure 14.1) and peculiar light-absorbing chromophore, which is responsible for their colorful presence in algae and other organisms. Due to the presence of these double bonds, the carotenoid molecules are sensitive to reactions such as oxidation and isomerisation and also to light, heat, acids, and oxygen (Amorim-Carrilho et al. 2014; Cardoso et al. 2017).

Among the carotenoids found in nature, β-carotene is perhaps the most important (Pisal and Lele 2005) and the best-known carotenoid of importance to human health (Shete and Quadro 2013). Consequently, clinical trials on the effects of carotenoids on human health have been focused mostly on β-carotene (de Carvalho and Caramujo 2017).

The anticancer role of natural β-carotene explains why it is highly valued by the health-products market. In principle, the benefits of this isoprenoid are associated with the mixture of *trans* and *cis*-isomers of β-carotene; although controversial in the origin of the *cis*-isomers, their mixture is only present in the natural β-carotene and not in the molecule produced by chemical synthesis (Ben-Amotz and Levy 1996).

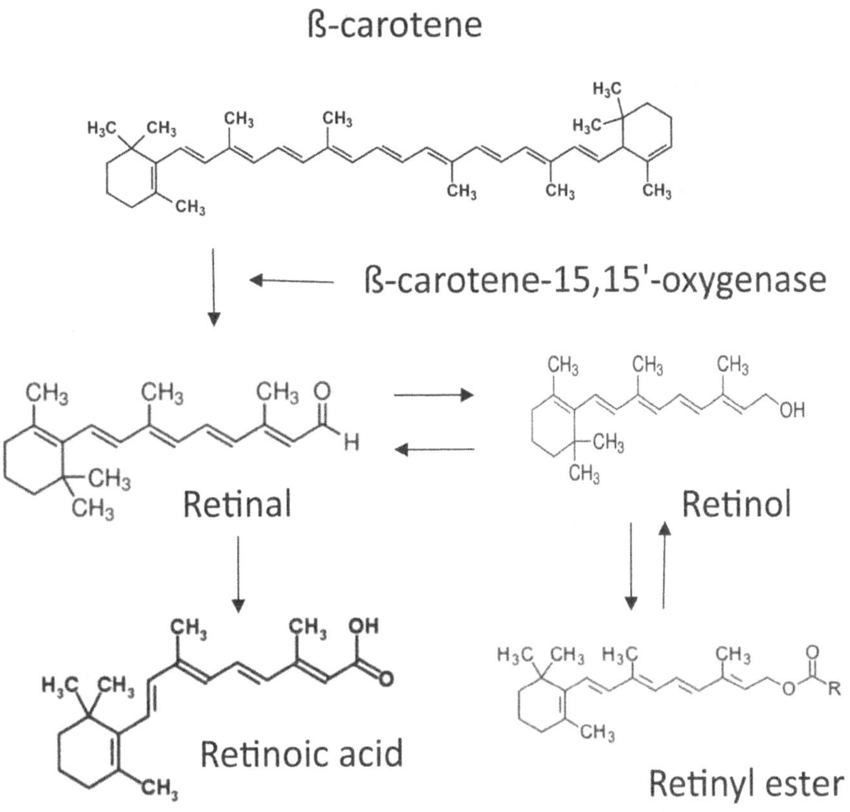

β-carotene

FIGURE 14.1 β-carotene bioconversion into vitamin A by human metabolism. (Modified after Shete and Quadro 2013.)

Several studies have indicated that excessive consumption of artificial pigments due to proclivity could be detrimental to health because of toxic effects such as allergic reactions, cancer, asthma, abdominal pain, nausea, and hepatic and renal damage reported in humans (Srivastava 2015; Cardoso et al. 2017). Natural carotenoids, however, show positive effects benefiting health (Chen et al. 2016). Because of these attributes, carotenoids such as β-carotene, lutein, zeaxanthine, α-carotene, β-cryptoxanthin, and lycopene are preferred in diets (Berman et al. 2015; Cardoso et al. 2017). In the case of β-carotene, its frequent consumption in foods has been shown to reduce the risk related to cardiovascular disease, which is an important cause of death globally, as well as other human and animal diseases. The suspected carcinogenic potential of synthetic food additives (El-Baky et al. 2003) are encouraging the acceptance of natural ingredients such as β-carotene of *D. salina* (Hsu et al. 2008; Guedes et al. 2011). Moreover, the public demand for natural sources of β-carotene has led to its incorporation in health foods and also in feed used in aquaculture, such as shrimp farming (Krinsky 2005; Spolaore et al. 2006; Guedes 2011; Paniagua-Michel 2015).

NATURAL β-CAROTENE AND HUMAN HEALTH

β-carotene is one of the most studied and utilized carotenoids because of its medicinal properties in therapeutic and preventive disease programs, mainly against diabetes, cancer, and cataracts (Shete and Quadro 2013). Its diverse use in food supplements, cosmetics, and pharmaceuticals is growing, and thus the demand for β-carotene is likely to increase over the next few years (Sathasivam and Kim 2018). Human and animal systems lack the enzymes to synthesize carotenoids, which is why they need to obtain the necessary carotenoids directly from the diet and to transform the dietary carotenoid precursors by metabolic reactions to meet their needs (Liñán-Cabello et al. 2002; Paniagua-Michel et al. 2002). In fact, among the metabolic bioconversion in both the human body and in animal systems, the following reactions can be classified as common: reduction, cyclization, oxidation, cleavage of double bonds or epoxy bonds, and translation of double bonds (Matsuno 1989; Carvalho and Caramujo 2017).

The importance of the antioxidant properties of carotenoids for human health derives from their potential to reduce the oxidative stress linked to various reactive oxygen species (ROS) related disorders, including various types of cancer and neurological and cardiovascular diseases (Voutilainen et al. 2006; Fiedor and Burda 2014; Carvalho and Caramujo 2017). Carotenoids (especially α-carotene, β-carotene, and β-cryptoxanthin) are also important precursors of vitamin A which is essential for normal growth and development, immune system function, and vision. On the other hand, β-carotene may serve either as an antioxidant or as a pro-oxidant, depending on its intrinsic properties as well as on the redox potential of the biological environment in which it acts (Palozza et al. 2003).

β-CAROTENE IN MARINE ORGANISMS

Marine and aquatic animals deposit carotenoids obtained mainly by feeding on autotrophs (phytoplankton and microalgae). They deposit the ingested carotenoids in their gonads, carapaces, muscles, and integuments. These carotenoids are either directly accumulated without modification or are converted into other carotenoids prior to deposition in tissues. Carotenoid-based skin coloration is an indication of quality of the flesh; the presence of β-carotene is ascribed health benefits as precursors for vitamin A and transcription regulators, antioxidants, free-radical scavengers, immune-system stimulants and cancer inhibitors. These pigments are also mobilized from muscle to ovaries, and have been associated with reproduction (Kodric-Brown and Brown 1984; Ando et al. 1986; Liñán-Cabello and Paniagua Michel 2004). Aquatic animals conspicuously accumulate carotenoids in their gonads, which is assumed to be essential for their reproduction and the successful development of their eggs and early larval stages (Paniagua-Michel and Liñán-Cabello 2002). In the case of the sea urchin, supplementation with β-carotene, which can be metabolized to echinenone, also increased reproduction and the survival of larvae (Tsushima et al. 1997). Color in shrimp and fish is a nutritive quality criterion demanded by consumers. The aesthetic appeal and nutritive value of the food from aquaculture organisms, *viz.*, salmonids, crustaceans, and several farmed fish (red porgy or red sea bream) are dependent on its color, which has a direct bearing on marketing and commercialization.

DUNALIELLA SALINA: AMONG THE RICHEST NATURAL SOURCES OF β-CAROTENE

Dunaliella salina was first described by Felix Dunal in 1838 and Clara Hamburger in 1905 (Teodoresco 1905; Jaenicke 1998; Oren 2005). They described *D. salina* as a eukaryotic unicellular microalgae that becomes red-colored when it thrives in the brines of salt lakes and salterns around the halophilic habitats of the French Mediterranean coast (Teodoresco 1905). Presently this genus has been reported in several hypersaline environments in various parts of the world. While Lerche (1937) conducted studies on development and reproduction in *Dunaliella*, it was Teodoresco (1866–1949) who named *D. salina* in honor of F. Dunal. Mil'ko (1963) and Massyuk (1966) were the first to propose that *D. salina* would be an ideal commercial source of natural β-carotene, and they conducted several trials of mass culture of this alga in Ukraine (Massyuk and Abdulla 1969; Borowitzka and Borowitzka 1988). Ben-Amotz et al. (1982) proposed this alga as source of glycerol as well. Actually, *Dunaliella salina* has been identified in all the hypersaline environments worldwide, where other oxygenic phototrophs fail to grow (Oren 2005). This flagellate alga accumulate the highest amount of β-carotene per cell of any organism, measuring up to 14% on the basis of dry weight (Mil'ko 1963; Aasen et al. 1969; Borowitzka and Borowitzka 1990). It is known that *D. salina* accumulates around 102 mg of β-carotene per 100 g, while carrots only accumulate 3 mg (USDA National Nutrient Database for Standard Reference Release 18 USA) (Al-Muhteseb and Emeish 2015). β-carotene from *D. salina* is currently produced on a commercial and pilot scale in several countries, among them Australia, Israel, the United States, India, China, and Spain. In fact, the market for natural β-carotene is increasing worldwide (Figure 14.2).

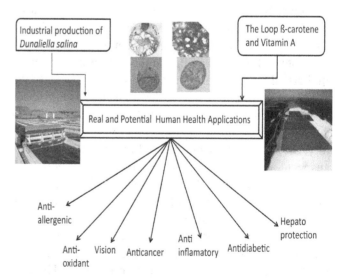

FIGURE 14.2 Examples of industrial production of β-carotene from *Dunaliella salina* as well as its uses and potential benefits to human health.

Recently, carotenoids from *D. salina* have been approved by USFDA as a food color and are recognized as safe natural color (Dufossé et al. 2005; Yang et al. 2013). Generally, carotenoids extracted from microalgae, such as β-carotene from *Dunaliella salina*, are considered to be eco-friendly colorants. The food industry has greatly benefited from the use of the β-carotene molecule, which can be safely used instead of the chemically synthesized C40 isoprenoid β-carotene (Basilya et al. 2018).

CAROTENOIDS AND β-CAROTENE MARKET PRICE

Microalgae pigments acquire relevance because of robust market demands. β-carotene price from natural sources, such as from microalgae, may range from US$800 kg−1 to US$1,500 kg−1 (Gellenbeck 2012), whereas its chemically synthesized molecule costs less than half this (Guedes et al. 2011). Even though the price of synthetic carotenoids is lower than that of natural ones, demand for the synthetic version started to decline because of their suspected toxic effects. Nowadays, consumer preference for sustainable production of natural products and green ingredients has led to a surge in development of the global market for natural products. The global carotenoid market was estimated at around ~1.24 billion USD in 2016, and according to conservative assessments, this market may reach ~1.53 billion USD by 2021 (Ambati et al. 2018). From 2016 to 2021, the compound annual growth rate (CAGR) could reach 3.78% at global level (www.bccresearch.com). Recently, a range of carotenoids from microalgae, mainly astaxanthin, β-carotene, fucoxanthin, and lutein, have become highly sought after for use in nutraceuticals, pharmaceuticals, food, animal feed, dietary supplements, and cosmetics (Lelyana 2016).

ATTRIBUTES OF *DUNALIELLA SALINA*: HEALTH AND NUTRACEUTICAL PRODUCTS

The motile chlorophyte *D. salina* may be found inhabiting the habitats lakes, oceans, and brackish waterbodies of all the continents. The content and composition of carotenoid pigments produced by microalgal species vary and are influenced by the culture conditions (Ranga Rao et al. 2013; Wells et al. 2017). Hence, the environmental conditions are the main factor that determines the massive accumulation of β-carotene in *D. salina* cells (Paniagua-Michel 2009). *D. salina* and *D. bardawil* both produce massive quantities of β-carotene during the carotenogenesis phase (Aasen et al. 1969; Borowitzka and Borowitzka 1988; Paniagua-Michel 1995). The process of the induction of carotenogenesis in *Dunaliella* proceeds under suboptimal growth conditions, such as nutrient limitations, mainly when nitrogen and phosphorous are exhausted. As a general rule, this photo-physiological process is inversely related to the specific growth rate, *viz.*, enhanced β-carotene accumulation occurs when the specific growth rate is low (Ben-Amotz et al. 1982; Borowitzka LJ et al. 1984). Carotenogenesis and predominant β-carotene accumulation is also influenced by light quality (wavelength) and/or intensity (Ben-Amotz 1973; Raja et al. 2006). However, optimal carotenogenesis and the highest amount of β-carotene in the cell is directly proportional to salinity (Borowitzka, MA and MAa Borowitzka 1989). One of the major advantages of β-carotene from *D. salina* is its ability to grow in conditions of extreme salinity in production systems, such as open raceways and ponds, which is a strategic step to arrest growth of competing organisms (Gellenbeck 2012). The salinity optimum for growth ranges between 18 and 22% NaCl, but for carotenogenesis, the optimum salinity that enhances carotenoid production is >27% NaCl (Borowitzka LJ et al. 1984). Carotenogenesis, as well as the optimum β-carotene yield, may be achieved in two stages. In the first stage, microalgae are grown in a low salinity, around 15% NaCl, in nutrient-rich medium for maximal biomass production. After achieving maximal growth. the biomass is transferred to a carotenogenic medium involving high salinity and low levels of nutrients, which lead to the accumulation of β-carotene in the biomass (Massyuk 1966; Borowitzka, LJ et al. 1984). Another approach is to select a mode of culture to grow the algae (batch or continuous) at an intermediate salinity, assuring the best overall yield of β-carotene over time (Borowitzka LJ et al. 1985). This alternative production strategy avoids the higher salinities that encourage the growth of predatory protozoa such as *Fabrea salina*, and the very low salinity conditions which could promote the growth of other algal forms (Borowitzka 1999). The large-scale *D. salina* cultures in outdoor conditions may achieve sustainable periods of productivity of 30–40 g dry weight m^{-2}. day^{-1} (Borowitzka 1999), while in smaller culture systems and over shorter periods, productivity may be as high as 60 g dry weight.m^{-2}.day^{-1} (Ben-Amotz 1980).

β-CAROTENE PROCESSING AND MANUFACTURING REQUIREMENTS IN *D. SALINA*

Carotenoids are extracted from *D. salina* cells with organic solvents using grinders and homogenizers that

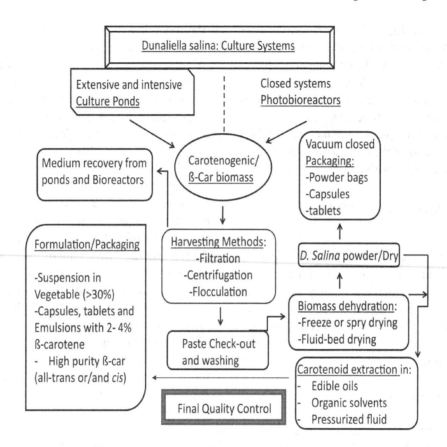

FIGURE 14.3 Flow-chart representation of *D. salina* processing and β-carotene extraction and packaging. (Modified after Tafreshi and Shariati 2009; Shankaranarayanan et al. 2018.)

in some cases are specially designed for *Dunaliella* (Sarada et al. 2006), aiming to suit the scaled-up processes of industrial applications (Figure 14.3). Cell wall breaking can be achieved by selecting the proper method. Hence milling, ultrasonication, microwaving, thawing, freezing, supercritical fluid extraction, and extraction of edible oils may be applied depending on the algal species (McMillan et al. 2013; Ranga Rao et al. 2018). But the main concern is to meet the GRAS status and safety requirements as per the regulatory demands. According to Gellenbeck (2012), for food-grade quality requirements, the following requirements must be met:

a) Efficacy: The new ingredient should exhibit the intended biological activity.
b) Sourcing: Production should be eco-friendly and should meet the ethical considerations as stipulated by the international norms.
c) Quality assurance: It should meet the international standards laid down by agencies such as USFDA, CODEX, INCI (International Nomenclature of Cosmetic Ingredients), the American Cosmetic Association, etc.
d) Regulatory requirements: It should meet the regulatory considerations of each country. GMP (good manufacturing practices) and the Fair Packaging and Labeling Act (Federal Trade Commission 2011) need to be complied with.

Hence, the existing methods used to separate the biomass and purify carotenoids to recover the pigments through downstream processing represent the major costs (Cardoso et al. 2017). The strategic development of low-cost processes for the purification of β-carotene will decrease the market price; no doubt this could be a leading theme in biotechnological research on β-carotene production in *Dunaliella salina*.

CONCLUSIONS

The higher consumer and market demand for natural sources of β-carotene has promoted biotechnological production from *Dunaliella salina* and *D. bardawil*. The high bioactivity of β-carotene is widely used in pharmaceuticals and medicinal formulations owing to its provitamin *A* activity. This molecule is also important as a coloring agent in various foods and health-food

FIGURE 14.4 Different commercial β-carotene presentations. A, food grade β-carotene crystals from *D. salina*; B, β-Carotene powder from *D. salina*; C, β-Carotene Emulsion (2-10%); D, β-carotene tablets; E, β-carotene capsules/pills.

formulations (Figure 14.4). It is also used extensively to impart coloration in aquaculture-farmed fishes and crustaceans. Consequently, the utilization of β-carotene from *Dunaliella* is in demand in food, animal feed, pharmaceutical, nutraceutical, and cosmeceutical applications. However, research is needed in the genomics, transcriptomics and metabolomics of β-carotene producing *Dunaliella* to obtain the carotenoid-rich biomass through autotrophic, heterotrophic and mixotrophic cultivation to maximize productivity and also to produce novel carotenoids on an industrial scale.

ACKNOWLEDGMENTS

The author is grateful for all the support provided by the internal funding for projects in Marine Biotechnology from the Center for Scientific Research and Higher Education of Ensenada BC (CICESE), Mexico.

REFERENCES

Aasen, A.J., Eimhjellen, K.E., Liaaen-Jensen, S. An extreme source of β-carotene. *Acta Chemica Scandinavica*. 1969. 23: 2544–2545.

Al-Muhteseb, S.I., Emeish, S. Producing natural mixed carotenoids from *Dunaliella salina*. *Journal of Natural Sciences Research*. 2015. 5: 53–59.

Ambati, R.R., Gogisetty, D., Aswathnarayana Gokare, R., Ravi, S., Bikkina, P.N., Su, Y., Lei, B. *Botryococcus* as an alternative source of carotenoids and its possible applications—An overview. *Critical Reviews in Biotechnology*. 2018. 38 (4): 541–558.

Amorim-Carrilho, K.T., Cepeda, A., Fente, C., Regal, P. Review of methods for analysis of carotenoids. *TrAC Trends in Analytical Chemistry*. 2014. 56: 49–73.

Basily, H., Nassar, M., Diwani, G.E., Abo El-Enin, S. Exploration of using the algal bioactive compounds for cosmeceuticals and pharmaceutical applications. *Egyptian Pharmaceutical Journal*. 2018. 17: 109–120.

Ben-Amotz, A., Avron, M. The role of glycerol in the osmotic regulation of the halophilic Alga *Dunaliella parva*. *Plant Physiology*. 1973. 51: 875–878.

Ben-Amotz, A., Avron, M. The biotechnology of cultivating the halotolerant alga *Dunaliella*. *Trends in Biotechnology*. 1990. 8: 121–126.

Ben-Amotz, A., Katz, A., Avron, M. Accumulation of β-carotene in halotolerant algae: Purification and characterization of β-carotene-rich globules from *Dunaliellabardawil* (Chlorophyceae). *Journal of Phycology*. 1982. 18: 529–537.

Ben-Amotz, A., Levy, Y. Bioavailability of a natural isomer mixture compared with synthetic all-trans β-carotene in human serum. *The American Journal of Clinical Nutrition*. 1996. 63 (5): 729–734.

Berman, J., Zorrilla-López, U., Farré, G., Zhu, C., Sandmann, G., Twyman, R.M., Capell, T., Christou, P. Nutritionally important carotenoids as consumer products. *Phytochemistry Reviews*. 2015. 14: 727–743.

Bogacz-Radomska, L., Harasym, J. β-carotene—Properties and production methods. *Food Quality and Safety*. 2018. 2: 69–74.

Borowitzka, M.A. Commercial production of microalgae: Ponds, tanks, tubes and fermenters. *Journal of Biotechnology*. 1999. 70: 313–321.

Borowitzka, L.J., Borowitzka, M.A. β-carotene (provitamin A) production with algae. In: Vandamme, E.J. (Ed.), *Biotechnology of Vitamins, Pigments and Growth Factors*. 1989: 15–26. Elsevier Applied Science, London.

Borowitzka, L.J., Borowitzka, M.A. Commercial production of β-carotene by Dunaliella salina in open ponds. *Bulletin of Marine Science*. 1990. 47: 244–252.

Borowitzka, M.A., Borowitzka, L.J. Dunaliella. In: Borowitzka, M. A. and Borowitzka, L. J. (Eds.), *Micro-Algal Biotechnology*. 1988. Cambridge.

Capa-Robles, W.R., Paniagua-Michel, J., Soto, J.O. The biosynthesis and accumulation of beta-carotene proceed via the glyceraldehyde 3-phosphate/pyruvate pathway. *Natural Product Research*. 2009. 23: 1021–1028.

Cardoso, L., Kanno, K.Y.F., Karp, S.G. Microbial production of carotenoids—A review. *African Journal of Biotechnology*. 2017. 16: 139–146.

Chen, J., Wang, Y., Benemann, J.R., Zhang, X., Hu, H., Qin, S. Microalgal industry in China: Challenges and prospects. *Journal of Applied Phycology*. 2016. 28: 715–725.

Christaki, E., Bonos, E., Giannenas, I., Florou-Paneri, P. Functional properties of carotenoids originating from algae. *Journal of the Science of Food and Agriculture*. 2013. 93: 5–11.

Cogdell, R.J. Carotenoids in photosynthesis. Philosophical Transactions of the Royal Society B: *Biological Sciences*. 1978. 284: 569–579.

De Carvalho, C.C.C.R., Caramujo, M.J. Carotenoids in aquatic ecosystems and aquaculture: A colorful business with implications for human health. *Frontiers in Marine Science*. 2017. 4: 93.

Dufossé, L., Galaup, P., Yaron, A., Arad, S.M., Blanc, P., Chidambara Murthy, K.N., Ravishankar, G.A. Microorganisms and microalgae as sources of pigments for food use: A scientific oddity or an industrial reality? *Trends in Food Science and Technology*. 2005. 16: 389–406.

Dunal, F. Extrait d'un mémoire sur les algues qui colorant en rouge certains eaux des Marais salants méditerranéens. *Annales des Sciences Naturelles Botanique*2 Séries. 1838. 9: 172.

Fiedor, J., Burda, K. Potential role of carotenoids as antioxidants in human health and disease. *Nutrients*. 2014. 6: 466–488.

Fuller, C.J., Butterfoss, D.N., Failla, M.L. Relative bioavailability of β-carotene from supplement sources. *Nutrition Research*. 2001. 21: 1209–1215.

Gellenbeck, K.W. Utilization of algal materials for nutraceutical and cosmeceutical applications—What do manufacturers need to know? *Journal of Applied Phycology*. 2012. 24: 309–313.

Hamburger, C. Zur Kenntnis der *Dunaliella salina* und einer Amöbe aus Salinenwasser von Cagliari. *Archiv für Protistenkunde*. 1905. 6: 111–130.

Hsu, Y.W., Tsai, C.F., Chang, W.H., Ho, Y.C., Chen, W.K., Lu, F.J. Protective effects of *Dunaliellasalina*—A carotenoid-rich alga, against carbon tetrachloride-induced hepatoxicity in mice. *Food and Chemical Toxicology*. 2008. 46: 3311–3317.

Jaenicke, L. Clara Hamburger and *Dunaliella salina* Teodoresco—A case study from the first half of the XXth century. *Protist*. 1998. 149: 381–388.

Kodric-Brown, A., Brown, J.H. Truth in advertising: The kinds of traits favored by sexual selection. *The American Naturalist*. 1984. 124: 309–323.

Krinsky, N.I., Johnson, E.J. Carotenoid actions and their relation to health and disease. *Molecular Aspects of Medicine*. 2005. 26: 459–516.

Larkum, A.W., Ross, I.L., Kruse, O., Hankamer, B. Selection, breeding and engineering of microalgae for bioenergy and biofuel production. *Trends in Biotechnology*. 2012. 30: 198–205.

Lelyana, R. Role of marine natural ingredient fucoxanthinon body's immune response of obesity. *Journal of Nanomedicine and Nanotechnology*. 2016. 07: 397.

Lerche, W. Untersuchungen über Entwicklung und Fortpflanzung in der Gattung Dunaliella. *Archiv für Protistenkunde*. 1937. 88: 236–268.

Liñán-Cabello, M.A., Paniagua Michel, J.J. Induction factors derived from Carotenoids and vitamin A during maturation of *Litopenaeus vannamei*. *Aquaculture International*. 2004. 12: 583–592.

Liñán-Cabello, M.A., Paniagua-Michel, J.J., Hopkins, P.M. Bioactive roles of carotenoids and retinoids in crustaceans. *Aquaculture Nutrition*. 2002. 8: 299–309.

Massyuk, N.P. Mass culture of the carotene containing alga *Dunaliella salina* Teod. *Ukrayins'kyi Botanichnyi Zhurnal*. 1966. 23: 12–19.

Massyuk, N.P., Abdulla, E.G. First experiment of growing carotene-containing algae under semi industrial conditions. *Ukraine Botanic Zhour*. 1969. 26: 21–27.

Matsuno, T. Animal carotenoids. In: Krinsky, N.I., Mathews-Roth, M.M. and Taylor, R.F. (Eds.). *Carotenoids: Chemistry and Biology*. 1989: 59–74. Springer, Boston, MA.

McMillan, J.R., Watson, I.A., Ali, M., Jaafar, W. Evaluation and comparison of algal cell disruption methods: Microwave, water bath, blender, ultrasonic and laser treatment. *Applied Energy*. 2013. 103: 128–134.

Mil'ko, E.S. Effect of various environmental factors on pigment production in the alga *Dunaliellasalina*. *Mikrobiologya*. 1963. 32: 299–307.

Oren, A.A. A hundred years of Dunaliella research: 1905–2005. *Saline Systems*. 2005. 1: 1–2.

Palozza, P., Serini, S., Di Nicuolo, F., Piccioni, E., Calviello, G. Prooxidant effects of β-carotene in cultured cells. *Molecular Aspects of Medicine*. 2003. 24: 353–362.

Paniagua-Michel, J., Capa-Robles, W.R., Gutierrez-Millan, L.E. The carotenogenesis pathway via the isoprenoid-β-carotene influence approach in a new strain of *Dunaliellasalina* isolated from Baja California, Mexico. *Marine Drugs*. 2009. 7: 45–56.

Paniagua-Michel, J., Dujardin, E., Sironval C. Growth of *Dunaliella bardawil* under carotenogenic conditions. *Journal of Marine Biotechnology*. 1995. 2: 101–104.

Paniagua Michel, J.J., Liñán-Cabello, M.A. Carotenoids, retinoids modulate ovarian development in *Litopenaeus vannamei*. *Global Aquaculture Advocate*. 2002: 25–26.

Paniagua Michel, J., Olmos, J., Acosta, R.M. Pathways of carotenoid biosynthesis in bacteria and microalgae. In: Barredo, J.-L. (Ed.). Microbial Carotenoids from Bacteria and Microalgae: Methods and protocols, *Methods in Molecular Biology*. 2012: 1–12. Springer.

Paniagua-Michel, J., Olmos, J., Morales-Guerrero, E.R. Microalgal Carotenoids: Bioactive roles, health foods and pharmaceuticals. In: Kim, S.K. (Ed.). *Marine Algae Extracts: Processes, Products and Applications*. 2015. 639–658.Wiley-VCH, Weinheim, Germany.

Paniagua Michel, J.J., Subramanian, V. Omics advances of biosynthetic pathways of isoprenoid production in microalgae. In: Kim, S.K. (Ed.). *Marine Omics: Principles and Applications*. 2016: 37–58. CRC Press.

Pisal, D.S., Lele, S.S. Carotenoid production from microalgae, *Dunaliella salina*. *Indian Journal of Biotechnology*. 2005. 4: 476–483.

Raja, R., Hemaiswarya, S., Rengasamy, R. Exploitation of *Dunaliella* for β-carotene production. *Applied Microbiology and Biotechnology*. 2007. 74: 517–523.

Ramos, A.A., Polle, J., Tran, D., Cushman, J.C., Jin, E.S., Varela, J.C. The unicellular green alga *Dunaliella salina* Teod. As a model for abiotic stress tolerance: Genetic advances and future perspectives. *Algae*. 2011. 26: 3–20.

Ranga Rao, A., Baskaran, V., Sarada, R., Ravishankar, G.A. *In vivo* bioavailability and antioxidant activity of carotenoids from micro algal biomass—a repeated dose study. *Food Research International*. 2013. 54: 711–717.

Ranga Rao, A., Deepika, G., Sarada Ravi, R.G.A., Narasimharao, P.B., Lei, B., Su, Y. Industrial potential of carotenoid pigments from microalgae: Current trends and future prospects. *Critical Reviews in Food Sciences and Nutrition*. 2018. 1: 1–22.

Sathasivam, R., Ki, J.-S. A review of the biological activities of microalgal carotenoids and their potential use in healthcare and cosmetic industries. *Marine Drugs*. 2018. 16: 26.

Schwender, J., Gemünden, C., Lichtenthaler, H.K. Chlorophyta exclusively use the 1-deoxyxylulose 5-phosphate/2-C-methylerythritol4-phosphate pathway for the biosynthesis of isoprenoids. *Planta*. 2001. 212: 416–423.

Shete, V., Quadro, L. Mammalian metabolism of β-carotene: Gaps in knowledge. *Nutrients*. 2013. 5: 4849–4868.

Spolaore, P., Joannis-Cassan, C., Duran, E., Isambert, A. Commercial applications of microalgae. *Journal of Bioscience and Bioengineering*. 2006. 101: 87–96.

Srivastava, S. Food adulteration affecting the nutrition and health of human beings. *Biol. Scientia Medica*. 2015. 1: 65–70.

Teodoresco, E.C. Organisation et développement du Dunaliella, nouveau genre de Volvocacée-Polyblepharidée. *Bot Centralbl, B.z.* 1905. XVIII: 215–232.

Tsushima, M., Kawakami, T., Mine, M., Matsuno, T. The role of carotenoids in the development of the sea urchin *Pseudocentrotus depressus*. *Invertebrate Reproduction and Development*. 1997. 32: 149–153.

Voutilainen, S., Nurmi, T., Mursu, J., Rissanen, T.H. Carotenoids and cardiovascular health. The *American Journal of Clinical Nutrition*. 2006. 83: 1265–1271.

Wells, M.L., Potin, P., Craigie, J.S., Raven, J.A., Merchant, S.S., Helliwell, K.E., Smith, A.G., Camire, M.E., Brawley, S.H. Algae as nutritional and functional food sources: Revisiting our understanding. *Journal of Applied Phycology*. 2017. 29: 949–982.

Yang, D.J., Lin, J.T., Chen, Y.C., Lu, F.J., Chang, T.J., Wang, T.J., Lin, H.W., Chang, Y.Y. Suppressive effect of carotenoid extract of *Dunaliella salina* alga on production of LPS-stimulated pro-inflammatory mediators in RAW264.7 cells via NF-kB and JNK inactivation. *Journal of Functional Foods*. 2013. 5: 607–615.

Zhang, J., Sun, Z., Sun, P., Chen, T., Chen, F. Microalgal carotenoids: Beneficial effects and potential in human health. *Food and Function*. 2014. 5: 413–425.

15 Exploring the Potential of Using Micro- and Macroalgae in Cosmetics

W.A.J.P. Wijesinghe and N.E. Wedamulla

CONTENTS

Abbreviations 149
Introduction 149
Microalgae and Macroalgae as Functional Ingredients 150
    Microalgae as a Source of Functional Ingredients 150
    Macroalgae as a Source of Functional Ingredients 151
Biological Activities of Microalgae and Macroalgae in Cosmeceutical Applications 154
    Antiaging Effects 154
    Moisturizing/Hydration Action 154
    Photo-Protective Action 154
    Whitening/Melanin-Inhibiting Effects 155
    Anticellulite and Slimming Effects 155
Isolation Techniques 155
Marketing Potential of Algal Cosmeceuticals 156
Conclusion 157
References 157

BOX 15.1 SALIENT FEATURES

Marine organisms are rich sources of structurally novel and biologically active metabolites. These metabolites possess various industrial applications, including pharmaceuticals, cosmeceuticals and functional foods. This chapter discusses the potential uses of macro- and microalgae in functional cosmetics. Macro- and microalgae produce a variety of bioactive components and they are diverse in their chemical structure. Hence, macro and microalgae have been identified as potential and easily accessible producers of a wide spectrum of natural substances that are vital for cosmeceutical products. Thus, the growing interest in the extraction of active compounds from these sources is obvious. The potent biological properties of certain active components isolated from macro- and microalgae may represent an interesting advance in the search for novel functional ingredients. Moreover, it is interesting to note that functional metabolites from macro- and microalgae offer a promising approach for use in cosmeceutical preparations such as whitening or de-pigmenting creams, antiwrinkle or antiaging lotions, deodorants and moisturizing creams.

ABBREVIATIONS

DHA: docosahexaenoic acid
EPA: eicosapentaenoic acid
ECM: extracellular matrix
NMF: natural moisturizing factor
ALA: α-linolenic acid
SFE: supercritical fluid extraction
SWE: subcritical water extraction
UAE: ultrasound-assisted extraction
MAE: microwave-assisted extraction

INTRODUCTION

The cosmetic industry is flourishing at a global level, and the use of natural ingredients in cosmeceuticals has drawn considerable attention owing to their beneficial effects on human health in addition to the desired properties (Thomas and Kim, 2013). Recently, customers have been constantly seeking novel bioactive compounds derived from natural sources to use in cosmeceuticals because of their beneficial effects compared to those of synthetic ingredients (Wijesinghe and Jeon, 2011; Kim, 2016). From this perspective, bioactive compounds derived from algae possess diverse functional roles as secondary metabolites and thus have potential applications in the development of novel cosmeceuticals (Thomas and Kim, 2013).

Marine organisms produce vast untapped sources of bioactive compounds which provide diverse leads in novel applications. Marine algae can be classified into two major groups, microalgae and macroalgae; blue-green algae, dinoflagellates and bacillariophyta (diatoms) are microalgae, while green, brown and red algae are classified as macroalgae (El Gamal, 2010). The cosmetic industry has already exploited the world of macroalgae; however, the information relevant to microalgae is still scanty (Ariede et al., 2017). There has been an influx of literature proving algae's biological properties, including its antioxidant, antibacterial and antiviral activity, and its positive effect on some types of cancer, heart disease, thyroid and immune functions, allergies and inflammation (Caccamese et al., 1981; Freile-Pelegrin and Morales, 2004; Da Costa et al., 2017). Algae have also been identified as a bio-sustainable source of ingredients (Pimentel et al., 2018) which are proven to be safe.

However, the development of an appropriate isolation technique is equally important in this regard to obtain safe and economically viable product suitable for industrial production. Considering the constraints associated with solvent-based extraction techniques, recent research is focusing more on solvent-free extraction techniques, which also deal with environmental concerns (Michalak et al., 2015).

MICROALGAE AND MACROALGAE AS FUNCTIONAL INGREDIENTS

Microalgae have been identified as one of the most diverse groups of microorganisms living in marine and fresh-water systems (Hu et al., 2008). Although microalgae are categorized as eukaryotic microorganisms in taxonomy, prokaryotic cyanobacteria, green algae and diatoms, along with many other eukaryotic groups, are also classified as microalgae in its broader definition (Metting, 1996). Green microalgae comprises common genera such as *Chlorella*, *Dunaliella* and *Haematococcus*, while diatoms include the genera *Phaeodactylum* (Fu et al., 2017). However, a few studies have also been carried out on microalgae and their commercial applications. Thus, only a few species of microalgae have been extensively studied, including *Spirulina*, *Chlorella*, *Haematococcus*, *Dunaliella*, *Botryococcus*, *Phaeodactylum*, and *Porphyridium* (Borowitzka, 2013; Christaki, 2014). Microalgae can grow rapidly and survive under harsh conditions and in the presence of environmental stressors owing to their unicellular or simple multicellular structure (Christaki, 2013). Therefore, microalgae serve as a source of valuable compounds such as polyunsaturated fatty acids, proteins, tocopherols, sterols and many other metabolites (Christaki, 2014).

Macroalgae, commonly known as seaweeds, are divided into four major classes: Rhodophyceae (red algae), Phaeophyceae (brown algae), Cyanophyceae (blue-green algae) and Chlorophyceae (green algae). Seaweeds also serve as a pool of unique biologically active components due to the wide diversity of their biochemical composition (Thomas and Kim, 2013). Dietary fiber, omega 3 fatty acids, essential amino acids and vitamins A, B, C and E are some of the health-promoting components present in seaweeds (Rajapakse and Kim, 2011).

MICROALGAE AS A SOURCE OF FUNCTIONAL INGREDIENTS

Recently, microalgae have drawn considerable attention from scientific communities (Shimizu, 1996; Barrera, et al., 2014). Microalgal extracts have been widely used in the formulation of cosmetic and skin-care products. These microalgal extracts serve as a rich source of bioactive proteins, vitamins, minerals and carotenoid pigments such as astaxanthin (Kim and Wijesekara, 2016; Ambati et al., 2018; Ranga Rao et al., 2018).

Among the active ingredients extracted from microalgae, polysaccharides demonstrate greater potential for use in cosmetic ingredients. Algal polysaccharides are used as gelling agents, thickeners and moisturizers in many cosmetic formulas (Jain et al., 2005). Moreover, biological activities of microalgae species also related with these polysaccharides. For example, the polysaccharides complexes from *Chlorella pyrenoidosa* and possibly *Chlorella ellipsoidea* are believed to have immune-stimulating properties. These complexes contain glucose and any combination of galactose, rhamnose, mannose, arabinose, N-acetylglucosamide and N-acetylgalactosamine. Further, the polysaccharide β-1,3-Glucan from *Chlorella* acts as active immune stimulator and free-radical collector (Hamed, 2016). These biological properties strengthen potential use of microalgae in skincare cosmetics, especially in retarding ageing (Mourelle et al., 2017). The *Chlorella* sps, the *Skeletonema* diatom and *Porphyridium* and *Nostoc flegelliforme* are identified as species rich in β–glucans (Hamed, 2016).

Microalgae contain 1–70% oil or fat content on a dry-weight basis. Microalgal lipids are categorized into two major groups: storage lipids, which consist of nonpolar lipids (triacylglycerides); and structural lipids, which consist of polar lipids such as phospholipids and sterols. Microalgae also contain prenyl derivatives (tocopherols, carotenoids, terpenes, quinines) and pyrrole derivatives (chlorophylls). Moreover, microalgae serve as an excellent source of polyunsaturated fatty acids (Sharma et al., 2012). Among these, essential fatty acids, primarily omega-3 and omega-6, are vital for the reliability of tissues. Linoleic acid is also used in the treatment of skin hyperplasia (Santhosh et al., 2016). Additionally, an influx of literature has provided evidence of the biological properties of the polyunsaturated fatty acids docosahexaenoic acid (DHA) and eicosapentaenoic acid (EPA), which are the most extensively studied. Studies have revealed the beneficial effect of algae-derived omega-3 fatty acids on visual and neural development. Further, these fatty acids have been found to prevent many diseases, *viz.*, heart conditions, hypertension, cancer, diabetes, cystic fibrosis, asthma, arthritis, depression and schizophrenia (Ohse et al., 2014). Microalgae-derived fatty acids provide an alternative source to fish oil, which has an unpleasant odour and is often contaminated with heavy metals (Robles Medina et al., 1998).

Microalgal-derived pigments with antioxidizing potential are generating incredible demand in the cosmetics industry as they act as antioxidants, as well as for their value as colorant. Carotenoids, phycobilins and chlorophylls are the major three classes of photosynthetic pigments present in autotrophs. Among carotenoids, β-carotene is also source of vitamin A. Carotenoids have been shown to protect against cancer, aging, ulcers, heart attacks and coronary artery disease. β-carotene is one of the most common careteniods extracted from *Dunaliella salina* that has wider applications in cosmetics. Astaxanthin is the second most important microalgal-derived carotenoid found in the green microalga *Haematococcus pluvialis* (Ranga Rao et al., 2013). Astaxanthin exhibits various biological properties including protection against cancer, inflammation and UV light. Further, it improves immune responses (Hamed, 2016). Since astaxanthin provides protection against UV-induced photo-oxidation, it can be used in natural sunscreen cosmetics (Mourelle et al., 2017; Ranga Rao et al., 2013). Other carotenoids include lutein, zeaxanthin and canthaxanthin, which are produced in low quantities (Hamed, 2016). Phycobiliproteins are protein-pigment complexes which are classified under another major class of pigments as phycocyanobilin (blue pigment) and phycoerythrobilin (red pigment). These pigments are being produced on a large scale from *Spirulina* (cyanobacterium) and *Porphyridium* (red microalgae). These pigments have wider applications in the decorative cosmetic industry and also act as good antioxidants. Chlorophylls are green pigments which can easily extracted from microalgae and used to mask odors in the cosmetic industry (Mourelle et al., 2017). In brief, microalgae-derived pigments exhibit vast diversity in terms of their biological properties, including antioxidant, anti-inflammatory and antiaging effects (Figure 15.1).

Microalgae represent an important source of vitamins, including A, B_1, B_2, B_6, B_{12}, C, E, nicotinate, biotin, folic acid and pantothenic acid. Owing its very strong antioxidant activity, vitamin E has potential applications in the cosmetic industry (Mourelle et al., 2017).

Spirulina is an excellent example of a microalgae with a pool of bioactive substances which include vitamins, carotenoids, fatty acids and minerals. *Spirulina* has also been identified as a rich source of protein with all the essential amino acids. Moreover, a surge of literature has proven the diverse biological properties of *Spiruilna* peptides, which are antimicrobial, antiallergic, antihypertensive and immunomodulatory. With these biological properties, some cosmetics tend to use peptides as ingredients among UV-protective, acne and dermatosis therapies and as a pro-inflammatory agent for skincare. Peptides are also used in age-defying skincare products, serving as an alternative to botulinum neurotoxin, a pigmentation modulator (Ovando et al., 2018).

Macroalgae as a Source of Functional Ingredients

Seaweed is one of the most important sources of bioactive compounds, which have wider applications in the food, pharmaceutical and cosmetic industries. Seaweed extracts are a rich source of cosmeceutical ingredients such as polysaccharides, proteins, lipids, minerals, and vitamins as well as secondary metabolites such as phenolic compounds, terpenoids, halogenated compounds and so on (Kim, 2014).

The cell walls of marine algae are considered to be a rich source of various bioactive polysaccharides. The total polysaccharide content of seaweed species ranged between 4% and 76% on a dry weight basis, and *Ascophyllum*, *Porphyra* and *Palmaria*-like species account for the highest polysaccharide concentrations. However, green seaweed species such as *Ulva* also have comparatively high contents, up to 65% on

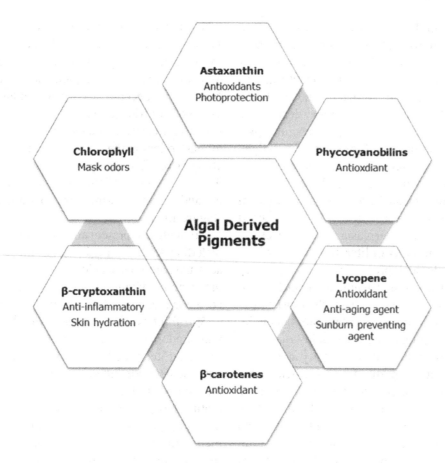

FIGURE 15.1 Microalgae-derived pigments and associated biological properties.

a dry weight basis (Holdt and Kraan, 2011). Fucoidans in brown algae, carrageenans in red algae and ulvans in green algae are good examples of these bioactive polysaccharides. Among polysaccharides, fucoidans are widely commercialized with the aid of brown seaweed sources which include *Laminaria japonica*, *Fucus vesiculosus*, *Undaria pinnatifida* and *Hizikia fusiformis*. Fucoidans have become one of the most promising research areas owing to their potent bioactive properties (Kim, 2014). Table 15.1 lists some significant macroalgal-derived polysaccharides and their functional properties.

Protein content is also utmost important in determining the value of biomass. This change with the species of interest; Rhodophyceae has the highest protein content (8%–50% dw) followed by Chlorophyceae (7–32% dw) and Phaeophyceae (6–24% dw) (Harnedy and FitzGerald, 2011). Macroalgae also serve as an excellent source of peptides and amino acids. The biofunctional properties of these peptides depend on their amino-acid composition and sequence in the parent protein (Harnedy and FitzGerald, 2011). Peptides possess functional ability to alter skin physiology. Peptides increase collagen production and compensate for the loss of extracellular matrix (ECM), thereby reducing the size and appearance of wrinkles. Further, peptides regulate the fibroblast production of ECM components (Katayama et al., 1991; Langholz et al., 1995). The free amino acid proportion of seaweed mainly consists of alanine, amino butyric acid, taurine, omithine, citruiline and hydroxyproline (Holdt and Kraan, 2011). Several species of red macroalgae such as *Palmaria* and *Porphyra* contain considerably high amounts of arginine. Arginine acts as a precursor of urea and serves as a component of natural moisturizing factor (NMF) and thus finds applications in cosmetic formulations (Bedoux et al., 2014).

Lipids include fats; waxes; sterols; fat-soluble vitamins; mono-, di- and triacylglycerols; diglycerides; and phospholipids. Some studies have proven phospholipids to be the main source of lipids in marine algae (Murata and Nakazoe, 2001), while others state that glycolipids are the major component (Narayan et al., 2005). Macroalgae account for comparable or higher polyunsaturated fatty acid content than that of terrestrial plants regardless of their low total lipid content (Pimentel et al., 2018).

TABLE 15.1
Biological Properties of Macroalgal-Derived Polysaccharides

Polysaccharide	Source	Biological Activity	References
Fucoidans	*Laminaria japonica*	Antioxidant activity	Kim, (2014); Holdt and Kraan, (2011); Rocha de Souza et al., (2007)
	Fucus vesiculosus	Anticoagulant activity	
	Undaria pinnatifida	Antiviral activity	
	Hizikia fusiformis	Anticancer activity	
Carrageenan	*Gracilaria*	Antiviral activity	Hardouin et al., (2014); Kim, (2014)
	Chondrus	Antioxidant activity	
	Gelidiella	Anticoagulant effect	
		Antitumor activity	
Porphyrans	*Porphyra*	Antitumor activity	Baweja et al., (2016); Noda, (1993)
		Significantly lower the artificially enhanced level of hypertension and blood cholesterol	
Alginate	*Ascophyllum*	Antibacterial effect	Holdt and Kraan, (2011)
	Laminaria	Decrease the concentration of cholesterol	
	Sargassum	Antihypertension effect	
		Prevent absorption of toxic chemical substances	
Ulvans	*Ulva* and *Enteromorpha* spp.	Antioxidant activity	Holdt and Kraan, (2011)
		Strain-specific anti-influenza activity	

Among these, long-chain polyunsaturated fatty acids are very important. Eicosapentaenoic acid (EPA; 20:5n-3) and docosahexaenoic acid (DHA; 22:6n-3) are the most important long-chain polyunsaturated fatty acids and are extensively studied for their health-promoting properties. Moreover, macroalgae serve as the source for α-linolenic acid (ALA; 18:3n-3) and docosapentaenoic-acid (22:5n-3). According to the reported studies, both brown and red species were richer in Arachidonic acid and EPA, while the green species showed high amounts of DHA (Pimentel et al., 2018).

Macroalgae also contain array of pigments which absorb light for photosynthesis. Most of these pigments are not commonly found in terrestrial plants. Three main classes of pigments are identified in macroalgae: chlorophylls, carotenoids and phycobiliproteins (Dumay and Morançais, 2016). These carotenoids are usually present as red, orange or yellow pigments with promising antioxidant activity. Moreover, β-carotene has identified as the most representative carotenoid present in all classes of macroalgae (Dumay and Morançais, 2016). Phycobiliproteins are a water-soluble photosynthetic pigment. Phycoerythrins, with a red pigment, are linked to the protein molecule; or phycocyanins, with a blue pigment in its place. Different wavelengths of the spectrum are absorbed by these different molecules, making those molecules very colorful and highly fluorescent in vivo and in vitro. This phenomenon generates a great deal of interest in biotechnological applications, in which they contribute to diverse biomedical diagnostic systems (e.g. immunochemical methods). Moreover, cosmetic applications of these natural colorants are well defined; pink and purple dyes in lipsticks, eyeliner and other cosmetic formulations are good examples of such applications (Pimentel et al., 2018).

Macroalgae are also recognized as promising source of vitamins, minerals and trace elements. Both fat-soluble vitamins and water-soluble vitamins such as thiamine, riboflavin, niacin, pantothenic acid, pyridoxine, cobalamin, biotin, folic acid and ascorbic acid are also contained in microalgae. Trace elements are also present in macroalgae and include calcium, sodium, potassium, magnesium, iron, copper, iodine and zinc (MacArtain et al., 2007).

Secondary metabolites such as phenolic compounds present in macroalgae do not directly intervene in primary metabolic processes; however, protection mechanisms therein, namely against oxidative stress or UV cytotoxic effects, are well defined (Pimentel et al., 2018).

BIOLOGICAL ACTIVITIES OF MICROALGAE AND MACROALGAE IN COSMECEUTICAL APPLICATIONS

The incorporation of algae-derived functional ingredients into cosmetic formulations has grown tremendously (Zhang et al., 2010; Freile-Pelegrin and Morales, 2004). Biological compounds derived from algal sources exhibit diverse functional roles intended for antiaging care, including protection against free radicals, prevention of skin flaccidity and wrinkles, antiphotoaging, photoprotection against UV radiation, skin whitening and moisturizing (Ariede et al., 2017;Wang et al., 2015).

ANTIAGING EFFECTS

Ageing can be simply introduced as an unavoidable, slow and complex phenomenon (Couteau and Coiffard, 2016). Two mechanisms are involved in aging: intrinsic and extrinsic aging. Atrophy, fibroblast reduction and thinning blood vessels result in intrinsic aging. The synthesis of collagen and elastin also steadily declines with age. On the other hand, UV damage is the primary cause of extrinsic aging. Environmental factors such as smoking, pollution and poor nutrition also contribute to this process. These types of damage lead to the increased degradation of collagen and elastin. Once the skin gets aged, it demonstrates a decrease in ECM proteins, increased collagen degradation and decreased fibroblasts. Further, extrinsic aging also leads to the generation of free radicals (Malerichand Berson, 2014).

The role of *Chlorella vulgaris* extracts in collagen repair mechanisms has been reported (Chandra et al., 2017). The ability of sulfated polysaccharides derived from marine microalgae to prevent free radicals and reactive chemical species activity is also widely acknowledged (de Jesus Raposo et al., 2015). Moreover, sulphated polysaccharides derived from macroalgae can exhibit interesting biological properties such as modulation of connective tissue proteolysis (Wijesinghe and Jeon, 2012) and anti-inflammatory effects (Kang et al., 2011). More importantly, these polysaccharides can also act as free-radical scavengers and antioxidants for the prevention of oxidative stress–induced damage in humans (Hu et al., 2001). In addition, studies have revealed the effectiveness of these sulfated polysaccharides in growth factor modulation, stimulating oxidative phenomena, increasing skin thickness and elasticity and inhibiting matrix metalloproteinases (Bedoux et al., 2014).

MOISTURIZING/HYDRATION ACTION

Moisturizing and hydration are vital for skincare, important for maintaining healthy looking skin as well as its elasticity. Hydration also helps by strengthening barrier capacity against harmful environmental factors (Bedoux et al., 2014). The amount of water present in the epidermis (60%) is determined by hygroscopic substances known as natural moisturizing factors (NMFs). These NMFs contain amino acids (40%), including serine (20–30%), lactic acid (12%), pyrrolidone carboxylic acid (12%), urea (8%), sugars and minerals (Couteau and Coiffard, 2016). The aforementioned components can act as humectants, and thus the topical application of those compounds can improve the skin's moisturizing ability and relieve dry skin conditions (Pimentel et al., 2018).

Polysaccharides can act as humectants and moisturizers. These properties are attributed to the fact that polysaccharides can be linked to keratin through hydrogen bonds and also exhibit high water storage capacity, thereby improving the moisturization of skin (Pimentel et al., 2018). Polysaccharides extracted from *Saccharina japonica* have been revealed to have better moisturizing properties than hyaluronic acid (Wang et al., 2013), thus suggesting potential applications as cosmetic ingredient. The study further revealed that sulphates act as the main active site for moisture absorption and moisture retention ability, and thus lower-molecular-weight polysaccharides showed the highest moisture absorption and moisture retention abilities (Wang et al., 2013). Microalgae-derived polysaccharides also have wider applications in cosmetics as a moisturizing agent; genus *Chlorella* is a good example of such applications (Mourelle et al., 2017).

PHOTO-PROTECTIVE ACTION

The most powerful environmental risk factor in skin-cancer pathogenesis is sunlight UV radiation. Thus, photoprotective products with UV filters are highly recommended for the prevention of several types of damage such as sunburn, photo-aging, photo-dermatoses or even skin cancer. Formulations containing sunscreen agents combined with antioxidants can increase the effectiveness and safety of the product in question (Ariede et al., 2017).

In addition to sunburn, UV radiation induces the formation of radical species; this leads to the destruction of proteins, DNA and other biomolecules as well as causing acute physiological stress. Moreover, this leads to proliferation of oncogenes, which can mutate and cause cancer. However, various natural compounds isolated from plant bases may have potential properties that can combat UV radiation (Bedoux et al., 2014). Bioactive compounds possess the ability to absorb UV radiation and thereby protect human fibroblast cells from UV-induced cell death and suppress UV-induced aging in human skin. Similarly, macroalgae have developed mechanisms which counteract the damaging effects of UV-B and UV-A by producing screen pigments such as carotenoids and phenolics (Bedoux et al., 2014).

Lycopene is the one of the most powerful antioxidants which scavenges reactive oxygen species. It also acts as sunburn-preventing agent and thus can be used as a sunscreen. Astaxanthin is another important microalgal-derived powerful antioxidant, thus effectively scavenging free radicals. This compound is of the utmost importance in skin protection against UV-induced photo-oxidation and thus has high potential for use in natural sunscreen cosmetics (Koller et al., 2014).

Whitening/Melanin-Inhibiting Effects

Melanin is the primary determinant of skin color. It absorbs UV radiation and prevents free-radical generation, thereby protecting skin from sun damage and aging (Hollinger et al., 2015). Melanin biosynthesis involves tyrosinase, which catalyzes melanin synthesis via two different pathways: the hydroxylation of L-tyrosine to 3,4-dihydroxy-l-phenylalanine (L-dopa) and the oxidation of L-dopa to dopaquinone, followed by further conversion to melanin. Therefore, the inhibition of tyrosinase is the easiest approach that can be used to achieve skin whiteness (Couteau and Coiffard, 2016). With this piece of information, more research is being directed towards the development of novel tyrosinase inhibitors from natural resources, and in recent years, marine algae have drawn considerable attention in the search for natural tyrosinase inhibitor agents (Thomas and Kim, 2013). For example, fucoxanthin extract of *L. japonica* acts as a tyrosinase inhibitor. Moreover, extracts of *Endarachne binghamiae, Schizymenia dubyi, E. cava* and *Sargassum siliquastrum* have been widely acknowledged as good candidates (Couteau and Coiffard, 2016). Phloroglucinol derivatives of brown algae also possess tyrosinase inhibitory activity owing to their ability to chelate copper in this enzyme (Thomas and Kim, 2013).

Anticellulite and Slimming Effects

Although cellulite is not identified as pathogenic concern, it still remains a matter of concern in cosmetics. Extensive research has been carried out to overcome symptoms and signs of cellulite, as well as the visual appearance of skin. Some species of macroalgae are already being used in cosmetic formulations with the objective of reducing cellulite; *Fucus vesiculosus* L. and *Laminaria digitata* (Huds.) are excellent examples of these applications (Bedoux et al., 2014).

Macroalgal extracts also demonstrate promising characteristics for use in slimming products; as revealed by research, they significantly decrease body weight gain, fat-pad weight and serum and hepatic lipid levels in high-fat-diet-induced Sprague Dawley male obese rats. Moreover, these protective effects are generated through the regulation of gene and protein expression involved in lipolysis and lipogenesis (Jang and Choung, 2013).

ISOLATION TECHNIQUES

The selection of proper solvents is vital in order to extract bioactive substances from raw biomass. Some disadvantages have been identified in relation to conventional extraction methods, which include extraction in Soxhlet apparatus, solid–liquid extraction and liquid–liquid extraction. One of the major constraints associated with these extraction techniques is the use of a high volume of solvents. Thus, these techniques are gradually being replaced by alternative methods like supercritical fluid extraction (Pan et al., 2012). Moreover, the use of eco-friendly and nontoxic solvents coupled with efficient and sustainable extraction techniques has gained considerable attention (Anastas and Warner, 1998).

Before bioactive compounds can be isolated from seaweeds, the algal raw material must be harvested and stored (by freeze drying) (Bedoux et al., 2014).

Alternative methods or methods commonly identified as green methods have shown several advantages over conventional methods; these include reduced amounts of solvent, shorter extraction time and performance at lower temperatures. Moreover, these methods exhibit better selectivity for the isolation of desired compounds and also avoid the formation of byproducts and unwanted reactions during the extraction process (Cikoš et al., 2018). There is a growing trend towards the application of innovative extraction techniques in the isolation of bioactive compounds, which include supercritical fluid extraction (SFE), subcritical water extraction (SWE), ultrasound-assisted extraction (UAE) and microwave-assisted extraction (MAE). SFE uses fluids in their supercritical conditions for the extraction; that is, temperature and pressure are raised above their critical point and thus the fluids have the characteristics of both liquids and gases (del Pilar Sánchez-Camargo et al., 2017). Carbon dioxide (CO_2) is the most commonly used solvent for SFE because of its nontoxicity, safety and low cost (Duarte et al., 2014). SWE has been identified as most promising technique for the extraction of bioactive compounds. This technique operates at high temperatures (50–200°C) and pressures (50–300 psi) for a short period of time (5–10 min) with a small amount of solvent. During the extraction, solvents are maintained near their critical region in the liquid state with the help of applied temperature and pressure, keeping the solvents below their boiling point (Duarte et al., 2014). Ultrasound waves with a frequency of 20–100 kHz are employed in UAE. These ultrasound waves create bubbles and zones of high and low pressure. Once the bubbles collapse in the strong ultrasound field, cavitation occurs. Ultimately, this implosive collapse, cavitation, near liquid–solid interfaces causes breakdown of particles. Thus, mass transfer is increased and bioactive compounds are released from the biological matrix (Kadam et al., 2013). Finally, MAE employs ionic conduction and dipole rotation, which act directly on the molecules and occur simultaneously. Microwave heating causes the absorption of energy by molecules in which loss of heat into the environment is kept to a minimum. Disruption of cells is take place due to absorption of energy by polar molecules. Destructed cells facilitate faster mass transfer and diffusion out of solid, where mass and heat transfer act synergistically and in the same direction (Seoane et al., 2017).

MARKETING POTENTIAL OF ALGAL COSMECEUTICALS

The algal cosmeceutical industry has identified as one of the most rapidly growing industries with a promising future. In the light of novel research outcomes relevant

TABLE 15.2
Examples of Algal-Based Cosmetic Products

Product Name	Algae	Product Description	References
Dove Regenerative Repair Shampoo + Conditioner	Red algae	Moisture retention properties of algae are well acknowledged and also assist proteins to maintain their structure. Thereby, they help to replenish nutrients, fortify the internal structure of damaged hair and give it a smoother appearance.	Nykaa (2018a)
L'Oreal Paris Pure Clay Mask (Red Algae) + Hydrafresh Anti-Ox Cream	Red algae	Exfoliates and refines pores; pure clay mask acts on rough skin and deep clogged pores on face to reveal a polished, pore-minimized and smooth complexion. Immediately skin texture appears refined and smooth. Ultimately, skin will be healthy looking with a beautiful glow.	Nykaa (2018b)
Superfood Skin Reset Mask	Microalgae	This mask acts, as reset button for the skin. It instantly helps balance and restore dull, tired skin. Spirulina, a sea-based superfood, and bioactive microalgae, a nutrient-rich source of exfoliating enzymes, are packed with phytonutrients and vitamins. Essential minerals deliver age-defying antioxidants, phytonutrients, essential amino acids, vitamins and minerals to skin.	Nordstrom (2018)
Ultra UV Defense Brightening Cream with Antioxidants— Broad Spectrum SPF 30 Sunscreen	Red algae—*Porphyra umbilicalis*	This product prevents premature skin aging and dark spots. Moreover, product is enriched with antioxidants, including red algae extract, for additional skin protection and antiaging benefits.	Jenelt (2018b)
Skin Renewal Radiance Lotion	Brown algae—*Dictyopteris membranacea*	Stimulates skin renewal, leaving skin healthier and younger-looking. Also, reduces dark spots and redness with a synergistic blend of natural brightening ingredients *viz.* algae, bearberry, licorice.	Jenelt (2018a)
Osea Eyes and Lips	Algae	Algae extract contains potent vitamins and minerals which assist in the hydration of dry skin and reinforce the appearance of a healthy, balanced, moisture-rich complexion.	OSEA (2018)

Applications of macro- and microalgae in cosmetics.

to algal bioactives and their potential cosmetic applications, a large number of small- and large-scale business entities have forced profusely towards using these algal metabolites in their cosmetic products. Table 15.2 gives a few examples of algal-based cosmetic products already on the market. Algal-derived ingredients have been used in many cosmetics because of their technological properties. However, there is a growing trend towards use of algae as a source of added value compounds owing to their health-promoting properties (Pimentel et al., 2018). Moreover, these applications are supported by an influx of scientific evidence directed towards the investigation of the biological effects of marine algae on skin, such as antiaging, moisturizing, whitening and photo-protective properties, which ultimately create a competitive advantage (Pimentel et al., 2018).

CONCLUSION

With growing consumer demand for skincare products based on natural ingredients, bioactive compounds derived from algal sources play a vital role in novel cosmetic formulations due to their unique biological properties, which ultimately have beneficial effects on health. Algae-derived compounds possess diverse biological activities, *viz.* antiaging, photoprotection, antiinflammatory effects, antioxidant effects, moisturizing effects and whitening effects. However, the comparatively high production cost and the constraints associated with ensuring a continuous supply of algal sources are major challenges identified in cosmetic applications of algal-derived biological substances. These challenges and growing consumer demand have created extra pressure to explore novel marine algal resources. In addition, the development of environmentally friendly extraction techniques is of the utmost importance from a sustainability point of view. Also, there is an urgent need to establish well organized procedures to commercialize algal-derived functional components based on the evidence and effectiveness of their biological activities.

REFERENCES

Ambati, R.R., Gogisetty, D., Aswathnarayana Gokare, R., Ravi, S., Bikkina, P.N., Su, Y., Lei, B., 2018. Botryococcus as an alternative source of carotenoids and its possible applications—An overview. *Critical Reviews in Biotechnology*, 38(4), pp. 541–558.

Ambati, R.R., Phang, S.M., Ravi, S., Aswathanarayana, R.G., 2014. Astaxanthin: Sources, extraction, stability, biological activities and its commercial applications – A review. *Marine Drugs*, 12(1), pp. 128–152.

Anastas, P.T., Warner, J.C., 1998. Principles of green chemistry. In: *Green Chemistry: Theory and Practice* (pp. 29–56).

Ariede, M.B., Candido, T.M., Jacome, A.L.M., Velasco, M.V.R., de Carvalho, J.C.M., Baby, A.R., 2017. Cosmetic attributes of alga—A review. *Algal Research*, 25, pp. 483–487.

Barrera, D., Gimpel, J., Mayfield, S., 2014. Rapid screening for the robust expression of recombinant proteins in algal plastids. In: *Chloroplast Biotechnology* (pp. 391–399). Humana Press, Totowa, NJ.

Baweja, P., Kumar, S., Sahoo, D., Levine, I., 2016. Biology of seaweeds. In: *Seaweed in Health and Disease Prevention*, Fleurence, J., Levine, I., Eds. (pp. 41–106). Elsevier Inc., Amsterdam, The Netherlands.

Bedoux, G., Hardouin, K., Burlot, A.S., Bourgougnon, N., 2014. Bioactive components from seaweeds: Cosmetic applications and future development. In: *Advances in Botanical Research* (Vol. 71, pp. 345–378). Academic Press.

Borowitzka, M.A., 2013. High-value products from microalga—their development and commercialisation. *Journal of Applied Phycology*, 25(3), pp. 743–756.

Caccamese, S., Azzolina, R., Furnari, G., Cormaci, M., Grasso, S., 1981. Antimicrobial and antiviral activities of some marine algae from eastern Sicily. *Botanica Marina*, 24(7), pp. 365–368.

Chandra, R., Parra, R., Iqbal, H.M., 2017. Phycobiliproteins: A novel green tool from marine origin blue-green algae and red algae. *Protein and Peptide Letters*, 24(2), pp. 118–125.

Chemat, F., 2011. Eco-extraction du végétal. Procédés Innovants et Solvants Alternatifs.

Christaki, E., 2014. Microalgae as a potential new generation of material for various innovative products. *Oceanography*, 1, p. e106.

Christaki, E., Bonos, E., Giannenas, I., Florou-Paneri, P., 2013. Functional properties of carotenoids originating from algae. *Journal of the Science of Food and Agriculture*, 93(1), pp. 5–11.

Cikoš, A.M., Jokić, S., Šubarić, D., Jerković, I., 2018. Overview on the application of modern methods for the extraction of bioactive compounds from marine macroalgae. *Marine Drugs*, 16(10), p. 348.

Couteau, C., Coiffard, L., 2016. Seaweed application in cosmetics. In: *Seaweed in Health and Disease Prevention*, Fleurence, J., Levine, I., Eds. (pp. 423–436). Elsevier Inc., Amsterdam, the Netherlands.

Da Costa, E., Melo, T., Moreira, A., Bernardo, C., Helguero, L., Ferreira, I., Cruz, M., Rego, A., Domingues, P., Calado, R., Abreu, M., Domingues, M., 2017. Valorization of lipids from Gracilaria sp. through lipidomics and decoding of antiproliferative and anti-inflammatory activity. *Marine Drugs*, 15(3), p. 62.

de Jesus Raposo, M.F., de Morais, A.M., de Morais, R.M., 2015. Marine polysaccharides from algae with potential biomedical applications. *Marine Drugs*, 13(5), pp. 2967–3028.

del Pilar Sánchez-Camargo, A., Ibáñez, E., Cifuentes, A., Herrero, M., 2017. Bioactives obtained from plants, seaweeds, microalgae and food by-products using pressurized liquid extraction and supercritical fluid extraction. In: *Green Extraction Techniques: Principles, Advances and Applications* (Vol. 76, p. 27).

Duarte, K., Justino, C.I.L., Gomes, A.M., Rocha-Santos, T., Duarte, A.C., 2014. Green analytical methodologies for preparation of extracts and analysis of bioactive compounds. In: *Comprehensive Analytical Chemistry* (Vol. 65, pp. 59–78). Elsevier.

Dumay, J., Morançais, M., 2016. Proteins and pigments. In: *Seaweed in Health and Disease Prevention*, Fleurence, J., Levine, I., Eds. (pp. 275–318). Elsevier Inc., Amsterdam, the Netherlands.

El Gamal, A.A., 2010. Biological importance of marine algae. *Saudi Pharmaceutical Journal*, 18(1), pp. 1–25.

Freile-Pelegrín, Y., Morales, J.L., 2004. Antibacterial activity in marine algae from the coast of Yucatan, Mexico. *Botanica Marina*, 47(2), pp. 140–146.

Fu, W., Nelson, D.R., Yi, Z., Xu, M., Khraiwesh, B., Jijakli, K., Chaiboonchoe, A., Alzahmi, A., Al-Khairy, D., Brynjolfsson, S., Salehi-Ashtiani, K., 2017. Bioactive compounds from microalgae: Current development and prospects. In: *Studies in Natural Products Chemistry* (Vol. 54, pp. 199–225). Elsevier.

Hamed, I., 2016. The evolution and versatility of microalgal biotechnology: A review. *Comprehensive Reviews in Food Science and Food Safety*, 15(6), pp. 1104–1123.

Hardouin, K., Burlot, A.S., Umami, A., Tanniou, A., Stiger-Pouvreau, V., Widowati, I., Bedoux, G., Bourgougnon, N., 2014. Biochemical and antiviral activities of enzymatic hydrolysates from different invasive French seaweeds. *Journal of Applied Phycology*, 26(2), pp. 1029–1042.

Harnedy, P.A., FitzGerald, R.J., 2011. Bioactive proteins, peptides, and amino acids from macroalgae. *Journal of Phycology*, 47(2), pp. 218–232.

Holdt, S.L., Kraan, S., 2011. Bioactive compounds in seaweed: Functional food applications and legislation. *Journal of Applied Phycology*, 23(3), pp. 543–597.

Hollinger, J.C., Kindred, C., Halder, R.M., 2015. Pigmentation and skin of color. *Cosmetic Dermatology: Products and Procedures*, pp. 23–32.

Hu, J.F., Gen, M.Y., Zhang, J.T., Jiang, H.D., 2001. An in vitro study of the structure–activity relationships of sulfated polysaccharide from brown algae to its antioxidant effect. *Journal of Asian Natural Products Research*, 3(4), pp. 353–358.

Hu, Q., Sommerfeld, M., Jarvis, E., Ghirardi, M., Posewitz, M., Seibert, M., Darzins, A., 2008. Microalgaltriacylglycerols as feedstocks for biofuel production: Perspectives and advances. *The Plant Journal*, 54(4), pp. 621–639.

Jenelt. 2018a. https://www.jenelt.com/search?q=Skin+Renewal+Radiance+Lotion (accessed December 30, 2018).

Jenelt. 2018b. https://www.jenelt.com/ultra-uv-defense-brightening-cream-with-antioxidants-broad-spectrum-spf-30-sunscreen (accessed December 30, 2018).

Jain, R., Raghukumar, S., Tharanathan, R., Bhosle, N.B., 2005. Extracellular polysaccharide production by thraustochytridprotists. *Marine Biotechnology*, 7(3), pp. 184–192.

Jang, W.S., Choung, S.Y., 2013. Antiobesity effects of the ethanol extract of *Laminaria japonica* Areshoung in high-fat-diet-induced obese rat. *Evidence-Based Complementary and Alternative Medicine*, 2013, p. 492807.

Kadam, S.U., Tiwari, B.K., O'Donnell, C.P., 2013. Application of novel extraction technologies for bioactives from marine algae. *Journal of Agricultural and Food Chemistry*, 61(20), pp. 4667–4675.

Kang, S.M., Kim, K.N., Lee, S.H., Ahn, G., Cha, S.H., Kim, A.D., Yang, X.D., Kang, M.C., Jeon, Y.J., 2011. Anti-inflammatory activity of polysaccharide purified from AMG-assistant extract of Ecklonia cava in LPS-stimulated RAW 264.7 macrophages. *Carbohydrate Polymers*, 85(1), pp. 80–85.

Katayama, K., Seyer, J.M., Raghow, R., Kang, A.H., 1991. Regulation of extracellular matrix production by chemically synthesized subfragments of type I collagen carboxypropeptide. *Biochemistry*, 30(29), pp. 7097–7104.

Kim, S.K., 2014. Marine cosmeceuticals. Journal of Cosmetic Dermatology, 13(1), pp. 56–67.

Kim, S.K., 2016. *Marine Cosmeceuticals: Trends and Prospects*. CRC Press: Boca Raton, FL.

Kim, S.K., Wijesekara, I., 2016. Cosmeceuticals from marine resources prospects and commercial trends. In: *Marine Cosmeceuticals: Trends and Prospects*, Kim, S.K., Ed. (pp. 5–6). CRC Press: Boca Raton, FL.

Koller, M., Muhr, A., Braunegg, G., 2014. Microalgae as versatile cellular factories for valued products. *Algal Research*, 6, pp. 52–63.

Langholz, O., Röckel, D., Mauch, C., Kozlowska, E., Bank, I., Krieg, T., Eckes, B., 1995. Collagen and collagenase gene expression in three-dimensional collagen lattices are differentially regulated by alpha 1 beta 1 and alpha 2 beta 1 integrins. *The Journal of Cell Biology*, 131(6 Pt 2), pp. 1903–1915.

MacArtain, P., Gill, C.I., Brooks, M., Campbell, R., Rowland, I.R., 2007. Nutritional value of edible seaweeds. *Nutrition Reviews*, 65(12 Pt 1), pp. 535–543.

Malerich, S., Berson, D., 2014. Next generation cosmeceuticals: The latest in peptides, growth factors, cytokines, and stem cells. *Dermatologic Clinics*, 32(1), pp. 13–21.

Metting, B., Pyne, J.W., 1986. Biologically active compounds from microalgae. *Enzyme and Microbial Technology*, 8(7), pp. 386–394.

Metting, F.B., 1996. Biodiversity and application of microalgae. *Journal of Industrial Microbiology and Biotechnology*, 17(5–6), pp. 477–489.

Michalak, I., Dmytryk, A., Wieczorek, P.P., Rój, E., Łęska, B., Górka, B., Messyasz, B., Lipok, J., Mikulewicz, M., Wilk,

R., Schroeder, G., Chojnacka, K., 2015. Supercritical algal extracts: A source of biologically active compounds from nature. *Journal of Chemistry*, 2015, pp. 1–14.

Mourelle, M., Gómez, C., Legido, J., 2017. The potential use of marine microalgae and cyanobacteria in cosmetics and thalassotherapy. *Cosmetics*, 4(4), p. 46.

Murata, M., Nakazoe, J.I., 2001.Production and use of marine algae in Japan. *Japan Agricultural Research Quarterly*, 35(4), pp. 281–290.

Narayan, B., Miyashita, K., Hosakawa, M., 2005. Comparative evaluation of fatty acid composition of different Sargassum (Fucales, Phaeophyta) species harvested from temperate and tropical waters. *Journal of Aquatic Food Product Technology*, 13(4), pp. 53–70.

Noda, H., 1993. Health benefits and nutritional properties of nori. *Journal of Applied Phycology*, 5(2), pp. 255–258.

Nordstrom. https://shop.nordstrom.com/s/youth-to-the-people-superfood-skin-reset-mask/4873716?origin=category personalizedsort&breadcrumb=Home%2FBeauty%2FSkin%20Care%2FNatural%20Skin%20Care&color=none (accessed December 30, 2018).

Nykaa. 2018a. https://www.nykaa.com/dove-regenerative-repair-shampoo-conditioner/p/91596?eq=desktop (accessed December 30, 2018).

Nykaa. 2018b. https://www.nykaa.com/l-oreal-paris-pure-clay-mask-red-algae-hydrafresh-anti-ox-cream/p/326814?ptype=product&productId=326814&skuId=326814 (accessed December 30, 2018).

Ohse, S., Derner, R.B., Ozório, R.Á., Corrêa, R.G., Furlong, E.B., Cunha, P.C.R., 2014. *Lipid Content and Fatty Acid Profiles in Ten Species of Microalgae*. IDESIA, Chile.

OSEA. https://oseamalibu.com/products/eyes-lips (accessed December 30, 2018).

Ovando, C.A., Carvalho, J.Cd, Vinícius de Melo Pereira, G., Jacques, P., Soccol, V.T., Soccol, C.R., 2018. Functional properties and health benefits of bioactive peptides derived from *Spirulina*: A review. *Food Reviews International*, 34(1), pp. 34–51.

Pan, J.L., Wang, H.M., Chen, C.Y., Chang, J.S., 2012. Extraction of astaxanthin from *Haematococcuspluvialis* by supercritical carbon dioxide fluid with ethanol modifier. *Engineering in Life Sciences*, 12(6), pp. 638–647.

Pimentel, F., Alves, R., Rodrigues, F., P. P. Oliveira, M., 2018. Macroalgae-derived ingredients for cosmetic industry—An update. *Cosmetics*, 5(1), p. 2.

Rajapakse, N., Kim, S.K., 2011. Nutritional and digestive health benefits of seaweed. In: *Advances in Food and Nutrition Research*, Kim, S.K., Ed. (Vol. 64, pp. 17–28). Academic Press, San Diego, CA, USA.

Ranga Rao, A., Deepika, G., Ravishankar, G.A., Sarada, R., Narasimharao, B.P., Lei, B., Su, Y., 2018. Industrial potential of carotenoid pigments from microalgae: Current trends and future prospects. *Critical Reviews in Food Science and Nutrition*, 25, pp. 1–22.

Ranga Rao, A., Sindhuja, H.N., Dharmesh, S.M., Sankar, K.U., Sarada, R., Ravishankar, G.A., 2013. Effective inhibition of skin cancer, tyrosinase, and antioxidant properties by astaxanthin and astaxanthin esters from the green alga *Haematococcuspluvialis*. *Journal of Agriculture and Food Chemistry*, 61, pp. 3842–3851.

Robles Medina, A., Molina Grima, E., Giménez Giménez, A., Ibañez González, M.J., 1998. Downstream processing of algal polyunsaturated fatty acids. *Biotechnology Advances*, 16(3), pp. 517–580.

Rocha de Souza, M.C., Marques, C.T., Guerra Dore, C.M., Ferreira da Silva, F.R., Oliveira Rocha, H.A., Leite, E.L., 2007. Antioxidant activities of sulfated polysaccharides from brown and red seaweeds. *Journal of Applied Phycology*, 19(2), pp. 153–160.

Santhosh, S., Dhandapani, R., Hemalatha, R., 2016. Bioactive compounds from microalgae and its different applications—A review. *Advances in Applied Science Research*, 7(4), pp. 153–158.

Seoane, P.R., Flórez-Fernández, N., Piñeiro, E.C., González, H.D., 2017. Microwave-assisted water extraction. In: *Water Extraction of Bioactive Compounds* (pp. 163–198). Elsevier.

Sharma, K.K., Schuhmann, H., Schenk, P.M., 2012. High lipid induction in microalgae for biodiesel production. *Energies*, 5(5), pp. 1532–1553.

Shimizu, Y., 1996. Microalgal metabolites: A new perspective. *Annual Review of Microbiology*, 50(1), pp. 431–465.

Thomas, N.V., Kim, S.K., 2013. Beneficial effects of marine algal compounds in cosmeceuticals. *Marine Drugs*, 11(1), pp. 146–164.

Wang, H.D., Chen, C.C., Huynh, P., Chang, J.S., 2015. Exploring the potential of using algae in cosmetics. *Bioresource Technology*, 184, pp. 355–362.

Wang, J., Jin, W., Hou, Y., Niu, X., Zhang, H., Zhang, Q., 2013. Chemical composition and moisture-absorption/retention ability of polysaccharides extracted from five algae. *International Journal of Biological Macromolecules*, 57, pp. 26–29.

Wijesinghe, W.A.J.P., Jeon, Y.J., 2011. Biological activities and potential cosmeceutical applications of bioactive components from brown seaweeds: A review. *Phytochemistry Reviews*, 10(3), pp. 431–443.

Wijesinghe, W.A.J.P., Jeon, Y.J., 2012. Biological activities and potential industrial applications of fucose rich sulfated polysaccharides and fucoidans isolated from brown seaweeds: A review. *Carbohydrate Polymers*, 88(1), pp. 13–20.

Zhang, Z., Wang, F., Wang, X., Liu, X., Hou, Y., Zhang, Q., 2010. Extraction of the polysaccharides from five algae and their potential antioxidant activity in vitro. *Carbohydrate Polymers*, 82(1), pp. 118–121.

16 Microalgae for Human Nutrition
Perspectives for the Future

Mariana F.G. Assuncao, Ana Paula Batista, Raquel Amaral, and Lília M.A. Santos

CONTENTS

Introduction 161
Microalgae as a Source of Nutrients for Human Consumption 162
    Macronutrients: Proteins, Carbohydrates and Lipids 162
    Micronutrients: Vitamins and Minerals 165
    Phytochemicals: Pigments and Polyphenols 167
Microalgae Already on the Market as a Source of Nutrients 169
    Arthrospira (*Spirulina*) and *Chlorella* 170
New Microalgae for the Market, Perspectives and Constraints 171
Conclusions 173
References 173

BOX 16.1 SALIENT FEATURES

Concerns with feeding an increasing world population and doing so with nutrient-rich foods, especially those which are naturally derived, are now included in the worldwide agenda. Microalgae have been considered as an answer to this search in the food sector since they can be cost-effective, large-scale production organisms that generate compounds with nutritional interest (macronutrients, micronutrients and phytochemicals) as well as having beneficial effects for human health. The European microalgae-derived food market is dominated by *Arthrospira* (*Spirulina*) *platensis* and *Chlorella vulgaris*, two microalgae indicated as a source of essential nutrients and health-promoting bioactive molecules. However, despite the studies that indicate their nutritional value and health-promoting effects, a question remains: how much of this nutritional and bioactive value is effectively assimilated by the human organism? Despite the disclosure of a large list of potential microalgae for use as food in recent years, the strict European legislation regarding novel foods remains as the main constraint to the use of microalgae as food in Europe. Efforts are being made to spread acceptance of microalgae for food consumption, namely the take-off collaborative actions between industrial producers, research bodies and other experts. With this approach, we hope that microalgae may soon be on everyone's plate, becoming one possible solution to hunger and malnutrition problems.

INTRODUCTION

At the dawn of the twenty-first century, hunger and malnutrition still persist as chronic, dramatic problems for humanity. Nutrition is a biological process essential for human development and health. The search for new sources of nutrients, especially those which are naturally derived, is now on the worldwide agenda. Cost-effective, large-scale production of organisms that produce compounds with nutritional interest is increasingly regarded as an answer to this search for novelty in the food sector. There are some records of the use of microalgae in human nutrition. Human consumption of microalgae goes back to the 1300s with the Aztecs' use of the blue-green biomass of *Arthrospira* (*Spirulina*) collected from lakes to prepare cakes (García et al., 2017). According to the revision by Borowitzka (2018), the first consideration given to cultivating microalgal biomass as commercial food source dates from 1947–1948, when solutions were being discussed to tackle the worldwide nutrition crisis. The production of *Chlorella* and *Scenedesmus* was announced as an alternative to the devising of nonagricultural production of nutrients (Geoghegan 1951). Studies into the large-scale production of microalgae led to the first plants for commercial production of *Chlorella* in Taiwan in 1956 and *Arthrospira (Spirulina)* in Japan, Thailand and the United States in 1970 (Borowitzka 2018). Currently there are many promising microalgal species proposed for commercial applications (Mobin and Alam 2017).

MICROALGAE AS A SOURCE OF NUTRIENTS FOR HUMAN CONSUMPTION

There is growing evidence that microalgae can be a significant source of a diverse number of essential nutrients to support human health, namely macronutrients, micronutrients and phytochemicals. The proximate chemical composition of different strains of microalgae have been published in the literature over the last 20 years (e.g. Rebolloso-Fuentes et al., 2000; Rebolloso-Fuentes et al., 2001a; Rebolloso-Fuentes et al., 2001b; Batista et al., 2013; Tokuşoglu and Ünal 2003; Kent et al., 2015; Tibbetts et al., 2015; Martins et al., 2016; Assunção et al., 2017; Diprat et al., 2017; Khantoon et al., 2018).

Macronutrients: Proteins, Carbohydrates and Lipids

Macronutrients are a group of nutrients needed in large quantities by the human body. They are classified as energy-providing nutrients and include proteins, lipids and carbohydrates (Caballero et al., 2005).

Microalgae as a source of protein dates back to the 1950s, when an increase in the world's population and the prediction of an insufficient protein supply led to the search for new sources of this macronutrient (Spolaore et al., 2006). The high protein content detected in some microalgal species compared with conventional foods (Table 16.1) is the main reason to consider these organisms as an alternative source of protein (Becker 2004). However, many of the published values on the protein content of microalgae are an estimation of crude protein, a measurement used to evaluate food and feed (Becker 2004). These calculations are based on the hydrolysis of the algal biomass and the evaluation of the total nitrogen (N) released. The application of an N-to-P (nitrogen-to-protein) conversion factor universally used for food labeling (N $\times$ 6.25) allows the calculation of total protein (Becker 2004; Wells et al., 2017). This conversion factor assumes that the protein source contains 16% of N, but it overlooks the content of nonprotein nitrogen in structural proteins found in microalgal cells, which are not nutritionally interesting: bioactive peptides, free amino acids, nucleic acids and ammonia (Lourenço et al., 2004; Tibbetts et al., 2015). Nevertheless, microalgae are still in line as a food source since only ~10% of the nitrogen detected in their biomass consists of nonprotein nitrogen (*Arthrospira* [*Spirulina*] 11.5% and *Dunaliella* 6%) (Becker 2004; Becker 2007). Some studies suggest the use of an overall mean N-to-P conversion factor of 4.78 since microalgae have nonprotein nitrogen in different amounts depending on the species, the cultivation method and the growth phase (Lourenço et al., 2004). The major nutritional value of microalgae is not their total protein content by itself, but rather their amino-acid profile and availability (Becker 2004; Becker 2007). The human organism is limited to the biosynthesis of some amino acids (nonessential amino acids), and it is through the diet that essential amino acids are procured (Becker 2004). Microalgae are capable of synthesizing all amino acids, including those essential for the human organism (Table 16.2) (Becker 2004; Becker 2007; Matos et al., 2017). Besides its nutritional value, microalgal protein has demonstrated some benefits for human health by preventing some diseases (e.g. Crohn's disease), cell/tissue damage and malabsorption after surgery (Matos et al., 2017).

Microalgae are also a good source of carbohydrates (Table 16.1), found in the form of starch, cellulose, sugars and other polysaccharides (Becker 2004; Chácon-Lee and González-Mariño 2010; Matos et al., 2017). These compounds provide energy used or stored by microalgal cells and as structural elements (Wells et al., 2017). Carbohydrates are important sources of energy in the human diet. Presently, in most cases, the microalgal food market is associated with the consumption of the whole biomass. This means that, in addition to the consumption of proteins, other components of the microalgal biomass such as lipids and carbohydrates are also consumed (Becker 2004). It is fair to say that algal polysaccharides are the least acknowledged algal-derived compound already being consumed as food and additives. Studies indicate that the human organism possesses enzymes capable of the hydrolysis of some carbohydrates in mono- and disaccharides, but it cannot digest the more complex ones such as cellulose, hemicellulose and pectin (Wells et al., 2017). A fraction of these indigestible carbohydrates has beneficial physiological effects for the human organism. It is therefore considered as functional food and is often referred to as *dietary fiber* (Wells et al., 2017). These carbohydrates resist digestion in the upper part of the gastrointestinal tract due to the presence of glycosidic bounds that are different

TABLE 16.1
Proximate Chemical Composition of Different Microalgae and Comparison with Conventional Foods. Values are Expressed as Percent of Dry Matter

Species	Ash	Protein	Carbohydrates	Lipids
Anabaena cylindrica	–	43–56	25–30	4–7
Aphanizomenon flos-aquae	–	62	23	3
Arthrospira maxima	–	46–71	8–16	4–11
Arthrospira platensis	7–10	46–63	8–22	3–14
Synechococcus sp.	–	63	15	11
Acutodesmus dimorphus	14	28	38	18
Botryococcus braunii	5–7	39	18–30	24–34
Chlamydomonas rheinhardtii	–	48	17	21
Chlorella pyrenoidosa	–	53–57	25–26	2–15
Chlorella vulgaris	5–6	12–58	8–33	5–28
Dunaliella salina	–	39–61	14–32	6–20
Dunaliella tertiolecta	0.08	18–43	31	17
Dunaliella sp.	19	34	14	14
Haematococcus pluvialis	30	48–51	27–43	5–15
Heterochlorella luteoviridis	0.14	13–41	14	11
Neochloris oleobundans	16	30	37	15
Scenesdesmus obliquus	–	6–56	10–64	11–21
Scenedesmus dimorphus	–	60–71	13–16	6–7
Scenedesmus sp.	15	30	27	15
Spirogyra sp.	–	6–45	21–64	4–21
Tetraselmis chuii	16	26–46	16–28	11–22
Porphyridium cruentum	17–24	8–57	21–57	6–40
Porphyridium aerugineum	8.9	31	45	13
Porphyridium purpureum	18	15	63	1
Euglena gracilis	–	10–61	14–40	14–41
Nannochloropsis granulata	6–7	17–33	27–36	23–47
Nannochloropsis oculata	–	46	–	–
Nannochloropsis spp.	8–11	28–30	9–35	18–21
Isochrisis galbana	16	16–56	10–17	12–17
Diacronema vlkianum	–	57	32	6
Hillea sp.	–	25	–	–
Ruttnera lamellosa	43	8	43	2
Chaetoceros calcitrans	–	36	27	15
Skeletonema costatum	–	15	–	–
Phaeodactilum tricornutum	9–23	39	25	18
Amphidinium carterae	–	24	–	–
Conventional foods				
Egg	–	47	4	41
Milk	–	26	38	28
Rice	–	8	77	2
Soya	–	37	30	20

Adapted from Chácon-Lee and González-Mariño 2010; Safi et al., 2014; Kent et al., 2015; Tibbetts et al., 2015; Martins et al., 2016; Assunção et al., 2017; Diprat et al., 2017; Khantoon et al., 2018.

TABLE 16.2
Amino Acid Composition of Protein from Microalgae

Amino acid	Aphanizomenon sp.	Arthrospira maxima	Arthrospira (Spirulina) platensis	Botryococcus braunii	Chlorella vulgaris	Dunaliella bardawil	Scenedesmus obliquus	Scenedesmus sp.	Tetraselmis chuii	Tetraselmis sp.	Porphyridium aerugineum	Phaeodactylum tricornutum	Nannochloropsis granulata	Egg	Soy
Isoleucine*	2.9 (a)	6.0 (a)	6.7 (a) 7.2 (b) 6.8 (b)	3.4 (a)	3.8 (a) 4.8 (a)	4.2 (a)	3.6 (a)	42 (c)	3.4 (a)	4.0 (a)	7.1 (a)	4.6 (a)	5.6 (a)	120 (c)	5.8 (a)
Leucine*	5.2 (a)	8.0 (a)	9.8 (a) 5.7 (b) 4.2 (b)	7.1 (a)	8.8 (a) 10.8 (a)	11.0 (a)	7.3 (a)	123 (c)	7.3 (a)	9.5 (a)	11.9 (a)	7.0 (a)	11.0 (a)	170 (c)	9.0 (a)
Valine*	3.2 (a)	6.5 (a)	7.1 (a) 4.9 (b)	4.4 (a)	5.5 (a) 7.9 (a)	5.8 (a)	6.0 (a)	61 (c)	4.8 (a)	5.7 (a)	7.3 (a)	5.1 (a)	7.1 (a)	130 (c)	7.2 (a)
Lysine*	3.5 (a)	4.6 (a)	4.8 (a) 4.2 (b) 3.3 (b)	4.7 (a)	8.4 (a) 7.7 (a)	7.0 (a)	5.6 (a)	230 (c)	5.6 (a)	6.5 (a)	8.0 (a)	6.4 (a)	8.5 (a)	160 (c)	6.7 (a)
Phenylalanine*	2.5 (a)	4.9 (a)	5.3 (a) 7.4 (b) 5.9 (b)	4.4 (a)	5.0 (a) 6.0 (a)	5.8 (a)	4.8 (a)	174 (c)	4.7 (a)	5.6 (a)	6.3 (a)	4.8 (a)	6.2 (a)	190 (c)	5.3 (a)
Tyrosine		3.9 (a)	5.3 (a)	2.8 (a)	3.4 (a) 3.0 (a)	3.7 (a)	3.2 (a)		3.0 (a)	3.6 (a)	5.8 (a)	3.4 (a)	4.2 (a)	–	4.3 (a)
Methionine*	0.7 (a)	1.4 (a)	2.5 (a) 1.7 (b) 1.9 (b)	2.5 (a)	2.2 (a) 1.6 (a)	2.3 (a)	1.5 (a)	286 (c)	2.4 (a)	2.8 (a)	3.7 (a)	2.7 (a)	3.5 (a)	60 (c)	3.0 (a)
Cysteine	0.2 (a)	0.4 (a)	0.9 (a)	1.4 (a)	1.4 (a) 0.5 (a)	1.2 (a)	0.6 (a)	–	2.8 (a)	1.4 (a)	2.2 (a)	1.5 (a)	1.6 (a)	–	2.1 (a)
Tryptophan*	0.7 (a)	1.4 (a)	0.3 (a) 3.8 (b) 4.2 (b)	2.2 (a)	2.1 (a) 1.1 (a)	0.7 (a)	0.3 (a)	–	2.3 (a)	–	3.3 (a)	2.6 (a)	2.8 (a)	15 (c)	1.7 (a)
Threonine*	3.3 (a)	4.6 (a)	6.2 (a) 3.8 (b) 4.2 (b)	3.7 (a)	4.8 (a) 5.6 (a)	5.4 (a)	5.1 (a)	84 (c)	4.0 (a)	5.2 (a)	5.8 (a)	4.8 (a)	5.4 (a)	85 (c)	5.3 (a)
Alanine	4.7 (a)	6.8 (a)	9.5 (a)	6.4 (a)	7.9 (a) 7.2 (a)	7.3 (a)	9.0 (a)		6.0 (a)	9.4 (a)	8.4 (a)	7.3 (a)	7.1 (a)	–	–
Arginine	3.8 (a)	6.5 (a)	7.3 (a)	20.5 (a)	6.4 (a) 8.0 (a)	7.3 (a)	7.1 (a)		9.4 (a)	5.0 (a)	8.6 (a)	5.7 (a)	7.4 (a)	–	6.4 (a)
Aspartic acid	4.7 (a)	8.6 (a)	11.8 (a)	8.7 (a)	9.0 (a) 10.4 (a)	10.4 (a)	8.4 (a)		14.1 (a)	–	15.0 (a)	11.6 (a)	11.4 (a)	–	10.7 (a)
Glutamine	7.8 (a)	12.6 (a)	10.3 (a)	12.7 (a)	11.6 (a) 11.6 (a)	12.7 (a)	10.7 (a)		12.0 (a)	–	15.6 (a)	18.8 (a)	14.1 (a)	–	12.3 (a)
Glycine	2.9 (a)	4.8 (a)	5.7 (a)	4.9 (a)	5.8 (a) 5.1 (a)	5.5 (a)	7.1 (a)		6.5 (a)	6.4 (a)	7.0 (a)	5.5 (a)	7.5 (a)	–	3.8 (a)
Histidine*	0.9 (a)	1.8 (a)	2.2 (a)	1.5 (a)	2.0 (a) 2.4 (a)	1.8 (a)	2.1 (a)		1.6 (a)	2.0 (a)	1.9 (a)	1.5 (a)	2.3 (a)	–	2.6 (a)
Proline	2.9 (a)	3.9 (a)	4.2 (a)	4.6 (a)	4.8 (a) 3.9 (a)	3.3 (a)	3.9 (a)		3.6 (a)	6.2 (a)	5.0 (a)	7.1 (a)	11.2 (a)	–	4.3 (a)
Serin	2.9 (a)	4.2 (a)	5.1 (a)	3.5 (a)	4.1 (a) 3.5 (a)	4.6 (a)	4.2 (a)		4.2 (a)	4.4 (a)	7.0 (a)	4.8 (a)	5.6 (a)	–	7.7 (a)

(*) Essential amino acids

(a) Expressed as g/100g of protein; (b) Expressed as g/16g of nitrogen; (c) Expressed as mg amino acid/g essential amino acid.

Adapted from Chácon-Lee and González-Mariño 2010; Schwenzfeier et al. 2011; Tibbetts et al. 2015.

from the digestive enzyme–susceptible α-1,4 and α-1,6 linkages (Aluko 2012; Wells et al., 2017).

They reach the colon and are fermented by prebiotic microflora that convert them into bioactive compounds, which are locally absorbed or enter the bloodstream to be distributed to target organs. They may also be used as growth promoters by probiotic organisms (Aluko 2012). The insoluble fiber (cellulose) can promote bowel movement, improving its health and the feeling of satiety (Aluko 2012). On the other hand, soluble fiber (oligosaccharides, pectin) has shown valuable antioxidant radical scavenging, anticancer, anticoagulant and antihypercholesterolemic properties (Aluko 2012; Wells et al., 2017). Microalgae also contain sulfated polysaccharides with important functional properties such as antioxidant and hypocholesterolemic effects (Gouveia et al., 2008; Marques et al., 2011; de Jesus Raposo et al., 2013).

Lipids and fatty acids are constituents of microalgal cell membranes, storage products, cell signaling molecules and energy storage (Becker 2004; Wells et al., 2017). Lipids from microalgae can be classified as *polar* lipids or *nonpolar* lipids. Many of the nonpolar lipids are triacylglycerols and free fatty acids, whereas the polar lipids are essentially glycolipids and phospholipids (Pignolet et al., 2013; D'Alessandro and Filho 2016). Lipids account for 1–40% of microalgal dry weight (Table 16.1) and, depending on the culture conditions, these values can reach up to 85% dry weight (Becker 2004; Chácon-Lee and González-Mariño 2010). Fatty acids are the structural components of most lipids so they constitute most of the nutritional interest in microalgae. The fatty acid composition differs according to the class of lipids. Nonpolar lipids are predominantly composed of saturated and monounsaturated fatty acids, whereas polar lipids are composed of polyunsaturated fatty acids (Pignolet et al., 2013). The human body needs to obtain the so-called essential fatty acids through diet. The polyunsaturated fatty acids linoleic (C18:2ω6), α-linolenic (C18:3ω3), arachidonic (C20:4ω6), eicosapentaenoic (C20:5ω3) and docosahexanoic (C22:6ω6) are not synthesized by the human body due to the lack of desaturases δ12 and δ15 (Gouveia et al., 2010; Khozin-Goldberg et al., 2011; Tvrzicka et al., 2011; de Jesus Raposo et al., 2013; Wells et al., 2017). They are of great importance to human health, being effective in treating or preventing cardiovascular disorders, lowering hypertension and preventing cancer, type 2 diabetes, inflammatory bowel disease, asthma, arthritis, kidney and skin disorders, depression and schizophrenia (Udayan et al., 2017; Matos et al., 2017; Wells et al., 2017). These fatty acids are present in microalgae in a slightly higher concentration than in fish (Table 16.3), with the advantage of having less potential to accumulate chemical contaminants frequently recorded for fish (Rajeshkumar and Li 2018). Low taste intensity and less off-odor are also advantages of microalgal lipids when compared to fish oils. But the main reason underlying the increasing interest in microalgal lipids is the fact that these organisms are able to synthetize ω3 fatty acids at levels exceeding 20% of their total lipid content. In Western diets, the ω6/ω3 ratio is high with unbalanced higher consumption of ω6 in relation to ω3 fatty acids. This nutritional behavior is linked to many diseases such as cancer, osteoporosis and inflammatory and autoimmune diseases, the same diseases which are suppressed by higher ω3 consumption (Simopoulos 2006). If we look to the ratio of these two classes of fatty acids found in most studied microalgae (Table 16.3), we find a low ω6/ω3 ratio, thus meeting overall indications for the healthy consumption of lipids in the diet.

Micronutrients: Vitamins and Minerals

Micronutrients, such as vitamins, minerals and metals, are required in minute amounts, yet they are extremely important for the normal functioning of the body, having critical functions in metabolic pathways (Caballero et al., 2005).

Microalgae also contain valuable micronutrients in high quantities, namely important vitamins (Table 16.4), which further increase their nutritional value. Vitamins are also essential organic micronutrients that the human body cannot synthesize directly in sufficient quantities and must obtain from the diet (Wells et al., 2017). They serve as precursors for essential enzyme cofactors and are needed for essential metabolic functions (Caballero et al., 2005). It is very well known that many human diseases are due to vitamin deficiency, including beriberi (lack of vitamin B_1), pellagra (lack of vitamin B_3), pernicious anemia (lack of vitamin B_{12}) and scurvy (lack of vitamin C) (Wells et al., 2017). Microalgae are a good source of B-group vitamins, and their concentrations can exceed those in conventional vitamin-rich foods such as liver, spinach and baker's yeast (Table 16.4). However, there is one exception that has generated some controversy on this topic, which is vitamin B_{12}. Some contradictory statements about the occurrence of this vitamin in different microalgae can be found in the literature (Becker 2004; Croft et al., 2005; Wells et al., 2017). Genomic studies proved that eukaryotic microalgae have a metabolic requirement for B_{12} but they cannot synthesize it (Croft et al., 2005). Vitamin B_{12} is synthesized by prokaryotes, so the concentrations detected in microalgae are presumably acquired in partnership with

TABLE 16.3
Essential Fatty Acid Profile of Microalgae

Microalgae	Linoleic Acid (C18:2ω6)	α-linolenic Acid (C18:3ω3)	Arachidonic Acid (C20:4ω6)	Eicosapentaenoic Acid (C20:5ω3)	Docosahexanoic Acid (C22:6ω6)
Ankistrodesmus falcatus	13.5–18.7 (a) 2.0 (b)	12.0–22.8 (a) 26.49 (b)	–	–	–
Ankistrodesmus fusiformis	12.23 (b)	26.28 (b)	–	–	–
Botryococcus braunii	5.2 (b)	5.3 (b)	–	–	–
Botryococcus terribilis	5.0 (b)	7.2 (b)	–	–	–
Chlamydocapsa bacillus	13.3 (b)	20.9 (b)	–	–	–
Chlamydomonas sp.	3.93 (b)	1.9 (b)	–	–	–
Chlorella vulgaris	8.6–9.7 (b) 3.39–20.9 (a)	0.20–47.0 (a) 1.6–1.9 (b)	–	3.23 (b)	20.94 (b)
Chlorococcum oleofaciens	9.8–16.3 (a)	5.1–11.6 (a)	–	–	–
Coelastrum microporum	8.58 (b)	–	–	–	–
Cylindrotheca fusiformis	2.8–1.2 (a)	–	1.1–4.9 (a)	0.6–2.9 (a)	–
Desmodesmus brasiliensis	12.0 (b)	4.5 (b)	–	–	–
Dunaliella primolecta	6.2–7.0 (a)	38.7–41.1 (a)	–	–	–
Emiliania huxleyi	0.9 (a)	5.5 (a)	–	–	9.2 (a)
Haematococcus pluvialis	3.24–20.99 (a)	0.13 (a) 15.0 (c)	–	–	–
Heterosigma akashiwo	1.6–4.5 (a)	4.2–6.7 (a)	3.5 (a)	8.7–14.8 (a)	0.7 (a)
Isochrysis sp.	2.9–5.1 (a)	3.0–4.9 (a) 29 (f)	–	1.4–1.7 (a) 2.8 (f)	46 (f)
Isochrisis galbana	1.14 (b)	0.46 (b)	1.07 (b)	1.93 (b)	18.79 (b)
Kirchneriella lunaris	4.5 (b)	39.7 (b)	–	–	–
Nannochloropsis sp.	0.6–10.8 (a)	0.8–28.2 (a)	2.3–5.9 (a)	4.5–10.0 (a)	–
Nannochloropsis gaditana		0.3 (f)		175 (f)	–
Nannochloropsis oculata	3.9–4.3 (a) 0.7 (f)	–	8.5–10.6 (a)	29.3–35.6 (a) 193 (f)	–
Neochloris oleobundans	22.4–29.0 (a)	17.3 (a)	–	–	–
Nostoc commune	12.5–19.3 (a)	16.3–38.1 (a)	–	–	–
Parietochloris incisa	14.3–17.0 (a)	3.0–14.3 (a)	14.0–46.0 (a)	1.0–4.3 (a)	–
Pavlova sp.	0.6 (a)	–	–	23.4–28.9 (a)	–
Pavlova lutheri	0.6 (a)	0.5 (a) 10 (f)	0.3 (a)	12.1–18.0 (a) 92 (f)	9.7 (a) 40.9 (f)
Pavlova viridis	0.73–1.9 (a)	0.15–2.4 (a)	1.8–2.5 (a)	9.5–15.7 (a)	2.4–7.2 (a)
Phaeodactylum tricornutum	0.5–4.4 (a)	0.4–3.2 (a) 0.8 (f)	–	9.0–14.4 (a) 111 (f)	0.7–1.1 (a) 8.3 (f)
Porphyridium cruentum	4.25–25.08 (a) 0.37–3.9 (d)	–	22.15–34.56 (a) 1.29–12.8 (d)	2.77–26.95 (a) 1.27–25.4 (d)	6.1 (d)
Porphyridium purpureum	1.10 (a)	–	2.7 (a)	0.5 (a)	–
Pseudokirchneriella subcapitata	7.5 (b) 1.7–4.4 (a)	9.3 (b) 8.4–23.1 (a)	–	–	–
Ruttnera lamellosa	1.7 (a)	2.1 (a)	0.2 (a)	0.5 (a)	6.7 (a)
Scenedesmus sp.	15.9–34.2 (a)	1.0–2.5 (a)	–	–	–
Scenedesmus obliquus	1.23–1.73 (c) 4.6 (b)	5.5–21.9 (a) 4.53–7.41 (c)	0.06–0.22 (c)	–	–
Arthrospira (Spirulina) platensis	10.37–14.45 (b) 21.2 (d) 48.1–23.7 (a)	0.62–0.68 (b)	0.34–0.41 (b)	2.21–2.91 (b)	3.30–3.51 (b)
Synechocystis sp.	0.2 (a)	4.7–14.2 (a)	–	–	–

(*Continued*)

TABLE 16.3 (CONTINUED)
Essential Fatty Acid Profile of Microalgae

Microalgae	Linoleic Acid (C18:2ω6)	α-linolenic Acid (C18:3ω3)	Arachidonic Acid (C20:4ω6)	Eicosapentaenoic Acid (C20:5ω3)	Docosahexanoic Acid (C22:6ω6)
Tetraselmis elliptica	2.7–3.4 (e)	8.9–13.7 (e)	–	–	–
Tetraselmis suecica	6.9–8.8 (a)	3.5–4.7 (a)	0.6–1.0 (a)	0.8 (a)	–
Tetraselmis subcordiformis	10.7–16.2 (a)	17.7–22.2 (a)	1.40–1.50 (a)	2.9–3.9 (a)	–
Thalassiosira weissflogii	–	0.3 (a)	–	4.0 (a)	0.1 (a)
Fish					
Boops boops (Bogue)	0.9 (a)	0.4 (a)	0.1 (a)	5.1 (a)	18.7 (a)
Mugil cephalus (Mullet)	1.8 (a)	1.4 (a)	0.1 (a)	10.5 (a)	7.7 (a)
Trachurus mediterraneous (Scad)	1.4 (a)	0.3 (a)	0.1 (a)	5.4 (a)	36.2 (a)
Sardinella aurita (Sardine)	2.0 (a)	0.3 (a)	0.6 (a)	11.7 (a)	13.3 (a)
Pagellus erythrinus (Pandora)	1.0 (a)	0.3 (a)	0.1 (a)	5.3 (a)	21.9 (a)
Scorpaena scrofa (Red scorpion fish)	4.0 (a)	0.3 (a)	0.1 (a)	4.7 (a)	28.0 (a)
Scopthalmus maeticus (Turbot)	1.9 (a)	0.5 (a)	0.2 (a)	5.3 (a)	30.3 (a)
Solea solea (Common sole)	1.4 (a)	0.1 (a)	0.3 (a)	7.7 (a)	18.7 (a)

(a) Expressed as percent of total fatty acids; (b) Expressed as percent of total lipids; (c) Expressed as μg/mg DW; (d) Expressed as percent of DW; (e) Expressed as mg/g DW; (f) Expressed as mg/g total lipid

Adapted from Özogul and Özogul 2007; Chácon-Lee and González-Mariño 2010; Lang et al., 2011; Griffiths et al., 2012; Huang et al., 2013; Nascimento et al., 2013; Hultberg et al., 2014; Ryckebosch et al., 2014; Martins et al., 2016; Assunção et al., 2017; Abomohra et al., 2017; Del Río et al., 2017.

bacteria closely associated with or grown on the algae (Croft et al., 2005; Wells et al., 2017). On the other hand, the detection of this vitamin in cyanobacteria is more likely to be authentic due to the close phylogenetic position of these organisms to bacteria (Becker 2004; Wells et al., 2017).

Microalgae have the ability to accumulate minerals gathered from the surrounding environment, exhibiting a balanced content in macrominerals (Na, K, Ca and Mg) and microminerals (Fe, Zn, Mn and Cu) (Matos et al., 2017). Human nutritional requirements demand at least 23 mineral elements and the required daily quantities of mineral nutrients are small, particularly when compared with nutrients such as carbohydrates and lipids (Quintaes and Diez-Garcia 2015). Minerals are involved in the formation of bones and teeth, muscle contraction, blood formation and synthesis of protein and production of energy (Quintaes and Diez-Garcia 2015). Since minerals are indispensable to functioning of the organism, they must be regularly present in the diet. The macrominerals are those present in greater proportions in body tissues, leading to greater amounts in the diet. Microminerals are equally essential to the human diet, although required in smaller amounts. Little is known about the mineral content in microalgae, with few studies focusing on elemental composition analysis (Wells et al., 2017). In proximate analysis, the ash represents the content of macro, trace and ultra-trace elements and corresponds to total mineral content. Ash values known for these organisms (Table 16.1, and references herein) reveal their mineral content potential.

Phytochemicals: Pigments and Polyphenols

Phytochemicals is a term not yet legally defined. It refers to nutrients/compounds with possible, but not yet fully established, effects on human health and that are not considered essential nutrients. Compounds such as pigments and polyphenols are included in this category (Beecher 1999; Caballero et al., 2005).

Pigments are at the center of photosynthetic activity in microalgal cells. Three classes of pigments are acknowledged in microalgae: chlorophylls, carotenoids and phycobiliproteins (e.g. Gouveia et al., 2008; Christaki et al., 2011; Begum et al., 2016; D'Alessandro and Filho 2016). Chlorophyll is the primary pigment, reaching 0.5–1.5% of microalgae's dry weight (Becker 2004; Christaki et al., 2011; D'Alessandro and Filho 2016). The nutritive value of chlorophyll is, however, quite controversial;

TABLE 16.4
Vitamin Content of Microalgae and Cyanobacteria Compared with Conventional Food Source of Vitamins in mg/kg DW

Vitamins	Cyanobacteria		Microalgae					Animal	Plant	Fungi
	Arthrospira (Spirulina) platensis	Aphanizomenon flos-aquae	Chlorella pyrenoidosa	Scenedesmus quadricauda	Tetraselmis suecica	Isochrysis galbana	Dunaliella tertiolecta	Chlorella stigmatophora	Liver	Spinach	Baker's yeast
Vit A	840	–	480	554	–	–	–	–	60.0	130.0	trace
Vit B_1	44	4.8	10	11.5	32.3	14.0	29.0	14.6	3.0	0.9	7.1
Vit B_2	37	57.3	36	27	19.1	30.0	31.2	19.6	29.0	1.8	16.5
Vit B_6	3	11.1	23	–	2.8	1.8	2.2	1.9	7.0	1.8	21.0
Vit B_{12}	7	8	–	1.1	0.5	0.6	0.7	0.6	0.65	–	–
Vit C	80	0.7	–	396	191.0	119.0	163.2	100.2	310.0	470.0	trace
Vit E	120	–	–	–	421.8	58.2	116.3	669	10	–	112.0
Nicotinate	–	0.1	240	108	89.3	77.7	79.3	82.5	136.0	5.5	4.0
Biotin	0.3	0.3	0.15		0.8	1.0	0.9	1.1	1.0	0.007	5.0
Folic acid	0.4	1.0			3.0	3.0	4.8	3.1	2.9	0.7	53.0
Pantothenic acid	13	6.8	20	46	37.7	9.1	13.2	21.4	73.0	2.8	–

Adapted from Fabregas and Herrero 1990; Becker 2004.

some studies report its anti-inflammatory, antimutagenic and anticarcinogenic health effects (Becker 2004; Koller et al., 2014; Begum et al., 2016; D'Alessandro and Filho 2016), while other studies identify the chlorophyll's derivates as a cause of some health problems, namely skin irritations (Becker 2004). Microalgal carotenoids are already used in foods—carotenes and xanthophylls reaching values up to 0.1–14% dry weight (Becker 2004; Christaki et al., 2011; Rajesh et al., 2017). The main carotenoids present in microalgae are β-carotene, lycopene, astaxanthin, zeaxanthin, violaxanthin and lutein (Becker 2004; Christaki et al., 2011; D'Alessandro and Filho 2016). The great effect on human health of carotenoids is related to their antioxidant activity, in addition to their ability to act as provitamin A (Koller et al., 2014; D'Alessandro and Filho 2016). The consumption of a diet rich in carotenoids has been epidemiologically correlated with a lower risk of diseases, such as cardiovascular and chronic diseases and cancer, which involve free-radical action on tissues (Rao and Rao 2007; Matos et al., 2017). β-carotene, astaxanthin, zeaxanthin and lutein are the most studied carotenoids extracted from microalgae (Udayan et al., 2017; Vidyashankar et al., 2017) with applications as nutritional supplements (Table 16.5). Phycobiliproteins are pigments present in rhodophytes, cryptophytes and cyanobacteria (Becker 2004; Koller et al., 2014; D'Alessandro and Filho 2016) and include allophycocyanin, phycocyanin, phycoerythrin and phycoerytrocyanin (Udayan et al., 2017; D'Alessandro and Filho 2016). Carotenoids and phycobilins extracted from microalgae have antioxidant activity as well as anti-inflammatory and hepatoprotective properties (Christaki et al., 2011). They are considered of interest for human nutrition for improving human health, and they can also be used as natural colorants for food products (Koller et al., 2014; Begum et al., 2016). Microalgal pigments therefore show interesting properties for human nutrition. However, some studies suggest that the threshold between the beneficial and adverse effects of some carotenoids, such as β-carotene, is low, and full understanding of their functional effects is needed (Van der Berg et al., 2000; Gouveia et al., 2008).

Polyphenols constitute one of the most numerous and ubiquitous group of phytochemicals and are an integral part of the human diet (Bravo 1998; Matos et al., 2017). Polyphenol-rich diets have been linked to many health benefits mainly due to their antioxidant, anti-inflammatory and antimicrobial properties (Matos et al., 2017). Microalgae have the ability to produce polyphenols with high antioxidant capacity, namely flavonoids and phenolic acids (Wells et al., 2017). Further research into these compounds is necessary since the extraction and characterization of polyphenols and the determination of their antioxidant activity are still challenging (Wells et al., 2017).

MICROALGAE ALREADY ON THE MARKET AS A SOURCE OF NUTRIENTS

The use of microalgae as source of nutrients started with and was confined to *Arthrospira* (*Spirulina*) and *Chlorella* for a long time, but it has recently expanded as new studies progressively reveal the potential of many other taxa (Chácon-Lee and González-Mariño 2010; Tibbetts et al., 2015; Assunção et al., 2017). Despite the disclosure of a large list of potential microalgae for food, the European market is still mainly confined to *Arthrospira* (*Spirulina*) and *Chlorella*.

TABLE 16.5
Carotenoids Extracted from Microalgae with Applications in Nutrition

Carotenoid	Microalgae	Yield (mg g^{-1})	Main Effect
β-carotene	*Dunaliella salina*	6–100	Reduces plasma cholesterol and atherogenesis, reduces fat accumulation and inflammation in liver
	Dunaliella bardawil	–	
Astaxanthin	*Haematococcus pluvialis*	1.30–30	Prevents obesity and fatty liver disease
	Chlorella zofingiensis	1	
Zeaxanthin	*Dunaliella salina*	6	Vision and overall antioxidant protection of body
	Nannochloropsis sp.	–	
Lutein	*Chlorella zofingiensis*	3.4	Protective antioxidative effect, vision
	Chlorella protothecoides	4.6	
	Murielliopsis sp.	4–6	

Adapted from Udayan et al., 2017; Vidyashankar et al., 2017.

ARTHROSPIRA (SPIRULINA) AND CHLORELLA

Arthrospira (*Spirulina*) and *Chlorella* are the two commercially available microalgae directly sold as food supplements in the form of dry biomass without any kind of processing (Enzing et al., 2014). *Arthrospira* (*Spirulina*) is a cyanobacteria with a long history of human consumption (Borowitzka 2018). The earliest records are from Spanish sailors who found that the Aztecs harvested this alga from Lake Texcoco and used it to make a cake named *Tecuilatl* (García et al., 2017). Also noteworthy is the traditional production and consumption of *Arthrospira* (*Spirulina*) in the region of Lake Chad (Africa), where *dihé*, a natural food prepared from Spirulina, still plays an important role in local communities' nutrition and income today (Carcea et al., 2015). In the 1960s, a renewed interest in the biochemical potential of *Arthrospira* (*Spirulina*) led to its mass production for commercial purposes that started in the late 1970s. *Arthrospira* (*Spirulina*) is indicated as source of protein (up to 60–70% dry weight), essential amino acids, fatty acids (namely palmitic, linoleic and γ-linolenic acids), minerals, pigments (chlorophyll a, β-carotene, zeaxanthin, cryptoxanthin, c-phycocyanin and allo-phycocyanin), vitamin B_{12} and sulphated polysaccharides (Jara et al., 2018). It is therefore considered as a "cell factory" of health-promoting bioactive molecules (Nicoletti 2016; Mathur 2018) and is sold in supermarkets as "Spirulina" powder and indicated as a "superfood." Its presence on the market is already generalized and accepted by consumers as an "alternative" source of vitamin B_{12} and proteins, particularly for non-animal-based diets such as vegetarianism. However, some studies indicate that a part of this vitamin B_{12} is in the form of its inactive analog, with no vitamin activity when consumed (Watanabe et al., 2002). Unexpectedly, the use of *Arthrospira* (*Spirulina*) for nutritional purposes is poorly studied. Most studies so far have focused on health-promoting effects and some are about the safety of consuming this cyanobacterium, and there is a gap in studies reporting the bioavailability of the ingested nutrients (Borowitzka 2018; Jara et al., 2018).

Chlorella is a green freshwater microalgae that is very easy to cultivate, which is probably the reason it was the first microalgae to be cultured on a large scale for the health-food market (Borowitzka 2018). The first records of the production of *Chlorella* date from 1950s, when it was produced in Taiwan and Japan as a source of health-promoting nutrients (Borowitzka 2018; Görs et al., 2010). Besides its effect on the lowering of cholesterol, preventive action against atherosclerosis or antitumor action, *Chlorella* produces a polysaccharide that acts as immunostimulator, free-radical scavenger and reducer of blood lipids (Suárez et al., 2006; Görs et al., 2010). *Chlorella* biomass is composed of 55–67% protein, 1–4% chlorophyll, 9–18% dietary fiber, minerals and vitamins (Bishop and Zubeck 2012), which results in its nutritional interest. *Chlorella* is sold in supermarkets mainly as powder and in tablet form. An extract of *Chlorella* is sold as "*Chlorella* Growth Factor" with a number of activities indicated, such as cancer prevention and antiaging, but no data about the composition can be found (Borowitzka 2018). Many claims for the benefits of consuming *Chlorella* have been made, but only a few are justified by scientific data and the identification of the active molecule (Borowitzka 2018).

Most studied microalgae with proven nutritional value as food are regarded as beneficial for human health. However, a question arises: how much of this nutritional and bioactive value is effectively assimilated by the human organism? The first factor to consider is the digestibility of microalgae in general. The need for a scientific validation of effective gastrointestinal processing and cell use of the nutritional value of flagged microalgae is the next essential step towards their release on the market. Fibrillary and mucilaginous components comprise the cell walls of most microalgae. The most common fibrillary component is cellulose, which is not digested by humans. The cell wall of *Arthrospira* (*Spirulina*) is devoid of cellulose and is not a barrier to proteolytic enzymes; however, in the case of *Chlorella*, a rigid indigestible cell wall composed of cellulose is present (Becker 2004). Studies confirmed the efficient metabolic protein utilization from *Arthrospira* (*Spirulina*) by humans (Cifferi 1983; Becker 2004; Mathur 2018); however, in the case of *Chlorella*, this is not proven.

Becker (2004) suggests a list of steps for enabling the utilization of microalgal biomass in human nutrition. After accurate determination of the nutritional value by proximate analysis, complemented by analysis of other compounds in the biomass, quality studies are in queue. Protein quality assessment by digestibility studies and metabolic uptake by human gastrointestinal cells are important determinations for calibrating the expectations and judgment of the true value of microalgae as food. Safety tests for ensuring the nontoxicity of biogenic and nonbiogenic substances present in the biomass are crucial for providing robust data for the whole process. Marketing assessments may help with the successful implementation of microalgal-derived products as food into the market.

NEW MICROALGAE FOR THE MARKET, PERSPECTIVES AND CONSTRAINTS

The potential use of microalgae in human nutrition has been widely discussed and recognized, especially as a natural source of highly valuable bioactive compounds for health-promoting functional foods (Matos et al., 2017).

Although the market size and amounts of nutrients produced from microalgae are still significantly smaller than various other commodity crops, demand for microalgal-based nutrients is expected to be significant in the predictable future. Several market reports for the global algal and microalgal products forecast a compound annual growth rate above 5% for the next 5–8 years (Credence Research 2016; Meticulous Research 2018; Persistence Market Research 2018). Nutraceuticals, followed by food and feed, are expected to prevail as the leading applications of microalgae, regarding revenues as well as sales expansion. For example, a growing trend towards plant-based algal proteins, considered free from allergens, gluten and GMO, is reflected by a European algal protein market valued at around €63 million in 2016, with 28% market share for Germany (Mordor Intelligence 2018).

Recently, the main constraints for using algae as food ingredient in Europe have been identified by the European Commission Joint Research Centre (Enzing et al., 2014) as: i) technical difficulties associated with their cultivation and high production costs; ii) low demand in European countries compared to Asian markets; and iii) strict European legislation regarding novel foods.

The former Regulation (EC) No. 258/97 (European Commission 1997), referring to the authorization of novel foods and novel food ingredients, has been recently revised and replaced by the Regulation (EU) No. 2015/2283 (European Union 2015) which is applicable since 1 January 2018. A novel food is defined as food that has not been consumed to any significant degree in the EU before 15 May 1997 (when the first novel food legislation came into force). *Arthrospira* (*Spirulina*) and *Chlorella*, which have been widely consumed since ancient times, do not fall under the novel food status and thus can be freely commercialized as a food ingredient. This is also applicable to a filamentous blue-green alga from Klamath Lake (Oregon, United States), *Aphanizomenon flos-aquae*, although some concerns regarding the potential toxicity of this natural bloom species have been raised (e.g. Niccolai et al., 2017; Roy-Lachapelle et al., 2017; Lyon-Colbert et al., 2018). In Australia, *A. flos-aquae* was considered as novel food, and safety assessments were required due to the potential presence of cyanobacterial toxins such as microcystins and nodularin (FSANZ 2016).

In order to ensure the highest level of protection for human health, novel foods must undergo a safety assessment before being placed on the European Union (EU) market. Some of the key aspects that should be considered are the history of use of the organism, predicted levels of ingestion, nutritional microbiological data, toxicological studies (animal and human), allergenic potential, substantial equivalence to other foods and possible implications for human nutrition.

Novel foods or novel food ingredient applications may follow a simplified procedure, only requiring notifications from the company when they are considered by a national food assessment body as "*substantially equivalent*" to existing foods or food ingredients (regarding their composition, nutritional value, metabolism, intended use and the level of undesirable substances contained therein). Table 16.6 presents examples of approved microalgal novel foods.

The new Regulation (EU) No. 2015/2283 (European Union 2015) presents some improvements in the regulatory conditions so that food businesses can more easily bring new and innovative foods to the EU market, while maintaining a high level of food safety for European consumers. Simplified and centralized authorization (by the European Community) and safety evaluation by European Food Safety Authority (EFSA) procedures are aimed at improving efficiency and transparency, as well as creating a faster, structured notification system for traditional foods from third-world countries on the basis of safe food use history. One of the main features is the promotion of innovation by granting an individual authorization for 5 years based on protected data. On the other hand, generic authorization of novel foods, as opposed to the previously restricted applicant-specific authorization, allows any food business operator to place an authorized novel food on the EU market, provided the authorized conditions of use, labeling requirements and specifications are respected. An EU-approved list of authorized novel foods (Table 16.6) has been published (European Union 2017) and will be updated for every novel food approval. This list includes the conditions under which the novel food may be used, that is, the specified food categories and maximum levels (e.g. maximum DHA level required for *Schizochytrium* oils); additional specific labeling requirements (e.g. "*Contains negligible amounts of iodine*" required for *Tetraselmis chuii*) and specifications (e.g. chemical composition, oil oxidative stability, DNA markers).

In the United States, *Chlorella*, *Arthrospira* (*Spirulina*), *Dunaliella*, *Haematococcus*, *Schizochytrium*,

TABLE 16.6
Microalgae Novel Foods That May Be Placed on the Market in the EU Pursuant to Regulation (EC) No. 258/97, and Indicated on the Union List of Novel Foods in Accordance with Regulation (EU) 2015/2283 (European Union 2017)

Microalgae Product	Applicant Company	Approval
Odontella aurita microalga	Fermentalg S.A.R.L. (France)	Dec 9, 2002
Schizochytrium sp. oil[a]	MartekBiosciencesCorporation (USA)	June 5, 2003
Schizochytrium sp. oil rich in DHA and EPA	DSM Nutritional Lipids (USA)	Jul 6, 2012
Schizochytrium sp. (ATCC PTA-9695) oil	DSM Nutritional Products (USA)	Mar 31, 2015
Schizochytrium sp. (T18) oil	Mara Renewables Corporation (Canada)	Jul 20, 2018
Astaxanthin-rich oleoresin from *Haematococcus pluvialis* algae	US Nutra (USA)	Jun 28, 2004
	AstaReal (Sweden)	May 17, 2006
	Cyanotech Corp. (USA)	Mar 7, 2007
	Alga Technologies Ltd. (Israel)	Apr 14, 2008
	FenchemBiotek Ltd. (China)	Feb 20, 2015
	InnoBio Ltd. (China)	Nov 10, 2016
	Beijing Gingko Group	Nov 23, 2016
Algal oil from the micro-algae *Ulkenia* sp.	Lonza Ltd (Switzerland)	Oct 21, 2009
Dried microalgae *Tetraselmis chuii*	Fitoplancton Marino S.L. (Spain)	Mar 4, 2014
Freeze-dried *Tetraselmis chuii* supplement	Fitoplancton Marino S.L. (Spain)	Nov 8, 2017
EPA-rich oil from *Phaeodactilum tricornutum*	Simris Alg AB (Sweden)	Pending

[a] Since 2011, 11 other companies have been authorized to commercialize DHA-rich oils from *Schizochytrium* sp., as they were considered equivalent to previously commercialized products (article 5 simplified procedure).

Porphyridium cruentum and *Crypthecodinium cohnii* are classified as food sources falling into the GRAS (generally recognized as safe) category (García et al., 2017). In recent years, the US Food & Drug Administration has received several applications for novel microalgae products for GRAS status (Table 16.7). It is worth mentioning an ongoing application for *Chlamydomonas reinhardtii*, a microalga which is not yet commercialized as food either in the United States or in Europe. From the European side, also noteworthy is the pending application of the Swedish company SimrisAlg AB for EPA-rich oil from *Phaeodactylum tricornutum* (Table 16.6). This marine diatom has been recognized as a rich source of polyunsaturated fatty acids, such as EPA, as well as of the carotenoid fucoxanthin, which is a valuable pigment with several biological properties (Gilbert-Lopez et al., 2017; Rodolfi et al., 2017). Batista and colleagues (2017) studied the incorporation of *P. tricornutum* biomass in traditional cookies which presented high phenolic and antioxidant content.

The regulatory progresses and successful authorization of novel microalgal based foods and food ingredients broaden perspectives for a wider inclusion of these valuable microorganisms in the human diet. Further efforts are being made to spread the use of microalgae for food use, with collaborative actions being taken between industrial producers, research bodies and other experts. The European Algal Biomass Association has a working group dedicated to novel foods, while in the United States, "The Future of Algae for Food & Feed (FAFF)" initiative was created to accelerate the development of cost-competitive algae-based food and feed in the global marketplace.

One of the current challenges that microalgae producers and biotechnologists are facing concerns the organic certification of food-grade microalgae to respond to the growing demand for organic and fully traceable plant-based ingredients. There are several practical issues to overcome; for instance, no detailed production rules have been defined for the production system, although it is clear that open cultivation systems involve risk of contamination and thus closed systems are possibly a better solution for controlled cultivation. Moreover, the addition of minerals hinders organic certification so there are many essential nutrients and micronutrients specific to microalgal growth that cannot be used, while the use of other organic fertilizers, such as manure, is not an option due to the risk of water and biomass contamination with harmful bacteria, toxins and heavy metals. Nevertheless, some microalgal producers are working

TABLE 16.7
Notifications of Microalgae Foods Generally Recognized as Safe (GRAS) by the US Food & Drugs Administration (FDA 2018)

GRN No.	Microalgal Product	Company	Approval
137	Algal oil (*Schizochytrium sp.*)	Martek Biosciences Corporation (USA)	Feb 12, 2004
319	Microalgal oil *Ulkenia* sp. SAM2179	Lonza Ltd. (Switzerland)	Aug 4, 2010
384	Algal oil derived from *Chlorella protothecoides* strain S106 (Cp algal oil)	Solazyme, Inc. (USA)	Jun 13, 2012
469	*Chlorella protothecoides* strain S106 flour with 40-70% lipid (algal powder)	Solazyme Roquette Nutritionals, LLC (USA)	Jun 7, 2013
527	Algal oil (87% oleic acid) derived from *Prototheca moriformis* strain S2532	Solazyme, Inc. (USA)	Feb 6, 2015
553	Algal oil (40% docosahexaenoic acid) derived from *Schizochytrium* sp.	DSM Nutritional Products (USA)	Jun 19, 2015
673	Algal fat derived from *Protothecamoriformis* (S7737)	TerraVia Holdings (USA) (Previously Solazyme, Inc.)	Oct 13, 2016
754	Algal oil (87% oleic acid) derived from *Prototheca moriformis* strain S6697	Corbion Biotech, Inc. (USA)	Jul 23, 2018
773	Dried biomass of *Chlamydomonas reinhardtii*	Triton Algae Innovations (USA)	Pending
776	Algal oil (35% docosahexaenoic acid) from *Schizochytrium* sp. FCC-1324	Fermentalg (France)	Pending
777	Algal oil (55% docosahexaenoic acid) from *Schizochytrium* sp. FCC-3204	Fermentalg (France)	Pending

towards overcoming these challenges; Allmicroalgae (Leiria, Portugal), for example, has recently received organic certification by the EU for a *Chlorella vulgaris* microalgal product.

CONCLUSIONS

The growing interest in microalgal-derived food and supplements is also accompanied by concern about the two main topics that arise in the case of every novel source of nutrients falling outside the conventional concept of food: the bioavailability of ingested nutrients and their safety. Many of the health benefits attributed to consuming microalgae as functional foods are still underestimated via scientific studies aiming to tangibly estimate the true value of microalgae in the food market. Once properly characterized, microalgae will possibly be on everyone's plate, and this colorful food will be the delight of chefs worldwide.

REFERENCES

Abomohra, A.E., El-Sheekh, M., Hanelt, D. Screening of marine microalgae isolated from the hypersaline Bardawil lagoon for biodiesel feedstock. *Renew. Energy.* 2017. 101: 1266–1272.

Aluko, R.E. *Functional Foods and Nutraceuticals.* 2012. New York, NY: Springer.

Assunção, M.F.G., Varejão, J.M.T.B., Santos, L.M.A. Nutritional characterization of the microalga *Ruttnera lamellosa* compared to *Porphyridium purpureum*. *Algal Res.* 2017. 26: 8–14.

Batista, A.P., Gouveia, L., Bandarra, N.M., Franco, J.M., Raymundo, A. Comparison of microalgal biomass profiles as novel functional ingredient for food products. *Algal Res.* 2013. 2: 164–173.

Batista, A.P., Niccolai, A., Fradinho, P., Fragoso, S., Bursic, I., Rodolfi, L., Biondi, N., Tredici, M.R., Sousa, I., Raymundo, A. Microalgae biomass as an alternative ingredient in cookies: Sensory, physical and chemical properties, antioxidant activity and in vitro digestibility. *Algal Res.* 2017. 26: 161–171.

Becker, E.W. Micro-algae as a source of protein. *Biotechnol. Adv.* 2007. 25: 207–210.

Becker, W. Microalgae in human and animal nutrition. In: Richmond, A. (editor), *Handbook of Microalgal Culture: Biotechnology and Applied Phycology.* 2004. 312–346. Blackwell Publishing Ltd.

Beecher, G.R. Phytonutrients' role in metabolism: Effects on resistance to degenerative processes. *Nutr. Rev.* 1999. 57: S3–S6.

Begum, H., Yusoff, F.M., Banerjee, S., Khatoon, H., Shariff, M. Availability and utilization of pigments from microalgae. *Crit. Rev. Food Sci. Nutr.* 2016. 56: 2209–2222.

Bishop, W.M., Zubeck, H.M. Evaluation of microalgae for use as nutraceuticals and nutritional supplements. *Nutr. Food.* 2012. 2: 1–6.

Borowitzka, M.A. Microalgae in medicine and human health: A historical perspective. In: Levine, I., Fleurence, J. (editors), *Microalgae in Health and Disease Prevention.* 2018. Academic Press.

Bravo, L. Polyphenols: Chemistry, dietary sources, metabolism and nutritional significance. *Nutr. Rev.* 1998. 56: 317–333.

Caballero, B., Allen, L., Prentice, A. *Encyclopedia of Human Nutrition.* 2005. Elsevier Ltd.

Carcea, M., Sorto, M., Batello, C., Narducci, V., Aguzzi, A., Azzini, E., Fantauzzi, P., Finotti, E., Gabrielli, P., Galli, V., Gambelli, L., Maintha, K.M., Namba, F., Ruggeri, S., Turfani, V. Nutritional characterization of traditional and improved *dihé*, alimentary blue-green algae from the Lake Chad region in Africa. *LWT—Food Sci. Technol.* 2015. 62: 753–763.

Chacón-Lee, T.L., González-Mariño, G.E. Microalgae for "healthy" foods—Possibilities and challenges. *Compr. Rev. Food Sci. Food Saf.* 2010. 9: 655–675.

Chisti, Y. Society and microalgae: Understanding the past and present. In: Levine, A., Fleurence, J. (editors), *Microalgae in Health and Disease Prevention.* 2018. Academic Press.

Christaki, E., Florou-Paneri, P., Bonos, E. Microalgae: A novel ingredient in nutrition. *Int. J. Food Sci.Nutr.* 2011. 62: 794–799.

Ciferri, O. Spirulina, the edible microorganism. *Microbiol. Rev.* 1983. 47: 551–578.

Credence Research. Algae Product Market by Application (Nutraceuticals, Food & Feed Supplements, Pharmaceuticals, Paints & Colorants, Pollution Control, Others)—Growth, Future Prospects, Competitive Analysis, and Forecast 2016-2023. 2017. Report Code: 57848-05-17.

Croft, M.T., Lawrence, A.D., Raux-Deery, E., Warren, M.J., Smith, A.G. Algae acquire vitamin B12 through a symbiotic relationship with bacteria. *Nature.* 2005. 438: 90–93.

D'Alessandro, E.B., Antoniosi Filho, N.R. Concepts and studies on lipid and pigments of microalgae: A review. *Renew. Sust. Energ. Rev.* 2016. 58: 832–841.

de Jesus Raposo, M.F., de Morais, R.M., de Morais, A.M. Health applications of bioactive compounds from marine microalgae. *Life Sci.* 2013. 93: 479–486.

Del Río, E., García-Gómez, E., Moreno, J., G. Guerrero, M., García-González, M. Microalgae for oil. Assessment of fatty acid productivity in continuous culture by two high-yield strains, *Chlorococcum oleofaciens* and *Pseudokirchneriella subcapitata. Algal Res.* 2017. 23: 37–42.

Diprat, A.B., Menegol, T., Boelter, J.F., Zmozinski, A., Rodrigues Vale, M.G.R., Rodrigues, E., Rech, R. Chemical composition of microalgae *Heterochlorella luteoviridis* and *Dunaliella tertiolecta* with emphasis on carotenoids. *J. Sci. Food Agric.* 2017. 97: 3463–3468.

Enzing, C., Ploeg, M., Barbosa, M., Sijtsma, L. Microalgae-based products for the food and feed sector: An outlook for Europe. In: Vigani, M., Parisi, C., Rodriguez-Cerezo, E. (editors), *JRC Scientific and Policy Reports.* 2014. European Commission.

European Commission. Regulation (EC) n° 258/97 of the European Parliament and of the Council of 27 January 1997 concerning novel foods and novel food ingredients. *Official Journal of the European Union*, L 43/1 of 14 February 1997, pp. 1.

European Union. Commission Implementing Regulation (EU) 2017/2470 of 20 December 2017 establishing the Union list of novel foods in accordance with Regulation (EU) 2015/2283 of the European Parliament and of the Council on Novel Foods. *Official Journal of the European Union*, L 351/72 of 30 December 2017.

European Union. Regulation (EU) n° 2015/2283 of the European Parliament and of the Council of 25 November 2015 on novel foods, amending Regulation (EU) n° 1169/2011 of the European Parliament and of the Council and repealing Regulation (EC) n° 258/97 of the European Parliament and of the Council and Commission Regulation (EC) n° 1852/2001. *Official Journal of the European Union*, L 327/1 of 11 December 2015, p. 1.

Fabregas, J., Herrero, C. Vitamin content of four marine microalgae. Potential use as source of vitamins in nutrition. *J. Ind. Microbiol.* 1990. 5: 259–264.

FDA. U.S. Food & Drugs Administration. Inventory of GRAS Notices. https://www.accessdata.fda.gov/scripts/fdcc/?set=GRASNotices&sort=GRN_No&order=DESC&startrow=1&type=basic&search=alga, accessed 15 September 2018.

FSANZ. Novel food-record of views. Record of views formed by the Food Standards Australia New Zealand (FSANZ). Novel Foods Reference Group or the Advisory Committee on Novel Foods, March 2016.

García, J.L., de Vicente, M., Galán, B. Microalgae, old sustainable food and fashion nutraceuticals. *Microb. Biotechnol.* 2017. 10: 1017–1024.

Geoghegan, M.J. Unicellular algae as food. *Nature.* 1951. 168: 426–427.

Gilbert-López, B., Barranco, A., Herrero, M., Cifuentes, A., Ibáñez, E. Development of new green processes for the recovery of bioactives from *Phaeodactylum tricornutum. Food Res. Int.* 2017. 99: 1056–1065.

Görs, M., Schumann, R., Hepperle, D., Karsten, U. Quality analysis of commercial *Chlorella* products used as dietary supplement in human nutrition. *J. Appl. Phycol.* 2010. 22: 265–276.

Gouveia, L., Batista, A.P., Sousa, I., Raymundo, A., Bandarra, N.M. Microalgae in novel food products. In: Papadopoulos, K.N. (editor), *Food Chemistry Research Developments.* 2008. 75–111. New York, NY: Nova Science Publishers, Inc.

Griffiths, M.J., van Hille, R.P., Harrison, S.T.L. Lipid productivity, settling potential and fatty acid profile of 11 microalgal species grown under nitrogen replete and limited conditions. *J. Appl. Phycol.* 2012. 24: 989–1001.

Huang, X., Huang, Z., Wen, W., Yan, J. Effects of nitrogen supplementation of the culture medium on the growth, total lipid content and fatty acid profiles of three microalgae (*Tetraselmis subcordiformis, Nannochloropsis oculata, Pavlova viridis*). *J. Appl. Phycol.* 2013. 25: 129–137.

Hultberg, M., Jönsson, H.L., Bergstrand, K.J., Carlsson, A.S. Impact of light quality on biomass production and fatty acid content in the microalga *Chlorella vulgaris*. *Bioresour. Technol.* 2014. 159: 465–467.

Jara, A., Ruano-Rodriguez, C., Polifrone, M., Assunção, P., Brito-Casillas, Y., Wägner, A.M., Serra-Majem, L. Impact of dietary *Arthrospira* (*Spirulina*) biomass consumption on human health: Main health targets and systematic review. *J. Appl. Phycol.* 2018. 30: 2403–2423.

Kent, M., Welladsen, H.M., Mangott, A., Li, Y. Nutritional evaluation of Australian microalgae as potential human health supplements. *PLOS ONE*. 2015. 10: e0118985.

Khantoon, H., Haris, H., Rahman, N.A., Zakaria, M.N., Begum, H., Mian, S. Growth, proximate composition and pigment production of *Tetraselmis chuii* cultured with aquaculture wastewater. *J. Ocean Univ. China*. 2018. 17: 641–646.

Khozin-Goldberg, I., Iskandarov, U., Cohen, Z. LC-PUFA from photosynthetic microalgae: Occurrence, biosynthesis, and prospects in biotechnology. *Appl. Microbiol. Biotechnol.* 2011. 91: 905.

Koller, M., Muhr, A., Braunegg, G. Microalgae as versatile cellular factories for valued products. *Algal Res.* 2014. 6: 52–63.

Lang, I., Hodac, L., Friedl, T., Feussner, I. Fatty acid profiles and their distribution patterns in microalgae: A comprehensive analysis of more than 2000 strains from the SAG culture collection. *BMC Plant Biol.* 2011. 11: 124.

Lourenço, S.O., Barbarino, E., Lavín, P.L., Lanfer Marquez, U.M., Aidar, E. Distribution of intracellular nitrogen in marine microalgae: Calculation of new nitrogen-to-protein conversion factors. *Eur. J. Phycol.* 2004. 39: 17–32.

Lyon-Colbert, A., Su, S., Cude, C. A systematic literature review for evidence of *Aphanizomenon flos-aquae* toxigenicity in recreational waters and toxicity of dietary supplements: 2000–2017. *Toxins*. 2018. 10: 254.

Marques, A.E., Miranda, J.R., Batista, A.P., Gouveia, L. Microalgae biotechnological applications: Nutrition, health and environment. In: Johnsen, M.N. (editor), *Microalgae: Biotechnology, Microbiology and Energy*. 2011. Nova Science Publishers, Inc.

Martins, C.B., Varejão, J.M.T.B., Santos, L.M.A. Biotechnological potential of *Haematococcus pluvialis* Flotow ACOI 3380. *Nova Hedwigia*. 2016. 103: 547–559.

Mathur, M. Bioactive molecules of Spirulina: A food supplement. In: Mérillon, J.M., Ramawat, K.G. (editors), *Bioactive Molecules in Food*. 2018. Springer International Publishing AG.

Matos, J., Cardoso, C., Bandarra, N.M., Afonso, C. Microalgae as healthy ingredients for functional food: A review. *Food Func.* 2017. 8: 2672–2685.

Meticulous Research. Algae Products Market– Global Opportunity Analysis and Industry Forecast (2017–2022). 2018. Report ID: MRFB-10441. 162 pp. https://www.meticulousresearch.com/product/algae-products-market-forecast-2022/.

Mobin, S., Alam, F. Some promising microalgal species for commercial applications: A review. *Energy Procedia*. 2017. 110: 510–517.

Mordor Intelligence. Europe Algae protein market—Growth, trends, and forecast (2018–2023). 2018. https://www.mordorintelligence.com/industry-reports/europe-algae-protein-market.

Nascimento, I.A., Marques, S.S.I., Cabanelas, I.T.D., Pereira, S.A., Druzian, J.I., de Souza, C.O., Vich, D.V., de Carvalho, G.C., Nascimento, M.A. Screening microalgae strains for biodiesel production: Lipid productivity and estimation of fuel quality based on fatty acids profiles as selective criteria. *Bioenerg. Res.* 2013. 6: 1–13.

Niccolai, A., Bigagli, E., Biondi, N., Rodolfi, L., Cinci, L., Luceri, C., Tredici, M.R. *In vitro* toxicity of microalgal and cyanobacterial strains of interest as food source. *J. Appl. Phycol.* 2017. 29: 199–209.

Nicoletti, M. Microalgae nutraceuticals. *Foods*. 2016. 5: 1–13.

Özogul, Y., Özogul, F. Fatty acid profiles of commercially important fish species from the Mediterranean, Aegean and Black Seas. *Food Chem*. 2007. 100: 1634–1638.

Persistence Market Research. Global Market Study on Microalgae: North America to be Largest Market for Microalgae during 2017 to 2026. 2018. Report code: PMRREP22360. 170 pp. https://www.persistencemarketresearch.com/market-research/microalgae-market.asp.

Pignolet, O., Jubeau, S., Vaca-Garcia, C., Michaud, P. Highly valuable microalgae: Biochemical and topological aspects. *J. Ind. Microbiol. Biotechnol.* 2013. 40: 781–796.

Quintaes, K.D., Diez-Garcia, R.W. The importance of minerals in human diet. In: Guardia, M., Garrigues, S. (editors), *Handbook of Mineral Elements in Food*. 2015. John Wiley & Sons, Ltd.

Rajesh, K., Rohit, M.V., Mohan, S.V. Microalgae-based carotenoids production. In: Rastogi, R.P., Madamwar, D., Pandey, A. (editors), *Algal Green Chemistry—Recent Progress in Biotechnology*. 2017. Elsevier.

Rajeshkumar, S., Li, X. Bioaccumulation of heavy metals in fish species from the Meiliang Bay, Taihu Lake, China. *Toxicol. Rep.* 2018. 5: 288–295.

Rao, A.V., Rao, L.G. Carotenoids and human health. *Pharmacol. Res.* 2007. 55: 207–216.

Rebolloso Fuentes, M.M., Acién-Fernández, G.G., Sánchez-Pérez, J.A., Guil-Guerrero, J.L. Biomass nutrient profiles of the microalga *Porphyridium cruentum*. *Food Chem*. 2000. 70: 345–353.

Rebolloso-Fuentes, M.M., Navarro-Pérez, A., García-Camacho, F., Ramos-Miras, J.J., Guil-Guerrero, J.L. Biomass nutrient profiles of the microalga *Nannochloropsis*. *J. Agric. Food Chem.* 2001b. 49: 2966–2972.

Rebolloso-Fuentes, M.M., Navarro-Pérez, A., Ramos-Miras, J.J., Guil-Guerrero, J.L. Biomass nutrient profiles of the microalga *Phaeodactylum tricornutum*. *J. Food Biochem.* 2001a. 25: 57–76.

Rodolfi, L., Biondi, N., Guccione, A., Bassi, N., D'Ottavio, M., Arganaraz, G., Tredici, M.R. Oil and eicosapentaenoic acid production by the diatom *Phaeodactylum tricornutum* cultivated outdoors in Green Wall Panel (GWP®) reactors. *Biotechnol. Bioeng.* 2017. 114: 2204–2210.

Roy-Lachapelle, A., Solliec, M., Bouchard, M.F., Sauvé, S. Detection of cyanotoxins in algae dietary supplements. *Toxins.* 2017. 9: E76.

Ryckebosch, E., Bruneel, C., Termote-Verhalle, R., Goiris, K., Muylaert, K., Foubert, I. Nutritional evaluation of microalgae oils rich in omega-3 long chain polyunsaturated fatty acids as an alternative for fish oil. *Food Chem.* 2014. 160: 393–400.

Safi, C., Ursu, A.V., Laroche, C., Zebib, B., Merah, O., Pontalier, P., Vaca-Garcia, C. Aqueous extraction of proteins from microalgae. *Algal Res.* 2014. 3: 61–65.

Schwenzfeier, A., Wierenga, P.A., Gruppen, H. Isolation and characterization of soluble protein from the green microalgae *Tetraselmis* sp. *Bioresour. Technol.* 2011. 102: 9121–9127.

Simopoulos, A.P. Evolutionary aspects of diet, the omega-6/omega-3 ratio and genetic variation: Nutritional implications for chronic diseases. *Biomed. Pharmacother.* 2006. 60: 502–507.

Spolaore, P., Joannis-Cassan, C., Duran, E., Isambert, A. Commercial applications of microalgae. *J. Biosci. Bioeng.* 2006. 101: 87–96.

Suárez, E.R., Syvitski, R., Kralovec, J.A., Noseda, M.D., Barrow, C.J., Ewart, H.S., Lumsden, M.D., Grindley, T.B. Immunostimulatory polysaccharides from *Chlorella pyrenoidosa*. A new galactofuranan. Measurement of molecular weight and molecular weight dispersion by DOSY NMR. *Biomacromolecules.* 2006. 7: 2368–2376.

Tibbetts, S.M., Milley, J.E., Lall, S.P. Chemical composition and nutritional properties of freshwater and marine microalgal biomass cultured in photobioreactors. *J. Appl. Phycol.* 2015. 27: 1109–1119.

Tokuşoglu, O., uunal, M.K. Biomass nutrient profiles of three microalgae: *Spirulina platensis*, *Chlorella vulgaris* and *Isochrisis galbana*. *J. Food Sci.* 2003. 68: 1144–1148.

Undayan, A., Arumugam, M., Pandey, A. Nutreaceuticals from algae and cyanobacteria. In: Rastogi, R.P., Madamwar, D., Pandey, A. (editors), *Algal Green Chemistry—Recent Progress in Biotechnology.* 2017. Elsevier.

Van den Berg, H., Faulks, R., Granado, H.F., Hirschberg, J., Olmedilla, B., Sandmann, G., Southon, S., Stahl, W. The potential for the improvement of carotenoid levels in foods and the likely systemic effects. *J. Sci. Food Agric.* 2000. 80: 880–912.

Vidyashankar, S., Daris, P.S., Mallikarjuna, K.G., Sarada, R. Microalgae as a source of nutritional and therapeutic metabolites. In: Siddiqui, M.W., Prasad, K., Bansal, V. (editors), *Plant Cell Biotechnology.* 2017. Apple Academic Press.

Watanabe, F., Takenaka, S., Kittaka-Katsura, H., Ebara, S., Miyamoto, E. Characterization and bioavailability of vitamin B12-compounds from edible algae. *J. Nutr. Sci. Vitaminol.* 2002. 48: 325–331.

Wells, M.L., Potin, P., Craigie, J.S., Raven, J.A., Merchant, S.S., Helliwell, K.E., Smith, A.G., Camire, M.E., Brawley, S.H. Algae as nutritional and functional food sources: Revisiting our understanding. *J. Appl. Phycol.* 2017. 29: 949–982.

17 Nutraceutical Aspects of Microalgae
Will Our Future Space Foods Be Microalgae Based?

Ceren Gurlek, Cagla Yarkent, Izel Oral, Ayse Kose, and Suphi S. Oncel

CONTENTS

Abbreviations 177
Introduction 177
Life Support Systems 178
Terraforming 179
History of Algae in Space 179
History of Space Food and Microalgae as a Novel Designer Food in Space 181
Future Scenarios 182
Conclusion 183
References 183

BOX 17.1 SALIENT FEATURES

Our curiosity about outer space and the quest to establish a civilization in space has led scientists to create self-sufficient systems or, as an even better solution, to create in outer space an *Earthlike* environment like the one we are familiar with. Photosynthetic organisms like plants and microalgae have the capability to provide nutritional needs and create an Earthlike atmosphere to allow humans to survive in outer space, and in particular on Mars. Microalgae have numerous advantages over plants, such as higher growth rate, higher photosynthesis rate, cheaper production (due to the fact that they do not require specific substrates), and higher adaptive capability to extreme conditions, and they can also be used as a fuel source, as biofertilizer, and for water recovery purposes; consequently, they are considered promising candidates for life-support systems. This chapter discusses the history of space missions and how and why microalgae can be used as space food and in biological life support systems based on this history; the future of space missions and space food; and the potential of microalgae for terraforming Mars.

ABBREVIATIONS

BLSS: Biological life support system
CO_2: carbon dioxide
DNA: deoxyribonucleic acid
ESA: European Space Agency
IR: infrared
ISS: International Space Station
LSS: Life support system
MELiSSA: Micro-ecological life support system alternative
NASA: National Aeronautics and Space Administration
UV: ultraviolet

INTRODUCTION

Futuristic space technology prepared us for the idea of using colorful food pills to ensure a balanced diet and satisfy our appetites. The introduction of certain nutraceutical raw materials and extracts and such single-cell proteins as yeast or microalgae complete the image of an all-in-one tablet for a healthier lifestyle. The use of microalgae for nutraceuticals is not a new concept. Society has been aware of this potential since the time of ancient civilizations. However, in industry, the biotechnological aspects of microalgal nutraceuticals are new concepts that are gaining acceptability in global market.

Beyond the Earth, in outer space, there is lack of breathable oxygen and thus aerobic organisms cannot survive (Bolonkin et al. 2017). To establish space civilization, a habitable and sustaining environment must be maintained in space (Daues 2006). Algae can provide the oxygen, food and fuel needed for human life (Chapman 2013).

Algae are chlorophyll-containing photosynthetic organisms, which makes them significantly important for providing oxygen. They are also nutritionally rich sources; thus, they can be very important for the human diet (Spolaore et al. 2006). With the increase in human populations, constant efforts are being made to find sustainable and cost-effective healthy and functional foods, which has led to the adoption of algal technologies and subsequently the production of highly nutritious algal forms in the food industry (Nicoletti 2016). *Spirulina* and *Chlorella* are the most popular strains and conventional diet supplements among the algae on the market (Mobin and Alam 2017); however, there are enormous numbers of species waiting to be discovered under extreme conditions, which may have potential food applications.

Algae can be used to produce a whole range of valuable primary products like proteins, carbohydrates and vitamins and secondary products like pigments, phenolics and flavonoids. These products have varied human health benefits such as antioxidant, antimicrobial and anti-inflammatory effects, reducing the risk of certain diseases and immunomodulation.

The development of oxygen in the Earth's atmosphere started with the photosynthesis of algae, now known as *cyanobacteria*. Algae changed the atmosphere of the planet and allowed the evolution of eukaryotic organisms by producing oxygen in the early stages of life (Chapman 2013). Algae living in water sources like rivers, lakes, lagoons and oceans contribute to about 50% of all the oxygen produced on the planet. A group of organisms capable of providing atmospheric oxygen on our planet would possibly be of same utility on another planet as well, but no-one can wait for millions of years for that to happen.

In their life cycle, microalgae take up CO_2, water and other nutrients to produce oxygen, biomass and metabolites while converting solar energy to chemical energy (Adeniyi et al. 2018; Pires 2017). Besides contributing to the buildup of oxygen in the environment, they are of importance because of their ability to produce lipids. The most notable feature of the oil they produce is its potential for conversion into different types of biofuels (Adeniyi et al. 2018) like bioethanol, biohydrogen, biomethanol, biodiesel and biogas (Koutra et al. 2018; Pires 2017). The huge potential of microalgae to tolerate high CO_2 concentration and produce oil approximately 60% offers valuable, sustainable and renewable fuel resource yet to be tapped economically (Adeniyi et al. 2018; NASA 2018;Rizwan et al. 2018,). However, the fuel produced by algal lipids has combustion properties that meet the needs of a range of applications, and it is thus a potential energy source at the International Space Station (ISS) (Adeniyi et al. 2018; https://www.nasa.gov/mission_pages/station/research/experiments/1810.html, 29.06.2018). In addition to the advantage as food and fuel is also its potential as an agent to produce potable water using the waste generated by humans (Adeniyi et al. 2018). In this chapter, we will focus on the potential utility of algae as space food for space missions and habitation.

LIFE SUPPORT SYSTEMS

A life support system (LSS) is a group of devices which are able to ensure favorable air pressure and temperature, protection from radiation, and the recovery of waste products for better cost-effectiveness and sustainability. These systems should be designed to maintain human life in space (Lehto et al. 2006).

Long-term missions in space require the supply of basic life-support components. In a traditional life-support system, the volume of water, oxygen and food storage and launch weight are directly proportional to the crew size and the length of the mission. Due to the limited spacecraft load, it is necessary to optimize the utility of the carbon and food recycling regime (Skoog 1984), warranting the creation of regenerative systems. This was the challenge for the development of the water recovery system on the International Space Station, used for converting urine to drinkable water. Though recycling systems are now in place for short duration missions of six months in space, a fully regenerative system using physicochemical and biological methods adopting biological life support systems (BLSS) needs to be developed.

An LSS should include some main parts, such as water supply, carbon dioxide control, oxygen supply, trace contaminant control and waste management. Also, having these parts does not necessarily mean that good closed gas and water loops are introduced. If an implementation scenario can be achieved successfully, there will be no need to use extra resupply in these systems. At this stage, a chemical CO_2 control system should be changed with CO_2 dump system, adding electrolysis for recovering oxygen from water, adding recovery system for condensate water, urine and wash water and increasing capabilities for CO_2 concentration (Standard et al. 1981). There is only one difference between LSS and BLSS aside from having the same parts; both LSSs and BLSSs have to include a closed food loop. This new design should be a balanced ecological and biotechnical system, affording cohabitation of human, animals, plants and microorganisms.

To achieve this, it is first necessary to consider the specification of the human diet, vitamins and trace minerals. After that, plant or algae species which have the ability to use waste water through photosynthesis and to produce food for humanity should be chosen (Standard et al. 1981). Even though a BLSS might meet the needs of the crew, it still involves a large initial volume and mass of consumables (about 30 metric tons) and high-power consumption for a 1000-day mission for a crew of six people, and thus resupplying is still a major issue. To create long-term self-sufficient BLSSs, the generation of resources from local materials is a prerequisite. Water, solar energy, carbon, nitrogen and other nutrients have been found on Mars, but they are not in a form ready to be used, which constrains their use by BLSS microorganisms. As a method for onsite habitat creation, *terraforming* is considered to be a suitable way to start colonization by plants and organisms and overcome the challenges posed by BLSSs (Farges et al. 2008).

TERRAFORMING

The best approach to ensuring a habitable environment is by changing the conditions of other planets, reducing costs, at the same time being safer. *Terraforming* is a term that implies changing the environments of other planets to make them Earthlike and thus suitable for human life, but it also may involve changing the surface of the planet so as to be able to produce food, and it still may not be suitable for humans. Mars is accepted as the optimum planet for human colonization; unfortunately, its conditions, such as high UV radiation, which may cause DNA damage; low temperatures; extremely dry conditions; and strong oxidative atmosphere are harsh for humans. Martian days do not have a significant difference compared to the days on Earth, but seasonal cycles are longer and unequal, and this affects the growth of terrestrial plants, which is an important part in balancing a terraformed ecosystem on Mars. In order to terraform Mars, it is important to keep organisms alive, preventing their loss or death due to extreme conditions. The first thing to do is to change the environment of Mars, then, constructing a habitable area with selected organisms to initiate life there (Alexandrov 2006). After terraforming Mars, it will be necessary to create a stable self-regulating ecosphere so as to achieve conditions in the range that humans can tolerate. For maintaining controlled conditions on Mars, automatic and continuous monitoring systems (Taylor 1998) will be required; in this way, the climate and seasonal cycles of Mars will become similar to those on Earth. The differences between Earth and Mars are shown in Table 17.1 (Cockell 2010; Lehto et al. 2006; Mackenzie et al. 2006; Madhavan Nair et al. 2008).

HISTORY OF ALGAE IN SPACE

Numerous studies have been carried out into how algae can be used in space for terraforming Mars and in spacecraft for regenerative life support systems from the beginning of the space journey, and algae are still seen as promising candidates for space applications.

On 19 August 1960, *Chlorella pyrenoidosa* was sent to space with two dogs (Belka and Strelka) and other organisms in the spacecraft *Korabl-Sputnik 2*. This spacecraft returned to Earth after a short duration (25 hours), and the results showed that conditions in the spacecraft do not affect the growth of the algae; however, photosynthetic activity was found to be lower than expected. On 22 February 1966, *Chlorella* strain cultures were sent into space on *Cosmos 110* and stayed in the space for 22 days. This space mission showed that algae can survive in space on a longer mission. On 15 September 1968, 10 November 1968 and 7 August 1969, the *Zond 5*, *Zond 6*, and *Zond 7* spacecraft flew to the Moon and back to Earth with *Chlorella* on board, and the results showed that while *Chlorella* survivability was lower on *Zond 5* and *Zond 6*, cell viability had not been affected on *Zond 7* (Alexandrov 2006). In 1971, *Chlorella vulgaris* was cultivated in the spacecraft *Salyut-1* for 72 days and there was no influence in the basic physiological parameters (Galkina and Meleshko 1975). Between 1977 and 1982, studies on the station *Salyut 6* showed that weightlessness has no effect on the growth of algae and their interaction with the environment. In 2008, *Chlorella*, *Rosenvingiellaradicans* and *Gloeocapsa*sp were kept on the ISS for 18 months and they preserved cell morphology, while cells were bleached and carotenoids were destroyed (Cockell et al. 2011). In 2011, *Chlamydomonasreinhardtii* showed higher photosynthetic activity in the spacecraft *STS-134* (Vukich et al. 2012).

Fifty-one research projects were carried out on algae adaptability in space from 1960 to 2018 (Figure 17.1). *Chlorella* species are the most investigated algae in space due to their high growth rate

TABLE 17.1
Comparison of Earth and Martian Conditions

Parameter	Earth	Mars
Distance from sun	150 million km	225 million km
Length of year	365 Earth days	687 Earth days
Length of days	24 hours	24 hours 37 minutes
Gravity	1 g	0.38 g
Surface temperature range	−60°C–50°C	−145°C–20°C
Mean surface temperature	+15°C	−60°C
Atmospheric pressure	1013 mb	5–11 mb
Atmospheric composition, average		
N_2	780 mbar, 78%	0.189 mbar, 2.7%
O_2	210 mbar, 21%	0.009 mbar, 0.13%
CO_2	0.38 mbar, 0.038%	6.67 mbar, 95.3%
Ar	10.13 mbar, 1%	0.112 mbar, 1.6%
UV radiation	>300 nm	>190 nm
Major rock type at the surface	Many volcanic, metamorphic and sedimentary rock types	Primarily basalts. Sulfate salts and phyllosilicates include sedimentary materials. Volcanic activities on Mars ended nearly 1 billion years ago.
Polar ice caps	Water ice	Water ice and dry ice (CO_2)

Cockell 2010; Lehto et al. 2006; Mackenzie et al. 2006; Madhavan Nair et al. 2008.

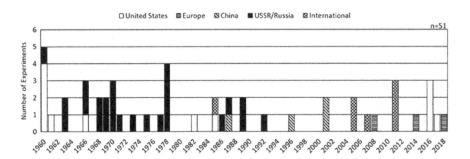

FIGURE 17.1 Researches about algae in space from 1960 to 2018 (Niederwieser et al. 2018).

and resistance to manipulations in cultivation, but *Chlamydomonasreinhardtii*, *Scenedesmusobliquus*, *Rosenvingiellaradicans* and *Sphaerocystissphaerocystiformis* have also been studied (Niederwieser et al. 2018).

One of the most important challenges for NASA's long-term flights is to have an efficient LSS that provides oxygen, food and water for human beings, and recycles waste (Javanmardian and Paisson 1992). This requirement can be met by using photosynthetic living creatures like higher plants, microalgae and cyanobacteria. These organisms make sure that gas and mineral cycles are closed (Wagner et al. 2015).

When we look at microalgae, they offer more advantages for use in LSSs than higher plants and cyanobacteria because of their properties. Filaments in cyanobacteria can block the building of suspension cultures, whereas the unicellular cells of microalgae can be easily cultivated in suspension cultures. Also, cyanobacteria can produce toxins which may increase under microgravity conditions. Compared to higher plants, microalgae have a higher photosynthesis rate, the efficiency of their conversion of light energy into biomass is five times higher, and they have an faster growth rate; moreover, it is cheaper to produce biomass from microalgae because they do not need specific substrates (Wagner et al. 2015).

Even though microalgae have advantages over other photosynthetic organisms, energy-efficient photobioreactors are needed in order to use microalgae successfully in LSSs. In order to establish an efficient cultivation system for microalgae in LSSs, it is necessary to consider the following important parameters, such as ensuring the scattering of light throughout the reactor, a high illuminated surface-to-volume ratio, a highly efficient gravity-independent gas–liquid exchange, and the removal of the harmful effects of UV light and the heating effects of IR light by selecting appropriate light-wave lengths before lighting in the reactor (Javanmardian and Paisson 1992).

To sum up this information, the MELiSSA (Micro-Ecological Life Support System Alternative) project, carried out by the European Space Agency (ESA), has been fundamental to understanding how to build life in space. In the MELiSSA project, *Arthrospiraplatensis* is cultivated in a photobioreactor and cyanobacteria is used for production of food, water and oxygen. The aim is to create a self-sustaining closed-loop system via five main compartments: waste liquefaction via thermophilic anaerobic bacteria; carbon transformation via photoheterotrophic bacteria; nitrification via nitrifying bacteria; and food, water and oxygen production with two sub-compartments (via higher plant and photoautotrophic bacteria, *Arthrospiraplatensis*) and crew compartment, as seen in Figure 17.2 (Farges et al. 2008).

Nutrients, food and water are supplied to the crew compartment. Wastes are collected from crew compartment and production compartment and sent to the waste compartment for fiber degradation. The residues from the degradation are collected from the waste compartment and are fed into the compartment for carbon transformation and the compartment for nitrification to produce nitrate and carbon dioxide to feed the organisms in the production compartment. Food is also sent from the compartment for carbon transformation to the crew compartment after the removal of volatile fatty acids. The production compartment produces water and oxygen and these are fed into the compartment for nitrification. Overall, this closed-loop system regenerates food, oxygen and water to reduce the demand on resources and make itself more cost-efficient and self-sufficient. The requirements for a single crew member per day are summarized in Table 17.2 (Farges et al. 2008).

HISTORY OF SPACE FOOD AND MICROALGAE AS A NOVEL DESIGNER FOOD IN SPACE

Gherman Titov, a Soviet cosmonaut, was the first person (in 1961) to eat in space, while John Glenn was the first American to do so (1962); until then, the eating and swallowing process had not been carried out in conditions of weightlessness. After the first eating process in space, NASA made efforts to design space food for improved spacecraft and longer missions. They aimed to

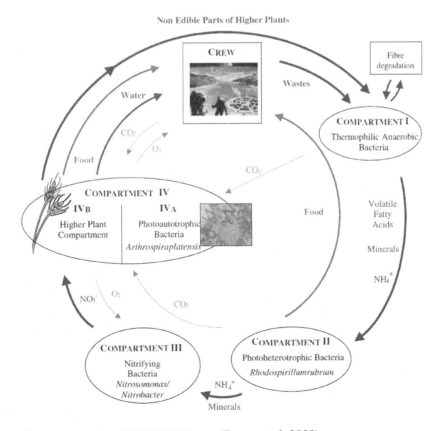

FIGURE 17.2 Schematic representation of MELiSSA Loop (Farges et al. 2008).

TABLE 17.2
The Quantity of the Requirements for Crew per Day in a Life Support System (Farges et al. 2008)

Requirements of the Crew	Quantity for Per Person Per Day
O_2 production	1 kg
CO_2 removal	1.2 kg
Drinking water production	2.8 kg
Food	2.7 kg
Hygiene water	13 kg
Energy to remove	15 kJ

produce easily consumable and nutritionally rich foods with a similar taste to those found on Earth. Longer missions, staying on another planet or even colonizing it, would require more improved foods. In the future, plants would be grown, harvested and served to humans after the planet had been successfully terraformed.

In the *Mercury* projects (1961–1963), the astronauts ate food from a tube without seeing or smelling it. Also, There was also food in cube form, but both of these forms were engineered foods and they did not have the same taste or texture as normal food. The tubes and cubes contained a high calorie mixture of protein, fat and sugar. For the *Gemini* missions (1965–1966), they did not change anything about the packaging other than improving them. The astronauts lost weight, and a lot of the food came back to Earth without being eaten. Hence, in the *Apollo* program (1968–1972), the scientists decided that the quality of the food was poor, and they tried to improve this. The astronauts in the Apollo program had hot water to rehydrate their food, which made it tastier, and canned foods allowed people to eat using utensils like on Earth. The *Skylab* program (1973–1974), offered 72 foods for astronauts along with freezers, refrigerators and food warmers. The US Space Shuttle (1981–present) tried a different and simplified automated food system to reduce the trash from meal trays with a menu of 350 different foods (Perchonok and Bourland 2002).

The progress in the development of space food shows that food must not only be nutritious; how it looks and how it is eaten must also be like it is on Earth for psychological reasons. Eating the same thing repeatedly is not good for the human body; all the necessary nutrients must be taken in via a balanced diet (Hinghofer-Szalkay and König 1992). Microalgae, such as *Arthrospiraplatensis*, can provide proteins, amino acids, minerals, vitamins and so on and meet human beings' nutritional needs; in addition to that, microalgae can be easily cultivated when compared to plants (Nicoletti 2016; Verseux et al. 2016). Even though *Arthrospiraplatensis* are nutritionally rich sources, they are not acceptable for humans taste-wise and they lack nutritional balance. Humans can only consume 10g of *Arthrospiraplatensis* per day because of this imbalance. However, these challenges can be overcome by mixing cyanobacteria with plants or other microalgae or with synthetic biology (Verseux et al. 2016).

Two types of microalgae and cyanobacteria are suitable candidates for LSSs. Because of its amino acid, vitamin and lipid content, the microalgae *Chlorella vulgaris* is a great candidate for a space food for humans. *Arthrospiraplatensis*, a cyanobacteria, has an appropriate nutritional composition for animal feed and human consumption. It also has health-promotion qualities (Lehto et al. 2006).

FUTURE SCENARIOS

Expeditions to Mars have changed how we see living in space. First of all, Mars may serve as a primitive biotic environment; the existence of water reserves and the quality of rocks and soils, along with the atmospheric conditions, could one day turn Mars into a green planet. Putting microalgae species as a supplement of food first, and blending the idea with biorefinery concepts as we are dealing on Earth right now, could be a logical understanding to produce food and prepare infrastructure, life gases, clean water and fertilizers for other plants to be grown. What will be dealt with here could face with cost and logistics barriers even if the idea fits well in space environment. The list given here offers some of the factors that are crucial for producing algae in space.

- Light energy, being the driving force of photosynthesis, and the supplementation of light, either naturally by sun via solar collectors or artificially with illumination systems. Heterotrophic and light-dark cycle production systems could also be integrated.
- Supplementation of macro- and micronutrients from the space environment, for which we know that Martian soil could be a good alternative.
- Choosing the right species. Do we want to use the species for food or to extract certain compounds from it? The adaptability of the species to extreme conditions is the key point, and cyanobacteria species are one of the best candidates for that; however, they also produce toxic compounds, mostly hepato- or neurotoxins, which cannot be used as an alternative food supplement. Ongoing bioprospecting studies on Earth will highlight

the right species to produce food in space, and using genetic modification and synthetic biology tools could also be helpful during this process. Cyanobacteria and some other organisms may live in extreme conditions on Earth (Cockell 2010), but the Martian atmosphere and surface has much more severe conditions that could negatively affect those organisms.

- Preservation, storage, cleaning, recycling of water and also sterilization of it, if needed, are the cost-required issues, so smart designs for this are emerging like those proposed in Project MELiSSA
- Construction of the materials, designing the production facility, determination of the scale and, if needed, scale-up criteria influence the cost and material requirements of the production facility. It should be cheap, resistant to outer environment conditions, long-lasting and protective. As it is in other LSS issues, bringing this equipment from Earth does not seem like an effective idea, so building the facility using the soil, water and other materials on site could be a novel design criteria. Algae itself can be used as construction material through the recycling of unused or "waste" biomass.
- Thinking about the nutritional facts of the algae new food formulation such as *Spirulina* power bars could be another point of the idea. High in nutrition, easy to prepare or cook, appetizing to the eye, these features are fundamental at the first glance.

CONCLUSION

The concerns of all, even the adventures in *Star Trek* and similar series, could be fascinating; the human body is not suitable for long-term space travel, and such travel could be tiring for the organism and some other drawbacks could be seen in human physiology. The concept requires a more targeted approach.

Space habitation and using microalgae as a part of it looks clever and promising at the first steps of colonization or using that "spot" as a transfer line. However, there are challenges and various physiological, psychological and technical limitations that should be borne in mind. The preliminary studies to mimic the idea on Earth could be inspiring, to later fit the conditions in space environment. Even though algae looks like a good source of quality food, it should be kept in mind that algae can also serve other aspects of technology. Algae as a space food is promising, and sustaining that idea definitely requires more investment and investigation. For these uses, algae are promising candidates with their photosynthetic properties and their ability to be used as a fuel and food source, as biofertilizers, and for water recovery purposes.

REFERENCES

Adeniyi, O. M., Azimov, U., and Burluka, A. Algae biofuel: Current status and future applications. *Renewable and Sustainable Energy Reviews*. 2018. 90: 316–335.

Alexandrov, S. Algal research in space: History, current status and future prospects. Innovare *Journal of Life Sciences*. 2006. 0(1): 1–4.

Bolonkin, A., Pensky, O., Krinker, O., Neumann, S., and Magarshak, Y. Man in outer space without space suit. *Biomedical Journal of Scientific & Technical Research*. 2017. 1(4): 1–6.

Chapman, R. L. Algae: The world's most important "plants"—an introduction. *Mitigation and Adaptation Strategies for Global Change*. 2013. 18(1): 5–12.

Cockell, C. S. Geomicrobiology beyond Earth: Microbe–mineral interactions in space exploration and settlement. *Trends in Microbiology*. 2010. 18(7): 308–314.

Cockell, C. S., Rettberg, P., Rabbow, E., and Olsson-Francis, K. Exposure of phototrophs to 548 days in low Earth orbit: Microbial selection pressures in outer space and on early earth. *The ISME Journal*. 2011. 5(10): 1671–1682.

Daues, K. *A History of Spacecraft Environmental Control and Life Support Systems*. NASA Johnson Space Center. 2006 January: 1–13.

Farges, B., Poughon, L., Creuly, C., Cornet, J. F., Dussap, C. G., and Lasseur, C. Dynamic aspects and controllability of the MELiSSA project: A bioregenerative system to provide life support in space. *Applied Biochemistry and Biotechnology*. 2008. 151(2–3): 686–699.

Galkina, T. B., and Meleshko, G. I. Investigation of the physiological activity of chlorella after exposure to spaceflight factors aboard the "Salyut" orbital station. *Space Biology and Aerospace Medicine*. 1975. 9: 36–42.

Hinghofer-Szalkay, H. G., and König, E. M. Human nutrition under extraterrestrial conditions. *Advances in Space Biology and Medicine*. 1992. 2: 131–179.

NASA. Nanoracks-Duchesne-algae production in microgravity with variable wavelengths of light. Available at: https://www.nasa.gov/mission_pages/station/research/experiments/1810.html. Accessed 29 June 2018.

Javanmardian, M., and Palsson, B.Ø. Design and operation of an algal photobioreactor system. *Advances in Space Research*. 1992. 12(5): 231–235.

Koutra, E., Economou, C. N., Tsafrakidou, P., and Kornaros, M. Bio-based products from microalgae cultivated in digestates. *Trends in Biotechnology*. 2018. 36(8): 819–833.

Lehto, K. M., Lehto, H. J., and Kanervo, E. A. Suitability of different photosynthetic organisms for an extraterrestrial biological life support system. *Research in Microbiology*. 2006. 157(1): 69–76.

Mackenzie, B., Leahy, B., and Fisher, G. The Mars homestead: A Mars base constructed from local materials. *American Institute of Aeronautics and Astronautics*. 2006. 501: 1–17.

Madhavan Nair, G., Sridhara Murthi, K. R., and Prasad, M. Y. S. Strategic, technological and ethical aspects of establishing colonies on Moon and Mars. *Acta Astronautica*. 2008. 63(11–12): 1337–1342.

Mobin, S., and Alam, F. Some promising microalgal species for commercial applications: A review. *Energy Procedia*. 2017. 110: 510–517.

Nicoletti, M. Microalgae nutraceuticals. *Foods*. 2016. 5(4): 54–66.

Niederwieser, T., Kociolek, P., and Klaus, D. A review of algal research in space. *Acta Astronautica*. 2018. 146: 359–367.

Perchonok, M., and Bourland, C. NASA food systems: Past, present, and future. *Nutrition*. 2002. 18(10): 913–920.

Pires, J. C. M. COP21: The algae opportunity? *Renewable and Sustainable Energy Reviews*. 2017. 79 (February): 867–877.

Rizwan, M., Mujtaba, G., Memon, S. A., Lee, K., and Rashid, N. Exploring the potential of microalgae for new biotechnology applications and beyond: A review. *Renewable and Sustainable Energy Reviews*. 2018. 92: 394–404.

Skoog, Å. I. BLSS: A contribution to future life support. *Advances in Space Research*. 1984. 4(12): 251–262.

Spolaore, P., Joannis-Cassan, C., Duran, E., and Isambert, A. Commercial applications of microalgae. *Journal of Bioscience and Bioengineering*. 2006. 101(2): 87–96.

Standard, H., Locks, W., and Orbiter, S. Trends in space life support. *Acta Astronautica*. 1981. 9: 1135–1146.

Taylor, R. L. S. Why Mars?—Even under the condition of critical factor constraint engineering technology may permit the establishment and maintenance of an inhabitable ecosystem on Mars. *Advances in Space Research*. 1998. 22(3): 421–432.

Verseux, C., Baqué, M., Lehto, K., De Vera, J. P. P., Rothschild, L. J., and Billi, D. Sustainable life support on Mars—The potential roles of cyanobacteria. *International Journal of Astrobiology*. 2016. 15(1): 65–92.

Vukich, M., Ganga, P. L., Cavalieri, D., Rivero, D., Pollastri, S., Mugnai, S., Mancuso, S., Pastorelli, S., Lambreva, M., Antonacci, A., Margonelli, A., Bertalan, I., Johanningmeier, U., Giardi, M. T., Rea, G., Pugliese, M., Quarto, M., Roca, V., Zanin, A., Borla, O., Rebecchi, L., Altiero, T., Guidetti, R., Cesari, M., Marchioro, T., Bertolani, R., Pace, E., De Sio, A., Casarosa, M., Tozzetti, L., Branciamore, S., Gallori, E., Scarigella, M., Bruzzi, M., Bucciolini, M., Talamonti, C., Donati, A., and Zolesi, V. BIOKIS: A model payload for multidisciplinary experiments in microgravity. *Microgravity Science and Technology*. 2012. 24: 397–409.

Wagner, I., Braun, M., Slenzka, K., and Posten, C. Photobioreactors in life support systems. *Advances in Biochemical Engineering*. 2015. 153: 143–184.

18 Microalgae and Cyanobacteria as a Potential Source of Anticancer Compounds

Wan-Loy Chu and Siew-Moi Phang

CONTENTS

Introduction 185
Potential Anticancer Compounds from Microalgae and Cyanobacteria 186
    Bioactive Peptides 186
    Polyketides 191
    Lipid Compounds 191
    Alkaloids 192
    C-phycocyanin 193
    Carotenoids 193
        Astaxanthin 194
        Fucoxanthin 195
        β-carotene 195
    Dinoflagellate Toxins 196
Future Directions for Research 197
Conclusion 198
Acknowledgments 198
References 199

BOX 18.1 SALIENT FEATURES

Microalgae and cyanobacteria are a potential source of a diverse range of the following bioactive compounds with anticancer activity: bioactive peptides, polyketides, lipid compounds, alkaloids, pigments (C-phycocyanin, astaxanthin [ATX], fucoxanthin and β-carotene) and dinoflagellate toxins. Despite the vast potential, only a limited number of the algal compounds have been brought to clinical trials and developed into anticancer drugs. The advent of "omic" tools and combinatorial chemistry is expected to accelerate the discovery and development of new cancer drugs derived from microalgae and cyanobacteria.

INTRODUCTION

Cancer is defined as a group of diseases characterized by the uncontrolled growth and spread of abnormal cells (American Cancer Society 2018). The etiology of cancer is multifactorial, including lifestyle factors such as tobacco use and excess body weight, nonmodifiable factors such as inherited genetic mutations, and immune conditions. These factors may act simultaneously or in sequence to cause cancer. According to the World Health Organization (WHO [2018]), cancer is the second leading cause of death worldwide, accounting for an estimated 9.6 million deaths in 2018. The most common cancers are lung (2.09 million cases), breast (2.09 million cases), colorectal (1.80 million cases), prostate (1.28 million cases), skin (nonmelanoma) (1.04 million cases) and stomach (1.03 million cases). There will be an ever-increasing cancer burden over the next decades, with over 20 million new cases expected annually as early as 2025 (Franceschi and Bray 2014).

There is an urgent need to search for new anticancer drugs because of the development of resistance in tumor cells to currently available drugs (Li et al. 2018; Nikolaou *et al.* 2018). Furthermore, the side effects caused by current cancer drugs, especially their toxicity in normal cells, has triggered the search for new bioactive compounds, particularly natural products, as anticancer agents (Pádua et al. 2015). In relation to this, the combination of anticancer drugs with natural compounds is regarded as a useful strategy to enhance the effectiveness of chemotherapeutic treatment of cancer (Pádua et al. 2015).

Microalgae and cyanobacteria represent a potential source of bioactive compounds, including anticancer compounds, that has yet to be fully explored. Furthermore, the bioactive compounds from microalgae are useful functional ingredients that can be incorporated into various food products to enhance their nutritional quality and for therapeutic actions on chronic diseases (Vaz et al. 2016). Microalgae and cyanobacteria have many advantages over terrestrial plants in terms of exploitation for bioactive compounds. For instance, algae can grow rapidly, can be cultured on a large scale, and do not compete directly with crops for agricultural lands (Fu et al. 2017). Furthermore, manipulating the growth conditions of microalgae or cyanobacteria, for instance, by imposing different stresses on the cells, could enhance production of biomass with valuable secondary metabolites, including those of pharmaceutical and/or industrial values (Martínez-Francés and Escudero-Oñate 2018). Another advantage of microalgae and cyanobacteria is their metabolic plasticity, which can be exploited to produce a variety of compounds with potential applications in various biotechnology sectors including the food industry, pharmaceutical, nutraceutical and cosmetic sectors (Lauritano et al. 2016).

There are an estimated 72,500 microalgal species, but only about 44,000 have been described (De Clerck et al. 2013). Cyanobacteria (blue-green algae) are the prokaryotic taxa of microalgae and are the most abundant phototrophic organisms on Earth. They are versatile and have successfully colonized a wide range of aquatic and terrestrial habitats, including extreme-environment habitats (Seckbach 2007). Cyanobacteria have an enormous diversity, with huge potential for biotechnological exploitation (Shah et al. 2017). *Spirulina* and *Chlorella* are among the most well-known microalgae that are currently being marketed as functional food, and they contain a variety of bioactive compounds with therapeutic potential, including anticancer effects (Andrade 2018). In addition, microalgae such as *Haematococcus* are a potential source of functional ingredients, particularly astaxanthin, which is a potent antioxidant that has been shown to have anticancer activity (Chu 2012).

About 60% of anticancer compounds are either natural products or their derivatives (Cragg et al. 2005). Marine organisms including algae have attracted much interest as a potential resource for discovery of new anticancer drugs due to their extremely high biodiversity (Simmons et al. 2005). There have been two reviews on the anticancer potential of microalgae and cyanobacteria, but the focus was mainly on marine species (Costa et al. 2012; Martinez Andrade et al. 2018). The primary aim of this chapter is to provide a comprehensive review of potential anticancer compounds from microalgae and cyanobacteria of both freshwater and marine origins. Another highlight of this review is on the anticancer potential of commercial species which are currently being consumed as functional food and are known to be a source of nutraceuticals.

POTENTIAL ANTICANCER COMPOUNDS FROM MICROALGAE AND CYANOBACTERIA

A great variety of bioactive compounds derived from microalgae and cyanobacteria are known to display anticancer activity. These compounds include primary metabolites such as pigments (carotenoids and C-phycocyanin) and novel secondary metabolites such as bioactive peptides, polyketides and dinoflagellate toxins that are produced by certain species of microalgae and cyanobacteria.

BIOACTIVE PEPTIDES

A wide range of biologically active peptides isolated from microalgae and cyanobacteria are known to possess great therapeutic potential, and these have attracted much interest from the pharmaceutical industries (Gerwick and Moore 2012). These peptides from microalgae usually consist of 10–20 amino-acid residues and can be released by solvent extraction, enzymatic hydrolysis or microbial fermentation (Giordano et al. 2018). There are distinct structural classes of peptides, which include linear peptides, linear depsipeptides, linear lipopeptides, cyclic peptides, cyclic depsipeptides and cyclic lipopeptides (Mi et al. 2017). In terms of biosynthesis, some peptides are synthesized by a multiple enzyme system, such as non-ribosomal peptide synthetase (NRPS) or polyketide synthase hybrid (NRPS/PKS) pathways, while other peptides are gene-coded, ribosomally synthesized and posttranslationally modified. Marine cyanobacteria represent a rich source of peptide metabolites in terms of structure and bioactivity. For instance, programs for drug discovery from marine cyanobacteria, such as the Panama International Cooperative Biodiversity Group

(ICGB) program, have discovered more than 400 new peptide compounds between 2007 and 2016 (Mi et al. 2017). Many of the bioactive peptides derived from cyanobacteria were found to display anticancer activity.

The various bioactive peptides, especially those isolated from marine cyanobacteria, are known to display significant anticancer activity against various cell lines (Table 18.1). Among the linear peptides, the dolastatins are well studied in terms of their anticancer activity. A member of this group, dolastatin 10, is a pentapeptide originally isolated from the sea hare *Dolabella auricularia*, which was later confirmed to be a cyanobacterial metabolite, following its isolation from the marine cyanobacterium *Symploca* sp. VPF642 from Palau (Luesch et al. 2001). In addition, a chemically related analogue, symploplastin 1, was isolated from Guamanian and Hawaiian varieties of *Symploca hydroides* in the same study (Luesch et al. 2001). Symplopastin 1, like dolstatin 10, was found to be a potent microtubule inhibitor that was effective against a drug-insensitive mammary tumor and a drug-insensitive colon tumor.

Dolastatin 10 binds to tubulin on the rhizoxin-binding site and affects microtubule assembly, arresting the cell in the G2/M phase (Singh et al. 2011). The bioactive compound was brought into clinical tests but was discontinued at clinical phase II trial because of the development of peripheral neuropathy in 40% of the subjects and the lack of activity as a single agent in patients with hormone refractory metastatic prostate adenocarcinoma (Vaishampayan et al. 2000). An analogue of dolastatin 10, TZT-1027 (auristatin PE or soblidotin), was found to display antitumor activity against two human xenograft models, MX-1 breast carcinoma and LX-1 lung carcinoma in mice (Kobayashi et al. 1997). The drug was brought into phase I clinical trials due to its good preclinical activity.

A drug, brentuximabvedotin, developed based on antibody conjugated to the peptide analog of dolastatin 10, has been approved by the Food and Drug Administration (FDA) for the treatment of Hodgkin's lymphoma and anaplastic large cell lymphoma (Gerwick and Moore 2012; Minich 2012). In addition, a conjugate of auristatin (analogue of dolastatin 10) with a monoclonal antibody directed to the adhesion molecule E-selectin was found to inhibit the growth of prostate cancer cells by up to 80% in treated mice (Bhaskar et al. 2003). Recently, Yokosaka et al. (2018) evaluated a series of novel dolastatin 10 analogues, with varying functional groups such as amines, alcohols and thiols, as payload for conjugated drugs. The novel analogues were found to display good potency in a tumor cell proliferation assay and proved to be suitable payloads in conjugated drugs. Another recent study, Akaiwa et al. (2018), showed that azide modifications to the P2 and P4 subunits (N-terminal) could enhance the cytotoxic activity of dolastatin 10 analogues. A patent related to the method of production and use of dolastatin 10 and auristatins in the treatment of cancer was recently filed (Perez et al. 2018).

Another member of the dolastatin family, dolastatin 15, was also found to be effective against various cancer cell lines, with an ED50 of 2.4×10^{-3} µg/mL (Singh et al. 2011). The compound bound directly to the vinca alkaloid site on tubulin and blocked the progression to M phase of the cell cycle. However, no clinical trials have been conducted because of the structural complexity, low synthetic yield and poor water solubility of this compound. A water-soluble analogue of dolastatin 15, cematodin (LU-103793), was found to be effective against breast and other cancers in a phase I trial but was discontinued in a phase II trial due to unexpected results. A third-generation analogue, ILX-651 (synthatodin) has successfully completed a phase I clinical trial, and a phase II trial has been recommended (Cunningham et al. 2005).

Recently, Liang et al. (2018) predicted FKBP1A (FK506 binding protein) as the potential target of another member of dolastatin family, dolastatin 16, using modelling analysis by computational virtual screening. A natural analogue of dolastatin 10, symplostatin 1, isolated from cyanobacteria of the genus *Symploca*, was found to be a potent inhibitor of cell proliferation of MDA-MB-435 breast cancer and SK-OV-3 human ovarian cancer cell lines (Mooberry et al. 2003). Symplostatin 1 caused cell-cycle arrest at G2/M phase, inducing apoptosis through phosphorylation of Bcl-2, formation of micronuclei and activation of caspase-3. In addition, the compound was found to inhibit tubulin polymerization and endothelial cell proliferation and invasion. Although symplostatin 1 was active against murine colon 38 and murine mammary 16/c in an animal model study, the compound was poorly tolerated and the mice were slow to recover from the toxicity.

Cyclic peptides from marine cyanobacteria have received much attention as a class of natural products with diverse structures and great pharmacological potential. This group of compounds can be further classified into cyclic depsipeptides, cyclic lipopeptides and cyclic peptides. Between 2007 and 2016, a total of 87 cyclic peptides isolated from marine cyanobacteria were described (Mi et al. 2017). Cyclic peptides which have been found to display anticancer activity include

TABLE 18.1
Anticancer Activity of Bioactive Peptides Derived from Cyanobacteria

Compound	Class of Compound	Source	Anticancer Activity	References
Tasiamide (16 analogues)	Linear peptide	*Symploca* sp.	Cytotoxic activity against human nasopharyngeal carcinoma (KB) and human non-small-cell lung tumor (A549) cell lines- IC_{50}=1.29–12.88 µM	Zhang et al. (2014b)
Symplostatin 1	Linear peptide	*Symploca* sp.	Antiproliferative activity against MDA-MB-435 breast cancer (IC_{50}=0.15 nM) and SK-OV-3 ovarian cancer (IC_{50}=0.09 nM) cells	Mooberry et al. (2003)
Dolastatin 10	Linear pentapeptide	*Symploca* sp.	Subcutaneous implanted solid tumors in mice-murine mammary 16/C—gross log cell kill 2.3 murine pancreatic 03—gross log cell kill 1.9	Luesch et al. (2001)
Jahayne	Acetylene-containing linear lipopetide	*Lyngbya* sp.	Cytotoxicity against KB (IC_{50}=0.052 nM) and LoVo (IC_{50}=0.066 nM) cells	Iwasaki et al. (2015)
Belamide A	Methylated linear tetrapeptide	*Symploca* sp.	Inhibited the growth of HeLa (IC_{50}=1.8 µM) and HL60 cells (IC_{50}=0.63 µM); induced apoptosis in HeLa cells Displayed disruption activity against microtubule network in A-10 cells at 20 µM; cytotoxic activity against HCT-116 colon cancer cell line IC_{50}=0.74 µM	Simmons et al. (2006)
AlmiramidesB and H	Linear lipopeptide	A mat dominated by *Oscillatoria nigroviridis*	Strong cytotoxicity against five human cancer cell lines (A549, MDA-MB231, MCF-7, HeLa and PC3); not selective as the compounds were also cytotoxic against a gingival fibroblast cell line	Quintana et al. (2014)
Somocystinamide A	Linear lipopeptide	Assemblage of *Lyngbya mujuscula Schizothrix*	Significant cytotoxicity against neuro-2a neuroblastoma cells—IC_{50}=1.4 µg/mL	Nogle and Gerwick (2002)
Maedamide	Linear depsipeptide	Marine assemblage of *Lyngbya* sp.	Inhibited the growth of HeLa (LC_{50}=4.2 µM) and HL60 cells (LC_{50}=2.2 µM); induced apoptosis in HeLa cells	Iwasaki et al. (2014)
Somocystinamide A	Linear lipopeptide	*Lyngbya mujuscula*	Antiproliferative activity against NB7 neuroblastoma cells (IC_{50}=810 nM) due to induction of apoptosis; systemic treatment of zebrafish with the compound resulted in dose-dependent inhibition of angiogenesis	Wrasidlo et al. (2008)
Malyngamide A	Linear lipopeptide	*Moorea producens*	Cytotoxicity against human lung cancer cells (NCI-H460) and mouse neuro-2a neuroblastoma—EC_{50} = 1.45×10^{-5} µM/mL	Sabry et al. (2017)
Pitiprolamide	Proline-rich depsipeptide	*Lyngbya musjuscula*	Weak cytotoxic activity against HCT116 colon and MCF7 breast cancer cell lines—IC_{50} = 33 µM	Montaser et al. (2011)
Aurilides B and C	Cyclic depsipeptides	*Lyngbya musjuscula*	Cytotoxicity against NCI-H460 human lung tumor and neuro-2a mouse neuroblastoma cell lines—LC_{50} = 0.01–0.13 µM Aurilide B—high level of cytotoxicity against leukaemia, renal and prostate cancer cell lines (GI_{50}<10 nM)	Han et al. (2006)

(Continued)

TABLE 18.1 (CONTINUED)
Anticancer Activity of Bioactive Peptides Derived from Cyanobacteria

Compound	Class of Compound	Source	Anticancer Activity	References
Odoamide	Cyclic depsipeptide	*Okeania* sp.	Potent cytotoxicity against HeLa S3 cells- IC_{50} = 26.3 nM	Sueyoshi et al. (2016)
Palau'amide	Cyclic depsipeptide	*Lyngbya* sp.	Strong cytotoxicity against KB oral cancer cell line- IC_{50} = 13 nM	Williams et al. (2003)
Lagunamides A and B	Cyclic depsipeptides	*Lyngbya majuscula*	Potent cytotoxicity against P388 murine leukemia cell lines – IC_{50} = 6.4 and 20.5 nM respectively	Tripathi et al. (2010)
Symplocamide A	Cyclic depsipeptide	*Symploca* sp.	Potent cytotoxicity against H-460 lung cancer cells (IC_{50} = 40 nM) and neuro-2a neuroblastoma cells (IC_{50} = 29 nM)	Linington et al. (2008)
Apratoxin A	Cyclic depsipeptide	*Lyngbya* sp.	Transcription of cell-cycle genes in HT29 cells—two-fold or more changes in transcription tumour suppressors (e.g. *RASSF4*), oncogenes (e.g. *RAB11B*), stress response (e.g. *DDIT4*) and cell cycle-related genes (e.g. *CDC25A*). Inhibited angiogenesis in vitro (human umbilical vein cells, HUVEC). Inhibited FGF signaling in zebrafish	Luesch et al. (2006)
Apratoxin E	Cyclic depsipeptide	*Lyngbya bouillonii*	Antiproliferative activity against HT 29 cells (IC_{50} = 21 nM), HeLa cells (IC_{50} = 72 nM) and U2OSc cells (IC_{50} = 59 nM)	Matthew et al. (2008)
Apratoxin H and apratoxin A sulfoxide	Cyclic depsipeptide	*Moorea prodecens*	Cytotoxicity against NCI-H460 lung cancer cells (IC_{50} = 3.4 and 89.9 nM respectively)	Thornburg et al. (2013)
Apratoxin S10 (Apra S10)	Cyclic depsipeptide	Analogue of apratoxin	Inhibition of pancreatic cancer cells—EC68 (GI_{50} = 0.35 nM); EC46 (GI_{50} = 0.32 nM) and PANC-1 (GI_{50} = 2.75 nM). Pancreatic patient-derived xenograft mouse model—retarded tumor growth without causing weight loss	Cai et al. (2018b)
Cryptophycin 52	Cyclic depsipeptide	Analogue of cryptophycin	Cytotoxic effect against LNCaP (androgen-dependent) and DU-145 (androgen-independent) pancreatic cancer cells at ≥ 0.1 pM	Drew et al. (2002)
Laxaphycins	Cyclic lipopeptide	*Hormothamnionenteromorphoides*	Cytotoxicity against human colon cancer HCT116 cells: Laxaphycin B4 (IC_{50} = 1.7 μM) Laxaphycin A2—weak effect	Cai et al. (2018a)

aurilides (Han et al. 2006; Sueyoshi et al. 2016), apratoxins (Luesch et al. 2006; Liu et al. 2009; Cai et al. 2018b), cryptophycins (Weiss et al. 2017) and laxaphycins (Cai et al. 2018a) (Table 18.1).

Apratoxins are a group of cyclic depsipeptide metabolites of mixed biosynthetic origin (i.e. peptide-polyketide hybrid) isolated mainly from *Lyngbya* spp. (=*Moorea* spp.), collected from areas such as Guam and Palau (Luesch et al. 2002; Luesch et al. 2006). A member of these depsipeptides, apratoxin A, displayed potent inhibitory activity against cancer cell growth by inducing G1 phase cell-cycle arrest and apoptosis (Luesch et al. 2006). This was partially initiated through antagonism of fibroblast growth receptor (FGFR) signaling via the transcription factor STAT3. In another study, Liu et al. (2009) reported that apratoxin A displayed potent antitumor activity against U2OS osteosarcoma cells. The antitumor activity of the agent was attributed to its action in inhibiting the cellular secretory pathway by preventing the N-glycosylation and subsequent cotranslational translocation of several cancer-associated receptors and secreted proteins.

Recently, Cai et al. (2018b) developed an analogue of apratoxin, named apratoxin S10 (Apra S10), as an anti-pancreatic-cancer agent which inhibited growth of both established as well as patient-derived primary pancreatic cancer cells. The mechanism of action involved down-regulation of multiple receptor tyrosine kinase and inhibition of growth factor and cytokine secretion. The study, using an orthotropic pancreatic-derived xenograft mouse model, also showed that Apra S10 was highly enriched in pancreas tissue. With such unusual mechanisms of action, the compound may have potential therapeutic use against cancers with active secretory pathways, such as pancreatic cancer (Cai et al. 2018b).

Cryptophycins are another type of cyclic depsipeptides, first isolated from *Nostoc* sp. ATCC 53789 in 1990 (Schwartz et al. 1990). The compounds displayed potent anticancer activity (Shin 2001), which was related to their ability to interact with tubulin and interfere with microtubule dynamics (Weiss et al. 2017). This prevents microtubules from forming correct mitotic spindles, causing cell-cycle arrest and apoptosis. An analogue of cryptophycin, cryptophycin-52, showed strong cytotoxic effect against prostate cancer cells by inducing apoptosis (Drew et al. 2002). The apoptotic effect of the compound was associated with proteolytic processing and activation of the caspase-3 and caspase-7 and cleavage of the substrate poly(ADP-ribose) polymerase (PARP). In another study, Shih and Teicher (2001) observed that cryptophycin 52 in combination with gembicitabine, cisplatine or carboplatin resulted in antitumor activity greater than either alone when tested against human non-small-cell lung carcinoma and human small-cell-carcinoma xenografts.

Cryptophycin-52 (LY355703) was brought into phase I clinical trial on patients with advanced solid tumors (Sessa et al. 2002). However, adverse effects associated with dose-dependent peripheral neuropathy and myalgia were reported. A multicenter phase II clinical trial of cryptophycin 52 was later conducted on patients with non-small-cell lung cancer (NSCLC) (Edelman et al. 2003). However, the results showed that cryptophycin 52 only had limited activity in NSCLC cells. Two other analogues with improved stability and water solubility, cryptophycins 249 and 309, were later considered as second-generation clinical candidates (Liang et al. 2005). Numerous structure–activity relationship studies have been carried out on other analogues of cryptophycin-52 (Kumar et al. 2015; Weiss et al. 2017). However, most of the analogues were not selective against cancer cells and thus were not better than the parent compound (Figueras et al. 2018).

One strategy to overcome systemic toxicity and undesired side effects is to develop selective delivery of antitumor agents to the tumor. The drug is linked directly or through a suitable linker system to a tumor-targeting moiety known as a *homing device* (Weiss et al. 2017). The use of homing devices such as monoclonal antibodies, hyaluronic acid, folic acid and peptides has been reported. Recently, Figueras et al. (2018) designed, synthesized and evaluated the antitumor activity of three novel cryptophycin analogues, with the aim of identifying functional groups for conjugation to homing devices. The study found that two of the analogues showed reduced activity but were still very cytotoxic against KB-3-1 cell line. In another development, a patent was recently filed on the invention related to antibody- or peptide-drug conjugate where crptophycin derivatives are covalently attached (Steinkuhler et al. 2018).

Another group of cyanobacterial cyclic lipopeptides that exhibit anticancer activity is the laxaphycins. These compounds are characterized by a rare fatty β-amino acid with a linear chain of up to 12 atoms and can be categorized into cyclic undeca- and dodecapeptides, the major representatives being laxaphycin A and laxaphycin B, respectively (Cai et al. 2018a). Laxaphycins A and B were first isolated from the terrestrial (mud sample) cyanobacterium *Anabaena laxa* (Frankmölle et al. 1992a; Frankmölle et al. 1992b). The compounds had also been isolated from a mixed assemblage of *Lyngbya majuscula* (Bonnard et al. 2007). Laxaphycin B was found to show much more potent antiproliferative activity ($IC_{50} < 2$ μM) than laxaphycin A ($IC_{50} > 20$ μM) when tested on a panel of human cancer cell lines (Bonnard et al. 2007). In addition, the combination of both laxaphycins showed synergistic antiproliferative effect against cancer cell lines. Recently, Cai et al. (2018a) isolated two new laxaphycins, namely laxaphycins B4 and A2 from the marine cyanobacterium *Hormothamnionenteromorphoides*. Laxaphycin B4 showed strong antiproliferative effect against human colon cancer HCT116 cells (IC$_{50}$=1.7 µM), while laxaphycins A and A2 only displayed weak activities. Laxaphycins A and B4 were also shown to act synergistically to inhibit the growth of HCT116 colorectal cancer cells.

There have also been several reported studies on bioactive peptides derived from eukaryotic microalgae, especially *Chlorella* species, which display anticancer activity. For instance, an undecapeptide isolated from the protein waste of *Chlorella vulgaris* was found to display strong dose-dependent antiproferative activity (IC$_{50}$ =70.7µg/mL) against gastric cancer AGS cells (Sheih et al. 2010). The compound induced a post-G1 cell-cycle arrest without causing cytotoxicity in normal lung fibroblast W1-38 cells. In addition, the peptide showed antioxidant activity against peroxyl radicals and low-density lipoprotein (LDL). In a separate study, Wang and Zhang (2013) isolated a bioactive polypeptide (named CPAP) from *Chlorella pyrenoidosa* using low-temperature high-pressure extraction, enzymatic hydrolysis, ion exchange and gel-filtration chromatography. The polypeptide showed antiproliferative activity against human liver cancer HepG2 cells (IC$_{50}$=426 µg/mL). The authors also demonstrated the resistance of the bioactive peptide to gastrointestinal enzymatic degradation. A similar antitumor peptide (Y2) was obtained from trypsin digest of *Spirulina platensis* (Zhang and Zhang 2013). The encapsulation of the peptide with chitosan did not cause any loss to the antitumor activity.

Polyketides

Polyketides are a diverse class of compounds that are synthesized through a series of modular enzymes that condense and then modify chains of acetate or propionate units via reduction, dehydration, cyclization and aromatization reactions (Tidgewell et al. 2010). A polyketide isolated from the marine cyanobacterium *Trichodemium thiebautii*, trichophycin A, was found to exhibit cytotoxic activity against neuro-2a neuroblastoma cell line (EC$_{50}$=6.5 µM) and human colon cancer cell line HCT-116 (EC$_{50}$=11.7 µM) (Bertin et al. 2017). The cytotoxicity of the compound could be related to its polyol character. In another study, nuiapolide isolated from *Okeaniaplumata* was found to display antichemotactic activity against Jurkat cells as well as slowing or blocking the G2/M phase of the cancerous cells (Mori et al. 2015). Another polyketide, polycavernoside D, isolated from *Okeania* sp, showed moderate activity against the H-460 human lung carcinoma cell line (Navarro et al. 2015). Andrianasolo et al. (2005) isolated another polyketide, swinholide A, from *Symploca* cf. sp collected from Fiji; and two related glycosylated derivatives, ankaraholides A and B, from *Geitlerinema* sp. collected from Madagascar. The swinholide-based compounds showed potent inhibition against cancer cell growth and exerted their cytotoxic effect by disrupting actin cytoskeleton. Teruya et al. (2009) tested biselyngbyaside, a macrolide glycoside isolated from *Lyngbya* sp., and found that it displayed broad-spectrum cytotoxicity in a human tumor cell line panel consisting of 39 cancer cell lines. The compound showed potent antiproliferative activity against the central nervous system cancer SNB-78 (GI$_{50}$= 0.036 µM) and lung cancer NCI H522 (GI$_{50}$ = 0.067 µM) cell lines.

Lipid Compounds

Curacin A is a lipid compound first isolated from *Lyngbya majuscula*, and it was shown to have antimitotic activity (IC$_{50}$=7–200 nM) by inhibiting microtubule assembly and the binding of colchicine to tubulin (Gerwick et al. 1994). Its structure is unique as it contains the sequential positioning of a thiazoline and cyclopropyl ring (Chang et al. 2004). However, the compound was so insoluble that its bioactivity could not be demonstrated in animal models (Singh et al. 2011). Several soluble semisynthetic derivatives of curacin A have been synthesized using combinatorial chemical techniques (Peter et al. 2004). In view of their potential as future anticancer drugs, these compounds are currently undergoing preclinical evaluation. The gene cluster involved in the synthesis of this compound has been identified, and the metabolic system is comprised of a nonribosomal peptide synthetase (NRPS) and multiple polyketide synthetases (PKSs) (Chen et al. 2003).

Isomalyngamide A1 and A2 are fatty-acid amides isolated from the marine cyanobacterium *Lyngbya majuscula* (Chang et al. 2011). The compounds were found to have therapeutic potential against tumor cell migration based on studies done on breast cancer cells. The synthetic analogs of the compounds were further shown to have an antimetastatic effect by inactivating the expression of p-FAK, FAK, p-Akt and Akt through the β1 integrin-mediated pathway.

In another study, Kim et al. (2014) found stigmasterol isolated from *Navicula incerta* showed potent apoptotic effect against HepG2 liver cancer cells. The compound was found to up-regulate the expression of proapoptotic genes (Bax, p53) but down-regulate the expression of antiapoptotic genes including Bcl-2. In addition, flowcytometric analysis revealed that the HepG2 cells were arrested at G2/M phase. In another study, Samarakoon et al. (2014) found that a novel fatty alcohol ester, nonyl 8-acetoxy-6-methyloctanoate (NAMC), isolated from *Phaeodactylum tricornutum*, showed strong suppression of the growth of HL-60 (human promyelocyticleukaemia) cells. The compound induced apoptosis by activating Bax and suppressing Bcl-xL and by up-regulating other inducers of apoptosis, particularly caspase-3 and p53.

Some diatoms are known to produce the teratogenic compounds polyunsaturated aldehydes (PUA) that induce abortions, birth defects, poor development and high offspring mortality in predatory planktonic and benthic invertebrates (Leflaive and Ten-Hage 2009). Polyunsaturated aldehydes (PUA) play an important role as an allelopathic agent to outcompete other phytoplankton or to reduce herbivory in some environments (Cózar et al. 2018). The PUA are the end products of the lipoxygenase/hydroperoxide lyase pathway, which is triggered due to damage of algal cells, as occurs through grazing by predators (Fontana et al. 2007). Cell damage activates lipase enzymes, which release polyunsaturated fatty acids (PUFA) from cell membranes that are immediately oxidized and cleaved to form PUA and a variety of other compounds collectively known as *oxylipins* (Pohnert 2005; Fontana et al. 2007).

The PUA first described in marine diatoms were 2-trans,4-cis,7-cis-decatrienal, 2-trans,4-trans,7-cis-decatrienal and 2-trans,4-trans-decadienal (DD) and were shown to display antiproliferative and apoptotic effects against the human adenocarcinoma CaCo$_2$ cell line (Miralto et al. 1999). Following this, the marine diatoms *Skeletonema marinoi* and *Thalassiosira rotula* were found to produce PUA other than DD, including 2-trans,4-trans-heptadienal (HD), 2-trans,4-trans-octadienal (OD) and 2-trans,4-trans,7-octatrienal (d'Ippolito et al. 2002a; 2002b). The PUA DD was shown to induce an apoptotic effect in both copepod and sea-urchin embryos (Romano 2003). Further research on similar effects on human cancerous cells has been reported. For instance, testing on three PUA showed that they exerted cytotoxic effect on both A549 lung cancer and COLO 205 colon cancer cell lines but not on BEAS-2B normal cells (Sansone et al. 2014). In addition, DD activated the death signaling pathway in the lung cancer cells, which involved Tumor Necrosis Factor Receptor 1 (TNFR1) and Fas associated death domain (FADD), leading to necroptosis via caspase-3 without activating the survival pathway receptor-interacting protein (RIP).

Alkaloids

Alkaloids consist of a diverse range of naturally containing compounds produced by marine organisms, frequently as potent toxins against predators (Shah et al. 2017). Many alkaloids are pharmacologically well characterized and are used in clinical applications, ranging from chemotherapeutics to analgesic agents. For instance, calothrixin A, an indolo[3,2-j]phenanthridine alkaloid isolated from *Calothrix* sp., was found to induce apoptotic killing of human Jurkat cancer cells (lymphoma) and also caused G2/M cell-cycle arrest, suggesting intracellular DNA damage (Chen et al. 2003). In addition, the compound appeared to be redox-active, causing cleavage of plasmid DNA as well as inducing ROS formation. The cytotoxic activity of calothrixin A was postulated to be related to its ring structure, which has the characteristics of a DNA intercalator. Bernardo et al. (2007) synthesized several analogues related to calothrixins A and B and tested the compounds against HeLa and normal CV-1 cells. Two of the products, indolophenanthrenedione and benzocarbazoledione, displayed the greatest selectivity, with potent cytotoxic activity against HeLa cancer cells (EC_{50}= 1.5 and 1.8 µM respectively) but no measurable activity against normal CV-1 cells.

A new hybrid thiazoline-containing alkaloid, laucysteinamide A, was isolated from *Caldora penicillata* collected from Lau Lau Bay Saipan, Northern Mariana Island (Zhang et al. 2017). This compound showed moderate cytotoxicity against H-460 human NSCLC cells (IC_{50}=11 µM). In a separate study, three hapaindole-type alkaloids, namely hapalindole X, deschlorohapalindole 1 and 13-hydroxy dechlorofontonamide, were isolated from the extract of two cultured cyanobacteria, *Westiellopsis* sp. and *Fischerellamuscicola* (Kim et al. 2012a). Of the three compounds, hapalindole X displayed moderate cytotoxicity against HT-29 (colon), MCF-7 (breast), NCI-H460 (lung) and SF268 (CNS) cancer cells with IC_{50} values of 24.8, 35.4, 23.0 and 23.5 µM respectively. Recently, Acuna et al. (2018) tested another alkaloid compound isolated from *Fischerella muscicola*, hapalindole H, against PC-3 androgen-insensitive prostate cancer cells. The compound showed selective cytotoxicity against PC-3 prostate cancer cells (EC_{50}=20 nM).

C-PHYCOCYANIN

Phycocyanin is one type of the accessory pigments, phycobiliproteins, found in cyanobacteria, which are important in capturing light and transferring energy in photosynthesis. There are three types of phycocyanin, namely C-phycocyanin, R-phycocyanin and allophycocyanin. C-phycocyanin has been reported to have various pharmacological activities, including anticancer activity, due to its β-subunit (Liu et al. 2016; Hao et al. 2018). In general, the anticancer activity of C-phycocyanin is due to its inhibition of tumor cell cycles, induction of tumor cell apoptosis and autophagy (Jiang et al. 2017). C-phycocyanin has no one specific target but acts in the membrane, cytoplasm and nucleus with diverse mechanisms of action (Fernandes e Silva et al. 2018). The cell targets of C-phycocyanin include MDR1 gene, cytoskeletal proteins and COX-2 enzymes, which make it capable of killing cancer cells resistant to chemotherapy.

Recently, Hao et al. (2018) showed that C-phycocyanin could significantly induce apoptosis and cell-cycle arrest, as well as suppress cell migration, proliferation, and colony formation ability of NSCLC cells through regulating multiple key genes. The pigment was also found to affect the cell phenotype by regulating the NF-κB signaling of NSCLC cells. C-phycocyanin was shown to have photodynamic effect in generating cytotoxic stress through ROS induction, which killed MDA-MB-231 breast cancer cells under 625-nm laser irradiation (Bharathiraja et al. 2016). Apoptotic cell death characteristics such as shrinking of cells, cytoplasmic condensation, nuclei cleavage and the formation of apoptotic bodies were observed in the treated cells. C-phycocyanin was also found to be effective against triple-negative MDA-MB-231 breast cancer cells (Ravi et al. 2015). Treatment of C-phycocyanin inhibited cell proliferation and reduced colony formation. It also caused G1 cell-cycle arrest, which could be attributed to decreased mRNA levels of cyclin E and CDK-2 and increased p21 levels.

C-phycocyanin from *Spirulina platensis* was shown to inhibit the proliferation of SKOV human ovarian cancer cells (Ying et al. 2016). Transcriptomic analysis showed that expression of genes involved in 18 classical pathways, including neutrophin signaling pathway, VEGF signaling pathway and p53 signaling pathway, were enhanced in cells treated with C-phycocyanin. In another study, C-phycocyanin was found to show antiangiogenic effect against colon cancer in a rat model (Saini and Sanyal 2014). Similarly, C-phycocyanin from *Spirulina platensis* was also shown to display antiangiogenesis effect on B16-F10 melanoma tumors in a C57BL/6 mouse model, although it was not able to reduce the proliferation of melanoma cells (Dibaei et al. 2018).

In another study, Gantar et al. (2012) found that C-phycocyanin from *Limnothrix* sp. could potentially improve the efficacy of available anticancer drugs against prostate cancer. The study showed that C-phycocyanin could enhance the efficacy of topotecan in killing prostate cells by increasing the level of ROS generated and activities of caspase-9 and caspase-3. In addition, C-phycocyanin was found to inhibit growth of the human hepatoma cell line HepG2, causing apoptosis, with the cells losing their nuclear entities and then becoming fragmented (Basha et al. 2008). C-phycocyanin was also found to induce apoptosis in human chronic myeloid leukemia cell line K562 by causing cytochrome c release from mitochondria into the cytosol, PARP cleavage and down-regulation of *Bcl-2* (Subhashini et al. 2004).

Recently, Liu et al. (2018) investigated the photodynamic therapy (PDT) effect of selenium-enriched phycocyanin (Se-PC) from *Spirulina platensis* against liver tumors based on both in vitro and in vivo models. The authors observed that Se-PC PDT showed antiproliferative effect against HepG2 cells, and at the cellular level, Se-PC was found to migrate from lysosomes to mitochondria in a time-dependent manner. In addition, in a mouse liver cancer model, tumor inhibition rate was significantly higher in the Se-PC PDT group than in the PC PDT group. Further, Se-PC PDT could induce cell death through free-radical production in tumors and enhance antioxidant enzymes with selenium in vivo.

CAROTENOIDS

Some 600–700 naturally occurring carotenoids have been identified from natural sources including plants, algae, fungi and bacteria (Jaswir 2011). The major carotenoids of commercial value found in microalgae include β-carotene, ATX, lutein, lycopene and canthaxanthin (Gong and Bassi 2016; Ranga Rao et al. 2018b). The health benefits of carotenoids are usually related to their antioxidant activities, which mediate the harmful effects of free radicals. This potentially helps to protect humans from compromised immune response, premature aging, cardiovascular diseases and certain cancers (Zhang et al. 2014a; Gong and Bassi 2016).

There have been many studies that evaluated the anticancer activity of carotenoid extracts from microalgae. For instance, Cha et al. (2008) showed that extracts from *Chlorella vulgaris* and *Chlorella ellipsoidea* containing xanthophylls (mainly violxanthin and lutein) inhibited growth of HCT116 human colon cells and enhanced apoptotic cell death. Carotenoid extract of *Chlorella ellipsoidea* (80 μg/mL) showed an effect 2.5 times stronger than that of *Chlorella vulgaris*, and the activity was comparable to the drug paclitaxel. Renju et al. (2014) evaluated the anticancer effect of lycopene from *Chlorella marina* against human prostate cancer line PC-3. The algal lycopene showed significantly higher inhibitory effect on the growth and colony formation of the cancer cells than lycopene from tomatoes. Algal lycopene caused an apoptotic effect on the cancer cells, which were arrested at the Go/G1 phase, as revealed by flow cytometry analysis. The carotenoids that have been widely studied for their anticancer activity include ATX, fucoxanthin and β-carotene, which will be described in detail in the following sections.

Astaxanthin

ATX is a keto-carotenoid (3,3'-dihydroxy-β,ß-carotene-4,4'-dione) produced mainly by the green microalga *Haematococcus pluvialis*, especially when it is under unfavorable growth conditions (Chu 2011). Accumulation of ATX occurs outside the chloroplast, and this can reach up to 4.5% dry weight (Boussiba et al. 1999). Recently, Hong et al. (2018) reported that ATX production by certain mutant strains of *Haematococcus pluvilais* was 1.59 times higher than that of the wild-type. ATX is a potent antioxidant, as its activity is 10 times higher than other carotenoids such as zeaxanthin, lutein, canthaxanthin and β-carotene, and 100 times more than α-tocopherol (Higuera-Cipara et al. 2006; Ranga Rao et al. 2018a). It has been suggested that ATX may be beneficial in reducing cancer due to its inflammation and oxidative-stress-reducing properties (McCall et al. 2018). Currently, ATX products are available for commercial applications in dosage forms as tablets, capsules, syrups, oils, soft gels, algal biomass and granulated powders (Ambati et al. 2014).

Several studies on the anticancer activity of ATX have been reported. For instance, Karimian et al. (2017) showed that ATX decreased the viability of T-47D and MDA-MB-231 cell lines by inducing apoptosis. In another study, Yan et al. (2017) showed that ATX was able to resensitize gemcitabine-resistant human pancreatic cancer cells (GR-HPCCs) to gemcitabine. The mechanism involved was via up-regulation of human equilibrative nucleoside transporter 1 (hENT1) and down-regulation of ribonucleoside diphosphate reductase (RRM) 1 and 2 to enhance gemcitabine-induced cell death in GR-HPCCs treated with the drug. In addition, cotreatment with ATX and gemcitabine in a tumor xenograft mouse model induced by GR-HPCCs supported the results from the in vitro study. The findings highlighted the potential use of ATX as a combination chemotherapy agent with gemcitabine in the treatment of human pancreatic cancer.

Rao et al. (2013) assessed the anticancer potency of total carotenoids, ATX, ATX monoester and ATX diester against UV–7,12-dimethylbenz(a)anthracene (DMBA)-induced skin cancer in rats. The authors showed that mono- and diester forms of ATX were more potent than ATX in reducing UV-DMBA-induced tumors in the rat model. In addition, the increase of tyrosinase and decrease of antioxidant levels were significantly normalized in the animals treated with ATX esters. The authors suggested that the higher anticancer potency of the esterified forms of ATX could be attributed to their higher bioavailability compared to the free form. In another study, 15-nitroastaxanthin, a major reaction product of ATX with peroxynitrite, displayed an inhibitory effect against two-stage carcinogenesis on mouse skin (Maoka et al. 2012). In the study, the mice were topically induced with 7,12-dimethylbenz[α]anthracene (DMBA) followed by formation of papilloma promoted by the application of 12-O-tetradecanoylphorbol-13- acetate (TPA).

There have been several reports on the anticancer effect of ATX against breast cancer based on both cell and animal models. For instance, McCall et al. (2018) demonstrated that treatment of ATX significantly reduced proliferation rates and inhibited breast cancer cell migration compared to the control normal breast epithelial cells. Inhibition of cancer cell migration is desirable as this would reduce the number of metastases formed. In addition, feeding of ATX was found to delay tumor growth and modulated immune response in a mouse cancer model (Nakao et al. 2010). The plasma levels of ATX as well as natural killer-cell subpopulation and plasma interferon-γ increased in mice that were fed ATX. However, the increase was only observed when ATX was given before tumor initiation, suggesting that an adequate blood ATX status is required to protect against tumor initiation. In another study, Yuri et al. (2016) compared the anticancer effects of ATX and canthaxanthin (CTX) in N-methyl-N-nitrosourea (MNU)-induced mammary cancer in a rat model. Feeding of ATX (0.4% diet) but not CTX reduced the incidence of palpable mammary carcinoma in the experimental animals. The study further showed that changes in adiponectin might be involved in the mechanism of action.

Apart from the purified forms of ATX, there have been studies that assessed the anticancer effect of ATX-rich extracts from *Haematococcus pluvialis*. Use of *Haematococcus* extract instead of purified ATX is attractive as the high costs involved in the production, isolation and purification of ATX may limit its application in cancer therapy (Palozza et al. 2009). For instance,

Palozza et al. (2009) evaluated the growth inhibitory effect of a lipid CO_2 extract from *Haematococcus pluvialis* on four human colon carcinoma cell lines, namely HCT-116, SW480, LS-174 and WiDR. The extract contained a mixture of 10.2 wt% ATX with minor amounts of β-carotene (0.1%), canthaxanthin (0.1% and lutein (0.05%). The extract inhibited the growth of all the colon cancer cell lines tested. In addition, the extract arrested cell-cycle progression and promoted apoptosis in HCT-116 colon cancer cells. The up-regulation of apoptosis was mediated via modification of the ratio of Bax/Bcl-2 and Bcl-XL and increase in the phosphorylation of p38, JNK and ERK1/2.

Recently, Chen et al. (2018) examined the combined effect of ATX and erlotinib, a selective epidermal growth factor receptor (EGFR) tyrosinase kinase inhibitor on cell proliferation and viability of NSCLC cells. The authors found that ATX synergistically enhanced cytotoxicity and cell growth inhibition by erlotinib in NSCLC cells, which were associated with the down-regulation of the expression of xeroderma pigmentosum complementation group C (XPC) and activation of p38 MAPK.

Fucoxanthin

Fucoxanthin is an orange-colored xanthophyll with a unique structure containing an allenic bond and oxygenic functional groups, such as hydroxyl, epoxy, carbonyl and carboxyl groups, in addition to polyene chain (Chu 2011; Kumar et al. 2013). The bioactivity of fucoxanthin is mainly attributed to its polyene chromophore with the allenic bond (Miyashita et al. 2011). Fucoxanthin is present mainly in brown seaweeds, but marine microalgae such as *Phaeodactylum tricornutm* and *Isochrysis galbana* have also been shown to have the potential for commercial production of the carotenoid (Kim et al. 2012b; Kim et al. 2012c).

The antiproliferative and cancer-preventing effects of fucoxanthin are exerted via different molecules and pathways including the Bcl-2 proteins, MAPK, NFκB, caspases, GADD45 and other molecules involved in cell-cycle arrest, apoptosis or metastasis (Kumar et al. 2013). For instance, fucoxanthin was shown to induce ROS production, promoting apoptosis in HL-60 human leukemia cells (Kim et al. 2010). The xanthophyll was also found to inhibit the growth of human leukemia (HL-60), colon cancer (Caco-2) and prostate cancer (PC-3 and LNCaP) cells, which was attributed to its apoptosis-inducing effect (Nakazawa et al. 2009). In another study, Hosokawa et al. (2004) found that fucoxanthin reduced the viability of human colon cancer cell lines (Caco-2, HT-29 and DLD-1), inducing DNA fragmentation and suppressing Bcl-2 protein level in the cancer cells. The authors also showed that fucoxanthin enhanced the anti-proliferative effect of poly (ADP-ribose)-polymerase (PPAR)-γ ligand, troglitazone, on colon cancer cells. Recently, Jin et al. (2018) showed that the combination of fucoxanthin with tumor necrosis factor-related apoptosis-inducing ligand (TRAIL) produced a strong synergistic effect on apoptosis in human cervical cancer cells by targeting the P13K/Akt/NF-κβ signaling pathway.

Fucoxanthin has also been shown to inhibit the metastatic potential of cancer cells. For instance, Chung et al. (2013) assessed the antimetastatic effect of fucoxanthin on murine B16-F10 melanoma cells based on assays for tumor cell invasion, migration, actin fiber organization and cancer cell-endothelial cell interaction. The study showed that fucoxanthin suppressed invasion of the highly metastatic B16-F10 melanoma cells. In addition, the carotenoid inhibited the expression of MMP-9, which plays a key role in tumor cell invasion and migration. Fucoxanthin also reduced the expression of the cell surface glycoprotein CD44 and CXC chemokine receptor-4 (CXCR4), which are involved in migration, invasion and cancer-endothelial cell adhesion.

There have also been several reports on the anticancer effect of fucoxanthin in in vivo models. For instance, Wang et al. (2012) showed that fucoxanthin could significantly inhibit the growth of sarcoma in xenografted sarcoma 180 (S180) mice. The study revealed that the effect was due to induction of apoptosis associated with down-regulation of STAT3/EGFR signaling in the S180 xenograft-bearing mice. In another study, Kim et al. (2013) showed that intraperitoneal administration of fucoxanthin significantly inhibited the growth of tumor mass in mice implanted with melanoma B16F10 cells. In addition, Chung et al. (2013) showed that fucoxanthin supplementation reduced the tumor nodules in the lungs due to metastasis in mice treated with B16-F10 melanoma cells.

In a recent study, Ravi et al. (2018) developed a nanogel system consisting of chitosan and glycolipid to increase the cellular uptake and anticancer efficacy of fucoxanthin against human colon cancer cells (CaCo-2). The authors found that fucoxanthin delivered via the nanogel exerted higher anticancer activity through ROS generation and via a caspase-dependent mechanism.

β-carotene

β-carotene is among those algal carotenoids that have been successfully produced on a commercial scale from *Dunaliella salina* grown in open ponds and photobioreactors (Chu 2011). In relation to the anticancer effects of β-carotene, early epidemiological evidence showed that there was an inverse correlation between dietary β-carotene intake and risk of cancer (Peto et al. 1981; Ziegler 1989). ß-carotene has also been shown to act synergistically with 5-fluorouracil in enhancing the anticancer effect against esophageal squamous carcinoma using in vivo and in vitro models (Zhang et al. 2016). In addition, Raja et al. (2007) reported that feeding of *Dunaliella salina* rich in β-carotene has a protective effect against fibrosarcoma in rats, with increase in catalase and SOD activities in the liver and kidneys. However, a study found that long-term β-carotene intake was associated with increased risk and overall mortality of lung cancer among a population of Finnish male smokers aged 50–69 years (Virtamo et al. 2014).

Dinoflagellate Toxins

Marine dinoflagellates are a large group of eukaryotic microalgae consisting of approximately 2000 living species (Taylor et al. 2008; Gallardo-Rodriguez et al. 2012). The dinoflagellate metabolites that display antitumor activity include amphidinolides (APDN) and amphidinols (APDL). Amphidinolides consist of unique macrolides produced by dinoflagellates of the genus *Amphidinium*, including those living as symbionts of marine flatworms (Kobayashi and Tsuda 2004; Kobayashi and Kubota 2007). The compounds, comprising over 40 members with a vast variety of backbone skeletons, have been shown to exhibit potent antitumor activity. For instance, APDN were found to display strong cytotoxicity against murine lymphoma L1210 and human epidermoid carcinoma KB cell lines (Kobayashi and Tsuda 2004). A related compound, caribenolide I, was also found to have strong cytotoxic effect against human colon tumor cell line HCT 116 and its drug-resistant variant, HCT116/VM46 cell line (Daranas et al. 2001). In another study, a novel polyhydroxy compound related to APDN, lingshuiol (1), was isolated from cultured *Amphidinium* sp. originated as epiphyte on seaweeds from Hainan Province, China (Huang et al. 2004). This compound was found to be highly cytotoxic against A-549 and HL-60 cells, with IC_{50} values of 0.21 and 0.23 µM respectively.

Espiritu et al. (2017) reported the anticancer activity of amphidinol 2 (AM2) isolated from *Amphidinium klebsii*. The metabolite displayed cytotoxic effect against HCT-116, HT-29, and MCF-7 cancer cells but not normal HDFn (human primary fibroblast) cells. The study also revealed that AM2 up-regulated the preapoptosis markers *cfos* and *cjun* in all the cancer lines tested. In another study, Echigoya et al. (2005) reported the cytotoxic activity of several APDL analogs isolated from *Amphidinium carterae* in New Zealand against mouse lymphoma P388 D1 cells.

Gymocin-A, a polyether isolated from the red tide dinoflagellate *Karenia* (=*Gymnodinium*) *mikimotoi* was found to display potent cytototoxic effect (IC_{50}= 1.3 µg/mL) against P388 mouse leukemia cells (Satake et al. 2005). Another group of polyether cyclic toxins is yessotoxins (YTX), which are produced by dinoflagellates of the genera *Protoceratium* and *Gonyaulax*, and are known to cause diarrhetic shellfish poisoning (Bowden 2006). Yessotoxin (YTX) was found to display anticancer effect by inducing apoptotic changes in BE(2)-M17 neuroblastoma cells, but with lower potency compared to okadaic acid, another dinoflagellate toxin (Leira et al. 2002). The toxin also showed potent effect in inducing death of HeLa cells at subnanomolar concentrations (Malaguti et al. 2002). The mode of action was indicative of proteolytic attack of the compound on PARP by caspases. However, another study showed that YTX could cause selective disruption of the E-cadherin-catenin system, which raises the concern that this compound might disrupt the tumor-suppressive functions of E-cadherin (Ronzitti et al. 2004).

Pectenotoxins (PTX) are another group of toxins that cause diarrhetic shellfish poisoning (PSP), believed to be produced by dinoflagellates from the genus *Dinophysis* (Miles et al. 2006). There are at least 14 different analogues of PTX with variations in structure and toxicity (Butler et al. 2012). Among these, PTX-2 was most well studied, especially regarding its inhibitory effects on polymerization of various actin isoforms. This compound was also found to be highly effective in activating an intrinsic pathway of apoptosis in p53-deficient tumor cells compared to those with functional p53 (Chae et al. 2005). PTX-2 triggered apoptosis through mitochondrial dysfunction, which was followed by the release of pro-apoptotic factors and caspase activation. In addition, Bax activation and Bim induction were only observed in p53 deficient cells after PTX-2 treatment.

In another study, Kim et al. (2008) demonstrated that PTX-2 induced its anticancer effects against leukemia cells through suppression of constitutive NF-κB activity. The study found that treatment of PTX-2 down-regulated NF-κB dependent expression of Cox2, IAP-1, IAP-2 and XIAP genes at the transcriptional and translational levels. In addition, the compound induced apoptosis by activating caspase-3 activity. In a separate study, Whitton et al. (2016) found that PTX-2 inhibited telomerase activity with down-regulation of hTERT expression in human leukemia cells. Treatment with PTX-2 also reduced c-Myc and Sp1 gene expression and DNA binding activity of the leukemia cells.

Another group of dinoflagellate toxins that has potential for anticancer drug development is karlotoxins (KTX). The toxins are produced by *Karlodinium veneficum*, are a group of polyketide compounds with haemolytic, cytotoxic and ichthyotoxic activity (Van Wagoner et al. 2008). The structure and physiological effects of KTX resemble that of APDL. The toxins are potentially useful for lowering cholesterol or targeting cancer cells high in cholesterol (Waters et al. 2010). The mechanism of toxicity of KTX is based on the ability of the toxin to form a pore in the cell membrane, which destroys the osmotic balance, leading to cell death (Deeds et al. 2015). In addition, the mechanism of cytotoxicity of KTX is linked to its perturbation of the lipid raft of the cell membrane through its interaction with cholesterol (Waters et al. 2010). Cholesterol is known to play an important role in mediating the function of the lipid raft and accumulation of cholesterol has been reported in some solid tumors. The link between cholesterol and cancer, and the potential of KTX as an anticancer agent targeting cholesterol is an area that is worth further investigations. Cytotoxicity testing of karlotoxin 2 on NCI-60 cell panel showed that the compound was active against a wide range of cell lines, including leukemia, NSCLC, ovarian cancer and breast cancer cell lines (Waters et al. 2015).

FUTURE DIRECTIONS FOR RESEARCH

Most of the novel metabolites produced by cyanobacteria and microalgae are in minute quantities. It is crucial to generate enough amounts of the lead compounds, using the synthetic chemistry approach for further characterization of their structure and pharmacological activity. Using this approach, synthetic analogues with enhanced biological activity compared to the parent compound can be produced. For instance, structural modification of dolastatin 10 has generated a series of analogues that are useful as payload for conjugated drugs (Yokosaka et al. 2018). Further, a large library of synthetic compounds based on the lead molecules from microalgae and cyanobacteria can be generated using a combinatorial chemistry approach (Prakash et al. 2018). This will be the way forward in attempts to discover new anticancer drugs based on microalgae and cyanobacteria.

Despite the fact that numerous metabolites from microalgae and cyanobacteria have been identified, the biosynthetic pathways involved are yet to be fully elucidated. More studies are needed to address knowledge gaps in this regard. A metabolomics approach will be a useful tool to do a comprehensive analysis of the secondary metabolites and to elucidate the pathways involved in the synthesis of the novel compounds (Prakash et al. 2018). Genetic engineering can then be applied to enhance production of the desired product. In a recent study, the carotenoid biosynthesis pathway of *Haematococcus pluvialis* was genetically modified to overproduce ATX, which resulted in a 67% higher ATX accumulation in transformed strains than in wild-type (Galarza et al. 2018). Another tool that is worth further exploring is the use of genome mining to discover new anticancer compounds. This approach is used instead of sequencing the whole algal genome, as it allows rapid identification of gene clusters that encode bioactive compounds. In-silico modelling can then be used to predict the chemical structure based on the sequence information, which can provide further guides to compound purification and structure confirmation. Using this approach, more than 21,000 novel cyanobacterial proteins have been discovered under the CyanoGEBA (Genomic Encyclopedia of Bacteria and Archaea) project (Shih et al. 2013).

An exciting area of development is the use of algal chloroplast as a synthetic biology platform for production of therapeutics (Dyo and Purton 2018). Using this approach, transgenes can be inserted at precise and predetermined locations within the chloroplast genome to allow for stable synthesis of a desired recombinant protein. Among the microalgal species, *Chlamydomonas reinhardtii* is one of the most advanced microalgal platforms for chloroplast transformation. The chloroplast genetic system of this microalga is well suited for synthetic biology as its genome is small (205 kb) and of low complexity (99 genes) (Maul et al. 2002). Recombinant proteins including anticancer immunotoxins have been successfully produced using this microalgal platform. For instance, Tran et al. (2013) successfully developed *Chlamyomonas reinhardtii* as a platform to produce a series of chimeric proteins. The recombinant protein consists of a single chain antibody (scFV) targeting the B-cell surface antigen CD22, genetically fused to the eukaryotic ribosome inactivating protein, gelonin, from *Gelonium multiform*. The antibody-toxin chimeric protein is useful as an immunotoxin that acts as a cytotoxic agent against B-cell lymphomas.

Research on natural products, especially those from cyanobacteria, has focused mainly on marine species. Novel microorganisms and natural products are continuously being discovered from terrestrial sources (Lam 2007). It is believed that up to 99% of microorganisms from terrestrial habitats, especially soils, have yet to be discovered. Cultured soil microbes have been the production source of cancer chemotherapeutics, including doxorubicin hydrochloride, bleomycin, daunorubicin and mitomycin (Pettit 2004). Furthermore, uncultured soil microbes represent another untapped resource of anticancer compounds. The screening of soil cyanobacteria has been focused on antifungal and antibacterial activity but not anticancer activity (Soltani et al. 2005). The potential of soil microalgae and cyanobacteria as a source of anticancer compounds is worth exploring.

Development of improved systems to enhance delivery and bioavailability of bioactives with anticancer properties is another potential area for further research. For instance, the limited bioavailability of ATX may reduce the effectiveness of the supplement, decreasing its assimilation into the body system (McCall et al. 2018). There have also been attempts to incorporate ATX into nanosystems using the emulsification-evaporation method to enhance its bioavailability (Anarjan et al. 2013). These nanosystems improve bioavailability by increasing the dissolution rate of ATX and the saturation solubility by reducing size and increasing surface area. Another system tested was encapsulation of ATX for delivery via high-pressure homogenization and microchannel emulsification (Khalid and Barrow 2018). In addition, Niizawa et al. (2019) developed an oleoresin encapsulation system for ATX using an external ionic gelation technique.

Combinatorial drug treatment has been regarded as an important therapeutic strategy for effective treatment of cancer (Chalakur-Ramireddy and Pakala 2018). Exploration of possible synergistic action of algal compounds with other therapeutic agents, especially anticancer drugs, is another potential area for further studies. Several algal compounds have been shown to exhibit synergistic action against cancer cells when combined with drugs. For instance, combination of ATX with erlotinib was found to act synergistically in inhibiting the growth of NSCLC cells (Chen et al. 2018). In addition, combination of ATX with gemcitabine caused synergistic effect in inducing cell death of pancreatic cancer cells (Yan et al. 2017). Another interesting area for further studies is to evaluate if combined algal compounds have a synergistic anticancer effect. For instance, the combinations of laxaphycins A and B (Bonnard et al. 2007) and, more recently, laxaphycins A2 and B4 (Cai et al. 2018a) have been shown to have synergistic antiproliferative effect against cancer cells. Interestingly, both the A- and B-type of laxaphycins were frequently observed to be produced by the same cyanobacterium.

Most of the experimental evidence on the anticancer properties of the bioactives, extracts or whole biomass of microalgae and cyanobacteria is based on preclinical studies involving cell lines or animal models. There have been very few clinical trials conducted to validate the efficacy of such products in cancer treatment. Some of the algal-based compounds that have been tested in clinical trials include dolastatin 10 (Vaishampayan et al. 2000) and crytophycin 52 (Edelman et al. 2003). ATX is among the algal compounds that has been tested in several clinical trials, which involved human subjects under various diseased conditions, including hypolipidemic and hypotensive effects and macular degeneration (Ambati et al. 2014). However, clinical trial of ATX on cancer patients has not been conducted. There has only been one drug developed from algal compound, brentuximabvedotin (dolastatin conjugated with antibody), that has been approved as a cancer therapeutic, particularly for the treatment of Hodgkin's lymphoma (Gerwick and Moore 2012). Multicenter clinical trials are urgently required before other algal compounds can be fully developed into therapeutic agents for cancer.

CONCLUSION

A diverse range of bioactive compounds with anticancer effects have been discovered from microalgae and cyanobacteria. Algal pigments such as ATX, fucoxanthin and C-phycocyanin are among the compounds which have been shown to exhibit anticancer activity. A diverse range of novel metabolites (e.g. bioactive peptides and polyketides) with anticancer effect have also been discovered from marine cyanobacteria and dinoflagellates. The potential use of algal compounds as lead or scaffold molecules for the development of anticancer drugs has yet to fully explored. The advent of "omic" tools and combinatorial chemistry is expected to accelerate the discovery and development of new anticancer drugs based on microalgae and cyanobacteria. Very few of the algal compounds have been tested in clinical trials to evaluate their efficacy and safety in cancer patients. Thus, it may take a long time before any anticancer drugs derived from microalgae and cyanobacteria can be brought to the market.

ACKNOWLEDGMENTS

The first author would like to acknowledge funding and support from the International Medical University for research on bioactive compounds from algae. Phang S. M. would like to acknowledge the following grants: UM Algae (GA003-2012); HICOE MOHE: IOES-2014 F; UM Grand Challenge-SBS No. GC002B-15SBS.

REFERENCES

Acuña UM, Mo S, Zi J, Orjala J, de Blanco EJ (2018) Hapalindole H induces apoptosis as an inhibitor of NF-κB and affects the intrinsic mitochondrial pathway in PC-3 androgen-insensitive prostate cancer cells. *Anticancer Res* 38(6): 3299–3307.

Akaiwa M, Martin T, Mendelsohn BA (2018) Synthesis and evaluation of linear and macrocyclic dolastatin 10 analogues containing pyrrolidine ring modifications. *ACS Omega* 3: 5212–5221.

Ambati RR, Phang SM, Ravi S, Aswathanarayana R (2014) Astaxanthin: Sources, extraction, stability, biological activities and its commercial applications—a review. *Mar Drugs* 12: 128–152.

American Cancer Society (2018) *Cancer Facts and 2018 Figures 2018*. American Cancer Society, Atlanta, GA.

Anarjan N, Nehdi IA, Tan CP (2013) Influence of astaxanthin, emulsifier and organic phase concentration on physicochemical properties of astaxanthin nanodispersions. *Chem Cent J* 7: 127.

Andrade LM (2018) *Chlorella* and *Spirulina* microalgae as sources of functional foods, nutraceuticals, and food supplements: An overview. *MOJFPT* 6.

Andrianasolo EH, Gross H, Goeger D, Musafija-Girt M, McPhail K, Leal RM, Mooberry SL, Gerwick WH (2005) Isolation of swinholide A and related glycosylated derivatives from two field collections of marine cyanobacteria. *Org Lett* 7: 1375–1378.

Basha OM, Hafez RA, El-Ayouty YM, Mahrous KF, Bareedy MH, Salama AM (2008) C-phycocyanin inhibits cell proliferation and may induce apoptosis in human HepG2 cells. *Egypt J Immunol* 15: 161–167.

Bernardo PH, Chai CL, Le Guen M, Smith GD, Waring P (2007) Structure–activity delineation of quinones related to the biologically active Calothrixin B. *Bioorg Med Chem Lett* 17: 82–85.

Bertin MJ, Wahome PG, Zimba PV, He H, Moeller PD (2017) Trichophycin a, a cytotoxic linear polyketide isolated from a *Trichodesmium thiebautii* bloom. *Mar Drugs* 15(1): 10.

Bharathiraja S, Seo H, Manivasagan P, Santha Moorthy M, Park S, Oh J (2016) In vitro photodynamic effect of phycocyanin against breast cancer cells. *Molecules* 21(11).

Bhaskar V, Law DA, Ibsen E, Breinberg D, Cass KM, DuBridge RB, Evangelista F, Henshall SM, Hevezi P, Miller JC, Pong M, Powers R, Senter P, Stockett D, Sutherland RL, von Freeden-Jeffry U, Willhite D, Murray R, Afar DE, Ramakrishnan V (2003) E-selectin up-regulation allows for targeted drug delivery in prostate cancer. *Cancer Res* 63: 6387–6394.

Bonnard I, Rolland M, Salmon JM, Debiton E, Barthomeuf C, Banaigs B (2007) Total structure and inhibition of tumor cell proliferation of laxaphycins. *J Med Chem* 50: 1266–1279.

Boussiba S, Bing W, Yuan JP, Zarka A, Chen F (1999) Changes in pigments profile in the green alga *Haeamtococcus pluvialis* exposed to environmental stresses. *Biotechnol Lett* 21: 601–604.

Bowden BF (2006) Yessotoxin—Polycyclic ethers from dinoflagellates: Relationships to diarrhetic shellfish toxins. *Toxin Rev* 25: 137–157.

Butler SC, Miles CO, Karim A, Twiner MJ (2012) Inhibitory effects of pectenotoxins from marine algae on the polymerization of various actin isoforms. *Toxicol In Vitro* 26: 493–499.

Cai W, Matthew S, Chen QY, Paul VJ, Luesch H (2018a) Discovery of new A- and B-type laxaphycins with synergistic anticancer activity. *Bioorg Med Chem* 26: 2310–2319.

Cai W, Ratnayake R, Gerber MH, Chen QY, Yu Y, Derendorf H, Trevino JG, Luesch H (2018b) Development of apratoxin S10 (Apra S10) as an anti-pancreatic cancer agent and its preliminary evaluation in an orthotopic patient-derived xenograft (PDX) model. *Invest New Drugs* 37(2): 367–374.

Cha KH, Koo SY, Lee DU (2008) Antiproliferative effects of carotenoids extracted from *Chlorella ellipsoidea* and *Chlorella vulgaris* on human colon cancer cells. *J Agric Food Chem* 56(22): 10521–10526.

Chae HD, Choi TS, Kim BM, Jung JH, Bang YJ, Shin DY (2005) Oocyte-based screening of cytokinesis inhibitors and identification of pectenotoxin-2 that induces Bim/Bax-mediated apoptosis in p53-deficient tumors. *Oncogene* 24: 4813–4819.

Chalakur-Ramireddy NKR, Pakala SB (2018) Combined drug therapeutic strategies for the effective treatment of Triple Negative Breast Cancer. *Biosci Rep* 38.

Chang TT, More SV, Lu IH, Hsu JC, Chen TJ, Jen YC, Lu CK, Li WS (2011) Isomalyngamide A, A-1 and their analogs suppress cancer cell migration in vitro. *Eur J Med Chem* 46: 3810–3819.

Chang Z, Sitachitta N, Rossi JV, Roberts MA, Flatt PM, Jia J, Sherman DH, Gerwick WH (2004) Biosynthetic pathway and gene cluster analysis of curacin A, an antitubulin natural product from the tropical marine cyanobacterium *Lyngbya majuscula*. *J Nat Prod* 67: 1356–1367.

Chen JC, Wu CH, Peng YS, Zheng HY, Lin YC, Ma PF, Yen TC, Chen TY, Lin YW (2018) Astaxanthin enhances erlotinib-induced cytotoxicity by p38 MAPK mediated xeroderma pigmentosum complementation group C (XPC) down-regulation in human lung cancer cells. *Toxicol Res* 7(6): 1247–1256.

Chen X, Smith GD, Waring P (2003) Human cancer cell (Jurkat) killing by the cyanobacterial metabolite calothrixin A. *J Appl Phycol* 15: 269–277.

Chu WL (2011) Potential applications of antioxidant compounds derived from algae. *Curr Top Nutraceutical Res* 9(3): 83–98.

Chu WL (2012) Biotechnological applications of microalgae. *IeJSME* 6: S24–S37.

Chung TW, Choi HJ, Lee JY, Jeong HS, Kim CH, Joo M, Choi JY, Han CW, Kim SY, Choi JS, Ha KT (2013)

Marine algal fucoxanthin inhibits the metastatic potential of cancer cells. *Biochem Biophys Res Commun* 439: 580–585.

Costa M, Costa-Rodrigues J, Fernandes MH, Barros P, Vasconcelos V, Martins R (2012) Marine cyanobacteria compounds with anticancer properties: A review on the implication of apoptosis. *Mar Drugs* 10: 2181–2207.

Cózar A, Morillo-García S, Ortega MJ, Li QP, Bartual A (2018) Macroecological patterns of the phytoplankton production of polyunsaturated aldehydes. *Sci Rep* 8: 12282.

Cragg GM, Kingston DGI, Newman DJ (2005) *Anticancer Agents from Natural Products*. Boca Raton, FL: CRC Press.

Cunningham C, Appleman LJ, Kirvan-Visovatti M, Ryan DP, Regan E, Vukelja S, Bonate PL, Ruvuna F, Fram RJ, Jekunen A, Weitman S, Hammond LA, Eder JP Jr. (2005) Phase I and pharmacokinetic study of the dolastatin-15 analogue tasidotin (ILX651) administered intravenously on days 1, 3, and 5 every 3 weeks in patients with advanced solid tumors. *Clin Cancer Res* 11: 7825–7833.

Daranas AH, Norte M, Fernández JJ (2001) Toxic marine microalgae. *Toxicon* 39: 1101–1132.

De Clerck O, Guiry MD, Leliaert F, Samyn Y, Verbruggen H (2013) Algal taxonomy: A road to nowhere? *J Phycol* 49: 215–225.

Deeds JR, Hoesch RE, Place AR, Kao JP (2015) The cytotoxic mechanism of karlotoxin 2 (KmTx 2) from *Karlodinium veneficum* (Dinophyceae). *Aquat Toxicol* 159: 148–155.

Dibaei F, Fazilati M, Moenzadeh F, Kafayat A, Jazayeri N, Talebi A (2018) Anti-angiogenesis effect of c-phycocyanin of *Spirulina platensis* on B16-F10 melanoma tumors in C57BL/6 mouse. *Pathobiol Res* 21(3): 141–146.

d'Ippolito G, Iadicicco O, Romano G, Fontana A (2002a) Detection of short-chain aldehydes in marine organisms: The diatom *Thalassiosira rotula*. *Tetrahedron Lett* 43: 6137–6140.

d'Ippolito G, Romano G, Iadicicco O, Miralto A, Ianora A, Cimino G, Fontana A (2002b) New birth-control aldehydes from the marine diatom *Skeletonema costatum*: Characterization and biogenesis. *Tetrahedron Lett* 43: 6133–6136.

Drew L, Fine RL, Do TN, Douglas GP, Petrylak DP (2002) The novel antimicrotubule agent cryptophycin 52 (LY355703) induces apoptosis via multiple pathways in human prostate cancer cells. *Clin Cancer Res* 8: 3922–3932.

Dyo YM, Purton S (2018) The algal chloroplast as a synthetic biology platform for production of therapeutic proteins. Microbiology 164: 113–121.

Echigoya R, Rhodes L, Oshima Y, Satake M (2005) The structures of five new antifungal and hemolytic amphidinol analogs from *Amphidinium carterae* collected in New Zealand. *Harmful Algae* 4: 383–389.

Edelman MJ, Gandara DR, Hausner P, Israel V, Thornton D, DeSanto J, Doyle LA (2003) Phase 2 study of cryptophycin 52 (LY355703) in patients previously treated with platinum based chemotherapy for advanced non-small-cell lung cancer. *Lung Cancer* 39: 197–199.

Espiritu RA, Tan MCS, Oyong GG (2017) Evaluation of the anti-cancer potential of amphidinol 2, a polyketide metabolite from the marine dinoflagellate *Amphidinium klebsii*. *Jordan J Biol Sci* 10(4): 297–302.

Fernandes e Silva E, Figueira FDS, Lettnin AP, Carrett-Dias M, Filgueira DMVB, Kalil S, Trindade GS, Votto APS (2018) C-phycocyanin: Cellular targets, mechanisms of action and multi drug resistance in cancer. *Pharmacol Rep* 70: 75–80.

Figueras E, Borbély A, Ismail M, Frese M, Sewald N (2018) Novel unit B cryptophycin analogues as payloads for targeted therapy. *Beilstein J Org Chem* 14: 1281–1286.

Fontana A, d'Ippolito G, Cutignano A, Romano G, Lamari N, Massa Gallucci A, Cimino G, Miralto A, Ianora A (2007) LOX-induced lipid peroxidation mechanism responsible for the detrimental effect of marine diatoms on zooplankton grazers. *Chembiochem* 8: 1810–1818.

Franceschi S, Bray F (2014) Chronic conditions rising in low- and middle-income countries: The case of cancer control. *Cancer Control* 17.

Frankmölle WP, Knübel G, Moore RE, Patterson GML (1992a) Antifungal cyclic peptides from the terrestrial blue-green alga Anabaena laxa. II. Structures of laxaphycins A, B, D and E. *J Antibiot* 45: 1458–1466.

Frankmölle WP, Larsen LK, Caplan FR, Patterson GM, Knübel G, Levine IA, Moore RE (1992b) Antifungal cyclic peptides from the terrestrial blue-green alga *Anabaena laxa*. I. Isolation and biological properties. *J Antibiot* 45: 1451–1457.

Fu W, Nelson DR, Yi Z, Xu M, Khraiwesh B, Jijakli K, Chaiboonchoe A, Alzahmi A, Al-Khairy D, Brynjolfsson S, Salehi-Ashtiani K (2017) Bioactive compounds from microalgae: Current development and prospects. In: *Studies in Natural Products Chemistry*. Amsterdam, the Netherlands: Elsevier, pp. 199–225.

Galarza JI, Gimpel JA, Rojas V, Arredondo-Vega BO, Henríquez V (2018) Over-accumulation of astaxanthin in *Haematococcus pluvialis* through chloroplast genetic engineering. *Algal Res* 31: 291–297.

Gallardo-Rodríguez J, Sánchez-Mirón A, García-Camacho F, López-Rosales L, Chisti Y, Molina-Grima E (2012) Bioactives from microalgal dinoflagellates. *Biotechnol Adv* 30: 1673–1684.

Gantar M, Dhandayuthapani S, Rathinavelu A (2012) Phycocyanin induces apoptosis and enhances the effect of topotecan on prostate cell line LNCaP. *J Med Food* 15: 1091–1095.

Gerwick WH, Moore BS (2012) Lessons from the past and charting the future of marine natural products drug discovery and chemical biology. *Chem Biol* 19: 85–98.

Gerwick WH, Proteau PJ, Nagle DG, Hamel E, Blokhin A, Slate DL (1994) Structure of curacin A, a novel antimitotic, antiproliferative and brine shrimp toxic natural product from the marine cyanobacterium *Lyngbya majuscula*. *J Org Chem* 59: 1243–1245.

Giordano D, Costantini M, Coppola D, Lauritano C, Núñez Pons L, Ruocco N, di Prisco G, Ianora A, Verde C (2018)

Biotechnological applications of bioactive peptides from marine sources. *Adv Microb Physiol* 73: 171–220.

Gong M, Bassi A (2016) Carotenoids from microalgae: A review of recent developments. *Biotechnol Adv* 34: 1396–1412.

Han B, Gross H, Goeger DE, Mooberry SL, Gerwick WH (2006) Aurilides B and C, cancer cell toxins from a Papua New Guinea collection of the marine cyanobacterium *Lyngbya majuscula*. *J Nat Prod* 69: 572–575.

Hao S, Yan Y, Li S, Zhao L, Zhang C, Liu L, Wang C (2018) The in vitro anti-tumor activity of phycocyanin against non-small cell lung cancer cells. *Mar Drugs* 16(6): 178.

Higuera-Ciapara I, Félix-Valenzuela L, Goycoolea FM (2006) Astaxanthin: A review of its chemistry and applications. *Crit Rev Food Sci Nutr* 46: 185–196.

Hong ME, Choi HI, Kwak HS, Hwang SW, Sung YJ, Chang WS, Sim SJ (2018) Rapid selection of astaxanthin-hyperproducing *Haematococcus* mutant via azide-based colorimetric assay combined with oil-based astaxanthin extraction. *Bioresour Technol* 267: 175–181.

Hosokawa M, Kudo M, Maeda H, Kohno H, Tanaka T, Miyashita K (2004) Fucoxanthin induces apoptosis and enhances the antiproliferative effect of the PPARγ ligand, troglitazone, on colon cancer cells. *Biochim Biophys Acta* 1675(1–3): 113–119.

Huang XC, Zhao D, Guo YW, Wu HM, Lin LP, Wang ZH, Ding J, Lin YS (2004) Lingshuiol, a novel polyhydroxyl compound with strongly cytotoxic activity from the marine dinoflagellate *Amphidinium* sp. *Bioorg Med Chem Lett* 14: 3117–3120.

Iwasaki A, Ohno O, Sumimoto S, Ogawa H, Nguyen KA, Suenaga K (2015) Jahanyne, an apoptosis-inducing lipopeptide from the marine cyanobacterium *Lyngbya* sp. *Org Lett* 17: 652–655.

Iwasaki A, Ohno O, Sumimoto S, Suda S, Suenaga K (2014) Maedamide, a novel chymotrypsin inhibitor from a marine cyanobacterial assemblage of *Lyngbya* sp. *Tetrahedron Lett* 55: 4126–4128.

Jaswir I (2011) Carotenoids: Sources, medicinal properties and their application in food and nutraceutical industry. *J Med Plants Res* 5(33): 7119–7131.

Jiang L, Wang Y, Yin Q, Liu G, Liu H, Huang Y, Li B (2017) Phycocyanin: A potential drug for cancer treatment. *J Cancer* 8: 3416–3429.

Jin Y, Qiu S, Shao N, Zheng J (2018) Fucoxanthin and tumor necrosis factor-related apoptosis-inducing ligand (TRAIL) synergistically promotes apoptosis of human cervical cancer cells by targeting PI3K/Akt/NF-κB signaling pathway. *Med Sci Monit* 24: 11–18.

Karimian A, Bahadori MH, Hajizadeh Moghaddam A, Mir Mohammadrezaei F (2017) Effect of astaxanthin on cell viability in T-47D and MDA-MB-231 breast cancer cell lines. *Multidiscip Cancer Investig*.

Khalid N, Barrow CJ (2018) Critical review of encapsulation methods for stabilization and delivery of astaxanthin. *J Food Bioactives* 1: 104–115.

Kim KN, Ahn G, Heo SJ, Kang SM, Kang MC, Yang HM, Kim D, Roh SW, Kim SK, Jeon BT, Park PJ, Jung WK, Jeon YJ (2013) Inhibition of tumor growth in vitro and in vivo by fucoxanthin against melanoma B16F10 cells. *Environ Toxicol Pharmacol* 35(1): 39–46.

Kim KN, Heo SJ, Kang SM, Ahn G, Jeon YJ (2010) Fucoxanthin induces apoptosis in human leukemia HL-60 cells through a ROS-mediated Bcl-xL pathway. *Toxicol In Vitro* 24: 1648–1654.

Kim MO, Moon DO, Heo MS, Lee JD, Jung JH, Kim SK, Choi YH, Kim GY (2008) Pectenotoxin-2 abolishes constitutively activated NF-kappaB, leading to suppression of NF-kappaB related gene products and potentiation of apoptosis. *Cancer Lett* 271: 25–33.

Kim H, Lantvit D, Hwang CH, Kroll DJ, Swanson SM, Franzblau SG, Orjalaa J (2012a) Indole alkaloids from two cultured cyanobacteria, *Westiellopsis* sp. and *Fischerella muscicola*. *Bioorg Med Chem* 20(17): 5290–5295.

Kim SM, Jung YJ, Kwon ON, Cha KH, Um BH, Chung D, Pan CH (2012b) A potential commercial source of fucoxanthin extracted from the microalga *Phaeodactylum tricornutum*. *Appl Biochem Biotechnol* 166: 1843–1855.

Kim SM, Kang SW, Kwon ON, Chung D, Pan CH (2012c) Fucoxanthin as a major carotenoid in *Isochrysis* aff. *galbana*: Characterization of extraction for commercial application. *J Korean Soc Appl Bio Chem* 55: 477–483.

Kim YS, Li XF, Kang KH, Ryu B, Kim SK (2014) Stigmasterol isolated from marine microalgae *Navicula incerta* induces apoptosis in human hepatoma HepG2 cells. *BMB Rep* 47(8): 433–438.

Kobayashi JI, Kubota T (2007) Bioactive macrolides and polyketides from marine dinoflagellates of the genus *Amphidinium*. *J Nat Prod* 70: 451–460.

Kobayashi JI, Tsuda M (2004) Amphidinolides, bioactive macrolides from symbiotic marine dinoflagellates. *Nat Prod Rep* 21: 77–93.

Kobayashi M, Natsume T, Tamaoki S, Watanabe J, Asano H, Mikami T, Miyasaka K, Miyazaki K, Gondo M, Sakakibara K, Tsukagoshi S (1997) Antitumor activity of TZT-1027, a novel dolastatin 10 derivative. *Jpn J Cancer Res* 88: 316–327.

Kumar A, Kumar M, Sharma S, Guru SK, Bhushan S, Shah BA (2015) Design and synthesis of a new class of cryptophycins based tubulin inhibitors. *Eur J Med Chem* 93: 55–63.

Kumar SR, Hosokawa M, Miyashita K (2013) Fucoxanthin: A marine carotenoid exerting anti-cancer effects by affecting multiple mechanisms. *Mar Drugs* 11: 5130–5147.

Lam KS (2007) New aspects of natural products in drug discovery. *Trends Microbiol* 15: 279–289.

Lauritano C, Andersen JH, Hansen E, Albrigtsen M, Escalera, L, Esposito F, Helland K, Hanssen KO, Romano G, Ianora A (2016) Bioactivity screening of microalgae for antioxidant, anti-Inflammatory, anticancer, anti-diabetes, and antibacterial activities. *Frontiers Mar Sci* 3: 68.

Leflaive J, Ten-Hage L (2009) Chemical interactions in diatoms: Role of polyunsaturated aldehydes and precursors. *New Phytol* 184: 794–805.

Leira F, Alvarez C, Vieites JM, Vieytes MR, Botana LM (2002) Characterization of distinct apoptotic changes induced by okadaic acid and yessotoxin in the BE (2)-M17 neuroblastoma cell line. *Toxicol In Vitro* 16: 23–31.

Li Y, Gao X, Yu Z, Liu B, Pan W, Li N, Tang B (2018) Reversing multidrug resistance by multiplexed gene silencing for enhanced breast cancer chemotherapy. *ACS Appl Mater Interfaces* 10: 15461–15466.

Liang J, Moore RE, Moher ED, Munroe JE, Al-awar, RS, Hay DA, Varie DL, Zhang TY, Aikins JA, Martinelli MJ, Shih C, Ray JE, Gibson LL, Vasudevan V, Polin L, White K, Kushner J, Simpson C, Pugh S, Corbett TH (2005) Cryptophycins-309, 249 and other cryptophycin analogs: Preclinical efficacy studies with mouse and human tumors. *Invest New Drugs* 23: 213–224.

Liang TT, Zhao Q, He S, Mu FZ, Deng W, Han BN (2018) Modeling analysis of potential target of dolastatin 16 by computational virtual screening. *Chem Pharm Bull* 66: 602–607.

Linington RG, Edwards DJ, Shuman CF, McPhail KL, Matainaho T, Gerwick WH (2008) Symplocamide A, a potent cytotoxin and chymotrypsin inhibitor from the marine cyanobacterium *Symploca* sp. *J Nat Prod* 71: 22–27.

Liu Q, Huang Y, Zhang R, Cai T, Cai Y (2016) Medical application of *Spirulina platensis*-derived C-phycocyanin. *Evid Based Complement Alternat Med* 2016: 7803846.

Liu Y, Law BK, Luesch H (2009) Apratoxin A reversibly inhibits the secretory pathway by preventing cotranslational translocation. *Mol Pharmacol* 76: 91–104.

Liu Z, Fu X, Huang W, Li C, Wang X, Huang B (2018) Photodynamic effect and mechanism study of selenium-enriched phycocyanin from *Spirulina platensis* against liver tumours. *J Photochem Photobiol B* 180: 89–97.

Luesch H, Chanda SK, Raya RM, DeJesus PD, Orth AP, Walker JR, Izpisúa Belmonte JC, Schultz PG (2006) A functional genomics approach to the mode of action of apratoxin A. *Nat Chem Biol* 2: 158–167.

Luesch H, Moore RE, Paul VJ, Mooberry SL, Corbett TH (2001) Isolation of dolastatin 10 from the marine cyanobacterium *Symploca* species VP642 and total stereochemistry and biological evaluation of its analogue symplostatin 1. *J Nat Prod* 64: 907–910.

Luesch H, Yoshida WY, Moore RE, Paul VJ (2002) New apratoxins of marine cyanobacterial origin from Guam and Palau. *Bioorg Med Chem* 10: 1973–1978.

Malaguti C, Ciminiello P, Fattorusso E, Rossini GP (2002) Caspase activation and death induced by yessotoxin in HeLa cells. *Toxicol In Vitro* 16: 357–363.

Maoka T, Tokuda H, Suzuki N, Kato H, Etoh H (2012) Antioxidative, anti-tumor-promoting, and anti-carcinogensis activities of nitroastaxanthin and nitrolutein, the reaction products of astaxanthin and lutein with peroxynitrite. *Mar Drugs* 10: 1391–1399.

Martínez Andrade KA, Lauritano C, Romano G, Ianora A (2018) Marine microalgae with anti-cancer properties. *Mar Drugs* 16(5): 165.

Martínez-Francés E, Escudero-Oñate C (2018) Cyanobacteria and microalgae in the production of valuable bioactive compounds. In: *Microalgal Biotechnology*. Intech Open.

Matthew S, Schupp PJ, Luesch H (2008) Apratoxin E, a cytotoxic peptolide from a Guamanian collection of the marine cyanobacterium *Lyngbya bouillonii*. *J Nat Prod* 71: 1113–1116.

Maul JE, Lilly JW, Cui L, dePamphilis CW, Miller W, Harris EH, Stern DB (2002) The *Chlamydomonas reinhardtii* plastid chromosome: Islands of genes in a sea of repeats. *Plant Cell* 14: 2659–2679.

McCall B, McPartland CK, Moore R, Frank-Kamenetskii A, Booth BW (2018) Effects of astaxanthin on the proliferation and migration of breast cancer cells in vitro. *Antioxidants* 7(10): 136.

Mi Y, Zhang J, He S, Yan X (2017) New peptides isolated from marine cyanobacteria, an overview over the past decade. *Mar Drugs* 15(5): 132.

Miles CO, Wilkins AL, Munday JS, Munday R, Hawkes AD, Jensen DJ, Cooney JM, Beuzenberg V (2006) Production of 7-epi-pectenotoxin-2 seco acid and assessment of its acute toxicity to mice. *J Agric Food Chem* 54: 1530–1534.

Minich SS (2012) Brentuximab vedotin: A new age in the treatment of Hodgkin lymphoma and anaplastic large cell lymphoma. *Ann Pharmacother* 46: 377–383.

Miralto A, Barone G, Romano G, Poulet SA, Ianora A, Russo GL, Buttino I, Mazzarella G, Laabir M, Cabrini M, Giacobbe MG (1999) The insidious effect of diatoms on copepod reproduction. *Nature* 402: 173–176.

Miyashita K, Nishikawa S, Beppu F, Tsukui T, Abe M, Hosokawa M (2011) The allenic carotenoid fucoxanthin, a novel marine nutraceutical from brown seaweeds. *J Sci Food Agric* 91: 1166–1174.

Montaser R, Abboud KA, Paul VJ, Luesch H (2011) Pitiprolamide, a proline-rich dolastatin 16 analogue from the marine cyanobacterium *Lyngbya majuscula* from Guam. *J Nat Prod* 74: 109–112.

Mooberry SL, Leal RM, Tinley TL, Luesch H, Moore RE, Corbett TH (2003) The molecular pharmacology of symplostatin 1: A new antimitotic dolastatin 10 analog. *Int J Cancer* 104: 512–521.

Mori S, Williams H, Cagle D, Karanovich K, Horgen FD, Smith R III, Watanabe CM (2015) Macrolactone nuiapolide, isolated from a Hawaiian marine cyanobacterium, exhibits anti-chemotactic activity. *Mar Drugs* 13: 6274–6290.

Nakao R, Nelson OL, Park JS, Mathison BD, Thompson PA, Chew BP (2010) Effect of dietary astaxanthin at different stages of mammary tumor initiation in BALB/c mice. *Anticancer Res* 30: 2171–2175.

Nakazawa Y, Sashima T, Hosokawa M, Miyashita K (2009) Comparative evaluation of growth inhibitory effects of stereoisomers of fucoxanthin in human cancer cell lines. *J Funct Foods* 1: 88–97.

Navarro G, Cummings S, Lee J, Moss N, Glukhov E, Valeriote FA, Gerwick L, Gerwick WH (2015) Isolation of polycavernoside D from a marine cyanobacterium. *Environ Sci Technol Lett* 2: 166–170.

Niizawa I, Espinaco BY, Zorrilla SE, Sihufe GA (2019) Natural astaxanthin encapsulation: Use of response surface methodology for the design of alginate beads. *Int J Biol Macromol* 121: 601–608. doi: 10.1016/j.ijbiomac.2018.10.044.

Nikolaou M, Pavlopoulou A, Georgakilas AG, Kyrodimos E (2018) The challenge of drug resistance in cancer treatment: A current overview. *Clin Exp Metastasis* 35: 309–318.

Nogle LM, Gerwick WH (2002) Somocystinamide A, a novel cytotoxic disulfide dimer from a Fijian marine cyanobacterial mixed assemblage. *Org Lett* 4: 1095–1098.

Pádua D, Rocha E, Gargiulo D, Ramos AA (2015) Bioactive compounds from brown seaweeds: Phloroglucinol, fucoxanthin and fucoidan as promising therapeutic agents against breast cancer. *Phytochem Lett* 14: 91–98.

Palozza P, Torelli C, Boninsegna A, Simone R, Catalano A, Mele MC, Picci N (2009) Growth-inhibitory effects of the astaxanthin-rich alga *Haematococcus pluvialis* in human colon cancer cells. *Cancer Lett* 283: 108–117.

Perez M, Rilatt I, Lamothe M (2018) Derivatives of dolastatin 10 and auristatins. Google Patents, US 2018/0222858 A1.

Peter W, Jonathan TR, Billy WD (2004) Chemistry and biology of curacin A. *Curr Pharmaceut Des* 10: 1417–1437.

Peto R, Doll R, Buckley JD, Sporn MB (1981) Can dietary beta-carotene materially reduce human cancer rates? *Nature* 290: 201–208.

Pettit RK (2004) Soil DNA libraries for anticancer drug discovery. *Cancer Chemother Pharmacol* 54: 1–6.

Pohnert G (2005) Diatom/copepod interactions in plankton: The indirect chemical defense of unicellular algae. *Chembiochem* 6: 946–959.

Prakash B, Kujur A, Yadav A (2018) Drug synthesis from natural products: A historical overview and future perspective. In: Tewari A, Tiwari S. (Eds) Synthesis of Medicinal Agents from Plants. Amsterdam, the Netherlands: Elsevier, pp. 25–46.

Quintana J, Bayona LM, Castellanos L, Puyana M, Camargo P, Aristizábal F, Edwards C, Tabudravu JN, Jaspars M, Ramos FA (2014) Almiramide D, cytotoxic peptide from the marine cyanobacterium *Oscillatoria nigroviridis*. *Bioorg Med Chem* 22: 6789–6795.

Raja R, Hemaiswarya S, Balasubramanyam D, Rengasamy R (2007) Protective effect of *Dunaliella salina* (Volvocales, Chlorophyta) against experimentally induced fibrosarcoma on wistar rats. *Microbiol Res* 162: 177–184.

Ranga Rao A, Deepika G, Ravishankar GA, Sarada R, Bikkina PN, Lei B, Su Y (2018b) Industrial potential of carotenoid pigments from microalgae: Current trends and future prospects. *Crit Rev Food Sci Nutr* 25: 1–22.

Ranga Rao A, Deepika G, Ravishankar GA, Sarada R, Narasimharao BP, Su Y, Lei B (2018a) *Botryococcus* as an alternative source of carotenoids and its possible applications—an overview. *Crt Rev Biotechnol* 38(4): 541–558.

Rao AR, Sindhuja HN, Dharmesh SM, Sankar KU, Sarada R, Ravishankar GA (2013) Effective inhibition of skin cancer, tyrosinase, and antioxidative properties by astaxanthin and astaxanthin esters from the green alga *Haematococcus pluvialis*. *J Agric Food Chem* 61: 3842–3851.

Ravi H, Kurrey N, Manabe Y, Sugawara T, Baskaran V (2018) Polymeric chitosan-glycolipid nanocarriers for an effective delivery of marine carotenoid fucoxanthin for induction of apoptosis in human colon cancer cells (Caco-2 cells). *Mater Sci Eng C* 91: 785–795.

Ravi M, Tentu S, Baskar G, Rohan Prasad S, Raghavan S, Jayaprakash P, Jeyakanthan J, Rayala SK, Venkatraman G (2015) Molecular mechanism of anti-cancer activity of phycocyanin in triple-negative breast cancer cells. *BMC Cancer* 15: 768.

Renju GL, Muraleedhara Kurup G, Bandugula VR (2014) Effect of lycopene isolated from *Chlorella marina* on proliferation and apoptosis in human prostate cancer cell line PC-3. *Tumour Biol* 35: 10747–10758.

Romano G, Russo GL, Buttino I, Ianora A, Miralto A (2003) A marine diatom-derived aldehyde induces apoptosis in copepod and sea urchin embryos. *J Exp Biol* 206: 3487–3494.

Ronzitti G, Callegari F, Malaguti C, Rossini GP (2004) Selective disruption of the E-cadherin-catenin system by an algal toxin. *Br J Cancer* 90: 1100–1107.

Sabry OM, Goeger DE, Gerwick WH (2017) Biologically active new metabolites from a Florida collection of *Moorea producens*. *Nat Prod Res* 31: 555–561.

Saini MK, Sanyal SN (2014) Targeting angiogenic pathway for chemoprevention of experimental colon cancer using c-phycocyanin as cyclooxygenase-2 inhibitor. *Biochem Cell Biol* 92: 206–218.

Samarakoon KW, Ko JY, Lee JH, Kwon ON, Kim SW, Jeon YJ (2014) Apoptotic anticancer activity of a novel fatty alcohol ester isolated from cultured marine diatom, *Phaeodactylum tricornutum*. *J Funct Foods* 6: 231–240.

Sansone C, Braca A, Ercolesi E, Romano G, Palumbo A, Casotti R, Francone M, Ianora A (2014) Diatom-derived polyunsaturated aldehydes activate cell death in human cancer cell lines but not normal cells. *PLOS ONE* 9: e101220.

Satake M, Tanaka Y, Ishikura Y, Oshima Y, Naoki H, Yasumoto T (2005) Gymnocin-B with the largest contiguous polyether rings from the red tide dinoflagellate, *Karenia* (formerly *Gymnodinium*) *mikimotoi*. *Tetrahedron Lett* 46: 3537–3540.

Schwartz RE, Hirsch CF, Sesin DF, Flor JE, Chartrain M, Fromtling RE, Harris GH, Salvatore MJ, Liesch JM, Yudin K (1990) Pharmaceuticals from cultured algae. *J Ind Microbiol* 5: 113–123.

Seckbach J (2007) *Algae and Cyanobacteria in Extreme Environments*. Berlin, Germany: Springer Science & Business Media.

Sessa C, Weigang-Köhler K, Pagani O, Greim G, Mora O, De Pas T, Burgess M, Weimer I, Johnson R (2002) Phase I and pharmacological studies of the cryptophycin analogue LY355703 administered on a single intermittent or weekly schedule. *Eur J Cancer* 38: 2388–2396.

Shah SAA, Akhter N, Auckloo BN, Khan I, Lu Y, Wang K, Wu B, Guo YW (2017) Structural diversity, biological properties and applications of natural products from cyanobacteria. A review. *Mar Drugs* 15(11): 354.

Sheih IC, Fang TJ, Wu TK, Lin PH (2010) Anticancer and antioxidant activities of the peptide fraction from algae protein waste. *J Agric Food Chem* 58: 1202–1207.

Shih C, Teicher BA (2001) Cryptophycins: A novel class of potent antimititotic antitumor depsipeptide. *Curr Pharmaceut Des* 13: 1259–1276.

Shih PM, Wu D, Latifi A, Axen SD, Fewer DP, Talla E, Calteau A, Cai F, Tandeau de Marsac N, Rippka R, Herdman M, Sivonen K, Coursin T, Laurent T, Goodwin L, Nolan M, Davenport KW, Han CS, Rubin EM, Eisen JA, Woyke T, Gugger M, Kerfeld CA (2013) Improving the coverage of the cyanobacterial phylum using diversity-driven genome sequencing. *Proc Natl Acad Sci USA* 110: 1053–1058.

Simmons TL, Andrianasolo E, McPhail K, Flatt P, Gerwick WH (2005) Marine natural products as anticancer drugs. *Mol Cancer Ther* 4: 333–342.

Simmons TL, McPhail KL, Ortega-Barría E, Mooberry SL, Gerwick WH (2006) Belamide A, a new antimitotic tetrapeptide from a Panamanian marine cyanobacterium. *Tetrahedron Lett* 47: 3387–3390.

Singh RK, Tiwari SP, Rai AK, Mohapatra TM (2011) Cyanobacteria: An emerging source for drug discovery. *J Antibiot* 64: 401–412.

Soltani N, Khavari-Nejad RA, Tabatabaei Yazdi M, Shokravi S, Fernández-Valiente E (2005) Screening of soil cyanobacteria for antifungal and antibacterial activity. *Pharmaceut Biol* 43: 455–459.

Steinkuhler MC, Gallinari MP, Osswald B, Sewald N, Ritzefeld M, Frese M, Figueras E, Pethö L (2018) Cryptophycin-based antibody-drug conjugates with novel self-immolative linkers. Google Patents, US2018/0078656 A1.

Subhashini J, Mahipal SV, Reddy MC, Mallikarjuna Reddy M, Rachamallu A, Reddanna P (2004) Molecular mechanisms in C-phycocyanin induced apoptosis in human chronic myeloid leukemia cell line-K562. *Biochem Pharmacol* 68: 453–462.

Sueyoshi K, Kaneda M, Sumimoto S, Oishi S, Fujii N, Suenaga K, Teruya T (2016) Odoamide, a cytotoxic cyclodepsipeptide from the marine cyanobacterium *Okeania* sp. *Tetrahedron* 72: 5472–5478.

Taylor FJR, Hoppenrath M, Saldarriaga JF (2008) Dinoflagellate diversity and distribution. *Biodivers Conserv* 17: 407–418.

Teruya T, Sasaki H, Kitamura K, Nakayama T, Suenaga K (2009) Biselyngbyaside, a macrolide glycoside from the marine cyanobacterium *Lyngbya* sp. *Org Lett* 11: 2421–2424.

Thornburg CC, Cowley ES, Sikorska J, Shaala LA, Ishmael JE, Youssef DTA, McPhail KL (2013) Apratoxin H and apratoxin A sulfoxide from the red sea cyanobacterium *Moorea producens*. *J Nat Prod* 76: 1781–1788.

Tidgewell K, Clark BR, Gerwick WH (2010) 2.06 The natural products chemistry of cyanobacteria. In: *Comprehensive Natural Products II. Chemistry and Biology* Vol 2. Oxford, UK: Elsevier, pp. 141–188.

Tran M, Henry RE, Siefker D, Van C, Newkirk G, Kim J, Bui J, Mayfield SP (2013) Production of anti-cancer immunotoxins in algae: Ribosome inactivating proteins as fusion partners. *Biotechnol Bioeng* 110: 2826–2835.

Tripathi A, Puddick J, Prinsep MR, Rottmann M, Tan LT (2010) Lagunamides A and B: Cytotoxic and antimalarial cyclodepsipeptides from the marine cyanobacterium *Lyngbya majuscula*. *J Nat Prod* 73: 1810–1814.

Vaishampayan U, Glode M, Du W, Kraft A, Hudes G, Wright J, Hussain M (2000) Phase II study of dolastatin-10 in patients with hormone-refractory metastatic prostate adenocarcinoma. *Clin Cancer Res* 6: 4205–4208.

Van Wagoner RM, Deeds JR, Satake M, Ribeiro AA, Place AR, Wright JL (2008) Isolation and characterization of karlotoxin 1, a new amphipathic toxin from *Karlodinium veneficum*. *Tetrahedron Lett* 49: 6457–6461.

Vaz BdS, Moreira JB, Morais MGd, Costa JAV (2016) Microalgae as a new source of bioactive compounds in food supplements. *Curr Opin Food Sci* 7: 73–77.

Virtamo J, Taylor PR, Kontto J, Männistö S, Utriainen M, Weinstein SJ, Huttunen J, Albanes D (2014) Effects of α-tocopherol and β-carotene supplementation on cancer incidence and mortality: 18-year postintervention follow-up of the alpha-tocopherol, beta-carotene cancer prevention study. *Int J Cancer* 135: 178–185.

Wang J, Chen S, Xu S, Yu X, Ma D, Hu X, Cao X (2012) In vivo induction of apoptosis by fucoxanthin, a marine carotenoid, associated with down-regulating STAT3/EGFR signaling in sarcoma 180 (S180) xenografts-bearing mice. *Mar Drugs* 10(9): 2055–2068.

Waters AL, Hill RT, Place AR, Hamann MT (2010) The expanding role of marine microbes in pharmaceutical development. *Curr Opin Biotechnol* 21: 780–786.

Wang X, Zhang X (2013) Separation, antitumor activities, and encapsulation of polypeptide from *Chlorella pyrenoidosa*. *Biotechnol Prog* 29: 681–687.

Waters AL, Oh J, Place AR, Hamann MT (2015) Stereochemical studies of the karlotoxin class using nmr spectroscopy and dp4 chemical-shift analysis: Insights into their mechanism of action. *Angew Chem Int Ed Engl* 54: 15705–15710.

Weiss C, Figueras E, Borbely AN, Sewald N (2017) Cryptophycins: Cytotoxic cyclodepsipeptides with potential for tumor targeting. *J Pept Sci* 23: 514–531.

Whitton RLA, Jefferson BDS, Villa RDS (2016) Algae reactors for wastewater treatment. Eng D thesis, Cranfield University. http://dspace.lib.cranfield.ac.uk/handle/1826/10289.

WHO (World Health Organization) Cancer. http://www.who.int/en/news-room/fact-sheets/detail/cancer. Accessed on 12 October 2018.

Williams PG, Yoshida WY, Quon MK, Moore RE, Paul VJ (2003) The structure of palau'amide, a potent cytotoxin from a species of the marine cyanobacterium *Lyngbya*. *J Nat Prod* 66: 1545–1549.

Wrasidlo W, Mielgo A, Torres VA, Barbero S, Stoletov K, Suyama TL, Klemke RL, Gerwick WH, Carson DA, Stupack DG (2008) The marine lipopeptide somocystinamide A triggers apoptosis via caspase 8. *Proc Natl Acad Sci USA* 105: 2313–2318.

Yan T, Li HY, Wu JS, Niu Q, Duan WH, Han QZ, Ji WM, Zhang T, Lv W (2017) Astaxanthin inhibits gemcitabine-resistant human pancreatic cancer progression through EMT inhibition and gemcitabine resensitization. *Oncol Lett* 14: 5400–5408.

Ying J, Wang J, Ji H, Lin C, Pan R, Zhou L, Song Y, Zhang E, Ren P, Chen J, Liu Q, Xu T, Yi H, Li J, Bao Q, Hu Y, Li P (2016) Transcriptome analysis of phycocyanin inhibitory effects on SKOV-3 cell proliferation. *Gene* 585: 58–64.

Yokosaka S, Izawa A, Sakai C, Sakurada E, Morita Y, Nishio Y (2018) Synthesis and evaluation of novel dolastatin 10 derivatives for versatile conjugations. *Bioorg Med Chem* 26: 1643–1652.

Yuri T, Yoshizawa K, Emoto Y, Kinoshita Y, Yuki M, Tsubura A (2016) Effects of dietary xanthophylls, canthaxanthin and astaxanthin on n-methyl-n-nitrosourea-induced rat mammary carcinogenesis. *In Vivo* 30: 795–800.

Zhang B, Zhang X (2013) Separation and nanoencapsulation of antitumor polypeptide from Spirulina platensis. *Biotechnol Prog* 29: 1230–1238.

Zhang C, Naman CB, Engene N, Gerwick WH (2017) Laucysteinamide a, a hybrid PKS/NRPS metabolite from a Saipan cyanobacterium, cf. *Caldora penicillata*. *Mar Drugs* 15(4): 121.

Zhang J, Sun Z, Sun P, Chen T, Chen F (2014a) Microalgal carotenoids: Beneficial effects and potential in human health. *Food Funct* 5: 413–425.

Zhang W, Sun T, Ma Z, Li Y (2014b) Design, synthesis and biological evaluation of tasiamide analogues as tumor inhibitors. *Mar Drugs* 12: 2308–2325.

Zhang Y, Zhu X, Huang T, Chen L, Liu Y, Li Q, Song J, Ma S, Zhang K, Yang B, Guan F (2016) β-carotene synergistically enhances the anti-tumor effect of 5-fluorouracil on esophageal squamous cell carcinoma in vivo and in vitro. *Toxicol Lett* 261: 49–58.

Ziegler RG (1989) A review of epidemiologic evidence that carotenoids reduce the risk of cancer. *J Nutr* 119: 116–122.

19 Macroalgae and Microalgae
Novel Sources of Functional Food and Feed

Shama Aumeerun, Joyce Soulange-Govinden, Marie Francoise Driver, Ambati Ranga Rao, Gokare A. Ravishankar, and Hudaa Neetoo

CONTENTS

Abbreviations 208
Introduction 208
Types of Algae 208
Diversity of Algae in the Mascarene Islands 209
    Macroalgae 209
Microalgae 209
Chemical Composition of Algae 209
    Carbohydrates/Polysaccharides 210
    Amino Acids and Proteins 210
    Lipids and Sterols 210
    Minerals and Vitamins 212
    Antioxidants 212
Health Benefits of Edible Algae as a Functional Food 212
    Anti-Obesity Activity 212
    Anti-Diabetic Activity 213
    Source of Dietary Fiber 213
Health Benefits of Edible Algae as a Functional Feed 213
    Growth Promoting Activity in Animals 213
    Growth Promoter of Beneficial Gut Bacteria 214
Potential Benefits of Algae to the Food and Feed Industry 214
    Food Industry 214
    Feed Industry 214
Conclusion 214
References 215

BOX 19.1 SALIENT FEATURES

Nowadays, there is an ever-increasing demand for new functional foods. This is due to the fact that consumers are becoming more conscious of the food they eat and its impact on their health. As a result, the food industry is actively prospecting for new functional food products to meet the demands of this niche market. Moreover, the feed industry has also garnered an interest in the research and development of functional feeds which can benefit the nutritional requirements and health of livestock animals. Algae are known to have various beneficial health-promoting properties such as reducing glycemic responses and colon cancer, significantly increasing stool volume and promote growth and protection of beneficial intestinal flora that can be exploited in the design and development of novel functional foods and feeds. This chapter starts with a general overview of macro- and microalgae, with reduced mention of specific algal species from the Mascarene Islands. It is then followed by an elaboration of the different biochemical constituents in macro- and microalgae, such as polysaccharides, amino acids and proteins, lipids, minerals, vitamins and antioxidants coupled with documentation of their health benefits. Finally, the advantages of algal-based ingredients to the food and feed sector are discussed.

ABBREVIATIONS

FAO: Food and Agriculture Organization
PUFA: Polyunsaturated fatty acid

INTRODUCTION

The marine environment holds an indubitable treasure of valuable resources that can prove to be highly beneficial to both humans and animals. Algae represent one of the primary marine resources that have been exploited for a very long time. They are currently being used by many different industries, be it by the food, dairy, pharmaceutical or cosmetic industry and in the production of biofuels and hydrogen gases (Pooja 2014; Suganya et al. 2016).

Algae are a heterogenous group of simple, non-cohesive, aquatic photosynthetic organisms classified under the kingdom of Protista (El Gamal 2010; Lewin and Andersen 2018). Their size varies from 3–10 μm for the unicellular species to up to a length of approximately 70 m for "Giant Kelps" (El Gamal 2010). Due to this variation, algae are classified into two main categories known as macroalgae and microalgae. Macroalgae are the macroscopic, multicellular type of algae while microalgae are microscopic and unicellular (Suganya et al. 2016). Algae are said to be planktonic, benthic or littoral organisms with a tolerance for a wide range of temperatures, pH, turbidity, oxygen and carbon-dioxide concentration (Barsanti and Gualtieri 2014).

Algae have different properties that can be very useful to humans. They are known to have a vast array of biological activities including anti-diabetic, antimicrobial, antifungal, anti-obesity and antioxidant (Ambati et al. 2014). They also have the potential to act as enzyme inhibitors and stimulants (Pal et al. 2014). All these properties are attributed to the presence of various bioactive compounds present in them (Ranga Rao et al. 2013).

Given these beneficial properties, algae may be exploited in the field of food and nutritional science especially in the development of new functional foods and feeds. Functional foods are defined as those foods that are expected to be consumed as part of a normal diet and which comprise of biologically active compounds that have the potential to reduce risk of diseases and provide better health (FAO 2007). Common examples include foods that consist of minerals, vitamins, fatty acids or dietary fibers or those that have biologically active compounds (FAO 2007). According to this definition, algae can be considered as a functional food.

Many algae are already being used in the food and feed industry. The microalgae *Spirulina* is often used as a feed supplement because of its excellent nutrient compounds and digestibility (Yaakob et al. 2014). The microalgae *Chlorella*, *Isochrysis*, *Pavlova*, *Phaeodactylum*, *Chaetoceros*, *Nannochloropsis*, *Skeletonema*, *Thalassiosira*, *Haematococcus* and *Tetraselmis* are used in fish farming to feed fish larvae (Madeira et al. 2017). Microalgae are not only used in animal nutrition but also as a food product. For example, the green algae *Spirogyra* and *Oedogonium* are consumed in Burma, Thailand, Vietnam and India (Garcia et al. 2017). The cyanobacterium *Aphanotheca sacrum* is considered as a delicacy in Japan and the species *Nostoc commune*, *Nostoc flagelliforme* and *Nostoc punctiforme* are used as dietary components in China, Mongolia, Tartaria and South America (Garcia et al. 2017). Likewise macroalgae are used as prebiotics for farm animals as they can increase the growth of beneficial gut bacteria, which in turn improves the intestinal health and growth performance of the animal (O'Sullivan et al. 2010). They are also widely being consumed as human food worldwide. In fact the total annual value of seaweed production is around US$6 billion and US$5 billion of this is used for producing food products for human consumption (Nayar and Bott 2014). Moreover among the 221 species of seaweeds that are being exploited worldwide, 145 of them are being used for food purposes (Nayar and Bott 2014).

TYPES OF ALGAE

Algae are classified into two major classes namely macroalgae and microalgae based on their size (Ranga Rao et al. 2018). These two classes are further sub-divided into different groups based on their morphological features such as pigmentation, chemical nature of their photosynthetic storage product and the organization of their photosynthetic membranes among others (Kilinç et al. 2013). Table 19.1 and 19.2 summarize some of the general features of both micro- and macroalgae.

TABLE 19.1
General Features of Macroalgae

Algae	General Features	References
Macroalgae/Seaweeds	Primary producer in aquatic food webs	Kilinç et al. (2013)
	Considered as marine plants as they are photosynthetic organisms with similar ecological functions as other plants	Diaz-Pulido and McCook (2008)
	Found in the intertidal and sublittoral to littoral areas of the sea and in brackish water	Hamed et al. (2018); Prasad et al. (2013)
	Classified under three main categories: *Phaeophyta*, *Rhodophyta* and *Chlorophyta* based on the nature of pigments present in them	Zamani et al. (2013); Misurcova (2012)

TABLE 19.2
General Features of Microalgae

Algae	General Features	References
Microalgae	Ubiquitous organism	Singh and Saxena (2015)
	Size ranges from about 5 to 50 μm	
	200,000 to 800,000 species present around the world	Venkatesan et al. (2015)
	Classified under four main classes: *Bacillariophyceae* (diatoms), *Chlorophyceae* (green algae), *Chrysophyceae* (golden algae) and *Cyanophyceae* (blue-green algae cyanobacteria)	Madeira et al. (2017)

DIVERSITY OF ALGAE IN THE MASCARENE ISLANDS

MACROALGAE

Mauritius, Rodrigues and Réunion, which represent the Mascarene Islands, are surrounded by the Indian Ocean and harbor a wide variety of marine algae. Mauritius as compared to the other two islands has a wide variety of macroalgae (Bolton et al. 2012). Its macroalgal flora consists of 435 species in which 59 are *Phaeophyceae*, 108 are *Chlorophyceae* and 268 are *Rhodophyceae* (Bolton et al. 2012). Réunion Island harbors 196 algal species among which 36 are *Phaeophyceae*, 31 are *Chlorophyceae* and 129 are *Rhodophyceae* (Mattio 2013). Algae from Rodrigues Island have not been studied much, thus it is only known to host 179 algal species and the exact number of the different classes of algae are yet to be described (Schils et al. 2004). As compared to the other algae *Rhodophyceae* represent the highest species diversity and the largest proportion of species with specific distribution pattern (Schils et al. 2004).

MICROALGAE

Information about the total diversity of microalgae in the Mascarene Islands is very limited as no studies have yet been conducted. Sadally et al. (2016) conducted a study to identify the microalgal diversity of 5 different ecosystems (coral reefs of blue bay, seagrass bed of Mahebourg, mangroves ecosystem at Pointe D'Esny, sandy beach of blue bay and estuarine areas of Le Goulet) around Mauritius Island. The authors recorded a total of 41 micro-phytoplankton genera and a total of 33 micro-phytobenthos genera in these 5 ecosystems. They also discovered that the most abundant microalgal species in these areas were diatoms, followed by dinoflagellates and cyanobacteria.

CHEMICAL COMPOSITION OF ALGAE

Both macro- and microalgae can be considered as very good sources of bioactive compounds comprising of carbohydrates, proteins, lipids, sterols, antioxidants, minerals and vitamins (Pérez et al. 2016). Macroalgae normally consist of 80–90% water and their dry weight encompasses 50% carbohydrates, 1–3% lipids and 7–38% minerals. Their protein content varies between 10–47% and consists of high proportions of essential amino acids (García-Casal et al. 2007). On the other hand, microalgae consist mostly of lipids (37.6 kJg^{-1}), followed by proteins (16.7 kJ g^{-1}) and carbohydrates (15.7 kJg^{-1}) (Markou et al. 2012). Algal biochemical compounds

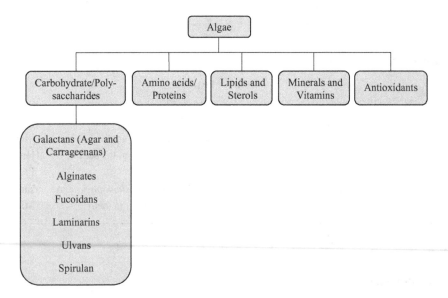

FIGURE 19.1 Chemical composition of algae.

can prove to be very beneficial to human and animals as they have a high nutritive value. The number of bioactive compounds present in different algae depends on factors such as season, growth environment, type and age of algae (Zubia et al. 2008; Connan et al. 2004). Figure 19.1 summarizes the bioactive compounds present in algae.

Carbohydrates/Polysaccharides

Algal carbohydrates consist mostly of various types of polysaccharides and a few monosaccharides and disaccharides (Stiger-Pouvreau et al. 2016). Research has shown that the hydrolysates of some brown algae have complex mixtures of monosaccharide which include galactose, glucose, mannose, fructose, xylose, fucose and arabinose (Stiger-Pouvreau et al. 2016). Different ratios of fructose, galactose, glucose, mannose and xylose are also found in microalgae (Yen et al. 2013). Polysaccharides are found in much larger amounts in algae (Ismail 2017). They are comprised of galactans, agars, alginates, carrageenans, fucoidans, laminarins, ulvans and other derivatives (Rajapakse and Kim 2011). As for the microalgal polysaccharides, not much is known about their common names, except for spirulan (de Jesus Raposo et al. 2015). Figure 19.2, Figure 19.3, Figure 19.4 and Figure 19.5 provide a summary of the different polysaccharides found in different algae.

Amino Acids and Proteins

Both seaweeds and microalgae are such good sources of proteins that some macro- and microalgae have the same protein content as some conventional sources of protein like milk, meat, egg and soybean (Bleakley and Hayes 2017). The protein content of algae varies according to factors such as species, season and geographical locations (Ismail 2017). The red algae *Palmaria palmata* and *Porphyra tenera* have a high protein content of 35% (w/w) and 47% (w/w) respectively, while on the other hand the green algae *Ulva pertuse* has a protein content of only 26% (w/w) (Harnedy and FitzGerald 2013). The microalgae *Chlorella vulgaris* has a high protein content ranging from 51% to 58% of its dry weight (Bleakley and Hayes 2017).

Most algae consist of all the essential amino acids and are excellent sources of the acidic amino acids, aspartic acid and glutamic acid (Admassu et al. 2015). For example, the microalgae *Spirulina* contain all the ten essential amino acids, with the exception of reduced methionine, cysteine and lysine (Allen 2016). Moreover, some amino acids are found in much higher levels in algae than in terrestrial plants (Admassu et al. 2015). For instance, 9 of the 10 essential amino acids were found in much higher levels in the algal species *Enteromorpha* than in an equal weight of soy beans (Chojnacka et al. 2012).

Lipids and Sterols

Lipid content of algae varies according to factors such as geographical location, season, temperature, light intensity, salinity and species (Ahmed and Ahmed 2016). Macroalgae usually have a low lipid content which is less than 5% of their dry weight (Marghraby and Fakhry 2015). On the other hand, microalgae have such a high level of lipids that they are considered as an alternative source for production of biodiesel (Zhu et al. 2016). In a

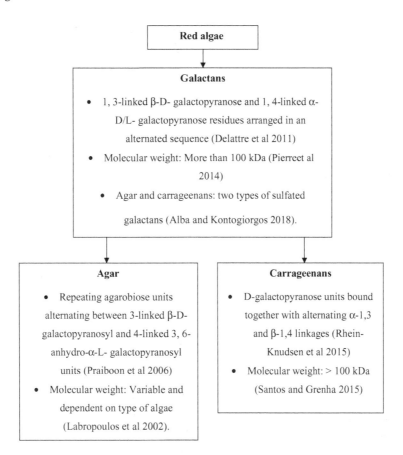

FIGURE 19.2 Polysaccharides of red algae.

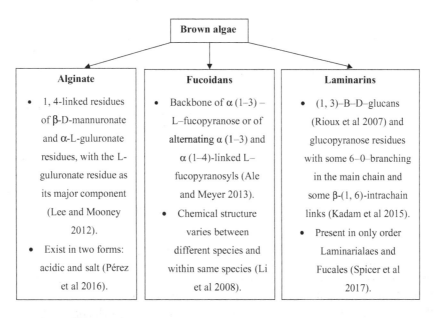

FIGURE 19.3 Polysaccharides of brown algae.

study conducted by Kendel et al. (2015), it was observed that 29% and 15% of the total lipids of *Ulva armoricana* and *Solieria chordalis* respectively were polyunsaturated fatty acids (PUFAs). Algae are composed of different types of PUFAs; microalgae contain the PUFAs eicosapentaenoic acid and docosahexaenoic acid while green seaweed consists of the PUFAs α-linolenic, stearidonic and linoleic acids (Handayania et al. 2011; Pérez et al. 2016). Red algae are composed of arachidonic and eicosapentaenoic acid and brown algae rich in the

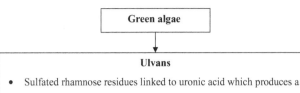

FIGURE 19.4 Polysaccharides of green algae.

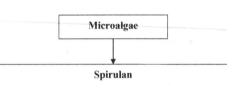

FIGURE 19.5 Polysaccharides of microalgae.

PUFAs present in both green and red algae (Pérez et al. 2016).

Sterols are a major nutritional component of algae (Pal et al. 2014). Cholesterol, fucosterol, isofucosterol and clionasterol are some examples of sterols found in macroalgae (Pérez et al. 2016). Sitosterol, campesterol, 24-ethylcholesterol, epibrassicasterol, dihydroxysterol, 24-methylenecholesterol, crinosterol, stigmasterol and 4-desmethyl-23,24-dimethyl steroid are microalgal sterols (Luo et al. 2015).

Minerals and Vitamins

Algae are good sources of minerals. Their mineral content varies according to factors such as species, season and geographical harvesting site (Bocanegra et al. 2009). Algae are rich in the macroelements P, K, Ca, Mg, Na, Fe and M and the trace elements Zn, Cu, Cr, Pb, Sr, As, Sd and Sc (Balina et al. 2016). These elements are found in much higher levels in algae than in terrestrial plants (Balina et al. 2016). Microalgae and macroalgae are also rich sources of vitamins. They consist of the water-soluble vitamins C and B1, B2, B12, fat-soluble vitamins E and carotenoids as provitamins of vitamins A (Misurcova 2012; Wells et al. 2017). Algal vitamins are of great importance due to their biochemical functions, antioxidant activity and health benefits (Škrovánková 2011).

Antioxidants

Algae are considered as good sources of antioxidants (Munir et al. 2013). Algal antioxidants consist of carotenoids and vitamin E (α-tocopherol), phycobiliproteins, polyphenols and vitamins (Munir et al. 2013). Carotenoids found in micro- and macroalgae actually represent the different pigments which together with chlorophyll give algae its characteristic color such as fucoxanthin, β-carotene, violaxantin, astaxanthin, lutein amongst others (Ranga Rao et al. 2018; Ambati et al. 2017). Phycobiliproteins are water-soluble pigment proteins normally present in cyanobacteria and some algae such as *Rhodophyta* and *Cryptomonas* (Sonani et al. 2017). They are normally classified as phycocyanin, phycoerythrins and allophycocyanin based on UV-visible absorption maxima (Yaakob et al. 2014). Polyphenols are the common secondary metabolites found in algae (Thomas and Kim 2011). They are said to be one of the most important classes of natural antioxidants (Machu et al. 2015). Seaweeds are rich in the polyphenolic compounds catechins, flavonols and phlorotannins (Gómez-Guzmán et al. 2018). Microalgae and cyanobacteria are composed of different classes of flavonoids such as isoflavones, flavanones, flavonols and dihydrochalcones (Goiris et al. 2012).

HEALTH BENEFITS OF EDIBLE ALGAE AS A FUNCTIONAL FOOD

Research has shown that the dietary intake of algae is beneficial for human health as they reduce a number of chronic diseases (Wan-Loy and Siew-Moi 2016). Some of the beneficial effects are summarized in the following.

Anti-Obesity Activity

Many studies have proved that algae have good anti-obesity potential. Hall et al. (2012) conducted a study to test the effect of *Ascophyllum nodosum* enriched bread on healthy, overweight and obese men. It was observed that as compared to the group of men that were fed the control bread, a significant reduction in the mean energy intake was observed in those men that were fed the *A. nodosum* enriched bread. Nakazono et al. (2016) also carried out a study where the anti-obesity effects of dietary acid-hydrolyzed (A-AO) and enzymatic-digested

(E-AO) alginate oligomers were tested in male mice fed a high-fat diet, and it was observed that E-AO had much stronger anti-obesity effects than A-AO. Paxman et al. (2008) investigated the ability of alginate to reduce increased uptake of cholesterol and glucose in overweight male subjects, and eventually concluded that alginates are very effective in controlling overweight and obesity problems. Moreover Jang and Choung (2013) showed that ethanolic extract of Laminaria japonica has excellent potential to act as an anti-obesity agent with no obvious toxicity. The algal sterol fucosterol and the carotenoid fucoxanthin have also been observed to have anti-obesity effects (Seca and Pinto 2018).

Anti-Diabetic Activity

Algae can be considered as a possible solution against diabetes. For instance, the sulfated polysaccharides alginate, fucoidan and laminarin of brown algae Sargassum have antioxidant potential which can prevent destruction of pancreatic β-cell that causes diabetes (Unnikrishnan et al. 2015). In a study conducted by Koneri et al. (2018), it was observed that methanolic extracts of algae Sargassum polycystum and Gracilaria edulis had a significant increase in beta-cell density showing insulin secretagogue activity. Studies have also shown that algal extracts are effective to control diabetes due to their action on the enzymes involved in carbohydrates digestion (Lji and Kadam 2013). Sun and Chen (2012) investigated the inhibitory effects of the microalgae Chlorella pyrenoidosa against the two key enzymes namely α-amylase and α-glucosidase relevant for type-2 diabetes. It was observed that Chlorella pyrenoidosa had a moderate inhibitory effect on α-amylase and strong inhibitory effect against α-glucosidase (Sun and Chen 2012). The authors stated that Chlorella pyrenoidosa is an optimal inhibitor of the enzymes as it does not show any excessive inhibitory effect on pancreatic α-amylase that can lead to abnormal bacterial fermentation of undigested carbohydrates in the colon causing flatulence, meteorism or diarrhea. Moreover, in another study the microalgae Isochrysis galbana has been observed to decrease glucose levels in diabetic rats (Nuño et al. 2013).

Source of Dietary Fiber

Algae can be considered as a good source of dietary fiber. Macroalgae consist of a number of soluble and insoluble fibers which includes agar, carrageenans, xylans, alginates, fucans, laminarans, sulfated rhamnoxyloglucurons, celluloses and mannans (McDermid et al. 2005). The level of dietary fiber present in seaweeds is said to be similar or higher than that present in terrestrial plants (Rajapakse and Kim 2011). The health-promoting effects of algal dietary fiber are that they reduce glycemic responses and colon cancer, significantly increase stool volume and promote growth and protection of beneficial intestinal flora (Gupta and Abu-Ghammam 2011). In fact, studies have shown that the fermentable fiber of brown algae has ability to maintain a favorable balance among colonic microflora and to promote growth of *Bifidobacteria and Lactobacteria* (Rajapakse and Kim 2011). Moreover, in a study conducted by Bhowmik et al. (2009), it was observed that the microalgae *Spirulina platensis* has the potential to increase the growth of the lactic acid bacteria *Lactobacillus casei, Lactobacillus acidophilus* and *Streptococcus thermophilus*.

HEALTH BENEFITS OF EDIBLE ALGAE AS A FUNCTIONAL FEED

In addition to the beneficial effects of algae on human health, there are many other evidences which prove that they also have the ability to provide better health to animals.

Growth Promoting Activity in Animals

The growth promoting activity of algae in animals can be a subject of debate. There are many studies that have shown that algae are unable to promote growth of animals. For instance, in a study conducted by Bach et al. (2008), it was observed that a sun-dried *Ascophyllum nodosum* supplemented diet had no effect on feed intake or growth of lambs. In another study, Gardiner et al. (2008) showed that there was a linear decrease in the daily weight gain of grower-finisher pigs that were fed *Ascophyllum nodosum*. Toyomizu et al. (2001) stated that there were no significant differences in the body weight of broilers that were fed *Spirulina*. However, there are many other studies proving that algae actually have growth promoting effects. El-Deekx and Brikaa (2009) showed that inclusion of the red seaweed *Polysiphonis* spp. in starter and finisher diets of ducks increases their weight gain. Choi et al. (2014) demonstrated that supplementation of broilers diet with *Undaria pinnatifida* by-product and *Hizikia fusiformis* by-product has a positive effect on their growth performance. Dadgar et al. (2011) concluded that the microalgae *Spirulina platensis* supplemented diet has a positive effect on the performance of broilers. A significant increase in the live weight, growth and body conformation was also observed in divergent Australian sheep that were fed Spirulina (Holman et al. 2012).

Growth Promoter of Beneficial Gut Bacteria

Algae have been observed to encourage growth and activity of probiotics. In a study conducted by Lynch et al. (2010), it was observed that supplementation of laminarin and fucoidan in pigs diet resulted in a reduction of intestinal *Enterobacteria* and increase of *Lactobacilli* spp. The authors suggested that seaweed polysaccharides can give a dietary mean of enhancing gut health of pigs. Other studies have shown that supplementation of seaweed in poultry diet increases growth of beneficial gut microbiota in the lower gastrointestinal tract of broilers (Kulshreshtha et al. 2014). Moreover, alginates have been observed to exert prebiotic activity in weanling pigs by increasing the number of Enterococci and enhancing bacterial diversity of the intestine (de Jesus Raposo et al. 2016). Microalgae are also known to have prebiotic activities (de Jesus Raposo et al. 2016). The microalgae *Spirulina platensis*, *Chlorella* spp., *Tetraselmis* spp., *Dunaliella salina*, *Scenedesmus* spp., *Chlorococcum* spp. and *Cylindrospermum* spp. are said to have prebiotic potentials (Gupta et al. 2017). In fact *Spirulina platensis* extract is considered to be the best algal source for prebiotic due to its growth promoting effect on the bacteria *L. lactis, B. longum* and *L. bulgaricus* (Gupta et al. 2017). In a study conducted by Nuño et al. (2013), it was also observed that the microalgae *Isochrysis galbana* increased lactic acid bacteria counts in rats.

POTENTIAL BENEFITS OF ALGAE TO THE FOOD AND FEED INDUSTRY

In addition to the health benefits on human and animals, algae can prove to be highly advantageous and profitable to both the food and feed industry.

Food Industry

One main advantage of algae as a food ingredient is that it is suitable for all types of eating styles, be it vegetarian, vegan or pescatarian among others. Vegan diets are usually considered to be nutrient-deficient due to the unbalanced protein sources and low intake of long chain omega-3 fatty acids, some vitamins and minerals (Nicoletti 2016). Seaweed thus can provide vegans and also vegetarians with the required nutrients that are only available in animal-based food. For instance, the dried seaweed *Porphyra yezoensis*, commonly known as Nori, consists of many nutrients that are absent in vegetarian and vegan diets, such as Vitamin A, Vitamin B_{12}, iron and *n*-3 polyunsaturated fatty acids (Watanabe et al. 2014). Algae thus can prove to be very beneficial to the food industries as they would be able to encompass a larger number of consumers which will prove to be profitable for them. Moreover, given the increasing food demand worldwide, algae can prove to be a potential solution to address this problem. In fact both seaweeds and microalgae are considered as promising and novel sources of many nutrients such as protein and omega-3 fatty acid (Henchion et al. 2017; Ji et al. 2015). Actually, with the complications associated with the use of fish oil as a source of omega-3, such as purification of the fatty acids, the significant taste, odor, and stability problems, algae is viewed as a potential alternative for production of omega-3 (Ji et al. 2015).

Feed Industry

Algae provide a myriad of advantages to the feed industry. It has been observed that inclusion of algae in animal feed results in better quality of animal-based products. For instance, studies have shown that Tasco® which is a product of the macroalgae *Ascophyllum nodosum* has the potential to increase productivity and quality of animal products such as milk, meat and carcass grade among others (Evans and Critchley 2014). In a study conducted by Ginzberg et al. (2000) it was observed that the egg yolk of chickens fed with the algae *Porphyridium* sp. had lower levels of cholesterol and higher levels of linoleic acid and arachidonic acid. Studies have also shown that seaweed can be used in pigs' diet to increase the iodine concentration in their meat (Makkar et al. 2016). In fact an increase in the concentration of iodine has been observed in the tissues of pigs fed the seaweed *Ascophyllum nodosum* (Makkar et al. 2016). In another study, it was observed that supplementation of the n-3-fatty-acid-enriched microalgae *Schizochytrium* spp. in poultry feed can enhance fatty acid composition in the breast meat of broilers (Yan and Kim 2013). In addition algae can be used as an alternative to the traditional feed ingredients. In a study conducted by Altmann et al. (2018) the microalgae Spirulina and Hermetia had been observed to have the potential to replace soybean in poultry diet. This can be of great use especially in harsh climatic conditions such as snow seasons whereby growing of soybeans or corn can be difficult (Evans and Critchley 2014).

CONCLUSION

Algae, be it macro- or microalgae, harbor a wide variety of biochemical compounds that have excellent nutritional and bioactive properties. The beneficial effects of these biochemical constituents and their potential applications in the food and feed industry have been enumerated in many studies. Both macro- and microalgae have a very promising future as a potential source of functional food and feed. However, there are some limitations that should be taken into consideration before incorporating algae in food or feed. For instance, special attention should be given when selecting the type of algae for consumption especially microalgae as some of them are highly toxic or contain harmful compounds such as toxic metals present within them. The bioavailability of the nutritional compounds should also be assessed beforehand. Moreover, industries should take special care when processing the algae for consumption in order not to damage or destroy the important nutritional compounds. Nevertheless, these few limitations cannot negate the fact that algae are very good sources of nutritional compounds. Thus more research is warranted to tackle these limitations and to develop functional algal food or feed with excellent nutritional properties that can benefit humans and animals.

REFERENCES

Admassu, H., W. Zhao, R. Yang, M.A.A. Gasmalla, and E. Alsir. "Development of Functional Foods: Sea Weeds (Algae) Untouched Potential and Alternative Resource – A Review." *International Journal of Scientific and Technology Research* 4, no. 9 (2015): 108–115.

Ahmed, Niaz, and Kashif Ahmed. "Chemical and Different Nutritional Characteristics of Brown Seaweed Lipids." *Advances in Science, Technology and Engineering Systems Journal* 1, no. 1 (2016): 23–25.

Alba, Katerina, and Vassilis Kontogiorgos. "Seaweed Polysaccharides (Agar, Alginate Carrageenan)." In: *Reference Module in Food Science*, 2018. doi:10.1016/b978-0-08-100596-5.21587-4.

Ale, Marcel Tutor, and Anne S. Meyer. "Fucoidans from Brown Seaweeds: An Update on Structures, Extraction Techniques and Use of Enzymes as Tools for Structural Elucidation." *RSC Advances* 3, no. 22 (2013): 8131–8141.

Allen, Kathleen. "Evaluating Spirulina as a Protein Source in Nile Tilapia (*Oreochromis niloticus*) Grow-Out Diets," 2016. https://qspace.library.queensu.ca/bitstream/handle/1974/14452/Allen_Kathleen__201605_MSc.pdf?sequence=1.

Altmann, Brianne A., Carmen Neumann, Susanne Velten, Frank Liebert, and Daniel Mörlein. "Meat Quality Derived from High Inclusion of a Micro-Alga or Insect Meal as an Alternative Protein Source in Poultry Diets: A Pilot Study." *Foods* 7, no. 3 (2018): 34.

Ambati, R.R., D. Gogisetty, R. Aswathnarayana Gokare, S. Ravi, P.N. Bikkina, Y. Su, and B. Lei. "*Botryococcus* as an Alternative Source of Carotenoids and Its Possible Applications – An Overview". *Critical Reviews in Biotechnology* 38, no. 4 (2017): 541–558.

Ambati, Ranga Rao, Siew Moi Phang, Sarada Ravi, and Ravishankar Gokare Aswathanarayana. "Astaxanthin: Sources, Extraction, Stability, Biological Activities and Its Commercial Applications – A Review" *Marine Drugs* 12, no. 1 (2014): 128–152.

Bach, S.J., Y. Wang, and T.A. Mcallister. "Effect of Feeding Sun-Dried Seaweed (*Ascophyllum nodosum*) on Fecal Shedding of *Escherichia coli* O157:H7 by Feedlot Cattle and on Growth Performance of Lambs." *Animal Feed Science and Technology* 142, no. 1–2 (2008): 17–32.

Baky, Abd el H, K.F. Hanaa el Baz, and S.A. EL-Latife. "Induction of Sulfated Polysaccharides in *Spirulina platensis* as Response to Nitrogen Concentration and Its Biological Evaluation" *Journal of Aquaculture Research & Development* 5, no. 1 (2014): 1.

Balina, Karina, Francesco Romagnoli, and Dagnija Blumberga. "Chemical Composition and Potential Use of Fucus vesiculosus from Gulf of Riga." *Energy Procedia* 95 (2016): 43–49.

Barsanti, Laura, and Paolo Gualtieri. "General Overview." In: *Algae: Anatomy, Biochemistry, and Biotechnology*, 1–46. 2nd ed. Boca Raton, FL: CRC Press, 2014. https://books.google.mu/books?id=AZClAgAAQBAJ&printsec=frontcover&source=gbs_ge_summary_r&cad=0#v=onepage&q&f=false.

Becker, E.W. "Micro-Algae as a Source of Protein." *Biotechnology Advances* 25, no. 2 (2007): 207–210.

Bhowmik, D., J. Dubey, and S. Mehra. "Probiotic Efficiency of *Spirulina platensis* – Stimulating Growth of Lactic Acid Bacteria." *World Journal of Dairy and Food Sciences* 4 (2009): 160–163. https://pdfs.semanticscholar.org/89ed/88688f523ac4dcbc91d57e6895a516692d76.pdf.

Bleakley, Stephen, and Maria Hayes. "Algal Proteins: Extraction, Application, and Challenges Concerning Production." *Foods* 6, no. 5 (2017): 33.

Bocanegra, Aránzazu, Sara Bastida, Juana Benedí, Sofía Ródenas, and Francisco J. Sánchez-Muniz. "Characteristics and Nutritional and Cardiovascular-Health Properties of Seaweeds." *Journal of Medicinal Food* 12, no. 2 (2009): 236–258.

Bolton, J.J., R. Bhagooli, and L. Mattio. "The Mauritian Seaweed Flora: Diversity and Potential for Sustainable Utilisation." *University of Mauritius Research Journal* 18A (2012): 6–27.

Choi, Y.J., S.R. Lee, and J.W. Oh. "Effects of Dietary Fermented Seaweed and Seaweed Fusiforme on Growth Performance, Carcass Parameters and Immunoglobulin Concentration in Broiler Chicks." *Asian-Australasian Journal of Animal Sciences* 27, no. 6 (2014): 862–870.

Chojnacka, Katarzyna, Agnieszka Saeid, Zuzanna Witkowska, and Lukasz Tuhy. "Biologically Active Compounds in Seaweed Extracts – The Prospects for the Application." *The Open Conference Proceedings Journal* 3, no. 1 (2012): 20–28.

Connan, Solène, Fabienne Goulard, Valérie Stiger, Eric Deslandes, and Erwan Ar. Gall. "Interspecific and Temporal Variation in Phlorotannin Levels in an

Assemblage of Brown Algae." *Botanica Marina* 47, no. 5 (2004). doi:10.1515/bot.2004.057.

Dadgar, H., M. Toghyani, and M. Dadgar. "Effect of Dietary Blue-Green-Alga (*Spirulina platensis*) as a Food Supplement on Cholesterl, HDL, LDL Cholesterol and Triglyceride of Broiler Chicken." *European Journal of Pharmacology* 668 (2011). doi:10.1016/j.ejphar.2011.09.281.

de Jesus Raposo, Maria Filomena, Alcina Maria Bernardo de Morais, and Rui Manuel Santos Costa de Morais. "Marine Polysaccharides from Algae with Potential Biomedical Applications." *Marine Drugs* 13, no. 5 (2015): 2967–3028.

de Jesus Raposo, M., A. de Morais, and R. de Morais. "Emergent Sources of Prebiotics: Seaweeds and Microalgae." *Marine Drugs* 14, no. 2 (2016): 27.

Delattre, Cedric, T.A. Fenoradosoa, and Philippe Michaud. "Galactans: An Overview of Their Most Important Sourcing and Applications as Natural Polysaccharides." *Brazilian Archives of Biology and Technology* 54, no. 6 (2011): 1075–1092.

Diaz-Pulido, G., and L. McCook. "Macroalgae (Seaweeds)." In: A. Chin., (ed.) *The State of the Great Barrier Reef On-Line*, Great Barrier Reef Marine Park Authority, Townsville, 2008. http://www.gbrmpa.gov.au/__data/assets/pdf_file/0019/3970/SORR_Macroalgae.pdf.

Dos Santos, Marlise A., and Ana Grenha. "Polysaccharide Nanoparticles for Protein and Peptide Delivery." *Advances in Protein Chemistry and Structural Biology* 98 (2015): 223–261.

El-Deekx, A.A., and A. Mervat Brikaa. "Effect of Different Levels of Seaweed in Starter and Finisher Diets in Pellet and Mash Form on Performance and Carcass Quality of Ducks." *International Journal of Poultry Science* 8, no. 10 (2009): 1014–1021.

Evans, F.D., and A.T. Critchley. "Seaweeds for Animal Production Use." *Journal of Applied Phycology* 26, no. 2 (2014): 891–899.

ElGamal, Ali A. "Biological Importance of Marine Algae." *Saudi Pharmaceutical Journal* 18, no. 1 (2010): 1–25.

ElMaghraby, Dahlia M., and Eman M. Fakhry. "Lipid Content and Fatty Acid Composition of Mediterranean Macro-Algae as Dynamic Factors for Biodiesel Production." *Oceanologia* 57, no. 1 (2015): 86–92.

Food and Agriculture Organization, 2007. "Report on Functional Foods ." Rome, Italy. www.hranomdozdravlja.com/slatkis/file.php?file=Functional_Foods...Nov2007.pdf.

García, José L., Marta De Vicente, and Beatriz Galán. "Microalgae, Old Sustainable Food and Fashion Nutraceuticals." *Microbial Biotechnology* 10, no. 5 (2017): 1017–1024.

García-Casal, Maria N., Ana C. Pereira, Irene Leets, José Ramírez, and Maria F. Quiroga. "High Iron Content and Bioavailability in Humans from Four Species of Marine Algae." *Journal of Nutrition* 137, no. 12 (2007): 2691–2695.

Gardiner, G.E., A.J. Campbell, J.V. O'Doherty, E. Pierce, P.B. Lynch, F.C. Leonard, C. Stanton, R.P. Ross, and P.G. Lawlor. "Effect of *Ascophyllum nodosum* Extract on Growth Performance, Digestibility, Carcass Characteristics and Selected Intestinal Microflora Populations of Grower–Finisher Pigs." *Animal Feed Science and Technology* 141, no. 3–4 (2008): 259–273.

Ginzberg, A., M. Cohen, U.A. Sod-Moriah, S. Shany, A. Rosenshtrauch, and S. Arad. "Chickens Fed with Biomass of the Red Microalga Porphyridium sp. Have Reduced Blood Cholesterol Level and Modified Fatty Acid Composition in Egg Yolk." *Journal of Applied Phycology* 12, no. 3/5 (2000): 325–330. https://www.researchgate.net/publication/226275678_Chickens_fed_with_biomass_of_the_red_microalga_Porphyridium_sp_have_reduced_blood_cholesterol_level_and_modified_fatty_acid_composition_in_egg_yolk.

Goiris, K., K. Muylaert, I. Fraeye, I. Foubert, J. De Brabanter, and L. De Cooman. "Antioxidant Potential of Microalgae in Relation to Their Phenolic and Carotenoid Content." *Journal of Applied Phycology* 24, no. 6 (2012): 1477–1486.

Gómez-Guzmán, Manuel, Alba Rodríguez-Nogales, Francesca Algieri, and Julio Gálvez. "Potential Role of Seaweed Polyphenols in Cardiovascular-Associated Disorders." *Marine Drugs* 16, no. 8 (2018): 250.

Gupta, Shilpi, and Nissreen Abu-Ghannam. "Bioactive Potential and Possible Health Effects of Edible Brown Seaweeds." *Trends in Food Science and Technology* 22, no. 6 (2011): 315–326.

Gupta, Sneh, Amar P. Garg, and Dhan Prakash. "Prebiotic Efficiency of Blue Green Algae on Probiotics Microorganisms." *Journal of Microbiology & Experimentation* 4, no. 4 (2017).

A.C. Fairclough, K. Mahadevan, and J.R. Paxman. "*Ascophyllum nodosum* Enriched Bread Reduces Subsequent Energy Intake with No Effect on Post-Prandial Glucose and Cholesterol in Healthy, Overweight Males. A Pilot Study". *Appetite* 58, no. 1 (2012): 379–386.

Hamed, Seham M., Amal A. Abd El-Rhman, Neveen Abdel-Raouf, and Ibraheem B.M. Ibraheem. "Role of Marine Macroalgae in Plant Protection & Improvement for Sustainable Agriculture Technology." *Beni-Suef University Journal of Basic and Applied Sciences* 7, no. 1 (2018): 104–110.

Handayania, Noer Abyor, Dessy Ariyantib, and Hady Hadiyanto. "Potential Production of Polysunsaturated Fatty Acids from Microalgae." *International Journal of Science and Engineering* 2, no. 1 (2011): 13–16.

Harnedy, Pádraigín A., and Richard J. Fitzgerald. "In Vitro Assessment of the Cardioprotective, Anti-Diabetic and Antioxidant Potential of Palmaria Palmata Protein Hydrolysates." *Journal of Applied Phycology* 25, no. 6 (2013): 1793–1803.

Henchion, Maeve, Maria Hayes, Anne Maria Mullen, Mark Fenelon, and Brijesh Tiwari. "Future Protein Supply and Demand: Strategies and Factors Influencing a Sustainable Equilibrium." *Foods* 6, no. 7 (2017): 53.

Holman, B.W.B., A. Kashani, and A.E.O. Malau-Aduli. "Growth and Body Conformation Responses of Genetically Divergent Australian Sheep to Spirulina (Arthrospira Platensis) Supplementation." *American Journal of Experimental Agriculture* 2, no. 2 (2012): 160–173.

Ismail, Gehan Ahmed. "Biochemical Composition of Some Egyptian Seaweeds with Potent Nutritive and Antioxidant Properties." *Food Science and Technology* 37, no. 2 (2017): 294–302.

Jang, Woong Sun, and Se Young Choung. "Antiobesity Effects of the Ethanol Extract of Laminaria Japonica Areshoung in High-Fat-Diet-Induced Obese Rat." *Evidence-Based Complementary and Alternative Medicine* 2013 (2013): 492807.

Jaulneau, Valérie, Claude Lafitte, Christophe Jacquet, Sylvie Fournier, Sylvie Salamagne, Xavier Briand, Marie-Thérèse Esquerré-Tugayé, and Bernard Dumas. "Ulvan, a Sulfated Polysaccharide from Green Algae, Activates Plant Immunity through the Jasmonic Acid Signaling Pathway." *Journal of Biomedicine and Biotechnology* 2010 (2010): 525291.

Ji, Xiao-Jun, Lu-Jing Ren, and He Huang. "Omega-3 Biotechnology: A Green and Sustainable Process for Omega-3 Fatty Acids Production." *Frontiers in Bioengineering and Biotechnology* 3 (2015). doi:10.3389/fbioe.2015.00158.

Kadam, Shekhar U., Colm P. O'Donnell, Dilip K. Rai, Mohammad B. Hossain, Catherine M. Burgess, Des Walsh, and Brijesh K. Tiwari. "Laminarin from Irish Brown Seaweeds *Ascophyllum nodosum* and Laminaria Hyperborea: Ultrasound Assisted Extraction, Characterization and Bioactivity." *Marine Drugs* 13, no. 7 (2015): 4270–4280.

Kendel, Melha, Gaëtane Wielgosz-Collin, Samuel Bertrand, Christos Roussakis, Nathalie Bourgougnon, and Gilles Bedoux. "Lipid Composition, Fatty Acids and Sterols in the Seaweeds Ulva armoricana, and Solieria Chordalis from Brittany (France): An Analysis from Nutritional, Chemotaxonomic, and Antiproliferative Activity Perspectives." *Marine Drugs* 13, no. 9 (2015): 5606–5628.

Kilinç, Berna, Semra Cirik, Gamze Turan, Hatice Tekogul, and Edis Koru. "Seaweeds for Food and Industrial Applications." In: Innocenzo Muzzalupo (ed.) *Food Industry*, 735–748. 1st ed. Intech, 2013.

Koneri, R., D.K. Jha, and M.G. Mubasheera. An Investigation on the Type I Antidiabetic Activity of Methanolic Extract of *Marine Algae*, *Gracilaria edulis* and *Sargassum polycystum*. *International Journal of Pharmaceutical Sciences and Research* 9 (2018): 2952–2959.

Kraan, Stefan. "Algal Polysaccharides, Novel Applications and Outlook." In: *Carbohydrates – Comprehensive Studies on Glycobiology and Glycotechnology*, 2012. doi:10.5772/51572.

Kulshreshtha, Garima, Bruce Rathgeber, Glenn Stratton, Nikhil Thomas, Franklin Evans, Alan Critchley, Jeff Hafting, and Balakrishnan Prithiviraj. "Feed Supplementation with Red Seaweeds, Chondrus Crispus and Sarcodiotheca Gaudichaudii, Affects Performance, Egg Quality, and Gut Microbiota of Layer Hens." *Poultry Science* 93, no. 12 (2014): 2991–3001.

Labropoulos, K.C., D.E. Niesz, S.C. Danforth, and P.G. Kevrekidis. "Dynamic Rheology of Agar Gels: Theory and Experiments. Part I. Development of a Rheological Model." *Carbohydrate Polymers* 50, no. 4 (2002): 393–406.

Lee, Kuen Yong, and David J. Mooney. "Alginate: Properties and Biomedical Applications." *Progress in Polymer Science* 37, no. 1 (2012): 106–126.

Lewin, Ralph A., and Robert A. Andersen. "Algae." In: *Encyclopædia Britannica*, 2018. https://www.britannica.com/science/algae.

Li, Bo, Fei Lu, Xinjun Wei, and Ruixiang Zhao. "Fucoidan: Structure and Bioactivity." *Molecules* 13, no. 8 (2008): 1671–1695.

Lji, P.A., and M.M. Kadam. "Prebiotic Properties of Algae and Algae-Supplemented Products." In: Herminia Domínguez (ed.) *Functional Ingredients from Algae for Foods and Nutraceuticals*, 658–670. Woodhead Publishing, 2013. doi: https://doi.org/10.1533/9780857098689.4.658.

Luo, Xuan, Peng Su, and Wei Zhang. "Advances in Microalgae-Derived Phytosterols for Functional Food and Pharmaceutical Applications." *Marine Drugs* 13, no. 7 (2015): 4231–4254.

Lynch, M.B., T. Sweeney, J.J. Callan, J.T. O'Sullivan, and J.V. O'Doherty. "The Effect of Dietary Laminaria Derived Laminarin and Fucoidan on Intestinal Microflora and Volatile Fatty Acid Concentration in Pigs." *Livestock Science* 133, no. 1–3 (2010): 157–160.

Machu, Ludmila, Ladislava Misurcova, Jarmila Vavra Ambrozova, Jana Orsavova, Jiri Mlcek, Jiri Sochor, and Tunde Jurikova. "Phenolic Content and Antioxidant Capacity in Algal Food Products." *Molecules* 20, no. 1 (2015): 1118–1133.

Madeira, Marta S., Carlos Cardoso, Paula A. Lopes, Diogo Coelho, Cláudia Afonso, Narcisa M. Bandarra, and José A.M. Prates. "Microalgae as Feed Ingredients for Livestock Production and Meat Quality: A Review." *Livestock Science* 205 (2017): 111–121.

Makkar, Harinder P.S., Gilles Tran, Valérie Heuzé, Sylvie Giger-Reverdin, Michel Lessire, François Lebas, and Philippe Ankers "Seaweeds for Livestock Diets: A Review." *Animal Feed Science and Technology* 212 (2016): 1–17.

Markou, Giorgos, Irini Angelidaki, and Dimitris Georgakakis. "Microalgal Carbohydrates: An Overview of the Factors Influencing Carbohydrates Production, and of Main Bioconversion Technologies for Production of Biofuels." *Applied Microbiology and Biotechnology* 96, no. 3 (2012): 631–645.

Mattio, Lydiane, Mayalen Zubia, Ben Loveday, Estelle Crochelet, Nathalie Duong, Claude E. Payri, Ranjeet

Bhagooli, and John J. Bolton. "Sargassum (Fucales, Phaeophyceae) in Mauritius and Réunion, Western Indian Ocean: Taxonomic Revision and Biogeography Using Hydrodynamic Dispersal Models." *Phycologia* 52, no. 6 (2013): 578–594.

Mcdermid, Karla J., Brooke Stuercke, and Owen J. Haleakala. "Total Dietary Fiber Content in Hawaiian Marine Algae." *Botanica Marina* 48, no. 5–6 (2005). doi:10.1515/bot.2005.057.

Misurcova, Ladislava. "Chemical Composition of Seaweeds." In: Se-Kwon Kim (ed.) *Handbook of Marine Macroalgae: Biotechnology and Applied Phycology*, 171–192. 1st ed. John Wiley & Sons, Ltd, 2012. doi:10.1002/9781119977087.ch7.

Mišurcová, Ladislava, Jana Orsavová, and Jarmila Vávra Ambrožová. "Algal Polysaccharides and Health." *Polysaccharides* (2014): 1–29. doi:10.1007/978-3-319-03751-6_24-1.

Munir, N., N. Sharif, S. Naz, and F. Manzoor. "Algae: A Potent Antioxidant Source." *SKY Journal of Microbiology Research* 1 (2013): 22–31. http://skyjournals.org/sjmr/pdf/2013pdf/Apr/Munir%20et%20al%20pdf.pdf.

Nakazono, Satoru, Kichul Cho, Shogo Isaka, Ryogo Abu, Takeshi Yokose, Masakazu Murata, Mikinori Ueno, Katsuyasu Tachibana, Katsuya Hirasaka, Daekyung Kim, and Tatsuya Oda. "Anti-Obesity Effects of Enzymatically-Digested Alginate Oligomer in Mice Model Fed a High-Fat-Diet." *Bioactive Carbohydrates and Dietary Fibre* 7, no. 2 (2016): 1–8.

Nayar, Sasi, and Kriston Bott. "Current Status of Global Cultivated Seaweed Production and Markets." *World Aquaculture* 45, no. 2 (2014): 32–37. https://www.researchgate.net/publication/265518689_Current_status_of_global_cultivated_seaweed_production_and_markets.

Nicoletti, Marcello. "Microalgae Nutraceuticals." *Foods* 5, no. 3 (2016): 54.

Nuño, K., A. Villarruel-López, A.M. Puebla-Pérez, E. Romero-Velarde, A.G. Puebla-Mora, and F. Ascencio. "Effects of the Marine Microalgae *Isochrysis galbana* and *Nannochloropsis oculata* in Diabetic Rats." *Journal of Functional Foods* 5, no. 1 (2013): 106–115.

O'Sullivan, Laurie, Brian Murphy, Peter Mcloughlin, Patrick Duggan, Peadar G. Lawlor, Helen Hughes, and Gillian E. Gardiner. "Prebiotics from Marine Macroalgae for Human and Animal Health Applications." *Marine Drugs* 8, no. 7 (2010): 2038–2064.

Pal, Archana, Mohit Chandra Kamthania, and Ajay Kumar. "Bioactive Compounds and Properties of Seaweeds – A Review." *OALibJ* 01, no. 4 (2014): 1–17.

Paxman, Jenny R., J.C. Richardson, Peter W. Dettmar, and Bernard M. Corfe. "Alginate Reduces the Increased Uptake of Cholesterol and Glucose in Overweight Male Subjects: A Pilot Study." *Nutrition Research* 28, no. 8 (2008): 501–505.

Pérez, María José, Elena Falqué, and Herminia Domínguez. "Antimicrobial Action of Compounds from Marine Seaweed." *Marine Drugs* 14, no. 3 (2016): 52.

Pierre, G., C. Delattre, C. Laroche, and Philippe Michaud. "Galactans and Its Applications." *Polysaccharides* (2014): 1–37. doi:10.1007/978-3-319-03751-6_69-1.

Pooja, Shetty. "Algae Used as Medicine and Food-A Short Review." *Journal of Pharmaceutical Sciences and Research* 6, no. 1 (2014): 33–35.

Praiboon, J., A. Chirapart, Y. Akakabe, O. Bhumibhamon, and T. Kajiwara. "Physical and Chemical Characterization of Agar Polysaccharides Extracted from the Thai and Japanese Species of Gracilaria." *ScienceAsia* 32, s1 (2006): 11–17.

Prasad, M.P., Sushant Shekhar, and A.P. Babhulkar. "Antibacterial Activity of Seaweed (*Kappaphycus*) Extracts against Infectious Pathogens." *African Journal of Biotechnology* 12 (2013): 2968–2971.

Rajapakse, Niranjan, and Se-Kwon Kim. "Nutritional and Digestive Health Benefits of Seaweed." In: Se-Kwon Kim (ed.) *Advances in Food and Nutrition Research* (Vol. 64), 17–28. Elsevier, 2011. doi: 10.1016/B978-0-12-387669-0.00002-8.

Ranga Rao, A., G. Deepika, G.A. Ravishankar, R. Sarada, B.P. Narasimharao, L. Bo, and Y. Su. "Industrial Potential of Carotenoid Pigments from Microalgae: Current Trends and Future Prospects." *Critical Reviews in Food Science and Nutrition* 0 (2018): 1–22.

Ranga Rao, A., H.N. Sindhuja, S.M. Dharmesh, K. Udaya Sankar, R. Sarada, and G.A. Ravishankar. "Effective Inhibition of Skin Cancer, Tyrosinase and Antioxidative Properties by Astaxanthin and Astaxanthin Esters-fromgreen Alga *Haematococcus pluvialis*." *Journal of Agriculture and Food Chemistry* 61 (2013): 3842–3851.

Rhein-Knudsen, Nanna, Marcel Tutor Ale, and Anne S. Meyer. "Seaweed Hydrocolloid Production: An Update on Enzyme Assisted Extraction and Modification Technologies." *Marine Drugs* 13, no. 6 (2015): 3340–3359.

Rioux, L.-E., S.L. Turgeon, and M. Beaulieu. "Characterization of Polysaccharides Extracted from Brown Seaweeds." *Carbohydrate Polymers* 69, no. 3 (2007): 530–537.

Sadally, Shamimtaz Bibi, Nawsheen Taleb Hossenkhan, and Ranjeet Bhagooli. "Microalgal Distribution, Diversity and Photo-Physiological Performance across Five Tropical Ecosystems around Mauritius Island." *Western Indian Oceanographical Journal of Marine Science* 15, no. 1 (2016): 49–68.

Schils, Tom, Eric Coppejans, Heroen Verbruggen, Olivier De Clerck, and Frederik Leliaert. "The Marine Flora of Rodrigues (Republic of Mauritius, Indian Ocean): An Island with Low Habitat Diversity or One in the Process of Colonization?." *Journal of Natural History* 38, no. 23 (2004): 3059–3076.

Seca, Ana, and Diana Pinto. "Overview on the Antihypertensive and Anti-Obesity Effects of Secondary Metabolites from Seaweeds." *Marine Drugs* 16, no. 7 (2018): 237.

Singh, Jasvinder, and Rakesh Chandra Saxena. "An Introduction to Macroalgae: Diversity and Significance." In: Se-Kwon Kim (ed.) *Handbook of Marine Microalgae Biotechnology Advances*, 11–24. Academic Press, 2015.

Škrovánková, Soňa. "Seaweed Vitamins as Nutraceuticals.". *Advances in Food and Nutrition Research* 64 (2011): 357–369.

Sonani, R.R., R.P. Rastogi, and D. Madamwar. "Natural Antioxidants From Algae: A Therapeutic Perspective." In: Rajesh Prasad Rastogi, Datta Madamwar, and Ashok Pandey (eds) *Algal Green Chemistry*, 91–120. Elsevier, 2017.

Spicer, S.E., J.M.M. Adams, D.S. Thomas, J.A. Gallagher, and Ana L. Winters. "Novel Rapid Method for the Characterisation of Polymeric Sugars from Macroalgae." *Journal of Applied Phycology* 29, no. 3 (2017): 1507–1513.

Stiger-Pouvreau, V., N. Bourgougnon, and E. Deslandes. "Carbohydrates From Seaweeds." In: Joel Fleurence, and Ira Levine (eds) *Seaweed in Health and Disease Prevention*, 223–274. Academic Press, 2016. doi:10.1016/b978-0-12-802772-1.00008-7.

Suganya, T., M. Varman, H.H. Masjuki, and S. Renganathan. "Macroalgae and Microalgae as a Potential Source for Commercial Applications along with Biofuels Production: A Biorefinery Approach." *Renewable and Sustainable Energy Reviews* 55 (2016): 909–941.

Sun, Z., and F. Chen. "Evaluation of the Green Alga *Chlorella pyrenoidosa* for Management of Diabetes." *Journal of Food and Drug Analysis* 20 (2012): 246–249.

Thomas, Noel Vinay, and Se-Kwon Kim. "Potential Pharmacological Applications of Polyphenolic Derivatives from Marine Brown Algae." *Environmental Toxicology and Pharmacology* 32, no. 3 (2011): 325–335.

Toyomizu, M., K. Sato, H. Taroda, T. Kato, and Y. Akiba. "Effects of Dietary Spirulina on Meat Colour in Muscle of Broiler Chickens." *British Poultry Science* 42, no. 2 (2001): 197–202.

Unnikrishnan, P.S., K. Suthindhiran, and M.A. Jayasri. "Antidiabetic Potential of Marine Algae by Inhibiting Key Metabolic Enzymes." *Frontiers in Life Science* 8, no. 2 (2015): 148–159.

Venkatesan, Jayachandran, Panchanathan Manivasagan, and Se-Kwon Kim. "Marine Microalgae Biotechnology: Present Trends and Future Advances." In: Se-Kwon Kim (ed.) *Handbook of Marine Microalgae Biotechnology Advances*, 1–9. Academic Press, 2015.

Wan-Loy, Chu, and Phang Siew-Moi. "Marine Algae as a Potential Source for Anti-Obesity Agents." *Marine Drugs* 14, no. 12 (2016): 222.

Watanabe, Fumio, Yukinori Yabuta, Tomohiro Bito, and Fei Teng. "Vitamin B12-Containing Plant Food Sources for Vegetarians." *Nutrients* 6, no. 5 (2014): 1861–1873.

Wells, Mark L., Philippe Potin, James S. Craigie, John A. Raven, Sabeeha S. Merchant, Katherine E. Helliwell, Alison G. Smith, Mary Ellen Camire, and Susan H. Brawley. "Algae as Nutritional and Functional Food Sources: Revisiting Our Understanding." *Journal of Applied Phycology* 29, no. 2 (2017): 949–982.

Yaakob, Zahira, Ehsan Ali, Afifi Zainal, Masita Mohamad, and Mohd Sobri Takriff. "An Overview: Biomolecules from Microalgae for Animal Feed and Aquaculture." *Journal of Biological Research* 21, no. 1 (2014): 6.

Yan, L., and I.H. Kim. "Effects of Dietary ω-3 Fatty Acid-Enriched Microalgae Supplementation on Growth Performance, Blood Profiles, Meat Quality, and Fatty Acid Composition of Meat in Broilers." *Journal of Applied Animal Research* 41, no. 4 (2013): 392–397.

Yen, Hong-Wei, I.-Chen Hu, Chun-Yen Chen, Shih-Hsin Ho, Duu-Jong Lee, and Jo-Shu Chang. "Microalgae-Based Biorefinery – From Biofuels to Natural Products." *Bioresource Technology* 135 (2013): 166–174.

Zamani, S., S. Khorasaninejad, and B. Kashefi. "The Importance Role of Seaweeds of Some Characters of Plant." *International Journal of Agriculture and Crop Sciences* 5 (2013): 1789–1793.

Zhu, L.D., Z.H. Li, and E. Hiltunen. "Strategies for Lipid Production Improvement in Microalgae as a Biodiesel Feedstock." *BioMed Research International* 2016 (2016): 8792548.

Zubia, Mayalen, Claude Payri, and Eric Deslandes. "Alginate, Mannitol, Phenolic Compounds and Biological Activities of Two Range-Extending Brown Algae, Sargassum Mangarevense and Turbinaria Ornata (Phaeophyta: Fucales), from Tahiti (French Polynesia)." *Journal of Applied Phycology* 20, no. 6 (2008): 1033–1043.

20 Angiogenic Actions of Anionic Polysaccharides from Seaweed

Celina Maria Pinto Guerra Dore, Monique G. das Chagas Faustino Alves, Luiza Sheyla Evenni P. Will Castro, Luciana Guimaraes Alves Figueira, and Edda Lisboa Leite

CONTENTS

Abbreviations 221
Introduction 222
Algal Polysaccharides: Structure and Pharmacological Actions 222
Angiogenesis and Tubulogenesis 223
Action of Anionic Polysaccharides on Angiogenesis 224
Conclusions 227
References 227

BOX 20.1 SALIENT FEATURES

Marine organisms such as brown and red seaweeds are rich sources of sulfated polysaccharides fucans or fucoidans and galactans, respectively. These polysaccharides are structural components of the extra-cellular matrix. Their structures vary among species of seaweeds. Usually, fucoidans are composed essentially of sulfated α-L-fucose. They may also contain galactose, mannose, xylose, uronic acids and acetyl groups. The biological activities of these polysaccharides depend on the degree of sulfation, molecular structure, types of constituent monosaccharides and degree of branching. Sulfated galactans of red algae contain sulfated β-galactoses and a regular repetitive structure. However, structural variations in sulfated galactans occur among different species of red algae. Anti-inflammatory, anticoagulant, immune-modulating, antioxidant, antiproliferative, antitumor and antiangiogenic actions are described for these polymers. The blood vessel formation can occur as vasculogenesis and angiogenesis, in which different cellular mechanisms such as migration and differentiation are involved. In this chapter, the effect of anionic polysaccharides extracted from brown and red algae, fucans and galactans, on angiogenesis are discussed. Angiogenesis is the formation of new capillary vessels from pre-existing vasculature, being regulated by complex interactions between stimulatory and inhibitory factors, including growth factors, cytokines, proteolytic enzymes, integrins and extra-cellular matrix components. The lack of control in angiogenesis has been implicated in the growth of tumors and metastasis. The action of polysaccharides from brown and red seaweed may directly interfere with the binding of growth factors to their specific cognate receptors, such as vascular endothelial growth factor (VEGF). These polysaccharides could modulate the activity of heparin-binding vascular such as basic fibroblast growth factors (bFGF) interfering with heparan sulfate, a proteoglycan, during microvessel formation. We used chick-embryo chorioallantoic membrane (CAM) assay for the *in vivo* study of angiogenesis using these polysaccharides.

ABBREVIATIONS

bFGF: basic fibroblast growth factors
CAM: chorioallantoic membrane
CD34+: cluster of differentiation 34
FGF2: fibroblast growth factors 2
FRF: fraction rich in fucans
GAG: glycosaminoglycans
GFP08: agaran-type polysaccharide
HUVECs: human umbilical vein endothelial cells
LMWF: low molecular weight fucoidan

MMWF1: molecular weight fucoidan (15–20 kDa)
MMWF2: fucoidan with (30 kDa)
MMWF3: fucoidan (30 kDa)
L.s.-P: parental fucoidan of *Laminaria saccharina*
L.s.-1.0: sulfated polysaccharides eluted by DEAE-Sephacel by 1.0 M NaCl
L.s.-1.25: sulfated polysaccharides eluted by DEAE-Sephacel by 1.1.25 M NaCl
NF: natural fucoidan
OSF: oversulfated fucoidan
PSV1: purified fucans of *Sargassum vulgare* precipitated with 1,0 v of acetone
RAEC: rabbit aortic endothelial cells
Sarg: polysaccharides *Sargassum integerrimum*
SP1: polysaccharides (fucans) of *Sargassum vulgare* precipitated with 1,0 v acetona
SPs: sulfated polysaccharides of *Sargassum stenophyllium*
TF: tissue factor
VEGF: vascular endothelial growth factor
VEGFR2: vascular endothelial growth factor receptor 2

INTRODUCTION

The first study about angiogenesis, or new blood vessel formation, was conducted by John Hunter in 1794. His observations suggested a proportionality between vascularity and metabolic requirements in health and disease (Choi and Moon 2018). Subsequently, Judah Folkman in 1971 hypothesized that tumor growth is angiogenesis-dependent. Tumor angiogenesis provides not only oxygen and nutrients for tumor cells but also the necessary anchorage to facilitate tumor metastasis (Choi and Moon 2018). Thus, the control of angiogenesis could lead to cancer therapies, stimulating intensive research in the field. Antiangiogenic therapy has become an effective strategy inhibiting tumor growth.

The long history of the use of algae for a variety of objectives has led to many studies exploring their constituents, which are distinctly different from that of land plants. Marine algae are organisms that are rich sources of sulfated polysaccharides, sulfated galactans from red algae and fucoidans or/and fucans from brown algae (Folkman et al. 1983). Thus, there is a growing interest in developing new pharmacological strategies using algal constituents (Groth *et al.* 2009). In our laboratory, we have investigated the influence of anionic polysaccharides from brown fucans or fucoidans from *Sargassum vulgare*, *Lobophora variegate* and sulfated galactans from red algae of *Amansia multifida* and their action on angiogenesis and antitumoral effects.

ALGAL POLYSACCHARIDES: STRUCTURE AND PHARMACOLOGICAL ACTIONS

The quantity of carbohydrate synthesized by plants in the oceans is very high and probably exceeds that found on land and in fresh water. The total polysaccharide concentrations in the seaweed species of interest range from 4% to 76% of the dry weight. They are polymers of sugars, *monosaccharides*, linked by glycosidic bonds. Brown algae contain large amounts of polysaccharides, fucoidan or/and fucans, alginate and laminarin that are the major storage carbon compounds that have many novel biological effects on animals and human cells (Anastasakis et al. 2011). Fucoidan and/or fucan is a fucose-containing sulfated polysaccharide that is extracted from the cell walls of several brown macroalgae. Percival and McDowell (1967) and Painter et al. (1993) observed that polymers containing sulfated polysaccharides named fucoidan predominantly occur in *Fucus vesiculosus*. Usually, theses polymers are composed essentially of sulfated α-L-fucose residues. They may also contain galactose, mannose, xylose, uronic acids and acetyl groups. However, Leite et al. (1998), studying a polysaccharide of brown seaweed *Spatoglossum scroederi*, observed that it was composed of a core of $\beta(1-3)$ glucuronic acid–containing oligosaccharide of 4.5 kDa with branches at C_4 of the fucose chains $\alpha(1-3)$-linked. The fucose was mostly substituted at C_4 with sulfate groups and at C_2 with chains of $\beta(1-4)$ xylose, which, in turn, was also partially sulfated (Figure 20.1).

Dore et al. (2013) described two anionic polysaccharides from brown seaweed *Sargassum vulgare* with antiangiogenic action. The crude extract of polysaccharides was fractionated with acetone to obtain four fractions. The 1,0 v fraction was named as SV1 and was fractionated by gel-filtration chromatography, and the eluted with 160 kDa was named as PSV1. These polysaccharides comprised L-fucose, D-mannose, D-galactose, D-xylose, D-glucose and D-glucuronic acid. Besides their well-tested anticoagulant and antithrombotic activities, some fucoidans also act on the inflammation and immune systems. We have shown that the different profiles of biological activities exhibited by these polysaccharides depend on variations of their structural features (Medeiros et al. 2008; Castro et al. 2015). NMR spectra of the polysaccharide were too complex to allow direct structure elucidation, probably due to structural heterogeneity. Fucoidans appear to play a role in algal cell wall organization and could be involved in the cross-linkage of alginate and cellulose (Berteau and Mulloy 2003). Polysaccharides extracted from brown seaweed also exhibit antioxidant, antiproliferative and

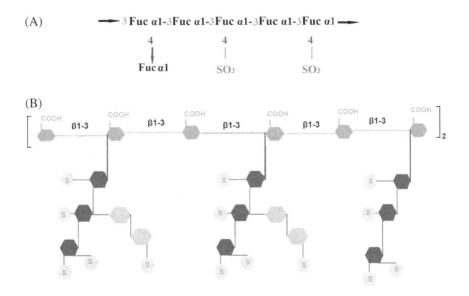

FIGURE 20.1 Two types of fine structures of fucoidans/fucans from (A) *Fucus vesiculosus* (Painter et al. 1983) and (B) *Spatoglossum scroederi*. (Adapted from Leite et al. 1998.)

antivasculogenic activity (Cumashi et al. 2007; Dias et al. 2008; Dore et al. 2013). The anticancer effects of fucoidans have been shown to vary depending on their structure, charge density, distribution of sulfate group and purity. Further, they can target multiple receptors or signaling molecules in various cell types, including tumor cells and immune cells. Fucoidan-mediated apoptosis of cancer cells would involve up-regulation or down-regulation of multiple signaling pathways. Signaling pathways leading to the apoptosis of cancer cells by fucoidan have not been fully understood (Kwak 2014).

ANGIOGENESIS AND TUBULOGENESIS

The formation of blood vessels can occur via several processes, such as vasculogenesis and angiogenesis, in which different cellular mechanisms such as migration and differentiation are involved (Risau 1997). *Vasculogenesis* is a process defined as the development of blood vessels from *in situ* differentiation of endothelial cell progenitor cells (angioblasts and hemangioblasts). These precursor cells are recruited from mesoderm areas adjacent to the embryo and originated by local cell division, organizing blood islets and establishing a primordial vascular plexus. *Angiogenesis* is the formation of new capillary vessels from pre-existing vasculature, being regulated by complex interactions between stimulatory and inhibitory factors, including growth factors, cytokines, proteolytic enzymes, integrins and extra-cellular matrix components (Bürgermeister et al. 2002; Kalluri and Zeisberg 2006). The importance of angiogenesis is based on the fact that this process is the key player in a series of physiological events such as ovulation, corpus luteum formation and wound healing and in diseases such as chronic arthritis, psoriasis, diabetic proliferative retinopathy, tumor growth and metastatic dissemination (Folkman 2008), which makes all the mechanisms that participate in angiogenesis promising targets of therapy. Tumors with strong angiogenic activity are related to the lower survival rates of patients (Giatromanolaki et al. 2004). Antiangiogenic substances produced in the tumor are able to halt the development of metastasis in some types of tumors, which confirmed the previous observations related to the increase of the aggressiveness of the metastases after the excision of the primary tumor (O'Relly et al. 1994).

One of the most specific and important factors involved in angiogenesis is vascular endothelial growth factor (VEGF), which may specifically bind to various endothelial cell membrane receptors (Ferrara 2002). In addition to stimulating angiogenesis, VEGF is also important for maintaining the integrity and permeability of blood vessels (Keck et al. 1989; Senger 2010). Members of the VEGF family stimulate cellular responses by binding to VEGFR tyrosine kinase receptors on the cell surface, causing them to dimerize and become activated via transphosphorylation (Shibuya 2013; Simons et al. 2016). Fibroblast growth factors (FGF) are also a family of factors involved in angiogenesis and wound healing. There is evidence supporting its action on embryonic development. FGFs are of great importance in the proliferation and differentiation of a wide variety of cells and tissues (Ferrara 2002; Tao et al. 2017).

ACTION OF ANIONIC POLYSACCHARIDES ON ANGIOGENESIS

Several in vitro and *in vivo* experimental methods have been adopted to study of the steps of angiogenesis, as well as for the determination of the influence of test compounds on this process (Tahergorabi and Khazaei 2012). Cumashi et al. (2007) evaluated the antiangiogenic action of polysaccharides from nine species: *Ascophyllum nodosum, Fucus evanescens, Fucus distichus, Fucus serratus, Fucus spiralis, Laminaria saccharina* and *Laminaria digitata*. The polysaccharides from these seaweeds were fucoidans composed by L-fucose, uronic acids and sulfate, which have shown their efficacy in inhibiting tubulogenesis in the *in vitro* assay of human umbilical vein endothelial cells (HUVEC). The HUVECs reorganize into tube-like structures when plated on matrigel *in vitro* tubulogenesis. This tube formation was blocked by 99% when challenged with 100 µg/mL offucoidans from *L. saccharina, L. digitata, F. evanescens, F. serratus,* and *F. distichus*. Fucoidans from *F. spiralis* and *A. nodosum* partially inhibited the tubulogenesis at 100 µg/mL and so were less effective for antiangiogenic activity, whereas the polysaccharides from *C. okamuranus* and *F. vesiculosus* were not able to impair the tubulogenesis. The difference of activity of these fucoidans in angiogenesis can be explained by the variable structure of the polysaccharides. Cumashi et al. (2007) suggested that the sulfate content and the presence of 2-O-α-D-glucuronyl substituents along the linear polysaccharide backbone are important factors for the antiangiogenic action of these fucoidans.

The antiangiogenic potential of sulfated polysaccharides from brown algae *Laminaria saccharina* was evaluated by Croci et al. (2011). In this study, the polysaccharide L.s.-P (Parental or crudefucoidan) was fracionated by ion-exchange chromatography to obtain two purified fractions: fraction L.s.-1.0 (mannoglucuronofucans sulfated) and L.s.-1.25 (fucans sulfated). L.s.-P and L.s.-1.25 showed inhibitory action on matrigel HUVEC capillary-like tube formation in all tested concentrations (1, 10 and 100 µg/mL). When polysaccharide fractions (100 µg/mL) were administered with bFGF, a modulator of angiogenesis, before plating the HUVEC onto matrigel, L.s.-P and L.s.-1.25, polysaccharides prevented around 99% of the bFGF-induced HUVEC tube formation, while L.s.-1.0 had no effect. These authors suggested that the particular chemical structure and conformation of fucoidans suggested the possibility that these polysaccharides may directly interfere with the binding of growth factors to their specific cognate receptors. L.s.-P and L.s.-1.25 interfered with bFGF in the interaction with receptor of BALB/c 3T3 fibroblasts. When polysaccharides were added to fibroblast suspensions on bFGF-coated plates, the adhesion of fibroblasts to bFGF was completely overcome. Sulfated polysaccharide from *L. saccharina* was studied for its influence on endothelial cells *in vitro* and also to evaluate its influence on tumor growth and tumor-related angiogenesis under *in vivo*. B6 mice were subcutaneously injected with matrigel plugs enriched with melanoma B16-F10 cells and 100 µg of L.s.-P polysaccharides and its purified fractions. The result reported was a reduction in angiogenesis in the tumor, microvessel density and tumor weight without any toxicity at the tested concentration.

The anionic polysaccharides from the brown seaweed *Sargassum stenophyllium*, which is composed of L-fucose, manose, galactose, xylose, glucuronic acid and sulfate, altered the vessel formation in the vascular membrane of a chick yolk sac. The yolk sac membrane treated with this fucoidan at 6-1500 µg/disk exhibited formation of fewer vessels compared to control (92 vessels). Similarly, a higher antivasculogenic activity was observed at the same concentration compared to hydrocortisone (Dias et al. 2008). In the presence of bFGF, the fucoidan fully prevented the vessels' formation due to its polyanionic characteristics, which are similar to those seen in the heparin; it was suggested that this polysaccharide should modulate the activity of heparin-binding vascular such as bFGF, interfering with heparan sulfate, a proteoglycan, during microvessel formation.

The sulfated polysaccharides from brown algae *Sargassum integerrimum* composed of L-fucose (SPs) exhibited a growth-inhibiting effect on HUVECs in a time- and dose-dependent manner. Moreover, this fucoidan presented inhibitory action on HUVEC migration in scratch-wound assay and transwell assay in which a concentration-dependent reduction in the number of migrated cells was observed. In a zebrafish embryo model, the SPs promoted a reduction in the total number of mature intersegmental vessels and subintestinal vessels when treated for 48 h compared with the control (Liu et al. 2016). A matrigel tube formation assay by rabbit aortic endothelial cells (RAEC) showed inhibition of capillary tube formation compared to the inhibitory action of heparin (positive control). Anionic polysaccharides did not interfere in RAEC cell proliferation. However, SV1 and PSV1 at 25, 50 and 100 µg/mL inhibited around 70% VEGF secretion in cell medium by RAEC (see Figure 20.2). It is important to consider that tumor vasculature is not a simple nutrient supply line for tumors; it governs the pathophysiology of the tumor and thus growth, metastasis and response to various

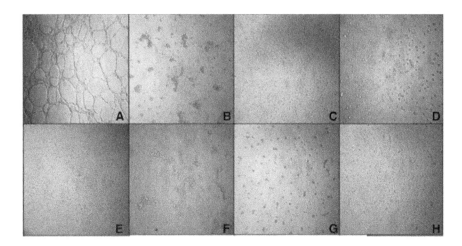

FIGURE 20.2 (See color insert.) Effects of SV1 and PSV1 on capillary tube formation of rabbit aortic endothelial cells (RAEC). RAECs (1 × 10⁵ per well) in medium containing 10% serum were seeded into Matrigel pre-coated 24-well plates and treated with different concentrations of SV1 and PSV1 (A—Control; B—Heparin (10 µg/mL); C—SV1 25; D —SV1 50; E—SV1 100; F—PSV1 25; G—PSV1 50; H — PSV1 100). Representative phase contrast photomicrographs (100× magnification). (Adapted from Dore et al. *Microvas* Res. 2013. 88:12–18.)

therapies. In this context, PSV1 and SV1 polysaccharides from *S. vulgare* stand out as being compounds that act directly *in vivo* and *in vitro* on the inhibition of angiogenesis, which is related to the ability of these polymers to inhibit the secretion of VEGF in endothelial cells (Dore et al. 2013).

Fraction rich in fucans (FRF) with 34.1% of sulfate obtained from *Lobophora variegata* showed antiangiogenic action in chorioallantoic membrane assay. It was determined that FRF at 10, 100 and 1000 µg egg−1 caused inhibition of around 50% at the higher concentration tested as evaluated by parameters such as the disk area, adjacent area and amount of neo-vessels to show relative inhibition compared to the blank. Spirolactone, phenanthrolin and heparin at 10 µg egg⁻¹ were used as positive controls as observed in Figure 20.3 (Castro et al. 2015). Research was done with chemically modified polysaccharide (fucans/fucoidans). Thus, Koyanagi et al. 2003 evaluated the antiangiogenic potential of a natural (NF) and oversulfated (OSF) fucoidan from the brown seaweed *Fucus vesiculosus*. Commercial fucoidan was fractionated by size, and a fraction between 100 and 130 kDa was named as natural fucoidan (NF), which was chemically sulfated and named as oversulfated fucoidan (OSF). NF and OSF possess 32.6% and 56.7% of sulfate content, respectively. NF at 100 µg/mL inhibited the VEGF-induced HUVEC proliferation, while OSF at 10 µg/mL already inhibited cell proliferation. Only OSF at 100 µg/mL showed inhibitory action on VEGF-induced HUVEC migration. This activity was caused by an altered signal transduction decreasing the VEGF/VEGFR2 phosphorylation. The sulfate introduced in OSF may promote a spatial orientation of negative charges related to the potency of VEGF binding. NF and OSF at 5 mg/kg administered intravenously also suppressed neovascularization in mice in which Sarcoma 180 cells were implanted.

Further evidence is emerging that ionic charges and molecular weight may be important in these processes. Anionic polysaccharide of low molecular weight (LMW) from brown seaweed shows proangiogenic activity. LWM fucoidan alone or in combination with bFGF at the site of the injury improved revascularization of ischemic areas (Luyt et al. 2003). These results were confirmed by phosphorylase activity, showing muscle regeneration in rats treated with the combination of bFGF and LMW fucoidan (Chabut et al. 2003). Matsubara et al. (2005) evaluated the action of fucoidans from *Laminaria japonica* of middle molecular weight of 15–30 kDa. The homogenate from seaweed was mildly hydrolyzed at 60°C with 0.01 HCl for 60 min. Then three fractions were obtained MMWF1 (15–20 kDa), MMWF2 with (30 kDa) and MMWF3 (30 kDa) with 8.2%, 32.2% and 33.2% of sulfate, respectively. MMWF-1 at 200 µg/mL had no effect on HUVEC tube formation on reconstituted basement membrane gel. However, MMWF-3 significantly showed inhibition at the same concentration. However, the inhibitory action was not observed at 50 µg/mL concentration. The authors opine that fucoidans of molecular weight of 20-30 kDa is critical for exhibiting antiangiogenic or proangiogenic potential (Matsubara et al. 2005). MMWF1 and MMWF3 at 100 µg/ml also stimulated VEGF-induced migration of HUVEC but

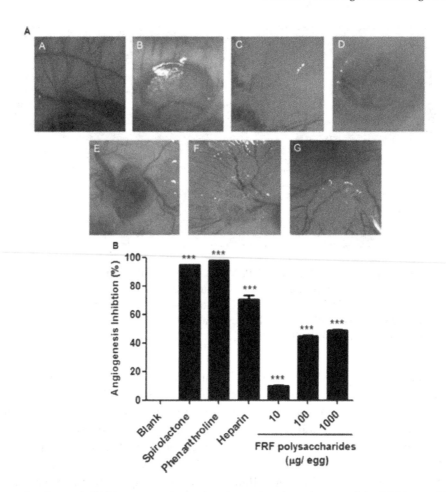

FIGURE 20.3 (See color insert.) CAM assay to determine the antiangiogenic potential of FRF. A: Chorioallantoic membrane assay (CAM) of embryonated chicken eggs: blank A agarose; negative controls: B phenantroline 10 µg egg^{-1} and C spironolactone 10 µg egg^{-1}. D Heparin 10 µg egg^{-1} (positive control). FRF polysaccharides 10 µg egg^{-1}, E 100 µg egg^{-1}, and F 1000 µg egg^{-1} G. B: Antiangiogenic activity was obtained by score of quantitative vessels in the region of application of polysaccharides, with 0 = no antiagiogenic effect; 0.5–1 = very low to medium antiagiogenic effect (no capillary-free area); 1 = medium antiagiogenic effect (reduction in the capillary density: action equivalent to double that of disk area); and 2 = hard antiangiogenic effect (more than double that of disk area). ***$p<0.05$. (Adapted from Castro et al. J Appl Phycol. 2015. 27:1315–1325.)

had no effect on HUVEC proliferation at 10-200 µg/mL. Ustyuzhanina et al. (2014) suggested that high-molecular-weight fucoidans (MW >30 kDa) of *A. nodosum* with a high sulfate content are antiangiogenic, whereas low molecular weight fucoidans (MW <15 kDa) from the same source are proangiogenic. However, it was not evaluated according to their sulfate content and sulfation pattern, both of which may have a stronger influence on their effect on angiogenesis. The influence of other fine structural details of fucoidans on angiogenesis remains to be established.

An agaran-type polysaccharide (GFP08) from the red seaweed *Grateloupia filicina* is composed of sulfated galactan and anhydrogalactose and not fucose, with small quantities of xylose and pyruvic acid. The GFP08 blocked the formation of tubes in Matrigel in vitro. Furthermore, suppression in the CAM neovascularization in eggs treated with GFP08 at 2–50 µg egg^{-1} was evident in *ex vivo* chorioallantoic membrane (CAM) assay, wherein the GFP08 did not inhibit HUVEC proliferation induced by EGF, bFGF and VEGF. GFP08 caused a downregulation on the protein expression levels of TF (tecidual factor) in HUVEC (Yu et al. 2012). The literature describe the tissue factor (TF) as a factor involved in regulating angiogenesis and as new therapeutic target for cancer (Fernandez and Rickles 2002; Kasthuri et al. 2009).

The red seaweed *Amansia multifida* is a source of sulfated galactans composed of minor quantities of glucose, xylose, mannose and glucuronic acid. The fraction (FT) with antiangiogenic activity. FT, total factor or crude polysaccharides presented antiangiogenic activity in all tested concentrations (Souza et al. 2012).

CONCLUSIONS

Bioproducts from algae as fucans, galactans and other anionic polysaccharides have attracted considerable attention due to their pharmacological effect. Several *in vitro* and *in vivo* models have indicated the importance of these polymers as antiangiogenic agents in different mechanisms of action. We have elucidated the potent antiangiogenic actions of two anionic polysaccharides, sulfated fucans from brown seaweed *Sargassum vulgare* and *Lobophora variegata*. These results suggest that seaweed fucoidans effectively suppress angiogenesis.

REFERENCES

Anastasakis, K., Ross, A. B., and Jones, J. M. Pyrolysis behaviour of the main carbohydrates of brown macroalgae. *Fuel*. 2011. 90: 598–607.

Berteau, O., and Mulloy, B. Sulfated fucans, fresh perspectives: Structures, functions, and biological properties of sulfated fucans and an overview of enzymes active toward this class of polysaccharide. *Glycobiology*. 2003. 13: 29R–40R.

Bürgermeister, J., Paper, D. H., Vogl, H., Linhardt, R. J., and Franz, G. LaPSvS1, a (1 → 3)-beta-galactan sulfate and its effect on angiogenesis in vivo and in vitro. *Carbohydr Res*. 2002. 337(16): 1459–1466.

Carmeliet, P. Blood vessels and nerves: Common signals, pathways and diseases. *Nat Rev Genet*. 2003. 4: 710–720.

Castro, L. S. E. P. W., de Sousa Pinheiro, T., Castro, A. J. G., da Silva Nascimento Santos, M., Soriano, E. M., and Leite, E. L. Potential anti-angiogenic, antiproliferative, antioxidant, and anticoagulant activity of anionic polysaccharides, fucans, extracted from brown algae *Lobophoravariegata*. *J Appl Phycol*. 2015. 27: 1315–1325.

Chabut, D., Fischer, A.-M., Colliec-Jouault, S., Laurendeau, I., Matou, S., Bonniec, B. L., and Helley, D. Low molecular weight fucoidan and heparin enhance the basic fibroblast growth factor-induced tube formation of endothelial cells through heparan sulfate-dependent α6 overexpression. *Mol Pharmacol*. 2003. 64(3): 696–702.

Choi, H., and Moon, A. Crosstalk between cancer cells and endothelial cells: Implications for tumor progression and intervention. *Arch Pharm Res*. 2018. 41(7): 711–724.

Coultas, L., Chawengsaksophak, K., and Rossant, J. Endothelial cells and VEGF in vascular development. *Glycobiology*. 2012. 22(10): 1343–1352.

Croci, D. O., Cumashi, A., Ushakova, N. A., Preobrazhenskaya, M. E., Piccoli, A., Totani, L., Ustyuzhanina, N. E., Bilan, M. I., Usov, A. I., Grachev, A. A., Morozevich, G. E., Berman, A. E., Sanderson, C. J., Kelly, M., Di Gregorio, P., Rossi, C., Tinari, N., Iacobelli, S., Rabinovich, G. A., and Nifantiev, N. E. Fucans, but not fucomannoglucuronans, determine the biological activities of sulfated polysaccharides from *Laminaria saccharina* brown seaweed. *PLOS ONE*. 2011. 6(2): e17283.

Cumashi, A., Ushakova, N. A., Preobrazhenskaya, M. E., D'Incecco, A., Piccoli, A., Totani, L., Tinari, N., Morozevich, G. E., Berman, A. E., Bilan, M. I., Usov, A. I., Ustyuzhanina, N. E., Grachev, A. A., Sanderson, C. J., Kelly, M., Rabinovich, G. A., Iacobelli, S., and Nifantiev, N. E. A comparative study of the anti-inflammatory, anticoagulant, antiangiogenic, and antiadhesive activities of nine different fucoidans from brown seaweeds. *Glycobiology*. 2007. 17(5): 541–552.

Dias, P. F., Siqueira, J. M., Maraschin, M., Ferreira, A. G., Gagliardi, A. R., and Ribeiro-do-Valle, R. M. A polysaccharide isolated from the brown seaweed *Sargassum stenophyllum* exerts antivasculogenic effects evidenced by modified morphogenesis. *Microvasc Res*. 2008. 75: 34–44.

Dore, C.M.P.G., Alves, M. G. C. F., Will, L. S., Costa, T. G., Sabry, D.A., Rêgo, S. L.A., Accardo, C.M., Rocha, H.A.O., Filgueira, L. G. A., and Leite, E. L. A sulfated polysaccharide, fucans, isolated from brown algae *Sargassum vulgare* with anticoagulant, antithrombotic, antioxidant and anti-inflammatory effects. *Carbohydr Polym*. 2013a. 91(1): 467–475.

Dore, C. M. P. G., Alves, M. G. F., Santos, N. D., Cruz, A. K. M., Câmara, R. B. G., Castro, A. J. G., Alves, L. G., Nader, H. B., and Leite, E. L. Antiangiogenic activity and direct antitumor effect from a sulfated polysaccharide isolated from seaweed. *Microvasc Res*. 2013b. 88: 12–18.

Fernandez, P. M., and Rickles, F. R. Tissue factor and angiogenesis in cancer. *Curr Opin Hematol*. 2002. 9: 401–406.

Ferrara, N. VEGF and the quest for tumour angiogenesis factors. *Nat Rev Cancer*. 2002. 2: 795–803.

Folkman, J. History of angiogenesis. In: *Angiogenesis—An Integrative Approach from Science to Medicine*. Figg WD, Folkman J (Eds). New York: Springer, 2008. pp. 1–14.

Folkman, J. Tumor angiogenesis. Therapeutic implications. *N Engl J Med*. 1971. 285: 1182–1186.

Folkman, J., Langer, R., Linhardt, R. J., Haudenschild, C., and Taylor, S. Angiogenesis inhibition and tumor regression caused by heparin or a heparin fragment in the presence of cortisone. *Science*. 1983. 221: 719–725.

Giatromanolaki, A., Sivridis, E., and Koukourakis, M. I. Tumour angiogenesis: Vascular growth and survival. *APMIS*. 2004. 112(7–8): 431–440.

Groth, I., Grünewald, I. N., and Alban, S. Pharmacological profiles of animal- and nonanimal-derived sulfated polysaccharides—Comparison of unfractionated heparin, the semisynthetic glucan sulfate PS3, and the sulfated polysaccharide fraction isolated from *Delesseria sanguine*. *Glycobiology*. 2009. 19(4): 408–417.

Hunter, J. *The Works of John Hunter, F.R.S*. 2015. Cambridge, UK: Cambridge University Press.

Hunter, J. *A Treatise on the Blood, Inflammation and Gunshot Wounds*. Palmer J. F. (Ed.). Philadelphia: Raswell, Barrington, and Haswell, 1794, 1840. p. 195.

Kalluri, R., and Zeisberg, M. Fibroblasts in cancer. *Nat Rev Cancer*. 2006. 6: 392–401.

Kasthuri, R. S., Taubman, M. B., and Mackman, N. Role of tissue factor in cancer. *J Clin Oncol.* 2009. 27: 4834–4838.

Keck, P. J., Hauser, S. D., Krivi, G., Sanzo, K., Warren, T., Feder, J., and Connolly, D. T. Vascular permeability factor, an endothelial cell mitogen related to PDGF. *Science.* 1989. 246: 1309–1312.

Koyanagi, S., Tanigawa, N., Nakagawa, H., Soeda, S., and Shimeno, H. Oversulfation of fucoidan enhances its anti-angiogenic and antitumor activities. *Biochem Pharmacol.* 2003. 65: 173–179.

Kwak, J. Y. Fucoidan as a marine anticancer agent in preclinical development. *Mar Drugs.* 2014. 12: 851–870.

Liu, G., Kuang, S., Wu, S., Jin, W., and Sun, C. A novel polysaccharide from *Sargassum integerrimum* induces apoptosis in A549 cells and prevents angiogensis *in vitro* and *in vivo. Sci Rep.* 2016. 6: 26722.

Luyt, C. E., Meddahi-Pellé, A., Ho-Tin-Noe, B., Colliec-Jouault, S., Guezennec, J., Louedec, L., Prats, H., Jacob, M. P., Osborne-Pellegrin, M., Letourneur, D., and Michel, J. B. Low-molecular-weight fucoidan promotes therapeutic revascularization in a rat model of critical hindlimb ischemia. *J Pharmacol Exp Ther.* 2003. 305(1): 24–30.

Matsubara, K., Xue, C., Zhao, X., Mori, M., Sugawara, T., and Hirata, T. Effects of middle molecular weight fucoidans on *in vitro* and *ex vivo* angiogenesis of endothelial cells. *Int J Mol Med.* 2005. 15: 695–699.

Medeiros, V. P., Queiroz, K. C., Cardoso, M. L., Monteiro, G. R., Oliveira, F. W., Chavante, S. F., Guimarães, L. A., Rocha, H. A., and Leite, E. L. Sulfated galactofucan from *Lobophoravariegata*: Anticoagulant and anti-inflammatory properties. *Biochemistry.* 2008. 73: 1018–1024.

O'Relly, M. S., Holmgren, L., Shing, Y., Chen, C., Rosenthal, R. A., Moses, M., Lane, W. S., Cao, Y., Sage, E. H., and Folkman, J. Angiostatin: A novel angiogenesis inhibitor that mediates the suppression of metastasis by a Lewis lung carcinoma. *Cell.* 1994. 79: 315–328.

Patankar, M. S., Oehninger, S., Barnett, T., Williams, R. L., and Clark, G. F. A revised structure for fucoidan may explain some of its biological activities. J Biol Chem. 1993. 268: 21770–21776.

Percival, E., and McDowell, R. H. *Chemistry and Enzymology of Marine Algal Polysaccharides.* New York, NY: Academic Press, 1967. p. 219.

Ribatti, D. Chicken chorioallantoic membrane angiogenesis model. *Methods Mol Biol.* 2012. 843: 47–57.

Ribatti, D. J. Judah Folkman, a pioneer in the study of angiogenesis. *Angiogenesis.* 2008. 11(1): 3–10.

Risau, W. Mechanisms of angiogenesis. *Nature.* 1997. 386(6626): 671–674.

Senger, D. R. Vascular endothelial growth factor: Much more than an angiogenesis factor. *Mol Biol Cell.* 2010. 21(3): 377–379.

Shibuya, M. Vascular endothelial growth factor and its receptor system: Physiological functions in angiogenesis and pathological roles in various diseases. *J Biochem.* 2013. 153: 13–19.

Simons, M., Gordon, E., and Claesson-Welsh, L. Mechanisms and regulation of endothelial VEGF receptor signalling. *Nat Rev Mol Cell Biol.* 2016. 17: 611–625.

Soeda, S., Kozako, T., Iwata, K., and Shimeno, H. Oversulfated fucoidan inhibits the basic fibroblast growth factor-induced tube formation by human umbilical vein endothelial cells: Its possible mechanism of action. *Biochim Biophys Acta.* 2000. 1497: 127–134.

Souza, L. A. R., Dore, C. M. P. G., Castro, A. J. G., Azevedo, T. C. G., Oliveira, M. T. B., Moura, M. F. V., Benevides, N. M. B., and Leite, E. L. Galactans from the red seaweed *Amansia multifida* and their effects on inflammation, angiogenesis, coagulation and cell viability. *Biomed Prev Nutrit.* 2012. 2: 154–162.

Tahergorabi, Z., and Khazaei, M. A review on angiogenesis and its assays. *Iran J Basic Med Sci.* 2012. 15: 1110–1126.

Tao, L., Huang, G., Song, H., Chen, Y., and Chen, L. Cancer associated fibroblasts: An essential role in the tumor microenvironment. *Oncol Lett.* 2017. 14(3): 2611–2620.

Ustyuzhanina, N. E., Bilan, M. I., Ushakova, N. A., Usov, A. I., Kiselevskiy, M. V., and Nifantiev, N. E. Fucoidans: Pro- or antiangiogenic agents? *Glycobiology.* 2014. 24(12): 1265–1274.

Yu, Q., Yan, J., Wang, S., Ji, L., Ding, K., Vella, C., Wang, Z., and Hu, Z. Antiangiogenic effects of GFP08, an agaran-type polysaccharide isolated from *Grateloupia filicina. Glycobiology.* 2012. 22(10): 1343–1352.

21 Platform Molecules from Algae by Using Supercritical CO_2 and Subcritical Water Extraction

Nidhi Hans, S.N. Naik and Anushree Malik

CONTENTS

Abbreviations 229
Introduction 230
Biochemical Composition of Microalgae 230
Extraction Techniques 230
    Conventional Extraction 230
    Modern Extraction Techniques 230
        Supercritical CO_2 Extraction 231
        Pressurized Fluid Extraction 237
        Ultrasound-Assisted Extraction 237
        Microwave-Assisted Extraction 237
        Enzyme-Assisted Extraction 238
Commercial and Industrial Applications of Microalgae 238
Economic Potential of Microalgae 239
Conclusion 239
Acknowledgment 240
References 240

BOX 21.1 SALIENT FEATURES

Currently, algal-based bioactive compounds such as lipids, proteins, carbohydrates, carotenoids, vitamins and minerals are increasingly being used worldwide. Photobioreactor and open-pond methods have been used to platform molecules in microalgae production for feedstock. Conventionally they are extracted using organic solvents from microalgae which have many limitations and drawbacks related to human health and environment. Due to low yield and less selectivity of bioactive compounds from conventional methods, greener and more efficient technology is required to extract these compounds. Therefore, green processes such as supercritical fluid extraction, pressurized fluid extraction, microwave-assisted extraction, ultrasound-assisted extraction and enzyme-assisted extraction are used as alternative methods for separating these molecules. Products extracted using green methods are regarded as safe by the United States Food and Drug Administration. This chapter highlights different techniques for extraction of high-value products from microalgae, which have commercial and industrial applications in the nutraceutical, cosmeceutical and pharmaceutical industries. It is reported that these compounds have antimicrobial, antifungal, antioxidant, anti-inflammatory and anticancerous properties, and can be used to treat life-threatening diseases. Therefore, the extraction techniques mentioned here are the most promising approach to extracting biologically active compounds from microalgae in the future.

ABBREVIATIONS

AA: Arachidonic acid
CO_2: Carbon dioxide
CVD: Cardiovascular disease
DHA: Docosahexaenoic acid

EPA: Eicosapentaenoic acid
FAME: Fatty acid methyl ester
FDA: Food and Drug Administration
GLA: Gamma-linolenic acid
GRAS: Generally recognized as safe
PLE: Pressurized liquid extraction
PUFAs: Polyunsaturated fatty acids
SWE: Supercritical water extraction

INTRODUCTION

The ever-increasing market for functional foods is always demanding new bioactive ingredients that can be used by food industries for the development of functional products. In this regard, much attention has recently been paid to natural compounds like polyunsaturated fatty acids (PUFAs), carotenoids, proteins, polyphenols and their associated derivatives. A marine habitat including microalgae and macroalgae is considered as a rich and underexploited source of bioactive compounds. Under different cultivation conditions, microalgae can synthesize high-value added compounds in significant quantities. Some species are able to accumulate high PUFA content, such as eicosapentaenoic acid (EPA), docosahexaenoic acid (DHA) (Ackman, 1981) antioxidant-like β-carotene (Herrero et al., 2006) and antimicrobial and antiviral molecules that are of interest to the food and pharmaceutical industries (Supamattaya et al., 2005).

A conventional method such as Soxhlet can be used to extract compounds of interest from microalgae dried biomass using different organic solvents, but this method has low selectivity and is toxic and inflammable. In order to overcome this, supercritical CO_2 is widely employed to obtain bioactive compounds from microalgae. This is generally recognized as safe (GRAS) by the U.S. Food and Drug Administration (FDA) as it is nontoxic, nonflammable, does not react with bioactive compounds and can extract thermally labile compounds even at low temperatures. Under ambient conditions, CO_2 has a density similar to liquid, that is, 0.470 g/ml; and viscosity and diffusivity like gas, which allows it to penetrate inside solid biomass and results in the separation of solvent-free extract (Berk, 2009).

In this chapter, we discuss different techniques used to efficiently extract bioactive compounds present in microalgae. Algal-based products and their economic value worldwide is also examined. Due to the huge biodiversity of algae, discussion is limited and we have tried to give an overview of technologies being used to extract valuable compounds from algae, their use in food and therapeutic applications and their market applications for better development in the future.

BIOCHEMICAL COMPOSITION OF MICROALGAE

Microalgae contain lipids (15–60%), polyunsaturated fatty acids, proteins (5–50%), carbohydrates, carotenoids, polysaccharides (4–76%), dietary fiber (25–75%), antioxidants, photosynthetic pigments, vitamins and minerals (Figure 21.1). The yield of production of these bioactive compounds depends on conditions of culture and also relies on the extraction method, degree of pretreatment and type of solvent used.

EXTRACTION TECHNIQUES

Different methods are used to extract biological active compounds from algal biomass to achieve a high yield of the desired compounds. An appropriate method should be selected for extraction with optimized conditions (temperature, pressure, pH), minimal energy consumption, less waste generation to increase the scale. Extraction of bioactive compounds from some microalgae and macroalgae with antibacterial, antiviral, antifungal, antioxidative and anti-inflammatory activity is summarized in Tables 21.1 and 21.2 respectively.

Conventional Extraction

Traditionally, Soxhlet extraction apparatus is used for the isolation of compounds from microalgae. In this process, different nonpolar organic solvents such as Hexane, petroleum ether, benzene, toluene, diethyl ether and chloroform and polar organic solvents such as acetonitrile, ethanol, methanol, acetone and water are used for extraction.

Modern Extraction Techniques

Due to the low yield of bioactive compounds with less selectivity achieved by conventional methods, novel techniques have been developed for the extraction of platform molecules such as polyunsaturated fatty acids, lipids, carotenoids, proteins, carbohydrates, vitamins, minerals, polyphenols and many more from algal biomass. New extraction techniques which have been successfully used to extract biologically active compounds are supercritical fluid extraction, pressurized fluid extraction, microwave-assisted extraction, ultrasound-assisted extraction and enzyme-assisted extractions, which are also used in the food and pharmaceutical industries.

Supercritical CO_2 Extraction

Supercritical fluids are used for separating and extracting bioactive compounds from algae. In this method, different solvents are used at their critical temperature and pressure (Taylor, 1996). At critical temperature and pressure, fluid can diffuse through solid biomass and dissolve molecules of interest. CO_2 has moderate critical temperature (31.2°C) and pressure (7.4 MPa) and is commonly used in supercritical fluid extraction as it is generally recognized as safe (GRAS) by the FDA and is non-toxic, cheap, environmentally friendly, inexpensive, nonflammable and odorless. It is inert as it does not react and can easily be removed from the product, and it can also extract thermally labile compounds without degradation at low temperatures (Macias-Sanchez et al., 2007). In this method, microalgae biomass is tightly packed and placed inside a cylindrical extraction vessel which maintains a specific temperature. The extraction vessel is supplied with CO_2 at a pressure greater than critical pressure. As critical temperature is reached, CO_2 mixes with the sample and performs lipid extraction.

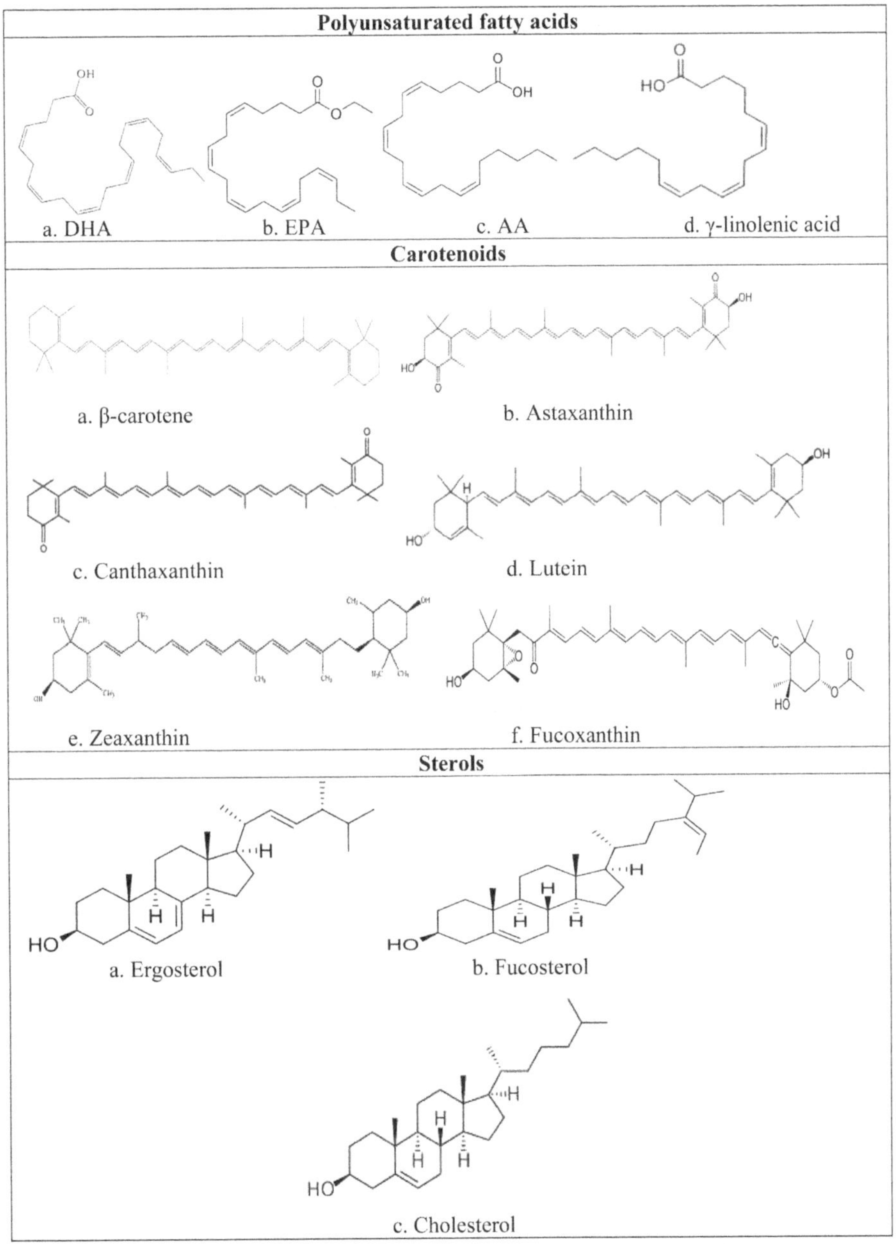

FIGURE 21.1 Chemical structure of main bioactive compounds from microalgae.

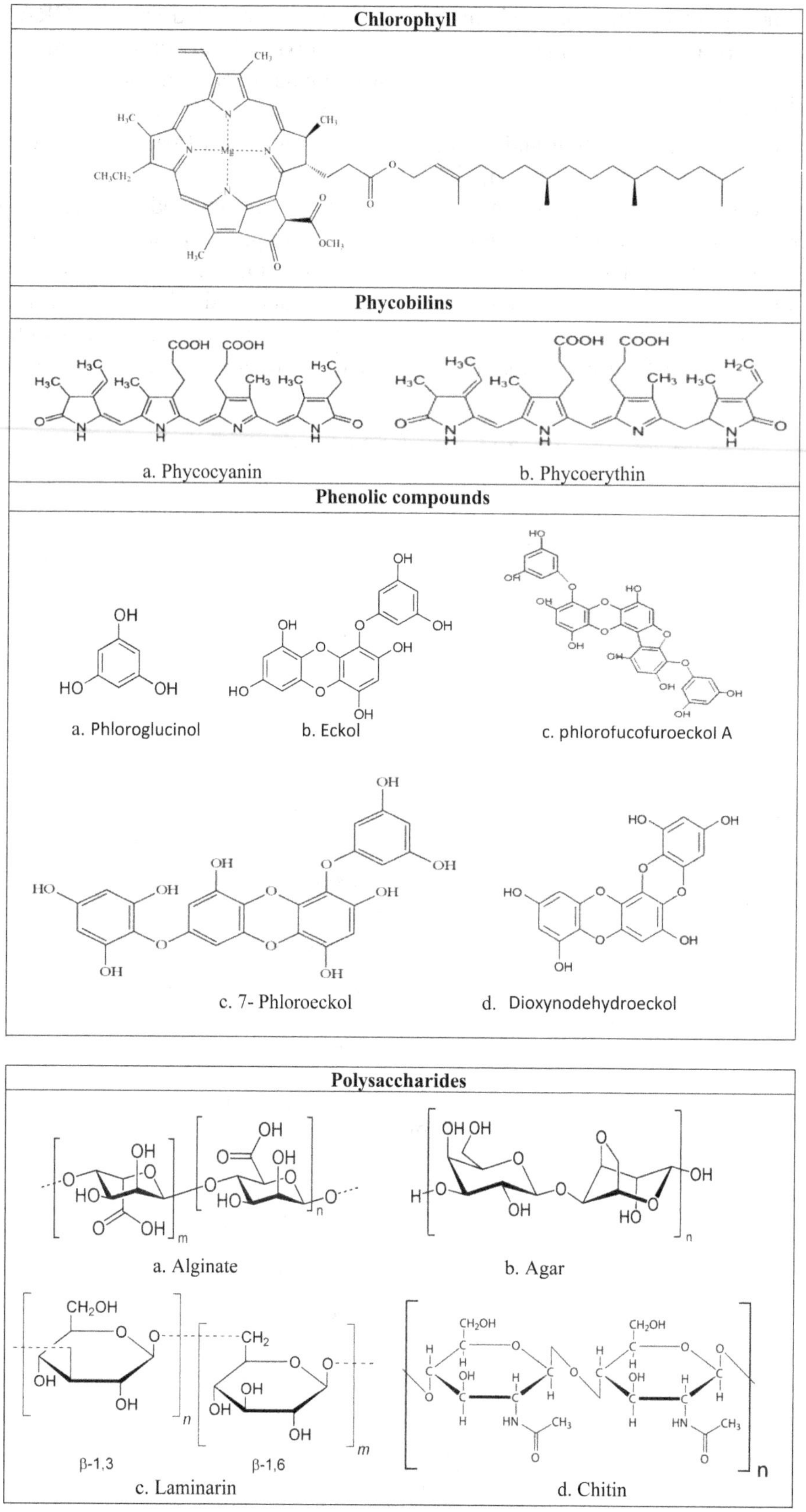

FIGURE 21.1 (CONTINUED) Chemical structure of main bioactive compounds from microalgae.

TABLE 21.1
Extraction of Bioactive Compounds from Some Microalgae and Their Area of Application

Species	Method	Product	Activity	Yield	Application	References
Botryococcus braunii	Solvent extraction	Lipids Xanthophylls Alkadeins	Antioxidant Anticancer	25–75%dw 85%	Cosmetic mask Energy Food additives	Lee et al. (2010) Mendes at al., (2003); Ranga Rao et al. (2006); (2018)
Chaetocerosmuelleri	SC-CO_2	Antioxidant extracts	Antioxidant		Health	Mendiola et al. (2007)
Phormidiumvalderianum	SC-CO_2	Antioxidant extracts	Antioxidant	93% reduction in toxins like Anatoxin-a	Food Pharmaceuticals	Chatterjee and Bhattacharjee (2014)
Chlorococcum sp.	Solvent extraction	Lipids Astaxanthin	Antioxidant	6.8% 7.09mg/g DW	Food Cosmetics	Halim et al. (2011); Ma and Chen (2001)
Chlorella pyrenoidosa	SC-CO_2	Lutein	Prevent macular degeneration	93.1%	Health	Inbaraj et al. (2006)
Haematococcus pluvialis	SC-CO_2	Astaxanthin Fatty acids phenols	Antioxidant Antiviral	87%	Natural pigment Dietary supplement	Sarada et al. (2006); Jaime et al. (2010)
Synechoccus 833	SC-CO_2	Allophycocyanin Carotenoid	Antioxidant	85-88%DW 87.6%	Pharmaceuticals Food and beverages	Viskari and Colyer (2003); Macías-Sanchez et al. (2005)
Chlorella vulgaris	SC-CO_2 Ultrasonication	Lipids Carotenoids Polyphenol Flavanoids Chlorophyll A	Antioxidant Anticancerous anti-inflammatory	5–58% dw 3.5mg/g 13.4mg/g 3.18mg/g Degrade chlorophyll	Food supplements Cosmetics Pharmaceuticals	Gouveia et al. (2007); Aluç et al. (2018); Wang et al., 2010)
Euglena gracilis	N_2 sources	α-tocopherol	Antioxidant	283.6µg/g	Cosmetics Food	Durmaz (2007)
Limnotherix sp.	SC-CO_2	C-phycocyanin	Antioxidant Antitumor	18% DW	Florescent markers, Cosmetics	Gantar et al. (2012)
Dunaliellasalina	SC-CO_2	β-carotene Indolic derivative	Antioxidant	0.2–27.7 µg/mg DW	Food Energy Pro-vitamin A	Mendiola et al. (2008)
Scenedesmusdimorphus	Ultrasonication Acid hydrolysis	Lipids β-carotene sugars	Antioxidant Anti-inflammatory	21% 0.370-0.74 95.6%	Food	Abrahamsson et al. (2012); Miranda et al. (2012)

(*Continued*)

TABLE 21.1 (CONTINUED)
Extraction of Bioactive Compounds from Some Microalgae and Their Area of Application

Species	Method	Product	Activity	Yield	Application	References
Spirulina platensis	SC-CO_2	Lipids	Antimicrobial	7.8%	Food	Furuki et al. (2003)
	SC-CO_2	β-carotene	Antioxidant	70%	Health	Mendiola et al. (2007)
	Ultrasonication	γ-linolenic acid		90% purity	Cosmetics	Herrero et al. (2005)
	SC-CO_2	C-phycocyanin		29.4mg/g	Immunofluorescence technique	
	PLE	Vitamin E		20%	Antibody labels	
		Phycobiliproteins				
Spirulina maxima	SC-CO_2	Lipids	Antioxidant	3.1%	Food supplements	Mendes et al. (2006, 2003); Mao et al. (2005)
		γ-linolenic acid phycocyanin	Anti-inflammatory	45%		
Nannochloropsis sp.	SC-CO_2	Lipids Carotenoids	Antioxidant, phagocytotic activity	25%	Health	Andrich et al. (2005)
Ochromonasdanica	SC-CO_2	Lipids	Antioxidant	28%	Food	Polak et al. (1989)
Spirulina pacifica	SC-CO_2	β-carotene	Antioxidant	118 mg/100g	Food	Careri et al. (2001)
		β-cryptoxanthin	Anti-inflammatory	7.5 mg/100g	Health	
		Zeaxanthin		48 mg/100g		
Leptolyngbya sp. KC45	Ammonium sulfate precipitation	Phycoerythrin	Anti-inflammatory	1.36%	Immunofluorescence technique Antibody labels	Pumas et al. (2012)
Skeletonemacostatum	SC-CO_2	Lipids Carotenoids	Antioxidant	8.6%	Food	Polak et al. (1989)
Porphyridiumcruentum	SC-CO_2	Phycoerythrin Phycocyanin	Anti-inflammatory	32.7% DW 3.48%	Nutritional supplements	Bermejo Román et al. (2002)
Synechoccus sp.	SC-CO_2	β-carotene β-cryptoxanthin zeaxanthin	Antioxidant	71.6% 7.9% 9.3%	Food Health	Mendiola et al. (2008)
Phaeodactylumtricornutum	Nitrogen starvation	PUFAs	Antimicrobial Anti-inflammatory	60%	Baby and health food application	Pulz and Gross (2004)

TABLE 21.2
Extraction of Bioactive Compounds from Some Macroalgae and Their Area of Application

Species	Method	Product	Activity	Yield	Application	References
Ulva rigida	Endo-protease (EAE)	Lipids Endo-protease, sugar, protein Polysaccharide Dietary fiber	Antiviral Antioxidant Antitumor	16.3 mg/g 4–9% 15–65% 38%	Food supplements Pharmaceuticals	Hardouin et al. (2016); Sathivel et al. (2008); Dawczynski et al. (2007)
Bangia atropurpurea	SC-CO_2	Lipids	Antioxidant	13.3%	Food	Chen and Chou (2002)
Chaetomorpha linum	SC-CO_2	Lipids	Antioxidant	53%	Energy	Aresta et al. (2005)
Caulerpa sp.	Pepsin (EAE)	Lipids	Antitumor	4.7–15.3 mg/g	Health	Lin et al. (2012)
Dictyopteris membranacea	SC-CO_2	Volatile metabolites (C_{11} Hydrocarbons)	Antimicrobial	7.5–40.5%	Health	El Hattab et al. (2007)
Dilophus ligulatus	SFE	Antifungal extract	Antifungal	2.80%	Health	Subra et al. (1991)
Galaxaura cylindrica	SFE	Lipids	Prevent CVD	19.8%	Food Health	Chen and Chou (2002)
Galaxaura filicina	SFE	Lipids	Reduce blood cholesterol	13.6%	Food Health	Chen and Chou (2002)
Helmintocladia australis	SFE	Lipids	Prevent blood-platelet aggregation	19.7%	Food	Chen and Chou (2002)
Gracilaria sp.	Proteolytic enzyme	Lipids Protein Sulphated polysaccharide	Anticoagulant Antioxidant	2.9–9.7 mg/g 5% 36%	Food supplements Health	Fidelis et al. (2014)
Hypnea charoides	SC-CO_2	Lipids	Antioxidant	67 mg/g DW	Food	Cheung (1999)
Laminaria sp.	SC-CO_2	Lipids Protein Polysaccharide Dietary fibre	Antibacterial, Anticancer, Anticoagulant	1.8% 3–14% 38% DW 36% DW	Food Health	Wen et al. (2006); Dawczynski et al. (2007)
Liagora boergesenii	SFE	Lipids	Anticoagulant	21.5%	Food	Chen and Chou (2002)
Liagora orientalis	SFE	Lipids	Reduce Blood pressure	17.6%	Food	Chen and Chou (2002)
Plocamium cartilagineum	SFE	Halogenated monoterpenes	Antifungal Antimicrobial	95%	Health	Gao et al. (2001)
Hypnea sp.	SC-CO_2	Lipids	Antioxidant	7–9%	Food	Cheung (1999)
Porphyra angusta	SC-CO_2	Lipids Polysaccharide Dietary fiber	Antitumor	12.4% 40% 35–49%	Food	Chen and Chou (2002); Murata and Nakazoe (2001); Dawczynski et al. (2007)
Porphyra dentata	SFE	Lipids Protein	Antioxidant Prevent CVD	11.2% 7%	Food, Health	Chen and Chou (2002)

(Continued)

TABLE 21.2 (CONTINUED)
Extraction of Bioactive Compounds from Some Macroalgae and Their Area of Application

Species	Method	Product	Activity	Yield	Application	References
Sargassum hemiphyllum	SC-CO_2	Lipids	Anticoagulant, anticancer	5.39% DW	Food	Cheung et al. (1998); Marinho-Soriano et al. (2006); Dawczynski et al. (2007)
		Protein		9%		
		Polysaccharide		4%		
		Dietary fiber		49–62%		
Scenedesmus obliqquus	SFE	Lipids	Antioxidant	3.97%	Food	Choi et al. (1987)
Scinaia monoliformis	SFE	Lipids	Antioxidant	17%	Food	Chen and Chou (2002)
Undaria pinnatifida	SC-CO_2	Fucoxanthin	Antioxidant Prevention of CVD	7.53 mg/g	Energy	Crespo and Yusty, (2005); Je et al. (2009); Dawczynski et al. (2007)
		Aliphatic hydrocarbons		9.28%		
		Polychlorinated biphenyls		9.2%		
		Polysaccharide		35–45%		
		Dietary fiber		30%		

An extracted lipid with CO_2 enters the collection vessel where CO_2 returns to a gaseous state under depressurization, and crude lipids remain precipitated. The collected lipid is free from solvents and does not need to undergo any solvent-removal step.

Supercritical CO_2 acts as a nonpolar solvent and is unable to interact with polar or neutral lipids. Therefore, the addition of a co-solvent such as ethanol, methanol or toluene is used to alter the polarity and increase the interaction of solvents with polar lipids. Therefore, operating variables like temperature, pressure, fluid flow rate and modifier addition can be optimized to maximize lipid yield with more unsaturated fatty acids (Pourmortazavi and Hajimirsadeghi, 2007). One major constraint to supercritical CO_2 extraction of microalgae is the level of moisture content in the sample. It obstructs diffusion of CO_2 into the sample and reduces contact time with solvent, thus reducing the diffusion of lipids out of the algal cell. That is why samples are dried before supercritical extraction.

Pressurized Fluid Extraction

In this method, organic solvents are used at high temperatures (from 20 to 200°C) and pressure (30–200 bars) so that it remains in a liquid state during extraction. This technique is also known as pressurized liquid extraction (PLE), pressurized solvent extraction (PSE), accelerated solvent extraction (ASE) or supercritical water extraction ([SWE], when extraction solvent is water). The combined effect of high temperature and pressure allows for a faster extraction process and requires a smaller amount of organic solvents (~10–30% w/v). This enhances mass transfer rate, decreasing the solvent viscosity and hence increasing the solubility of analytes present in the biomass. This facilitates the penetration of the solvent deeper into the matrix more easily, dissolve analytes and get collected in bulk solvent. Briefly, the instrumentation comprises of an extraction cell where biomass is loaded, which is placed in an oven to maintain the temperature; and high-pressure pump, which transfers solvent into the cell. Solvent flows through the extraction cell, and at the end of flow path, it is collected in a tube. After that, the extraction cell is purged with nitrogen gas for 2–3 minutes to remove any remaining residue from the biomass (Fernández, 2009). A variety of polar and nonpolar solvents are used in this method. In SWE, a high-temperature (200–250°C) dielectric constant of water (ε ~ 80) is decreased to value similar to some of the organic solvents such as ethanol and methanol (ε ~ 30–25), and this can be employed in various applications as an alternative to many organic solvents (Herrero et al., 2006). High-value metabolites are extracted from microalgae like lipophilic compounds (3–39.31% w/w), carotenoids (16–30% w/w), phycobiliproteins (~ 20% w/w), antioxidants and fatty acids. Parameters like pressure, temperature, type and composition of solvent and time should be optimized to obtain a high yield. One limitation of this technique is that extraction is exhaustive and it sometime produces nonselective compounds, which can be reduced by using adsorbents. Also, their high-cost technologies, which can be equilibrated with reduction in solvent volume and time.

Ultrasound-Assisted Extraction

In this method, when solvent is irradiated by ultrasound waves of a frequency above 20 KHz, it generates sound waves that circulate in the liquid media and results in alternate high-pressure and low-pressure cycles. During a high-pressure cycle, microbubbles are formed; they grow, oscillate extremely fast and collapse when acoustic pressure is too high. If bubbles collapse near the cell wall of the biomass, it results in the emission of shock waves; these disrupt the cell wall mechanically to form cavities and allow the solvent to penetrate inside the biomass to release intracellular contents. It increases the contact surface area between solvent and targeted compounds, leading to increased mass transfer. Lipids can be recovered by increasing the amplitude of the exposure time and further enhanced (50–500%) by using a mixture of polar and nonpolar solvents and a 10-fold reduced extraction time (Suali and Sarbatly, 2012). There is no thermal denaturation of essential biomolecules as it is run at low temperature (Ranjith et al., 2015). Free radicals can be generated if cells are subjected to sonication for a long time, which can degrade the quality of lipids by oxidation. This can be reduced by using a nonpolar solvent like hexane, which is resistant to peroxide formation (Chemat et al., 2004). This method is inexpensive, requires less operational time and greatly improves the yield and purity of the final product.

Microwave-Assisted Extraction

Microwaves are nonionizing electromagnetic waves of frequency ranges from 300MHz to 300GHz which can easily penetrate inside the biomass. The energy of the microwave is transferred to the solvent via two mechanisms: dipole rotation and ionic conduction. During microwave heating, dipole rotation of molecules takes place, which breaks weak hydrogen bonds, and migration of dissolved ions increases penetration of the solvent into the biomass to extract targeted compounds. It generates the vibration of water molecules which leads to an increase in the temperature of the intracellular water molecules of the microalgae cells. Then evaporation of water occurs, which exerts pressure on cell wall; it breaks and releases intracellular content into the medium. In this process, an ample amount of pressure is created inside the biomass which alters the physical properties of microalgae tissues, leading to the formation of pores which let the solvent penetrate inside the matrix. Parameters like time, temperature, dielectric constant of solvent, biomass-to-solvent ratio and concentration of solvent should be considered in this method. Water is a suitable solvent for use in microwave-assisted extraction as it generates appropriate energy for the extraction of various metabolites from microalgae. For high lipid extraction, nonpolar solvents should be mixed with polar solvents as the transfer of microwave energy take place from polar to nonpolar molecules, which enhances its interaction with dissolved molecules. The total consumption of solvent is reduced in microwave extraction (Chen et al., 2007). With less extraction time and a high microwave power level (30–1200 watts), yield is increased without degradation of final products (Hu et al., 2008; Wang et al., 2008). Also at high temperature (60–200°C), the viscosity of the solvent decreases, which increases its solubility and mobility and thus increases the extraction efficiency (Khajeh et al., 2010). Both polar and nonpolar compounds can be extracted simultaneously using microwaves, but this technique is not good for extraction of heat-sensitive metabolites. Also it requires additional step to remove unwanted or solid biomass from solvent. It is a cost-effective method for the extraction of high-quality lipids from wet microalgae in less reaction time, but its maintenance cost at an industrial level is a constraint factor.

Enzyme-Assisted Extraction

Enzymes are the biological catalyst that enhances the rate of conversion of substrate to product under mild conditions. Mixtures of enzymes are used to enhance the extraction of targeted bioactive compounds from algae, like proteins, phenols, carotenoids and lipids, by degrading its cell wall, which is chemically and structurally heterogeneous. Enzymes used in this method are nontoxic, ecofriendly and of food grade (Michalak and Chojnacka, 2014). Enzymes used frequently are amyloglucosidase, agarase, alcalase (for hydrolysis of proteins), carragenanase, celluclast (to break cellulosic material), kojizyme (protease), neutrase (metallo-proteinase), termamyl (amylase), ultraflo (hydrolyze 1,3- or 1,4- linkages in β-D-glucans xylanase), umamizyme (proteolytic enzyme), xylanase and viscozyme (break cell wall) (Kadam et al., 2013). This technology is highly specific and is executed under mild conditions to protect bioactive compounds from degradation. This technique is difficult to scale up to industrial scale as enzyme activities vary with different environmental conditions.

COMMERCIAL AND INDUSTRIAL APPLICATIONS OF MICROALGAE

Microalgae are a potential source for high-value products at commercial level. Nutritional supplements like capsules and tablets which are produced from microalgae (such as *Chlorella* species, *Spirullina*, *Dunaliellasalina*, etc.) are available on the market.

DHA produced commercially from *Crypthecodinium cohnii* is used in infant formulas and DHA from *Schizochytrium* is used in dietary supplements for adults and pregnant women (Ward and Singh, 2006; Spolaore et al., 2006). Carotenoids produced from microalgae have nutritional and therapeutic values and are used as colorants in the food industry and in cosmetics, dietary supplements and animal feed. β-carotene extract and dried powder of *Dunaliella* is commercially available for human and animal feed use (Wijffels, 2008). Astaxanthin produced from *Hematococcus* is widely used in pharmaceutical industries.

Florescent pigments isolated from cryptomonads and dinoflagellates are used as intracellular markers and serve as labels for antibodies and receptors in immunolabelling experiments (Glazer and Alexander, 1994). Phycobiliprotein produced from cyanobacteria *Arthosporia* and rhodophyte *Porphyridium* is commercially available and is widely used as colorants in candy bars, sweets, cold drinks, chewing gums, dairy products and so on (Spolaore et al., 2006).

It is reported that bioactive compounds from algae have antibacterial, antiviral, anticancer properties and therefore it is used commercially in drug screening. Antibacterial compounds from algae have various applications in agriculture; compounds like tjipanozoles extracted from cyanobacteria show a cytotoxic and fungicidal effect on *candida albicans* and *Magnaporthegrisea* (Bonjouklian et al., 1991), and algaecides that are produced help in controlling algal blooms (Bagchi et al., 1990). Extracts from cyanobacteria also show antiviral activity against *Herpes simplex* virus type II (Patterson, 1993). Compounds such as tubericidin, scytophycin B, toyocamycin and tolytoxin extracted from different microalgae show anticancerous activities (Lau et al., 1993).

Microalgae also have important applications in cosmetics. *Chlorella vulgaris* extract helps in the regeneration of tissue and reduction of wrinkles. Extract from *Arthosporia* species act as antiaging product, and *Nannochloropsis oculata* extract helps in skin tightening.

Also *Dunaliellasalina* enhances proliferation of cells and protects skin from UV radiation (Stolz, 2005).

Microalgae such as *Nostoc*, *Anabaena*, *Tolypothrix* and *Aulosira* are exploited industrially as biofertilizers which help in fixing atmospheric nitrogen and improving the physicochemical properties of soil (Marris, 2006). It is also used as feed for a variety of animals, from fish to farm animals. Many other high-value products are produced from microalgae with outstanding commercial impact.

ECONOMIC POTENTIAL OF MICROALGAE

Algae have been exploited since the 1950s for the use in food and beverages, personal care, pharmaceutical products, dietary supplements and feed. The algal market is estimated to be US$ 3.98 billion in 2018 and is projected to reach a value of US$ 5.17 billion by 2023, at a CAGR of 5.4% (PRNewswire, 2018). Production of the microalgae *Chlorella* at an industrial scale for human consumption was begun in Japan in 1960. From the 1980s, in India, the United States, Asia, Israel and Australia, algae-production facilities were established on a large scale (Enzing et al., 2014). Products from microalgae are available in two forms. The first is a dried form, particularly *Chlorella* (5000 tons/year, 2012) and *Spirilluna* (2000 tons/year, 2003), which are a rich source of protein and carbohydrate and are directly sold as dietary supplements; the second form is where compounds are extracted from microalgae such as astaxanthin (300 tons/year, turnover of US$ 200, 2004), phycobiliprotein (turnover of US$ 50, 2004), ω-3 fatty acids (240 tons/year in 2003, turnover of US$ 14,390, 2009), β-carotene (1200 tons/year in 2010, turnover of US$ 285, 2012) and added to food and feed to enhance its nutritional value (Spolaore et al., 2006; Ismail, 2010; Norsker et al., 2011; Milledge, 2012). More than 125 million dollars/year worth of algae products were exported by China, Indonesia and South Korea and 13 million dollars/year by Ireland, France and the Netherlands in the period 2010–2012.

The cost of producing dry algal biomass feed varies from US$ 80/kg–US$ 800/kg in Australia. Of the algae biomass produced, 30% is used for animal feed, especially *Arthosporia*, and more than 50% of it is used in food supplements. DHA-rich *C. cohinii*, used for infant formula, is estimated to have world wholesale market of US$ 10 billion per annum (Ward and Singh, 2005). The price of β-carotene extracted from *Dunaliella* varies from US$ 300 to US$ 3000/kg in year 2004 (Ben-Amotz, 2004). The aquaculture market for astaxanthin isolated from *Hematococcus* is estimated to be US$ 200 million/annum in 2004 (Hejazi and Wijffels, 2004). The estimated market for lutein, a carotenoid, in the pharmaceutical industry is to be US$ 190 million/annum, while in the food and nutraceutical industries it is US$ 110 million/annum. Labeled isotope compounds from algae have a market value of US$ 13 million/year (Spolaore et al., 2006). Growth of algal product marketing is increasing as the adoption level of algal-based products by consumers is increasing, which helps new emerging companies to innovate and develop different products from algae. Companies including Cargill (United States), BlueBioTech Int. GmbH (Germany), DowDuPont (United States), AlgaeCan Biotech Ltd. (Canada), DIC Corporation (Japan), Algatechologies ltd. (Israel), DSM (the Netherlands), Cellna Inc. (United States), BASF (Germany), Tianjin Norland Biotech Co., Ltd. (China), Taiwan Chlorella Manufacturing Company (China), Cyanotech Corporation (United States), Sun Chlorella Corporation (Japan), Algaetech International SdnBhd (Malaysia), Kerry (Ireland), Ingredion (United States), CP Kelco (United States), Corbion (the Netherlands), Roquette Frères (France), FenchemBiotek (China), Pond Technologies Inc. (Canada) and E.I.D Parry (India) focus on increasing their research and development expenditure and manufacturing capacities to develop new algae products with better food-grade quality.

CONCLUSION

Algae is considered an emerging biological source for its potential applications in different fields including functional food, feed, pharmaceuticals, nutraceuticals, personal care, dietary supplements and many more. The global market for algae-based products is well developed, and dried algal biomass or extracted bioactive compounds are currently used for enhancing the nutritional value of foods. Valuable commercial algal-based products face the biggest challenges in downstream processing to separate out particular products. Optimized methods are used to extract compounds efficiently from algae which are accumulated within cells. To get the maximum yield, pretreatment methods like ultrasonication and microwave-assisted extraction can be coupled with solvent extraction. All the extraction techniques discussed here have their own advantages and disadvantages, but supercritical CO_2 extraction is the most promising approach for the future as it yields solvent-free, biologically active compounds without degrading them and is harmless for human consumption and in animal feed. Carotenoids (astaxanthin, fucoxanthin, β-carotene, lutein and zeaxanthin) and unsaturated fatty acids (DHA, EPA etc.) extracted from algae have antioxidant, antimicrobial, antiviral and anticancerous activities and can be used to cure many diseases.

With the increased awareness among consumers of the influence of algae products on health, the market demand for carotenoids (natural pigment) and PUFAs has grown, leading to increased large-scale production of algae, and therefore the production cost is predicted to reduce significantly. Government and industries focus on the bioavailability and bioaccessibility of extracted compounds before and after incorporating them into food by taking regulatory/ethical issues into account. Marketing of algae derivative products requires a Good Manufacturing Practice (GMP) certificate and other ISO quality clearances.

Therefore a large-scale sustainable, efficient and economical viable downstream process must be developed for the sake of the economy, the environment and human health.

ACKNOWLEDGMENT

The authors would like to acknowledge Indo-Sri Lanka project for the financial support through project no. DST/INT/SL/P-20/2016.

REFERENCES

Abrahamsson, V., Rodriguez-Meizoso, I., & Turner, C. Determination of carotenoids in microalgae using supercritical fluid extraction and chromatography. *J Chromatogr A*. 2012. 1250:63–68.

Ackman, R. G. Algae as sources for edible lipids. *N Sour Fats Oils*. 1981:189–220.

Aluç, Y., Başaran Kankılıç, G., & Tüzün, İ. Determination of carotenoids in two algae species from the saline water of Kapulukaya reservoir by HPLC. *J Liq Chromatogr Relat Technol*. 2018. 41(2):93–100.

Andrich, G., Nesti, U., Venturi, F., Zinnai, A., & Fiorentini, R. Supercritical fluid extraction of bioactive lipids from the microalga *Nannochloropsis* sp. *Eur J Lipid Sci Technol*. 2005. 107(6):381–386.

Aresta, M., Dibenedetto, A., Carone, M., Colonna, T., & Fragale, C. Production of biodiesel from macroalgae by supercritical CO2 extraction and thermochemical liquefaction. *Environ Chem Lett*. 2005. 3(3):136–139.

Bagchi, S. N., Palod, A., & Chauhan, V. S. Algicidal properties of a bloom-forming blue-green alga, Oscillatoria sp. *J Basic Microbiol*. 1990. 30(1):21–29.

Barrow, C., & Shahidi, F. *Marine Nutraceuticals and Functional Foods*. CRC Press: Boca Raton, FL. 2008.

Behrens, P. W., Bingham, S. E., Hoeksema, S. D., Cohoon, D. L., & Cox, J. C. Studies on the incorporation of CO2 into starch by Chlorella vulgaris. *J Appl Phycol*. 1989. 1(2):123–130.

Behrens, P. W., & Kyle, D. J. Microalgae as a source of fatty acids. J Food Lipids. 1996. 3(4):259–272.

Berk, Z. Extraction. In: Z. Berk (Ed.), *Food Process Engineering and Technology*. Academic Press: San Diego, CA. 2009. 259–277.

Bermejo Román, R., Alvárez-Pez, J. M., Acién Fernández, F. G., & Molina Grima, E. Recovery of pure B-phycoerythrin from the microalga Porphyridiumcruentum. *J Biotechnol*. 2002. 93(1):73–85.

Bonjouklian, R., Smitka, T. A., Doolin, L. E., Molloy, R. M., Debono, M., Shaffer, S. A., Moore, R. E., Stewart, J. B., & Patterson, G. M. L. Tjipanazoles, new antifungal agents from the blue-green alga Tolypothrixtjipanasensis. *Tetrahedron*. 1991. 47(37):7739–7750.

Borowitzka, M. A. High-value products from microalgae—their development and commercialisation. *J Appl Phycol*. 2013. 25(3):743–756.

Careri, M., Furlattini, L., Mangia, A., Musc, M., Anklam, E., Theobald, A., & Von Holst, C. Supercritical fluid extraction for liquid chromatographic determination of carotenoids in Spirulina pacifica algae: A chemometric approach. *J Chromatogr A*. 2001. 912(1):61–71.

Chatterjee, D., & Bhattacharjee, P. Supercritical carbon dioxide extraction of antioxidant rich fraction from Phormidiumvalderianum: Optimization of experimental process parameters. *Algal Res*. 2014. 3:49–54.

Chemat, F., Grondin, I., Costes, P., Moutoussamy, L., Sing, A. S. C., & Smadja, J. High power ultrasound effects on lipid oxidation of refined sunflower oil. *Ultrason Sonochem*. 2004. 11(5):281–285.

Chen, C. Y., & Chou, H. N. Screening of red algae filaments as a potential alternative source of eicosapentaenoic acid. *Mar Biotechnol*. 2002. 4(2):189–192.

Chen, L., Ding, L., Yu, A., Yang, R., Wang, X., Li, J., Jin, H., & Zhang, H. Continuous determination of total flavonoids in *Platycladus orientalis* (L.) Franco by dynamic microwave-assisted extraction coupled with on-line derivatization and ultraviolet–visible detection. *Anal Chim Acta*. 2007. 596(1):164–170.

Cheung, P. C. Temperature and pressure effects on supercritical carbon dioxide extraction of n-3 fatty acids from red seaweed. *Food Chem*. 1999. 65(3):399–403.

Cheung, P. C. K., Leung, A. Y. H., & Ang, P. O. Comparison of supercritical carbon dioxide and Soxhlet extraction of lipids from a brown seaweed, *Sargassum hemiphyllum* (Turn.) C. Ag. *J Agric Food Chem*. 1998. 46(10): 4228–4232.

Choi, K. J., Nakhost, Z., Krukonis, V. J., & Karel, M. Supercritical fluid extraction and characterization of lipids from algae *Scenedesmus obliquus*. Food Biotechnol. 1987. 1(2):263–281.

Crespo, M., & Yusty, M. Comparison of supercritical fluid extraction and Soxhlet extraction for the determination of PCbs in seaweed samples. *Chemosphere*. 2005. 59(10):1407–1413.

Crespo, M. P., & Yusty, M. L. Comparison of supercritical fluid extraction and Soxhlet extraction for the determination of aliphatic hydrocarbons in seaweed samples. *Ecotoxicol Environsaf*. 2006. 64(3):400–405.

Dellert, S. F., Nowicki, M. J., Farrell, M. K., Delente, J., & Heubi, J. E. The 13C-xylose breath test for the diagnosis of small bowel bacterial overgrowth in children. *J Pediatr Gastroenterol Nutr*. 1997. 25(2):153–158.

Durmaz, Y. Vitamin E (α-tocopherol) production by the marine microalgae *Nannochloropsis oculata* (Eustigmatophyceae) in nitrogen limitation. *Aquaculture*. 2007. 272(1–4):717–722.

El Baky, H. A., El Baz, F. K., El Baroty, G. S., Asker, M. S., & Ibrahim, E. A. Phospholipids of some marine macroalgae: Identification, antivirus, anticancer and antimicrobial bioactivities. *Pharm Chem*. 2014. 6:370–382.

El Hattab, M., Culioli, G., Piovetti, L., Chitour, S. E., & Valls, R. Comparison of various extraction methods for identification and determination of volatile metabolites from the brown alga Dictyopterismembranacea. *J Chromatogr A*. 2007. 1143(1–2):1–7.

Enzing, C., Ploeg, M., Barbosa, M., & Sijtsma, L. Microalgae-based products for the food and feed sector: An outlook for Europe. *JRC Sci Policy Rep*. 2014:19–37.

Fernández Álvarez, M. Study of the Photochemical Behavior and Determination of Phytosanitary Compounds in Environmental and Agro-Alimentary Matrices by Means of Advanced Techniques of Extraction and Microextraction. Univ Santiago de Compostela. 2009.

Fidelis, G. P., Camara, R. B. G., Queiroz, M. F., Santos Pereira Costa, M. S., Santos, P. C., Rocha, H. A. O., & Costa, L. S. Proteolysis, NaOH and ultrasound-enhanced extraction of anticoagulant and antioxidant sulfated polysaccharides from the edible seaweed, *Gracilaria birdiae*. *Molecules*. 2014. 19(11):18511–18526.

Furuki, T., Maeda, S., Imajo, S., Hiroi, T., Amaya, T., Hirokawa, T., Ito, K., & Nozawa, H. Rapid and selective extraction of phycocyanin from *Spirulina platensis* with ultrasonic cell disruption. *J Appl Phycol*. 2003. 15(4):319–324.

Gantar, M., Simović, D., Djilas, S., Gonzalez, W. W., & Miksovska, J. Isolation, characterization and antioxidative activity of C-phycocyanin from Limnothrix sp. strain 37-2-1. *J Biotechnol*. 2012. 159(1–2):21–26.

Gao, D., Okuda, R., & Lopez-Avila, V. Supercritical fluid extraction of halogenated monoterpenes from the red alga *Plocamium cartilagineum*. *J AOAC Int*. 2001. 84(5):1313–1331.

Glazer, A. N. Phycobiliproteins—A family of valuable, widely used fluorophores. *J Appl Phycol*. 1994. 6(2):105–112.

Gouveia, L., Nobre, B. P., Marcelo, F. M., Mrejen, S., Cardoso, M. T., Palavra, A. F., & Mendes, R. L. Functional food oil coloured by pigments extracted from microalgae with supercritical CO2. *Food Chem*. 2007. 101(2):717–723.

Halim, R., Gladman, B., Danquah, M. K., & Webley, P. A. Oil extraction from microalgae for biodiesel production. *Bioresour Technol*. 2011. 102(1):178–185.

Hardouin, K., Bedoux, G., Burlot, A. S., Donnay-Moreno, C., Bergé, J. P., Nyvall-Collén, P., & Bourgougnon, N. Enzyme-assisted extraction (EAE) for the production of antiviral and antioxidant extracts from the green seaweed Ulva armoricana (Ulvales, Ulvophyceae). *Algal Res*. 2016. 16:233–239.

Hejazi, M. A., & Wijffels, R. H. Milking of microalgae. Trends Biotechnol. 2004. 22(4):189–194.

Herrero, M., Cifuentes, A., & Ibañez, E. Sub- and supercritical fluid extraction of functional ingredients from different natural sources: Plants, food-by-products, algae and microalgae: A review. *Food Chem*. 2006. 98(1):136–148.

Herrero, M., Simó, C., Ibáñez, E., & Cifuentes, A. Capillary electrophoresis-mass spectrometry of *Spirulina platensis* proteins obtained by pressurized liquid extraction. *Electrophoresis*. 2005. 26(21):4215–4224.

Hu, Z., Cai, M., & Liang, H. H. Desirability function approach for the optimization of microwave-assisted extraction of saikosaponins from Radix Bupleuri. *Sep Purif Technol*. 2008. 61(3):266–275.

Inbaraj, B. S., Chien, J. T., & Chen, B. H. Improved high performance liquid chromatographic method for determination of carotenoids in the microalga *Chlorella pyrenoidosa*. *J Chromatogr A*. 2006. 1102(1–2):193–199.

Ismail, A. Marine lipids overview: Markets, regulation, and the value chain. *Ocl-Ol- Corps Gras Li*. 2010. 17(4):205–208.

Jaime, L., Rodríguez-Meizoso, I., Cifuentes, A., Santoyo, S., Suarez, S., Ibáñez, E., & Señorans, F. J. Pressurized liquids as an alternative process to antioxidant carotenoids' extraction from Haematococcuspluvialis microalgae. *LWT—Food Sci Technol*. 2010. 43(1):105–112.

Je, J. Y., Park, P. J., Kim, E. K., Park, J. S., Yoon, H. D., Kim, K. R., & Ahn, C. B. Antioxidant activity of enzymatic extracts from the brown seaweed *Undaria pinnatifida* by electron spin resonance spectroscopy. *LWT—Food Sci Technol*. 2009. 42(4):874–878.

Kadam, S. U., Tiwari, B. K., & O'Donnell, C. P. Application of novel extraction technologies for bioactives from marine algae. *J Agric Food Chem*. 2013. 61(20):4667–4675.

Kainosho, M. Isotope labelling of macromolecules for structural determinations. *Nat Struct Biol*. 1997. 4:858–861.

Khajeh, M., Reza Akbari Moghaddam, A., & Sanchooli, E. Application of Doehlert design in the optimization of microwave-assisted extraction for determination of zinc and copper in cereal samples using FAAS. *Food Anal Methods*. 2010. 3(3):133–137.

Kladi, M., Vagias, C., & Roussis, V. Volatile halogenated metabolites from marine red algae. *Phytochem Rev*. 2004. 3(3):337–366.

Lattimer, J. M., & Haub, M. D. Effects of dietary fiber and its components on metabolic health. *Nutrients*. 2010. 2(12):1266–1289.

Lau, A. F., Siedlecki, J., Anleitner, J., Patterson, G. M., Caplan, F. R., & Moore, R. E. Inhibition of reverse transcriptase activity by extracts of cultured blue-green algae (cyanophyta). *Planta Med*. 1993. 59(2):148–151.

Lee, J. Y., Yoo, C., Jun, S. Y., Ahn, C. Y., & Oh, H. M. Comparison of several methods for effective lipid extraction from microalgae. *Bioresour Technol.* 2010. 101(1):S75–S77.

Li, Q., Du, W., & Liu, D. Perspectives of microbial oils for biodiesel production. *Appl Microbiol Biotechnol.* 2008. 80(5):749–756.

Lin, H. C., Chou, S. T., Chuang, M. Y., Liao, T. Y., Tsai, W. S., & Chiu, T. H. The effects of Caulerpamicrophysa enzyme-digested extracts on ACE-inhibitory activity and in vitro anti-tumour properties. *Food Chem.* 2012. 134(4):2235–2241.

Lustbader, J. W., Birken, S., Pollak, S., Pound, A., Chait, B. T., Mirza, U. A., Ramnarain, S., Canfield, R. E., & Brown, J. M. Expression of human chorionic gonadotropin uniformly labeled with NMR isotopes in Chinese hamster ovary cells: An advance toward rapid determination of glycoprotein structures. *J Biomol NMR.* 1996. 7(4):295–304.

Ma, R. Y. N., & Chen, F. Enhanced production of free trans-astaxanthin by oxidative stress in the cultures of the green microalga *Chlorococcum* sp. *Proc Biochem.* 2001. 36(12):1175–1179.

Ma, W., Lu, Y., Dai, X., Liu, R., Hu, R., & Pan, Y. Determination of anti-tumor constitute mollugin from traditional Chinese medicine *Rubia cordifolia*: Comparative study of classical and microwave extraction techniques. *Sep Sci Technol.* 2009. 44(4):995–1006.

Macías-Sánchez, M. D., Mantell, C., Rodríguez, M., Martínez de la Ossa, E., Lubián, L. M., & Montero, O. Supercritical fluid extraction of carotenoids and chlorophyll A from *Synechococcus* sp. *J Supercrit Fluids.* 2007. 39(3):323–329.

Mao, T. K., Van de Water, J., & Gershwin, M. E. Effects of a Spirulina-based dietary supplement on cytokine production from allergic rhinitis patients. *J Med Food.* 2005. 8(1):27–30.

Marinho-Soriano, E., Fonseca, P. C., Carneiro, M. A. A., & Moreira, W. S. C. Seasonal variation in the chemical composition of two tropical seaweeds. *Bioresour Technol.* 2006. 97(18):2402–2406.

Marris, E. Putting the carbon back: Black is the new green. *Nature.* 2006. 442(7103):624–626.

Mendes, R. L., Nobre, B. P., Cardoso, M. T., Pereira, A. P., & Palavra, A. F. Supercritical carbon dioxide extraction of compounds with pharmaceutical importance from microalgae. *Inorg Chim Acta.* 2003. 356:328–334.

Mendes, R. L., Reis, A. D., & Palavra, A. F. Supercritical CO2 extraction of γ-linolenic acid and other lipids from *Arthrospira maxima* (*Spirulina*): Comparison with organic solvent extraction. *Food Chem.* 2006. 99(1):57–63.

Mendiola, J. A., García-Martínez, D., Rupérez, F. J., Martín-Álvarez, P. J., Reglero, G., Cifuentes, A., Barbas, C., Ibañez, E., & Señoráns, F. J. Enrichment of vitamin E from *Spirulina platensis* microalga by SFE. *J Supercrit Fluids.* 2008. 43(3):484–489.

Mendiola, J. A., Jaime, L., Santoyo, S., Reglero, G., Cifuentes, A., Ibañez, E., & Señoráns, F. J. Screening of functional compounds in supercritical fluid extracts from *Spirulina platensis*. *Food Chem.* 2007. 102(4):1357–1367.

Mendiola, J. A., Santoyo, S., Cifuentes, A., Reglero, G., Ibáñez, E., & Señoráns, F. J. Antimicrobial activity of sub- and supercritical CO2 extracts of the green alga *Dunaliella salina*. *J Food Prot.* 2008. 71(10):2138–2143.

Mendiola, J. A., Torres, C. F., Toré, A., Martín-Álvarez, P. J., Santoyo, S., Arredondo, B. O., Señoráns, F. J., Cifuentes, A., & Ibáñez, E. Use of supercritical CO2 to obtain extracts with antimicrobial activity from *Chaetoceros muelleri* microalga. A correlation with their lipidic content. *Eur Food Res Technol.* 2007. 224(4):505–510.

Michalak, I., & Chojnacka, K. Algal extracts: Technology and advances. *Eng Life Sci.* 2014. 14(6):581–591.

Milledge, J. J. Microalgae—commercial potential for fuel, food and feed. *Food Sci Technol.* 2012. 26(1):28–30.

Miranda, J. R., Passarinho, P. C., & Gouveia, L. Pre-treatment optimization of *Scenedesmus obliquus* microalga for bioethanol production. *Bioresour Technol.* 2012. 104:342–348.

Norsker, N. H., Barbosa, M. J., Vermuë, M. H., & Wijffels, R. H. Microalgal production—A close look at the economics. *Biotechnol Adv.* 2011. 29(1):24–27.

Patterson, G. M. L., Baker, K. K., Baldwin, C. L., Bolis, C. M., Caplan, F. R., Larsen, L. K., Lavine, I. A., Moore, R. E., Nelson, C. S., Tschappat, K. D., Tuang, G. D., Boyd, M. R., Cardellina, J. H., Collins, R. P., Gustafson, K. R., Snader, K. M., Weisloi, O. S., & Lewin, R. A. Antiviral activity of cultured blue-green algae (cyanophyta). *J Phycol.* 1993. 29(1):125–130.

Polak, J., Balaban, M., Peplow, A., & Philips, A. Supercritical carbon dioxide extraction of lipids from algae. In: K. Johnston; Penninger, J., (Eds), *Supercritical Fluid Science and Technology, ACS Symposium Series No. 406*. American Chemical Society: Washington, DC. 1989. 449–467.

Pourmortazavi, S. M., & Hajimirsadeghi, S. S. Supercritical fluid extraction in plant essential and volatile oil analysis. *J Chromatogr A.* 2007. 1163(1–2):2–24.

Priyadarshani, I., & Rath, B. Commercial and industrial applications of micro algae—A review. *J Algal Biomass Util.* 2012. 3(4):89–100.

PRNewswire. Algae products market worth 517 billion USD by 2023. Available from: Prnewswire.com/news-releases/algae-products-market-worth-517-billion-usd-by-2023 680567331, 2018.

Pulz, O., & Gross, W. Valuable products from biotechnology of microalgae. *Appl Microbiol Biotechnol.* 2004. 65(6):635–648.

Pumas, C., Peerapornpisal, Y., Vacharapiyasophon, P., Leelapornpisid, P., Boonchum, W., Ishii, M., & Khanongnuch, C. Purification and characterization of a thermostable phycoerythrin from hot spring cyanobacterium *Leptolyngbya* sp. KC45. *Int J Agric Biol.* 2012. 14(1).

Ranga Rao, A., Deepika, G., Ravishankar, G. A., Sarada, R., Narasimharao, B. P., Su, Y., & Lei, B. *Botryococcus* as an alternative source of carotenoids and its possible applications—An overview. *Crt Rev Biotechnol.* 2018. 38(4):541–558.

Ranga Rao, A., Sarada, R., Baskaran, R., & Ravishankar, G. A. Antioxidant activity of *Botryococcus braunii* extract elucidated in in vitro models. *J Agri Food Chem.* 2006. 54:4593–4599.

Ranjith Kumar, R., Hanumantha Rao, P., & Arumugam, M. Lipid extraction methods from microalgae: A comprehensive review. *Front Energy Res.* 2015. 2:61.

Rodríguez-Bernaldo de Quirós, A., Castro de Ron, C., López-Hernández, J., & Lage-Yusty, M. A. Determination of folates in seaweeds by high-performance liquid chromatography. *J Chromatogr A.* 2004. 1032(1–2):135–139.

Sarada, R., Vidhyavathi, R., Usha, D., & Ravishankar, G. A. An efficient method for extraction of astaxanthin from green alga *Haematococcuspluvialis*. J Agric Food Chem. 2006. 54(20):7585–7588.

Sathivel, A., Raghavendran, H. R. B., Srinivasan, P., & Devaki, T. Anti-peroxidative and anti-hyperlipidemic nature of *Ulva lactuca* crude polysaccharide on D-galactosamine induced hepatitis in rats. *Food Chem Toxicol.* 2008. 46(10):3262–3267.

Stolz, P. Manufacturing microalgae for skin care. Cosmet Toiletries. 2005. 120:99–106.

Suali, E., & Sarbatly, R. Conversion of microalgae to biofuel. *Renew Sust Energ Rev.* 2012. 16(6):4316–4342.

Subra, P., Tufeu, R., Garrabos, Y., & Boissinot, P. Supercritical fluid extraction from a Mediterranean brown alga. *J Supercrit Fluids.* 1991. 4(4):244–249.

Supamattaya, K., Kiriratnikom, S., Boonyaratpalin, M., & Borowitzka, L. Effect of a *Dunaliella*extract on growth performance, health condition, immune response and disease resistance in black tiger shrimp (*Penaeus monodon*). *Aquaculture.* 2005. 248(1–4):207–216.

Taylor, L. T. Supercritical Fluid Extraction. John Wiley & Sons: New York, NY. 1996.

Viskari, P. J., & Colyer, C. L. Rapid extraction of phycobiliproteins from cultured cyanobacteria samples. *Anal Biochem.* 2003. 319(2):263–271.

Wang, H. M., Pan, J. L., Chen, C. Y., Chiu, C. C., Yang, M. H., Chang, H. W., & Chang, J. S. Identification of anti-lung cancer extract from *Chlorella vulgaris* CC by antioxidant property using supercritical carbon dioxide extraction. *Proc Biochem.* 2010. 45(12):1865–1872.

Wang, Y., You, J., Yu, Y., Qu, C., Zhang, H., Ding, L., Zhang, H., & Li, X. Analysis of ginsenosides in Panax ginseng in high pressure microwave-assisted extraction. *Food Chem.* 2008. 110(1):161–167.

Ward, O. P., & Singh, A. Omega-3/6 fatty acids: Alternative sources of production. *Proc Biochem.* 2005. 40(12): 3627–3652.

Wen, X., Peng, C., Zhou, H., Lin, Z., Lin, G., Chen, S., & Li, P. Nutritional composition and assessment of *Gracilaria lemaneiformis* Bory. *J Integr Plant Biol.* 2006. 48(9):1047–1053.

Wijffels, R. H. Potential of sponges and microalgae for marine biotechnology. *Trends Biotechnol.* 2008. 26(1):26–31.

22 Biological Activities and Safety Aspects of Fucoxanthin

Naveen Jayapal, Madan Kumar Perumal, and Baskaran Vallikannan

CONTENTS

Highlights 245
Introduction 245
Sources of FX 246
Structure of Fucoxanthin 246
Bioavailability of Fucoxanthin 247
Metabolites of Fucoxanthin 247
Safety Aspects of Fucoxanthin 248
Biological Activity of Fucoxanthin 248
    Antioxidant Activity 248
    Hepatoprotective Effect 249
    Skin Protection 249
    Antiangiogenic Effect 250
    Ocular Protection 250
    Cerebrovascular Protection 250
    Bone-Protection 250
    Anti-Inflammatory Activity 250
    Anticancer Effect 251
    Antiobesity Effect 251
    Antidiabetic Effect 252
Conclusion 252
Acknowledgments 252
Author Contributions 252
References 252

HIGHLIGHTS

1. Carotenoids are tetraterpenoid organic pigments synthesized by plants, algae, fungi, yeasts and bacteria.
2. Among those carotenoids of marine origin, fucoxanthin (FX) acquires a major part when compared to other marine carotenoids.
3. FX with molecular formula $C_{42}H_{58}O_6$ contain a unique structure with an allelic bond and 5,6-monoepoxy group and six oxygen atoms, making it a potent antioxidant and antiobesity carotenoid.
4. FX's absorption and its bioavailability is higher than other carotenoids.
5. FX possesses numerous biological activities such as antioxidant, antidiabetic, antiobesity, antiangiogenic, anticancer, anti-inflammatory, bone-protective, cerebrovascular protective, skin protective, ocular protective and hepatoprotective effects.

INTRODUCTION

Carotenoids are tetraterpenoid organic pigments synthesized by plants, algae, fungi, yeasts and bacteria. Carotenoids have typical linear C40 molecular backbone encompassing 11 conjugated double bonds, which are found in photosynthetic organisms, including seaweeds

(Farré et al. 2010; Takaichi 2011). Carotenoids are the second most naturally occurring pigments on earth, with more than 750 members characterized to date. Among them, almost 50 carotenoids are consumed in the human diet (Krinsky 1993; Crews et al. 2001). In plants, carotenoids are synthesized in the form of color pigments reflecting red, orange and yellow in flowers, vegetables and fruits. While animals cannot synthesize carotenoids, they absorb carotenoids in their diet, which are responsible for the color of some animals like lobster, flamingo and fishes (Farré et al. 2010). Approximately 10–15 carotenoids represent most of its dietary intake, which are found in measurable concentrations in human blood and tissues (Britton 2004; Maiani et al. 2009). Carotenoids are present in both terrestrial plants and in marine sources like seaweed (red, green and brown seaweed) and fishes. Several researchers have proved that marine carotenoids have considerable potential and promising applications for human health. Among those carotenoids of marine origin, fucoxanthin (FX) acquired major part when compared to other marine carotenoids. In brown seaweeds, FX forms a complex with chlorophyll-protein and plays a vital role in photoprotection and light harvesting, i.e., for effective light utilization and up-regulation of photosynthesis. FX generally occurs in the blade of the brown seaweed thallus, which experiences extreme light exposure (Lobban and Wynne 1981). FX has been reported to show numerous biological activities such as antioxidant, hepatoprotective, anti-inflammatory, anticancer, antidiabetic, antiobesity and neuroprotective effects. Hence, in this article, the recent available scientific literature regarding the origin, metabolism, safety and biological activities of FX will be reviewed.

SOURCES OF FX

FX is a light harvesting orange xanthophyll pigment present in the chloroplasts of the Eukaryotic macroalgae, microalgae and diatoms (Cavalier Smith and Chao 2006). In the marine environment, FX contributes more than 10% of the total carotenoids production. FX was first isolated from the marine brown seaweeds viz *Fucus, Dictyota,* and *Laminaria* by Willstatter and Pag (1914). In addition to this, other marine sources of FX is tabulated in Table 22.1.

STRUCTURE OF FUCOXANTHIN

FX's systemic name is 3S,5R,6S,3'S,5'R,6'R)-5,6-Epoxy-3'-ethanoyloxy-3,5'- dihydroxy-6',7'-didehydro-5,6,7,8,5',6'-hexahydro-beta,-beta-caroten-8-one. The

TABLE 22.1
Marine Sources of Fucoxanthin

Marine Source	References
Brown Seaweed	
Alaria crassifolia	Airanthi et al. (2011a)
Cladosiphon okamuranus	Mise et al. (2011)
Cystoseira hakodatensis	Airanthi et al. (2011a); Airanthi et al. (2011b)
Eisenia bicyclis	Airanthi et al. (2011),
Fucus serratus	Strand et al. (1998)
Fucus vesiculosus	Zaragozá et al. (2008)
Hijikia fusiformis	Nishino, (1998); Yan et al. (1999)
Ishige okamurae	Kim et al. (2010)
Kjellmaniella crassifolia	Airanthi et al. (2011a)
Laminaria japonica	Das et al. (2008); Zhang et al. (2008); Miyata et al. (2009)
Laminaria ochotensis	Miyata et al. (2009)
Myagropsis myagroides	Heo et al. (2010)
Padina tetrastromatica	Sangeetha et al. (2010)
Petalonia binghamiae	Murakami et al. (2002)
Sargassum fulvellum	Yan et al. (1999)
Sargassum heterophyllum	Afolayan et al. (2008)
Sargassum horneri	Airanthi et al. (2011b)
Sargassum siliquastrum	Heo et al. (2008)
Undaria pinnatifida	Hosokawa et al. (1999); Yan et al. (1999); Ikeda et al. (2003); Hosokawa et al. (2004); Maeda et al. (2005); Sachindra et al. (2007); Asai et al. (2008); Khan et al. (2008a); Liu et al. (2009); Nakazawa et al. (2009); Airanthi et al. (2011b); Okada et al. (2011)
Diatoms	
Chaetoseros sp.	Iio et al. (2011a); Iio et al. (2011b)
Cylindrotheca closterium	Rijstenbil, (2003)
Odontella aurita	Moreau et al. (2006)
Phaeodactylum tricornutum	Nomura et al. (1997)

structure of FX (Figure 22.1) was first elucidated by Englert et al. (1990). FX contains a unique structure with an allelic bond and 5.6- monoepoxy group and six oxygen atoms (Chemspider 2015), with the molecular formula $C_{42}H_{58}O_6$. The unique functional groups hydroxyl, epoxy, acetate and allenic differentiate FX from other carotenoids like β-carotene and astaxanthin. FX extractable solvents are dimethyl sulfoxide (DMSO), ethanol, methanol, dimethyl ether, diethyl ether, petroleum ether, acetone, ethyl acetate and n-hexane (Kim 2011a; Kanda et al. 2014). FX captures a broader spectrum of light than chlorophyll A and C, (449–540nm), which increases the

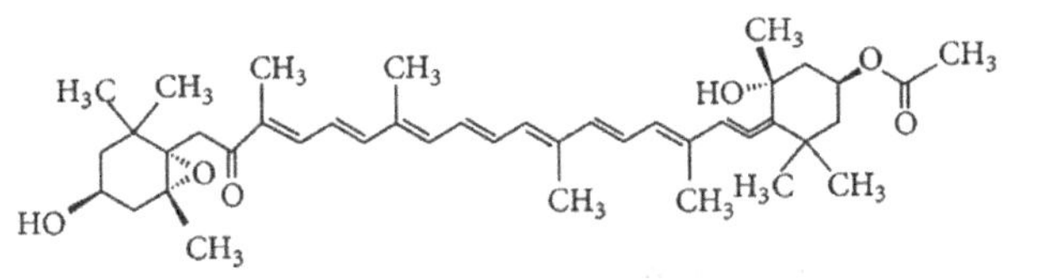

FIGURE 22.1 Molecular structure of FX.

efficiency of photosynthesis (Pyszniak and Gibbs 1992; Kim 2011b). In nature, FX exists in *trans* or cis configuration. The *trans* isomer of FX is the more common, comprising ~90% of that found in nature. The *trans* isomer is chemically more stable and acts as an active antioxidant cis isomer (Nakazawa et al. 2009; Holdt and Kraan 2011).

BIOAVAILABILITY OF FUCOXANTHIN

The absorption and metabolism of FX in the system are closely related to its bioavailability. Bioavailability of carotenoids requires numerous steps like disruption, dispersion in lipid emulsion, solubilization in micelles, movement across the unagitated water layer adjacent to the microvilli, uptake by the cells and incorporation into lymphatic lipoproteins (Furr and Clark 1997; Deming and Erdman 1999; van het Hof et al. 2000). But the absorption of carotenoids is compromised because of the presence of an unstirred water layer across the intestinal barrier. To overcome these limitations, mixed micelles formation of carotenoids along with the lipids has been studied to facilitate diffusion (active transport) (Borel et al. 2013). Micelles formation is dependent on the physicochemical nature of the carotenoids, and mixed micelles play a vital role in digestion, absorption and uptake of carotenoids in intestinal cells (Garrett et al. 1999). FX solubility is a critical factor for oral administration. Carotenoids in mixed micelles increase solubilization and aid in intestinal absorption (Sugawara et al. 2001). A study by Sugawara et al. (2001) showed phospholipids had enhanced effects on carotenoid absorption in cells, which are dependent on the lipophilicity of carotenoids. Intestinal lumen-containing mixed micelles play an essential role in the digestion and absorption of triacylglycerols and the uptake of lipophilic compounds. Combining FX with dietary edible oils or lipids increases the absorption of FX in obese premenopausal women (Abidov et al. 2010) and in KK-Ay mice (Maeda et al. 2007a; Maeda et al. 2007b; Okada et al. 2011). Maeda et al. (2007a; 2007b) studied the solubility of FX and found that fish oil and medium-chain triacylglycerols were more effective in solubilizing the FX than soybean oil and vegetable oils. Sugawara et al. (2001) showed that lysophosphatidylcholine and phospholipase A2 help in the diffusion of carotenoids in the intestinal epithelium and significantly increase the absorption of carotenoids in the digestive tract.

The rate of FX absorption is more than lutein (Yonekura et al. 2010) and the accumulation of FX metabolites was more than that of astaxanthin (Hashimoto et al. 2009). Even though the extraction, purification and structural elucidation of FX are well known, its digestion, absorption and metabolism are still poorly understood. Some reports related to FX digestion, metabolism and its metabolites are summarized below.

METABOLITES OF FUCOXANTHIN

It is also necessary to consider the conversion of FX into its metabolites and their characterization in different organs before the biological activity of FX can be understood. FX is quickly hydrolyzed in the gastrointestinal region, and as a result, no unchanged FX was noticed in the plasma or liver in mice (Asai et al. 2004). Asai et al. (2004) predominantly observed the conversion of fucoxanthinol into amarouciaxanthin A in liver microsomes of mice and in the HepG2 cell.

Fucoxanthinol was considered the primary active metabolite of FX, but amarouciaxanthin A was not observed in human plasma (Hashimoto et al. 2009). Sangeetha et al. (2010) proposed a FX metabolism pathway and projected that its metabolism might be due to enzymatic reactions such as dehydrogenation, oxidation, isomerization, demethylation and deacetylation. In this study, they found fucoxanthinol and amarouciaxanthin A to be important metabolites in FX in intubated rats. The metabolites of fucoxanthinol were first isolated from sea squirt (Nishino et al. 1998; Konishi et al. 2006). In marine animals like oysters and clams, halocynthiaxanthin is formed from metabolism of fucoxanthinol (Murakami et al. 2002; Maoka et al. 2005; Maoka 2011) (Figure 22.2).

The distribution of FX metabolites in tissue has been associated with lipophilicity, but the depletion was inverse to lipophilicity (Yonekura et al. 2010). Yonekura et al. (2010) studied the metabolism, tissue distribution and depletion of FX and observed that fucoxanthinol and amarouciaxanthin A partitioned more into adipose tissues than into plasma, liver and kidney of ICR mice. The half-life of the depletion (t1/2) of FX metabolites in adipose tissues (>41 days) was longer than that in kidneys (4.44 days), liver (2.63 days) and plasma (1.16 days). Conversion of FX into metabolites is given sequentially (Figure 22.3).

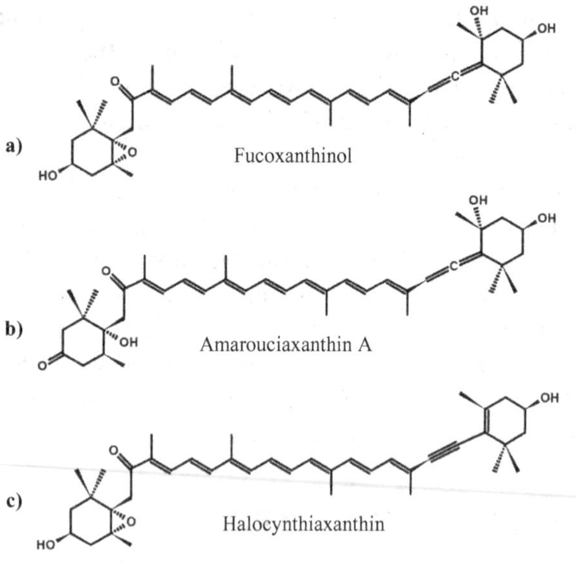

FIGURE 22.2 Molecular structures of FX metabolites: a) Fucoxanthinol; b) Amarouciaxanthin A; c) Halocynthiaxanthin.

FIGURE 22.3 Sequential order of FX metabolism and its metabolites.

SAFETY ASPECTS OF FUCOXANTHIN

Most studies recommend FX as a safe pharmaceutical ingredient (Ishikawa et al. 2008; Kadekaru et al. 2008). But some studies reported the adverse effect of FX on normal cells both *in vitro* and *in vivo* (Ikeda et al. 2003; Liu et al. 2011). FX showed subchronic toxicity and genotoxicity in mice (Iio et al. 2011). Woo et al. (2010) revealed that the dose of FX (50 mg/kg, daily for 28 days) markedly increased blood cholesterol levels in rats and ICR mice. Clinical studies showed that FX was easily metabolized (Riccioni et al. 2011) and repeated oral dosing of FX (95% purity) for 28 days did not show any obvious toxicity in rats (Kadekaru et al. 2008). Toxicity studies with extracts from *Fucus vesiculosus* (containing 0.0012% FX) indicated that a daily dosage of 750 mg/kg for 4 weeks did not show any relevant signs of acute toxicity in mice and rats (Zaragozá et al. 2008).

In vivo toxicity study of a single dose (1000 and 2000 mg/kg) and repeated oral dose (500 and 1000 mg/kg for 30 days) of purified FX (93% purity) to ICR mice did not showed any mortality and abnormalities in gross appearance. Also, no significant changes were observed in histological studies of different tissues like kidney, liver, spleen and gonadal tissues. Subsequently, FX treatment showed no genotoxicity and no mutagenic effects on the bone-marrow cells of mice. No mortality and appearance changes were observed in a single oral dose study. The LD 50 of FX (2000 mg/kg body weight) and 13-week oral dose (200 mg/kg body weight) study showed no adverse effect under subchronic dose condition. FX did not show toxicity and mutagenicity, hence it was concluded that FX was a safe compound for pharmaceutical usage (Beppu et al. 2009a; Beppu et al. 2009b; Iio et al. 2011).

BIOLOGICAL ACTIVITY OF FUCOXANTHIN

FX exhibited various pharmacological effects including antioxidant, hepatoprotective, antiangiogenic, anticancer, anti-inflammatory, antidiabetic, antiobesity, bone-protective and cerebrovascular protective and skin and ocular protective effects (Figure 22.4).

Antioxidant Activity

Oxidant stress has been reported to be involved in the pathogenesis of several human disorders. An increase in the oxidative stress is due to dysfunction of the antioxidant defense system. The carotenoids act as a primary antioxidant by quenching singlet oxygen and scavenging free radicals (Stahl and Sies 2012). FX primarily acts by donating an electron to reactive oxygen species (ROS) instead of a proton. FX structure and the presence of functional groups play a vital role in its antioxidant activity and was found higher when compared to other carotenoids. FX is shown to enhance the antioxidant capacity of blood serum levels in mammals. The key structural differences between FX and other carotenoids are the existence of an unusual allenic carbon (C-7'),

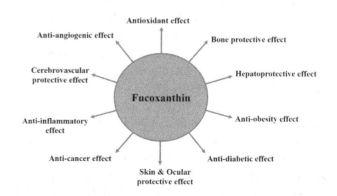

FIGURE 22.4 Biological activities of FX on different human complications. (The red arrow indicates the attenuation effect of FX.)

5,6-monoepoxide, a carbonyl group, an acetyl group and two hydroxyl groups in the terminal ring of FX (Nomura et al. 1997; Yan et al. 1999). The allenic bond of FX was responsible for its higher antioxidant activity (Sachindra et al. 2007), and the presence of six oxygen atoms might be more sensitive to radicals, exclusively under anoxic conditions. Nomura et al. (1997) suggested that the carotenoid antioxidant activity is well correlated with the existence of intramolecular oxygen atoms. In addition to FX, fucoxanthinol also showed *in vitro* free-radical scavenging and singlet oxygen quenching activity similar to α-tocopherol.

Li et al. (2000) showed that FX and β-carotene could minimize the oxidation of vitamin D2 by quenching singlet oxygen in the presence of light, riboflavin and oxygen. Sangeetha et al. (2008, 2009) compared the effects of FX and β-carotene on oxidative stress indicators (glutathione transferase, catalase, and Na$^+$K$^+$-ATPase) and its probable role in suppressing lipid peroxidation resulting from retinol deficiency in rats. The results conclude that FX more effectively downregulated lipid peroxidation in plasma and liver tissue. FX exerts its antioxidant effect by acting against pro-oxidant, as confirmed by the enhanced expression of HO-1 and NQO1 in murine hepatic BNL CL.2 cells through activation of the Nrf2/ARE system (Liu et al. 2011). Ha et al. (2013) found that FX supplementation activated the Nrf2 pathway and downregulated NQO1, resulting in an improved antioxidant effect observed in the serum of obese rats. Dang et al. (2018) found the antioxidant activity of FX extracted from brown seaweed was more than that in phenolic compound.

Hepatoprotective Effect

Park et al. (2011) demonstrated that FX significantly reduced the accumulation of hepatic lipid droplets in mice fed a high-fat diet (HFD) and protected against liver damage by inhibiting the activity of fatty acid synthesis–related enzymes in the liver. Woo et al. (2010) found that FX significantly lowered the hepatic lipid contents; meanwhile, fecal lipids and feces weight markedly increased in HFD-fed C57BL/6N mice due to the inhibition of lipid absorption. Liu et al. (2011) demonstrated that FX (1–20 μM) pretreatment for 24 hours attenuated the oxidative damage and decreased cell proliferation induced by ferric nitrilotriacetate in murine hepatic BNL CL.2 cells in a dose-dependent manner. Additionally, FX not only decreased the level of protein carbonyl contents and thiobarbituric acid-reactive substances but also increased the level of GSH in a concentration-dependent manner.

Also, FX did not show any dose-dependent (0.05% and 0.2%) effect, and the study concluded that the 0.05% might be sufficient to improve the hepatic lipid content. But Maeda et al. (2008) found no significant changes in plasma lipids in Wistar rats. Park et al. (2011) fond that FX normalized the hepatic glycogen content in HFD fed mice by upregulating glucokinase significantly in the liver, which meanwhile increased the ratio of glycogen and hepatic glucokinase/glucose-6-phosphatase content. An increase in the content of docosahexaenoic acid (DHA) reduced the activity of hepatic fatty acid synthesis enzymes and increased β-oxidation of hepatic fatty acid, due to which the hepatic lipid content was reduced. Tsukui et al. (2009) initially described that FX and fucoxanthinol improved the amount of DHA and arachidonic acid in the liver of KK-Ay mice, but in the small intestine, the DHA content remained unaltered. Subsequently, Tsukui et al. (2007) confirmed that FX improved the level of DHA in the liver of normal adult C57BL/6J mice. An increase in arachidonic acid (ω-6) levels was observed in FX-fed mice, demonstrating that FX might have altered the metabolic pathways of ω-6 and ω-3 highly unsaturated fatty acids (Tsukui et al. 2007; Maeda et al. 2008; Tsukui et al. 2009).

Skin Protection

Increased exposure to sunlight would lead to ROS generation, causing angiogenesis and inflammatory reaction of the skin, since sunlight contains UV radiation. Skin angiogenesis is the primary factor in damage to the cellular constituents, causing diseases like skin cancer pigmentation, erythema, laxity and wrinkling (Heo and Jeon 2009; Urikura 2011). FX significantly reduced the intracellular ROS generated by UV-B radiation in human fibroblasts. The increase in cell survival percentage and the inhibition of cell injury in FX pretreated human fibroblast cells suggested that FX protected skin against photodamage induced by UV-B irradiation from sunlight (Heo and Jeon 2009).

Shimoda et al. (2010) found that FX inhibited melanogenesis in melanoma, tyrosinase activity and UV-B induced skin pigmentation. The results revealed that FX significantly downregulated the expression of endothelin receptor A, p75 neurotrophin receptor, melanocortin 1 receptor, tyrosinase-related protein 1, prostaglandin E receptor 1 and cyclooxygenase-2. FX showed antipigmentary activity by oral or topical application in UV-B induced melanogenesis via the suppression of melanogenic stimulant receptors and prostaglandin E2 synthesis.

Urikura et al. (2011) showed that topical treatment with FX prevented skin-wrinkle formation and skin photoaging in UV-B irradiated hairless mice through its antioxidant and antiangiogenic effect. FX significantly suppressed UV-B induced epidermal hypertrophy, which causes wrinkle formation, the increase of thiobarbituric acid-reactive substances (TBARS) and matrix metalloproteinases-13 expression in the skin of hairless mice. Hence, the study suggested that FX may be used as an ingredient in cosmetics or as a skin-protective sunscreen.

Antiangiogenic Effect

FX showed an antiangiogenic effect by preventing angiogenesis-related diseases such as cancer, diabetic retinopathy, atherosclerosis and psoriasis. Sugawara et al. (2006) studied the anti-angiogenic effect of FX in rat aortic rings and cultured endothelial cells of the human umbilical vein, and they found that FX significantly suppressed the formation of new blood vessels and the differentiation of endothelial progenitor cells into the endothelial cell and significantly reduced the endothelial cells' tube length. An *ex vivo* angiogenesis assay showed that FX and fucoxanthinol inhibited microvessel outgrowth in a rat aortic ring.

Ocular Protection

Posterior capsule opacification of cataract is the critical long-term complication of extracapsular cataract extraction due to the migration of the epithelial cells of the lens and the proliferation left in the capsular bag after cataract surgery (Van Tenten et al. 2001). FX acted as an efficient and safe antiproliferative agent by inhibiting the growth of SRA 01/04, human lens epithelial cells. FX could be applied to the formulation of ocular implant products used after-cataract to prevent cataracts (Moreau et al. 2006). Shiratori et al. (2005) showed FX suppressed the development of the lipopolysaccharide-induced uveitis in male Lewis rats, showing the antiocular inflammatory effect of FX.

Cerebrovascular Protection

Ikeda et al. (2003) studied the effect of *Undaria pinnatifida* (Wakame) on the development of cerebrovascular diseases in stroke-prone spontaneously hypertensive rats. The brown seaweed wakame, containing FX, delayed stroke development by maintaining the normal blood pressure and it increased the life-span of hypertensive rats. Also, FX attenuated neuronal cell injury through its free-radical scavenging activity in reoxygenation and hypoxia conditions. This result showed FX's beneficial effect in cerebrovascular protection against ischaemic neuronal cell death in stroke-prone spontaneously hypertensive rat (SHRSP) with stroke.

FX significantly acted against intracellular ROS and H_2O_2-induced neuronal apoptosis, due to which the altered activities of PI3-K/Akt and ERK pathways was restored in SH-SY5Y cells. Subsequently, specific inhibitors of mitogen-activated protein kinase kinase and glycogen synthase kinase 3β significantly protected against H_2O_2-induced neuronal death (Yu et al. 2017).

Zhang et al. (2017) investigated the role of FX on the Nrf2-antioxidant-response element and Nrf2-autophagy pathways in the putative neuroprotection of traumatic brain injury in the mice model. FX alleviated traumatic brain injury–induced secondary brain injury, including neurological deficits, cerebral edema, brain lesion and neuronal apoptosis. Additionally, it was also observed that the upregulation of malondialdehyde, glutathione peroxidase activity increased neuron survival and reduced the ROS level.

Bone-Protection

Das et al. (2010) examined FX's effect on the differentiation of precursor RAW264.7 monocytes and its cytotoxic effect on differentiated RAW264.7 cells and the osteoblast-like cell line MC3T3-E1. The result showed that FX suppressed the differentiation of RAW264.7 cells significantly without any toxic effect on RAW264.7 cells. Suppression of differentiation was achieved by inducing apoptosis through caspase-3 activation and cleavage to the enzyme poly-ADP-ribose polymerase in RAW264.7 cells. But the cell viability of MC3T3-E1 cells was unchanged, indicating that the FX was cytotoxic to osteoclast-like RAW264.7 cells not for osteoblast-like MC3T3-E1 cells. The study revealed that the osteoclastogenesis suppression of FX was achieved by inhibiting osteoclast differentiation but did not show the antagonistic effect on bone formation. This result indicated that FX is helpful for the prevention of bone diseases such as rheumatoid arthritis and osteoporosis.

Anti-Inflammatory Activity

Inflammatory response is a self-defensive mechanism encountered with various pathogenic stimuli. Inflammatory cells are activated by inflammation mediators, which generate superoxide anion and nitric oxide radicals and become a destructive self-damaging process (Choi et al. 2008; Zaragozá et al. 2008).

FX inhibited the inflammatory cytokines and mediators in lipopolysaccharide-stimulated RAW 264.7 macrophages. Induced nitric oxide synthase and cyclooxygenase-2 protein expression was inhibited by FX, which in turn reduced the level of nitric oxide, tumor necrosis factor (TNF) alpha, prostaglandin E2, interleukin-6 and interleukin-1β through the phosphorylation of mitogen-activated protein kinases and the inhibition of NF-κB activation (Heo et al. 2008; Kim et al. 2010). Mast cells played an important role in inflammation and immediate-type allergic reactions, and FX inhibited the degranulation of mast cells by suppressing antigen-induced aggregation of high affinity IgE receptors by activating the degranulating signals of mast cells (Sakai et al. 2009).

Khan et al. (2008a, b) studied the anti-inflammatory activity of methanol extract of *U. pinnatifida* against mouse erythema and ear edema induced by phorbol myristate acetate. The extract suppressed erythema and edema with relative suppression of 78% and 85%, respectively (Khan et al. 2008b), and inhibited erythema by 50% when applied within 1 hour before or 15 minutes after application of phorbol myristate acetate (Khan et al. 2008a). In addition, the extract inhibited the acetic acid–induced writhing response in the analgesic test and also confirmed antipyretic activity in yeast-induced hyperthermic mice (Khan et al. 2009).

Sakai et al. (2011) examined the effect of FX on dinitrofluorobenzene-induced contact hypersensitivity in mice to elucidate their effect on mast-cell degranulation *in vivo*. FX significantly inhibited swelling of the ear and reduced the TNF-α and histamine level, which showed FX's anti-inflammatory effect.

ANTICANCER EFFECT

Oxidative stress is the major factor involved in the pathogenesis of several human cancers (Reuter et al. 2010). The anticancer effect of carotenoids is largely attributed to their antioxidant potential (Fiedor and Burda 2014). A vast number of studies have quantified the anticancer property of FX in various human cancers including those of the colon, breast, liver, leukemia and prostate.

In DLD-1 cells, FX suppressed protein expression and activation of integrin signals and in AOM/DSS mice mediated its chemopreventive effect by suppressing polyp formation and increasing the number of anoikis cells in normal colonic mucosa (Terasaki et al. 2019). For the first time, Ravi et al. (2018) reported that FX, when loaded with nanogels and glycolipids, showed higher anticancer activity in Caco-2 cells, which was mediated through ROS generation via a caspase-dependent mechanism (Figure 22.5). Carotenoids including FX in combination with low doses of doxorubicin selectively altered oxidative stress-mediated apoptosis in breast cancer cells (Vijay et al. 2018). FX in SGC-7901 (human

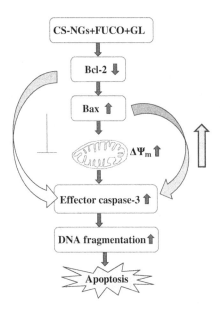

FIGURE 22.5 Anticancer activity of FX loaded in polymeric gel in Caco-2 cells mediated through ROS generation via caspase-dependent mechanism.

gastric cancer) cells induced both autophagy and apoptosis by upregulating the expressions of beclin-1, LC3 and cleaved caspase-3 and by downregulating Bcl-2 (Zhu et al. 2018).

Fucoxanthinol induced anoikis in human colorectal cancer cells through suppression of integrin signals (Terasaki et al. 2017). In human glioblastoma cells (U87 and U251), FX inhibited PI3K/Akt/mTOR pathways thereby activating apoptosis and reducing cell proliferation, migration and invasion (Liu et al. 2016). FX induced G0/G1 cell-cycle arrest and induced apoptosis in human bladder cancer (T24) cells via downregulating mortalin (Wang et al. 2014). Wang et al. (2014) showed that extracts from New Zealand *Undaria pinnatifida* contained FX as functional biomaterial for cancer *in vitro*. FX from brown alga *Saccharina japonica* showed antimetastatic effect both *in vitro* and *in vivo* of highly metastatic B16-F10 melanoma cells (Chung et al. 2013). A study performed in MDA-MB-231 and MCF-7 breast cancer cells showed the cytotoxicity induction potential of FX (de la Mare et al. 2013). FX from *Undaria pinnatifida* exhibited an antiproliferative effect in SK-Hep-1 cells by mainly upregulating Cx32 and Cx43 (Liu et al. 2009).

ANTIOBESITY EFFECT

FX isolated from brown algae showed increased lipolysis and inhibited lipogenesis in oleic acid–induced FL83B hepatocytes through promoting Sirt1/AMPK pathways

(Chang et al. 2018). Rebello et al. (2017) showed the anti-obesity effects of FX and fucoxanthinol, where these carotenoids did not induce browning in human adipocytes. FX when loaded with Chitosan-glycolipid nanogels did not show any acute or subacute toxicity in experimental rats (Ravi et al. 2015). Dietary treatment of FX to diabetic/obese KK-A(y) mice increased HDL-cholesterol and non-HDL-cholesterol levels by inducing sterol regulatory element-binding protein (SREBP) expression and also reduced cholesterol uptake in the liver via downregulation of low-density lipoprotein receptor (LDLR) and SR-B1 (Beppu et al. 2012). *Petalonia binghamiae* extract and its constituent FX exerted an antiobesity effect on HFD-fed mice by promoting β-oxidation and reducing lipogenesis (Kang et al. 2012). Hu et al. (2012) reported that FX, in combination with conjugated linoleic acid, exerted an antiobesity effect by regulating mRNA expression of enzymes related to lipid metabolism in the white adipose tissue of diet-induced obese rats. A *Undaria* lipid capsule containing n-3 polyunsaturated fatty acid rich scallop phospholipids with an incorporation of brown seaweed (*Undaria pinnatifida*) lipids containing FX exhibited enhanced antiobesity effects in KK-A(y) mice (Okada et al. 2011). A study by Maeda et al. (2007) showed that FX in combination with fish oil attenuated the weight gain of white adipose tissue of diabetic/obese KK-A(y) mice. FX from the edible seaweed *Undaria pinnatifida* exhibited antiobesity activity by upregulating UCP1 expression in white adipose tissues (Maeda et al. 2005).

ANTIDIABETIC EFFECT

Lin et al. (2017) reported the synergistic effect FX and low molecular weight Fucoidan (LMF) in db/db mice where the combination of FX and LMF showed better antidiabetic effect than that of FX or LMF alone. In diabetic/obese KK-A(y) mice, FX suppressed the development of hyperglycemia and hyperinsulinemia by activating the insulin signaling pathway, including GLUT4 translocation, and inducing GLUT4 expression in the soleus and EDL muscles (Nishikawa et al. 2012). FX from *Undaria pinnatifida* ethanol extract prevented insulin resistance and hepatic fat accumulation in HFD-induced obese C57BL/6J mice by modulating the hepatic glucose and lipid homeostasis (Park et al. 2011). Woo et al. (2010) showed that FX influenced the regulation of lipid metabolism in plasma and hepatic tissues, exhibiting an antidiabetic effect in HFD-fed C57BL/6N mice. FX regulated the expression of key inflammatory adipocytokines in the white adipose tissue of diabetic/obese KK-A(y) mice, but it did not affect lean C57BL/6J mice (Hosokawa et al. 2010). Maeda et al. (2007) reported that the antidiabetic effect of FX in combination with fish oil attenuated weight gain in the white adipose tissue of diabetic/obese KK-A(y) mice.

CONCLUSION

FX, a naturally occurring marine carotenoid, has shown biological activity against different human complication/disorders. FX acts on a range of biological complications, having antioxidant, antidiabetic, antiobesity, antiangiogenic, anticancer, anti-inflammatory, bone-protective, cerebrovascular protective, skin and ocular protective and hepatoprotective effects. The antioxidant properties of FX are due to its unique chemical structure. Even though the principal role of FX is attributable to its antioxidant activity (i.e., scavenging oxygen radicals), its specific mode of action in the alleviation of various biological disorders needs elucidation. Most of the studies revealed that the FX was not a toxic compound, showing no toxic effect in either *in vitro* or *in vivo* models. Mixed micelles of FX increase the bioavailability of FX and its metabolites (fucoxanthinol, amarouciaxanthin A and halocynthiaxanthin). Treating disease with nutraceuticals (FX) obviates the toxicity and side-effects of chemical drugs, and so research into the molecular mechanisms concerning the effect of FX on different complications is warranted. However, extensive studies like animal experiments and well-controlled human clinical trials are suggested to determine the safety and required daily dosage of FX. Also, technical alterations like encapsulation and sensory examinations must be undertaken before FX can be used effectively as a functional food constituent.

ACKNOWLEDGMENTS

The authors greatly acknowledge the University Grant Commission (UGC), SERB-DST (GAP-509) Government of India for financial assistance.

AUTHOR CONTRIBUTIONS

JN and PMK wrote the book chapter, VB edited the book chapter and helped review the literature. All authors approved the final version of the book chapter.

REFERENCES

Abidov, M., Ramazanov, Z., Seifulla, R., Grachev, S. The effects of Xanthigen in the weight management of obese premenopausal women with non-alcoholic fatty liver disease and normal liver fat. *Diabetes Obes. Metab.* 2010. 12: 72–81.

Afolayan, A. F., Bolton, J. J., Lategan, C. A., Smith, P. J., Beukes, D. R. Fucoxanthin tetraprenylated toluquinone and toluhydroquinone metabolites from *Sargassum heterophyllum* inhibit the in vitro growth of the malaria parasite *Plasmodium falciparum*. *Z. Naturforsch. C* 2008. 63: 848–852.

Airanthi, M. K., Hosokawa, M., Miyashita, K. Comparative antioxidant activity of edible Japanese brown seaweeds. *J. Food Sci.* 2011a. 76: C104–C111.

Airanthi, M. K. W. A., Sasaki, N., Iwasaki, S., Baba, N., Abe, M., Hosokawa, M., Miyashita, K. Effect of brown seaweed lipids on fatty acid composition and lipid hydroperoxide levels of mouse liver. *J. Agric. Food Chem.* 2011b. 59: 4156–4163.

Aldini, G., Yeum, K.J., Niki, E. and Russell, R.M. Eds. *Biomarkers for Antioxidant Defense and Oxidative Damage: Principle and Practical Applications.* Wiley-Blackwell Publishing: Ames, IA, 2010.

Asai, A., Sugawara, T., Ono, H., Nagao, A. Biotransformation of fucoxanthinol into amarouciaxanthin a in mice and HepG2 cells: Formation and cytotoxicity of fucoxanthin metabolites. *Drug Metab. Dispos.* 2004. 32(2): 205–211.

Asai, A., Yonekura, L., Nagao, A. Low bioavailability of dietary epoxyxanthophylls in humans. *Br. J. Nutr.* 2008. 100: 273–277.

Beppu, F., Hosokawa, M., Niwano, Y., Miyashita, K. Effects of dietary fucoxanthin on cholesterol metabolism in diabetic/obese KK-A(y) mice. *Lipids Health Dis.* 2012. 11: 112.

Beppu, F., Niwano, Y., Sato, E., Kohno, M., Tsukui, T., Hosokawa, M., Miyashita, K. In vitro and in vivo evaluation of mutagenicity of FX(FX) and its metabolite Fucoxanthinol. *J. Toxicol. Sci.* 2009b. 34: 693–698.

Beppu, F., Niwano, Y., Tsukui, T., Hosokawa, M., Miyashita, K. Single and repeated oral dose toxicity study of fucoxanthin, a marine carotenoid, in mice. *J. Toxicol. Sci.* 2009a. 34: 501–510.

Borel, P., Lietz, G., Goncalves, A., Szabo de Edelenyi, F., Lecompte, S., Curtis, P., Goumidi, L., Caslake, M. J., Miles, E. A., Packard, C., Calder, P. C., Mathers, J. C., Minihane, A. M., Tourniaire, F., Kesse-Guyot, E., Galan, P., Hercberg, S., Breidenassel, C., González Gross, M., Moussa, M., Meirhaeghe, A., Reboul, E. CD36 and SR-BI are involved in cellular uptake of provitamin A carotenoids by Caco-2 and HEK cells, and some of their genetic variants are associated with plasma concentrations of these micronutrients in humans. *J. Nutr.* 2013. 143(4): 448–456.

Britton, G., Liaaen Jensen, S., Pfander, H. E. *Carotenoids—Handbook.* Birkhauser Verlag: Basel, 2004.

Cavalier-Smith, T., Chao, E. E. Phylogeny and mega systematics of phagotrophic heterokonts (kingdom *Chromista*). *J. Mol. Evol.* 2006. 62(4): 388–420.

Chang, Y. H., Chen, Y. L., Huang, W. C., Liou, C. J. Fucoxanthin attenuates fatty acid–induced lipid accumulation in FL83B hepatocytes through regulated Sirt1/AMPK signaling pathway. *Biochem. Biophys. Res. Commun.* 2018. 495(1): 197–203.

Chemspider FUCO. Available at: http://www.chemspider.com/Chemical-Structure. 2015. 21864745.

Choi, S. K., Park, Y. S., Choi, D. K., Chang, H. I. Effects of astaxanthin on the production of NO and the expression of COX-2 and iNOS in LPS-simulated BV2 microglial cells. *J. Microbiol. Biotechnol.* 2008. 18: 1990–1996.

Chung, T. W., Choi, H. J., Lee, J. Y., Jeong, H. S., Kim, C. H., Joo, M., Choi, J. Y., Han, C. W., Kim, S. Y., Choi, J. S., Ha, K. T. Marine algal fucoxanthin inhibits the metastatic potential of cancer cells. *Biochem. Biophys. Res. Commun.* 2013. 439(4): 580–585.

Crews, H., Alink, G., Andersen, R., Braesco, V., Holst, B., Maiani, G., Ovesen, L., Scotter, M., Solfrizzo, M., van den Berg, R., Verhagen, H., Williamson, G. A critical assessment of some biomarker approaches linked with dietary intake. *Br. J. Nutr.* 2001. 86: S5–35.

Dang, T. T., Bowyer, M. C., Van Altena, I. A. V., Scarlett, C. J. Comparison of chemical profile and antioxidant properties of the brown algae. *Int. J. Food Sci. Technol.* 2018. 53: 174–181.

Das, S. K., Hashimoto, T., Kanazawa, K. Growth inhibition of human hepatic carcinoma HepG2 cells by FX is associated with down-regulation of cyclin D. *Biochim. Biophys. Acta.* 2008. 1780: 743–749.

Das, S. K., Ren, R. D., Hashimoto, T., Kanazawa, K. Fucoxanthin induces apoptosis in osteoclast-like cells differentiated from RAW264.7 cells. *J. Agric. Food Chem.* 2010. 58: 6090–6095.

de la Mare, J. A., Sterrenberg, J. N., Sukhthankar, M. G., Chiwakata, M. T., Beukes, D. R., Blatch, G. L., Edkins, A. L. Assessment of potential anti-cancer stem cell activity of marine algal compounds using an in vitro mammosphere assay. *Cancer Cell Int.* 2013. 13: 39.

Deming, D. M., Erdman, J. W. Jr. Mammalian carotenoid absorption and metabolism. *Pure Appl. Chem.* 1999. 71: 2213–2223.

Englert, G., Bjørnland, T., Liaaen-Jensen, S. 1D and 2D NMR study of some allenic carotenoids of the FXseries. *Magn. Reson. Chem.* 1990. 28: 519–528.

Farré, G., Sanahuja, G., Naqvi, S., Bai, C., Capell, T., Zhu, C., Christou, P. Travel advice on the road to carotenoids in plants. *Plant Sci.* 2010. 179: 28–48.

Fiedor, J., Burda, K. Potential role of carotenoids as antioxidants in human health and disease. *Nutrients.* 2014. 6: 466–488.

Furr, H. C., Clark, R. M. Intestinal absorption and tissue distribution of carotenoids. *J. Nutr. Biochem.* 1997. 8: 364–377.

Garrett, D. A., Failla, M. L., Sarama, R. J. Development of an in vitro digestion method to assess carotenoid bioavailability from meals. *J. Agric. Food Chem.* 1999. 47: 4301–4309.

Gerster, H. Anticarcinogenic effect of common carotenoids. *Int. J. Vitam. Nutr. Res.* 1993. 63: 93–121.

Ha, A. W., Na, S. J., Kim, W. K. Antioxidant effects of fucoxanthin rich powder in rats fed with high fat diet. *Nutr. Res. Pract.* 2013. 7(6): 475–480.

Hashimoto, T., Ozaki, Y., Taminato, M., Das, S. K., Mizuno, M., Yoshimura, K., Maoka, T., Kanazawa, K. The distribution and accumulation of fucoxanthin and its metabolites after oral administration in mice. *Br. J. Nutr.* 2009. 102: 242–248.

Heo, S. J., Jeon, Y. J. Protective effect of fucoxanthin isolated from *Sargassum siliquastrum* on UV-B induced cell damage. *J. Photochem. Photobiol. B.* 2009. 95: 101–107.

Heo, S., Ko, S., Kang, S., Kang, H., Kim, J., Kim, S., Lee, K., Cho, M., Jeon, Y. Cytoprotective effect of fucoxanthin isolated from brown algae *Sargassum siliquastrum* against H2O2-induced cell damage. *Eur. Food Res. Technol.* 2008. 228: 145–151.

Heo, S. J., Yoon, W. J., Kim, K. N., Ahn, G. N., Kang, S. M., Kang, D. H., Affan, A., Oh, C., Jung, W. K., Jeon, Y. J. Evaluation of anti-inflammatory effect of fucoxanthin isolated from brown algae in lipopolysaccharide-stimulated RAW 264.7 macrophages. *Food Chem. Toxicol.* 2010. 48: 2045–2051.

Holdt, S. L., Kraan, S. Bioactive compounds in seaweed: Functional food applications and legislation. *J. Appl. Phycol.* 2011. 23: 543–597.

Hosokawa, M., Kudo, M., Maeda, H., Kohno, H., Tanaka, T., Miyashita, K. Fucoxanthin induces apoptosis and enhances the antiproliferative effect of the PPARγ ligand, troglitazone, on colon cancer cells. *Biochim. Biophys. Acta.* 2004. 1675: 113–119.

Hosokawa, M., Miyashita, T., Nishikawa, S., Emi, S., Tsukui, T., Beppu, F., Okada, T., Miyashita, K. Fucoxanthin regulates adipocytokine mRNA expression in white adipose tissue of diabetic/obese KK-Ay mice. *Arch. Biochem. Biophys.* 2010. 504(1): 17–25.

Hosokawa, M., Wanezaki, S., Miyauchi, K., Kurihara, H., Kohno, H., Kawabata, J., Kawabata, J., Odashima, S., Takahashi, K. Apoptosis-inducing effect of fucoxanthin on human leukemia cell line HL-60. *Food Sci. Technol. Res.* 1999. 5: 243–246.

Hu, X., Li, Y., Li, C., Fu, Y., Cai, F., Chen, Q., Li, D. Combination of fucoxanthin and conjugated linoleic acid attenuates body weight gain and improves lipid metabolism in high-fat diet-induced obese rats. *Arch. Biochem. Biophys.* 2012. 519(1): 59–65.

Iio, K., Okada, Y., Ishikura, M. Single and 13-week oral toxicity study of fucoxanthin oil from microalgae in rats. *J. Food Hyg. Soc. Jpn.* 2011a. 52: 183–189.

Iio, K., Okada, Y., Ishikura, M. Bacterial reverse mutation test and micronucleus test of fucoxanthin oil from microalgae. *Shokuhin Eiseigaku Zasshi* 2011b. 52: 190–193.

Ikeda, K., Kitamura, A., Machida, H., Watanabe, M., Negishi, H., Hiraoka, J., Nakano, T. Effect of *Undaria pinnatifida* (wakame) on the development of cerebrovascular diseases in stroke-prone spontaneously hypertensive rats. *Clin. Exp. Pharmacol. Physiol.* 2003. 30: 44–48.

Ishikawa, C., Tafuku, S., Kadekaru, T., Sawada, S., Tomita, M., Okudaira, T., Nakazato, T., Toda, T., Uchihara, J. N., Taira, N., Ohshiro, K., Yasumoto, T., Ohta, T., Mori, N. Anti-adult T-cell leukemia effects of brown algae fucoxanthin and its deacetylated product, FUCOol. *Int. J. Cancer.* 2008. 123(11): 2702–2712.

Kadekaru, T., Toyama, H., Yasumoto, T. Safety evaluation of fucoxanthin purified from *Undaria pinnatifida*. *Nippon Shokuhin Kagaku Kogaku Kaishi.* 2008. 55(6): 304–308.

Kanda, H., Kamo, Y., Machmudah, S., Wahyudiono, E. Y., Goto, M. Extraction of fucoxanthin from raw macroalgae excluding drying and cell wall disruption by liquefied dimethyl ether. *Mar. Drugs.* 2014. 12(5): 2383–2396.

Kang, S. I., Shin, H. S., Kim, H. M., Yoon, S. A., Kang, S. W., Kim, J. H., Ko, H. C., Kim, S. J. *Petalonia binghamiae* extract and its constituent fucoxanthin ameliorate high-fat diet-induced obesity by activating AMP-activated protein kinase. *J. Agric. Food Chem.* 2012. 60(13): 3389–3395.

Khan, M. N. A., Choi, J. S., Lee, M. C., Kim, E., Nam, T. J., Fujii, H., Hong, Y. K. Anti-inflammatory activities of methanol extracts from various seaweed species. *J. Environ. Biol.* 2008b. 29: 465–469.

Khan, M. N. A., Lee, M. C., Kang, J. Y., Park, N. G., Fujii, H., Hong, Y. K. Effects of the brown seaweed *Undaria pinnatifida* on erythematous inflammation assessed using digital photo analysis. *Phytother. Res.* 2008a. 22: 634–639.

Khan, M. N. A., Yoon, S. J., Choi, J. S., Park, N. G., Lee, H. H., Cho, J. Y., Hong, Y. K. Anti-edema effects of brown seaweed (*Undaria pinnatifida*) extract on phorbol 12-myristate 13-acetate-induced mouse ear inflammation. *Am. J. Chin. Med.* 2009. 37: 373–381.

Kim, K. N., Heo, S. J., Kang, S. M., Ahn, G., Jeon, Y. J. Fucoxanthin induces apoptosis in human leukemia HL-60 cells through a ROS-mediated Bcl-xL pathway. *Toxicol. In Vitro.* 2010. 24: 1648–1654.

Kim, K. N., Heo, S. J., Yoon, W. J., Kang, S. M., Ahn, G., Yi, T. H., Jeon, Y. J. Fucoxanthin inhibits the inflammatory response by suppressing the activation of NF-κB and MAPKs in lipopolysaccharide-induced RAW 264.7 macrophages. *Eur. J. Pharmacol.* 2010. 649: 369–375.

Kim, S. K. Marine medicinal foods: Implications and application of macro and microalgae. In: *Advances in Food and Nutrition Research (Vol. 64).* Academic Press: San Diego, CA, 2011a.

Kim, S. K. Marine Cosmeceuticals: Trends and Prospects. CRC Press: Boca Raton, FL, 2011b.

Krinsky, N. I. Actions of carotenoids in biological systems. *Annu. Rev. Nutr.* 1993. 13: 561–587.

Li, T. L., King, J. M., Min, D. B. Quenching mechanisms and kinetics of carotenoids in riboflavin photosensitized singlet oxygen oxidation of vitamin D2. *J. Food Biochem.* 2000. 24: 477–492.

Lin, H. V., Tsou, Y. C., Chen, Y. T., Lu, W. J., Hwang, P. A. Effects of low-molecular-weight fucoidan and high stability fucoxanthin on glucose homeostasis, lipid metabolism, and liver function in a mouse model of Type II Diabetes. *Mar. Drugs.* 2017. 15(4): E113.

Liu, C. L., Chiu, Y. T., Hu, M. L. Fucoxanthin enhances HO-1 and NQO1 expression in murine hepatic BNL CL.2 cells through activation of the Nrf2/ARE system partially by its pro-oxidant activity. *J. Agric. Food Chem.* 2011.

Liu, C. L., Huang, Y. S., Hosokawa, M., Miyashita, K., Hu, M. L. Inhibition of proliferation of a hepatoma cell line by fucoxanthin in relation to cell cycle arrest and enhanced gap junctional intercellular communication. *Chem. Biol. Interact.* 2009. 182(2–3): 165–172.

Liu, C. L., Liang, A. L., Hu, M. L. Protective effects of fucoxanthin against ferric nitrilotriacetate-induced oxidative stress in murine hepatic BNL CL.2 cells. *Toxicol. In Vitro.* 2011. 25(7): 1314–1319.

Liu, Y., Zheng, J., Zhang, Y., Wang, Z., Yang, Y., Bai, M., Dai, Y. Fucoxanthin activates apoptosis *via* inhibition of PI3K/Akt/mtor pathway and suppresses invasion and migration by restriction of p38-MMP-2/9 pathway in human glioblastoma cells. *Neurochem. Res.* 2016. 41(10): 2728–2751.

Maeda, H., Hosokawa, M., Sashima, T., Funayama, K., Miyashita, K. Fucoxanthin from edible seaweed, *Undaria pinnatifida*, shows antiobesity effect through UCP1 expression in white adipose tissues. *Biochem. Biophys. Res. Commun.* 2005. 332(2): 392–397.

Maeda, H., Hosokawa, M., Sashima, T., Funayama, K., Miyashita, K. Effect of medium-chain triacylglycerols on anti-obesity effect of fucoxanthin. *J. Oleo Sci.* 2007b. 56: 615–621.

Maeda, H., Hosokawa, M., Sashima, T., Miyashita, K. Dietary combination of fucoxanthin and fish oil attenuates the weight gain of white adipose tissue and decreases blood glucose in obese/diabetic KK-Ay mice. *J. Agric. Food Chem.* 2007a. 55: 7701–7706.

Maeda, H., Tsukui, T., Sashima, T., Hosokawa, M., Miyashita, K. Seaweed carotenoid, fucoxanthin as a multi-functional nutrient. *Asia Pac. J. Clin. Nutr.* 2008. 17: 196–199.

Maiani, G., Castón, M. J., Catasta, G., Toti, E., Cambrodón, I. G., Bysted, A., Granado-Lorencio, F., Olmedilla-Alonso, B., Knuthsen, P., Valoti, M., Böhm, V., Mayer-Miebach, E., Behsnilian, D., Schlemmer, U. Carotenoids: Actual knowledge on food sources, intakes, stability and bioavailability and their protective role in humans. *Mol. Nutr. Food Res.* 2009. 53: S194–S218.

Mise, T., Ueda, M., Yasumoto, T. Production of fucoxanthin-rich powder from *Cladosiphon okamuranus*. *Adv. J. Food Sci. Technol.* 2011. 3: 73–76.

Miyata, M., Koyama, T., Kamitani, T., Toda, T., Yazawa, K. Anti-obesity effect on rodents of the traditional Japanese food, tororokombu, shaved *Laminaria*. *Biosci. Biotechnol. Biochem.* 2009. 73: 2326–2328.

Moreau, D., Tomasoni, C., Jacquot, C., Kaas, R., Le Guedes, R., Cadoret, J. P., Muller-Feuga, A., Kontiza, I., Vagias, C., Roussis, V., Roussakis, C. Cultivated microalgae and the carotenoid fucoxanthin from *Odontella aurita* as potent anti-proliferative agents in broncho-pulmonary and epithelial cell lines. *Environ. Toxicol. Pharmcol.* 2006. 22: 97–103.

Murakami, C., Takemura, M., Sugiyama, Y., Kamisuki, S., Asahara, H., Kawasaki, M., Ishidoh, T., Linn, S., Yoshida, S., Sugawara, F., Yoshida, H., Sakaguchi, K., Mizushina, Y. Vitamin A-related compounds, all-trans retinal and retinoic acids, selectively inhibit activities of mammalian replicative DNA polymerases. *Biochim. Biophys. Acta.* 2002. 1574: 85–92.

Nakazawa, Y., Sashima, T., Hosokawa, M., Miyashita, K. Comparative evaluation of growth inhibitory effect of stereoisomers of fucoxanthin in human cancer cell lines. *J. Funct. Foods.* 2009. 1: 88–97.

Nishikawa, S., Hosokawa, M., Miyashita, K. Fucoxanthin promotes translocation and induction of glucose transporter 4 in skeletal muscles of diabetic/obese KK-A(y) mice. *Phytomedicine.* 2012. 9(5): 389–394.

Nishino, H. Cancer prevention by carotenoids. *Mutat. Res.* 1998. 402: 159–163.

Nomura, T., Kikuchi, M., Kubodera, A., Kawakami, Y. Proton-donative antioxidant activity of fucoxanthin with 1,1-diphenyl-2-picrylhydrazyl (DPPH). *Biochem. Mol. Biol. Int.* 1997. 42: 361–370.

Okada, T., Mizuno, Y., Sibayama, S., Hosokawa, M., Miyashita, K. Antiobesity effects of *Undaria* lipid capsules prepared with scallop phospholipids. *J. Food Sci.* 2011. 76: H2–H6.

Park, H. J., Lee, M. K., Park, Y. B., Shin, Y. C., Choi, M. S. Beneficial effects of *Undaria pinnatifida* ethanol extract on diet-induced-insulin resistance in C57BL/6J mice. *Food Chem. Toxicol.* 2011. 49: 727–733.

Pyszniak, A. M., Gibbs, S. P. Immunocytochemical localization of photosystem I and the fucoxanthin-chlorophyll A/C light-harvesting complex in the diatom *Phaeodactylum tricornutum*. *Protoplasma.* 1992. 166: 208–217.

Ravi, H., Arunkumar, R., Baskaran, V. Chitosan-glycolipid nanogels loaded with anti-obese marine carotenoid fucoxanthin: Acute and sub-acute toxicity evaluation in rodent model. *J. Biomater. Appl.* 2015. 30(4): 420–434.

Ravi, H., Kurrey, N., Manabe, Y., Sugawara, T., Baskaran, V. Polymeric chitosan-glycolipid nanocarriers for an effective delivery of marine carotenoid fucoxanthin for induction of apoptosis in human colon cancer cells (Caco-2 cells). *Mater. Sci. Eng. C Mater. Biol. Appl.* 2018. 91: 785–795.

Rebello, C. J., Greenway, F. L., Johnson, W. D., Ribnicky, D., Poulev, A., Stadler, K., Coulter, A. A. Fucoxanthin and its metabolite Fucoxanthinol do not induce browning in human adipocytes. *J. Agric. Food Chem.* 2017. 65(50): 10915–10924.

Reuter, S., Gupta, S. C., Chaturvedi, M. M., Aggarwal, B. B. Oxidative stress, inflammation, and cancer: How are they linked? *Free Radic. Biol. Med.* 2010. 49(11): 1603–1616.

Riccioni, G., D'Orazio, N., Franceschelli, S., Speranza, L. Marine carotenoids and cardiovascular risk markers. *Mar. Drugs.* 2011. 9(7): 1166–1175.

Rijstenbil, J. W. Effects of UVB radiation and salt stress on growth, pigments and antioxidative defence of the marine diatom *Cylindrotheca closterium*. *Mar. Ecol. Prog. Ser.* 2003. 254: 37–48.

Sachindra, N. M., Sato, E., Maeda, H., Hosokawa, M., Niwano, Y., Kohno, M., Miyashita, K. Radical scavenging and singlet oxygen quenching activity of marine carotenoid FX and its metabolites. *J. Agric. Food Chem.* 2007. 55: 8516–8522.

Sakai, S., Sugawara, T., Hirata, T. Inhibitory effect of dietary carotenoids on dinitrofluorobenzene-induced contact hypersensitivity in mice. Biosci. Biotechnol. Biochem. 2011. 75: 1013–1015.

Sakai, S., Sugawara, T., Matsubara, K., Hirata, T. Inhibitory effect of carotenoids on the degranulation of mast cells via suppression of antigen-induced aggregation of high affinity IgE receptors. *J. Biol. Chem.* 2009. 284: 28172–28179.

Sangeetha, R. K., Bhaskar, N., Baskaran, V. Comparative effects of β-carotene and fucoxanthin on retinol deficiency induced oxidative stress in rats. *Mol. Cell. Biochem.* 2009. 331: 59–67.

Sangeetha, R. K., Bhaskar, N., Baskaran, V. Fucoxanthin restrains oxidative stress induced by retinol deficiency through modulation of Na+K+-ATPase and antioxidant enzyme activities in rats. *Eur. J. Nutr.* 2008. 47: 432–441.

Sangeetha, R. K., Bhaskar, N., Divakar, S., Baskaran, V. Bioavailability and metabolism of fucoxanthin in rats: Structural characterization of metabolites by LC-MS (APCI). *Mol. Cell. Biochem.* 2010. 333: 299–310.

Shimoda, H., Tanaka, J., Shan, S. J., Maoka, T. Antipigmentary activity of fucoxanthin and its influence on skin mRNA expression of melanogenic molecules. *J. Pharm. Pharmacol.* 2010. 62: 1137–1145.

Shiratori, K., Ohgami, K., Ilieva, I., Jin, X. H., Koyama, Y., Miyashita, K., Yoshida, K., Kase, S., Ohno, S. Effects of fucoxanthin on lipopolysaccharide-induced inflammation *in vitro* and *in vivo*. *Exp. Eye Res.* 2005. 81: 422–428.

Stahl, W., Sies, H. Photoprotection by dietary carotenoids: Concept, mechanism, evidence and future development. *Mol. Nutr. Food Res.* 2012. 56: 287–295.

Strand, A., Herstad, O., Liaaen-Jensen, S. FX metabolites in egg yolks of laying hens. *Comp. Biochem. Phys. A Mol. Integr. Physiol.* 1998. 119: 963–974.

Sugawara, T., Kushiro, M., Zhang, H., Nara, E., Ono, H., Nagao, A. Lysophosphatidylcholine enhances carotenoid uptake from mixed micelles by Caco-2 human intestinal cells. *J. Nutr.* 2001. 131: 2921–2927.

Sugawara, T., Matsubara, K., Akagi, R., Mori, M., Hirata, T. Antiangiogenic activity of brown algae fucoxanthin and its deacetylated product, FUCOol. *J. Agric. Food Chem.* 2006. 54: 9805–9810.

Takaichi, S. Carotenoids in algae: Distributions, biosynthesis and functions. *Mar. Drugs.* 2011. 9: 1101–1118.

Terasaki, M., Iida, T., Kikuchi, F., Tamura, K., Endo, T., Kuramitsu, Y., Tanaka, T., Maeda, H., Miyashita, K., Mutoh, M. Fucoxanthin potentiates anoikis in colon mucosa and prevents carcinogenesis in AOM/DSS model mice. *J. Nutr. Biochem.* 2019. 64: 198–205.

Terasaki, M., Maeda, H., Miyashita, K., Mutoh, M. Induction of anoikis in human colorectal cancer cells by Fucoxanthinol. *Nutr. Cancer.* 2017. 69(7): 1043–1052.

Tsukui, T., Baba, N., Hosokawa, M., Sashima, T., Miyashita, K. Enhancement of hepatic docosahexaenoic acid and arachidonic acid contents in C57BL/6J mice by dietary fucoxanthin. *Fish. Sci.* 2009. 75(1): 261–263.

Tsukui, T., Konno, K., Hosokawa, M., Maeda, H., Sashima, T., Miyashita, K. Fucoxanthin and fucoxanthinol enhance the amount of docosahexaenoic acid in the liver of KKAy obese/diabetic mice. *J. Agric. Food Chem.* 2007. 55(13): 5025–5029.

Urikura, I., Sugawara, T., Hirata, T. Protective effect of fucoxanthin against UVB-induced skin photoaging in hairless mice. *Biosci. Biotechnol. Biochem.* 2011. 75: 757–760.

van het Hof, K. H., West, C. E., Weststrate, J. A., Hautvast, J. G. A. J. Dietary factors that affect the bioavailability of carotenoids. *J. Nutr.* 2000. 130: 503–506.

Van Tenten, Y., Schuitmaker, H. J., De Wolf, A., Willekens, B., Vrensen, G. F. J. M., Tassignon, M. J. The effect of photodynamic therapy with bacteriochlorin A on lens epithelial cells in a capsular bag model. *Exp. Eye Res.* 2001. 72: 41–48.

Vijay, K., Sowmya, P. R., Arathi, B. P., Shilpa, S., Shwetha, H. J., Raju, M., Baskaran, V., Lakshminarayana, R. Low-dose doxorubicin with carotenoids selectively alters redox status and upregulates oxidative stress-mediated apoptosis in breast cancer cells. *Food Chem. Toxicol.* 2018. 118: 675–690.

Wang, L., Zeng, Y., Liu, Y., Hu, X., Li, S., Wang, Y., Li, L., Lei, Z., Zhang, Z. Fucoxanthin induces growth arrest and apoptosis in human bladder cancer T24 cells by up-regulation of p21 and down-regulation of mortalin. *Acta Biochim. Biophys.* Sin. 2014. 46(10): 877–884.

Wang, S. K., Li, Y., White, W. L., Lu, J. Extracts from New Zealand *Undaria pinnatifida* containing fucoxanthin as potential functional biomaterials against cancer *in vitro*. *J. Funct. Biomater.* 2014. 5(2): 29–42.

Willstatter, R., Page, H. J. Chlorophyll. XXIV. The pigments of the brown algae. *Justus Liebig's Ann. Chem.* 1914. 404: 237–271.

Woo, M. N., Jeon, S. M., Kim, H. J., Lee, M. K., Shin, S. K., Shin, Y. C., Park, Y. B., Choi, M. S. Fucoxanthin supplementation improves plasma and hepatic lipid metabolism and blood glucose concentration in high-fat fed C57BL/6N mice. *Chem. Biol. Interact.* 2010. 186: 316–322.

Yan, X., Chuda, Y., Suzuki, M., Nagata, T. Fucoxanthin as the major antioxidant in *Hijikia fusiformis*, a common edible seaweed. *Biosci. Biotechnol. Biochem.* 1999. 63: 605–607.

Yonekura, L., Kobayashi, M., Terasaki, M., Nagao, A. Keto-carotenoids are the major metabolites of dietary lutein and fucoxanthin in mouse tissues. *J. Nutr.* 2010. 140: 1824–1831.

Yu, J., Lin, J. J., Yu, R., He, S., Wang, Q. W., Cui, W., Zhang, J. R. Fucoxanthin prevents H2O2-induced neuronal apoptosis *via* concurrently activating the PI3-K/Akt cascade and inhibiting the ERK pathway. *Food Nutr. Res.* 2017. 61(1): 1304678.

Zaragozá, M. C., López, D., P Sáiz, M. P., Poquet, M., Pérez, J., Puig-Parellada, P., Mármol, F., Simonetti, P.,

Gardana, C., Lerat, Y., Burtin, P., Inisan, C., Rousseau, I., Besnard, M., Mitjavila, M. T. Toxicity and antioxidant activity *in vitro* and *in vivo* of two *Fucus vesiculosus* extracts. J. Agric. Food Chem. 2008. 56: 7773–7780.

Zhang, L., Wang, H., Fan, Y., Gao, Y., Li, X., Hu, Z., Ding, K., Wang, Y., Wang, X. Fucoxanthin provides neuroprotection in models of traumatic brain injury via the Nrf2-ARE and Nrf2-autophagy pathways. *Sci. Rep.* 2017. 7: 46763.

Zhang, Z. Y., Zhang, P. J., Hamada, M., Takahashi, S., Xing, G. Q., Liu, J. Q., Sugiura, N. Potential chemoprevention effect of dietary fucoxanthin on urinary bladder cancer EJ-1 cell line. *Oncol. Rep.* 2008. 20: 1099–1103.

Zhu, Y., Cheng, J., Min, Z., Yin, T., Zhang, R., Zhang, W., Hu, L., Cui, Z., Gao, C., Xu, S., Zhang, C., Hu, X. Effects of fucoxanthin on autophagy and apoptosis in SGC-7901cells and the mechanism. *J. Cell. Biochem.* 2018. 119(9): 7274–7284.

Section II

Algal Genomics and Metabolomics

23 Functional Omics and Big Data Analysis in Microalgae
The Repertoire of Molecular Tools in Algal Technologies

Chetan Paliwal, Tonmoy Ghosh, Asha A. Nesamma, and Pavan P. Jutur

CONTENTS

Abbreviations 261
Introduction 262
Molecular Tools—Functional Omics 263
Genomics 263
Transcriptomics 263
Proteomics 264
Lipidomics and Metabolomics 264
Advanced Microscopy and Cell Dynamics 265
Integration of Multiomics Data 265
Transcriptional Engineering—A Novel Perspective 265
Big Data Analysis 266
Phylogenomics of Metabolic Pathways 266
Sub-Cellular Networks 266
Prediction of Transcriptional Factors 267
Genome-Scale Metabolic Models (GSMMs) 267
Reconstruction of Microalgal Cell Factories 268
Conclusions 268
References 268

BOX 23.1   SALIENT FEATURES

Integrated omics (iOMICS) generates huge data deciphering which would enable us to better understanding of algal cell machinery. iOMICS would lead to construct an algal genome scale metabolic model for enhancing the production of various high value metabolites. Also, iOMICS would help us to identify transcription factors/regulatory hubs controlling these metabolic pathways.

ABBREVIATIONS

ACCase: Acetyl coenzyme A carboxylase
BN-PAGE: Blue native polyacrylamide gel electrophoresis
CARS: Coherent anti-stokes Raman scattering
DAGs: Diacylglycerols
DGTS: Diacylglyceroltrimethylhomoserine
EDX: Energy dispersive X-ray
ESTs: Expressed sequence tags
FASP: Filter-aided sample preparation
GRNs: Gene regulatory networks
GSMMs: Genome-scale metabolic models
HUP: Hexose proton symporter
MGDG: Monogalactosyldiacylglycerol
MS: Mass Spectroscopy
NADPH: Nicotinamide adenine dinucleotide phosphate
NMR: Nuclear magnetic resonance
PSR 1: Pi Starvation response1
RNA-Seq: RNA sequencing
SDS-PAGE: Sodium dodecyl sulfate polyacrylamide gel electrophoresis

SIMS:	Secondary ion mass spectrometry
TAGs:	Triacylglycerols
TALENs:	Transcription activator-like effector nucleases
TFs:	Transcription factors
TRs:	Transcriptional regulators
ZFNs:	Zinc finger nucleases

INTRODUCTION

Microalgae are a heterogeneous group of photosynthetic green bio-factories that fix ~40–50% of atmospheric carbon dioxide and considered as a potential source for various bioactive compounds like pigments, fatty acids, vitamins, and so on (Cadoret et al. 2012; Jutur et al. 2016; Paliwal et al. 2017). In the past decade, microalgae have emerged as a robust platform for biofuel production as they are rich in lipids, particularly triacylglycerols (TAGs), and pose no competition to either food or feed (Kim et al. 2015). Microalgae have the exceptional ability to tolerate stress conditions and alter the cellular machinery to accumulate lipids and/or carbohydrates along with other bioactive compounds that may be used in cosmetic, food, or pharmaceutical industries (Skjånes et al. 2013). The development of a feasible microalgae-based biorefinery is dependent on successful genetic engineering and mathematical modeling under certain conditions for the production of multiple products including bio-actives of high value (Paliwal et al. 2017).

Genetic engineering has made it possible to produce natural products by engineering and/or introducing specific genes for the production of new or desired products (Keasling 2012). Interestingly, several microalgal genomes have been sequenced to understand the genetic codes for biosynthesis of value-added metabolites, allowing for their pathway engineering (Sasso et al. 2012). However, there is little knowledge about them because of lack of molecular tools and poor expression studies of heterologous nuclear genes due to rapid gene silencing (Singh et al. 2016). Tools like genome sequencing and microalgal omics, including metabolomics, proteomics, transcriptomics, and genomics, have led to an in-depth understanding of changes in cellular metabolism, interaction, and functions of different proteins leading to the production of high-value metabolites (Chen et al. 2017).

The big data obtained through integrated omics with advanced techniques like confocal microscopy would enrich our understanding of the cellular dynamics in microalgae. The phylogenomic analysis and identification of subcellular localization of industrially important bioactive metabolites would help us in identifying missing links in the biosynthesis of a metabolite of interest, which would help us identify certain switch on/off transcription hubs which could improve its production levels. Therefore, a concomitant approach of constructing the genome-scale metabolic models through multi-omics and engineering strategies would help us upregulate the metabolism without compromising growth.

This chapter emphasises the repertoire of molecular tools and techniques primarily focusing on microalgal-based studies, ultimately leading to an in-depth understanding of microalgal cell dynamics, which will provide insights for the modification of metabolic processes for further development of a sustainable multi-product biorefinery (Figure 23.1).

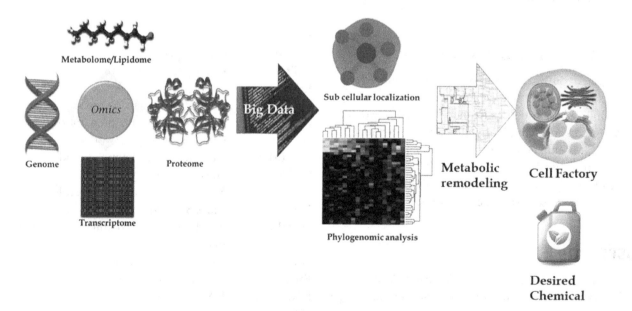

FIGURE 23.1 Integrated omic-engineering scheme for enhanced production of metabolites.

MOLECULAR TOOLS—FUNCTIONAL OMICS

Several advancements in molecular tools have paved the way for phylogenetic analysis and functional gene identification, but DNA mapping has provided insights into the metabolic pathways of the biological system (Jansson et al. 2012). Understanding the dynamics of an active cell needs an alternative approach such as the multi-omics platform, wherein transcriptomics will rationalize the RNA transcripts regulation by fold-change or analysis of the metabolites and / or proteins at certain conditions, thereby integrating all datasets to draw a functional omics pipeline.

GENOMICS

Genomics is the study of the whole genome of an organism, including genes, recombinant cDNA, and expressed sequence tags (ESTs), to obtain a hypothesis of the cellular machinery (Rai et al. 2016). Initially, only information about sequencing and gene assembly can be deduced from the genomic data, but functional genomics can infer the function of a gene helping in understanding and regulation of the metabolic pathways (Jamers et al. 2009). The year 2005 marked the beginning of next-generation sequencing, which has led to an unprecedented rise in the amount of microalgal genome data (Lu et al. 2016). High throughput analysis and systematic sequencing methodology have enabled an understanding of the molecular machinery of microalgae. By the end of 2010, only 7 microalgal genomes existed, but as of now >30 microalgal genomes have been sequenced (Brodie et al. 2017; Rismani-Yazdi et al. 2011).

The genomic revelation of *Chlorella protothecoides* discloses its lower genome size of 22 Mbp (compared to its close members) and hexose proton symporter (HUP)-like genes for glucose absorption, which plays an important role in heterotrophic growth (Gao et al. 2014). The same study has also concluded that the photosynthetic and CO_2 fixation enzymes were down-regulated during the heterotrophic cultivation during glucose assimilation, whereas the glycolysis and TCA cycle metabolism were upregulated.

The genomes of microalgae are structurally complicated and their sizes can range from 12.6 (*Ostreococcus tauri*) to 168 Mbp (*Emiliania huxleyi*), while the size of the dinophyte *Karenia brevis* is estimated to be 10 Gbp (Cadoret et al. 2012). One such study describing the family gene evolution in *C. reinhardtii* has revealed the pseudogenization of stress response genes and inferred that gene losses were more than gains (Wu et al. 2015). They identified 18,352 pseudogenes in nine different algal species and observed that with the increase in genome size, the gene numbers of the organism also increased. Furthermore, they found about 1,817 duplication events in *C. reinhardtii* lineage hypothesizing stress mitigation through gene duplications.

TRANSCRIPTOMICS

Transcriptomics provides insight into active genes expressed in certain culture conditions via identification of the RNA transcripts and their abundance. This has led to the discovery of transcriptional regulators (TRs) and regulatory proteins involved in selected biosynthetic pathways, allowing re-engineering of the cellular metabolism (Jamers et al. 2009; Rai et al. 2016). Earlier, our knowledge of transcriptome depended on ESTs of predicted genes, which is partial or biased. Therefore, high throughput next-generation RNA sequencing (RNA-Seq) was helpful in deciphering the complicated transcriptome structure and dynamics with precision and sensitivity (Martin and Wang 2011).

Transcriptomic data can be used in phylogenomic analysis besides studying the cellular metabolism, cell cycle analysis, and signal transduction under various environmental conditions (Cadoret et al. 2012; Sun et al. 2016). The transcriptome analysis of *Dunaliella tertiolecta* deduced that inositol phosphate metabolism to fatty acid biosynthesis pathway is linked to TAG accumulation (Yao et al. 2015). The transcriptome analysis of *Neochloris oleoabundans* revealed that nitrogen-replete conditions favored photosynthesis, carbon/protein metabolism, and cellular growth, while nitrogen stress shifted the carbon flux towards lipid and phospholipid biosynthesis with the activation of pentose phosphate pathway, NADPH regeneration, and lipid transport repressing the β-oxidation pathway (Rismani-Yazdi et al. 2012). Exposure of the model microalga *C. reinhardtii* to sub-nanomolar methylmercury concentrations in the short term severely affects the gene expression of energy and lipid metabolism, metal transport, and antioxidant enzymes (Beauvais-Flück et al. 2016). Another transcriptome study for enhancing lipid accumulation in *Nannochloropsis gaditana* deduced that there were 20 putative negative transcriptional regulators, and knocking down of fungal Zn(II)2Cys6-encoding genes homolog improves the lipid content from 20 to 50 % in nitrogen-replete conditions (Ajjawi et al. 2017).

Transcriptomics analysis of *C. reinhardtii* under phosphorus (P) starvation showed that transcriptional factor Pi starvation response 1 (PSR1) leads all the responses during P starvation and regulates the lipid and starch metabolism (Bajhaiya et al. 2017). Comparative analysis of *de novo* transcriptome and gene expression of *Scenedesmus acutus* TISTR8540 during N stress identified certain lipase genes which were down-regulated specifically along with glycolysis and starch synthesis, whereas gluconeogenesis, photosynthesis, TAG degradation and starch synthesis were up-regulated, confirming the channelling of carbon flux towards fatty acid and TAG synthesis (Sirikhachornkit et al. 2018). Therefore, transcriptome studies help in unraveling the molecular response during stress, thereby developing genetic manipulation strategies for the production of desired products.

PROTEOMICS

Proteomics is the study of all the proteins in a cellular system expressed during a given point of time under a particular set of physical, chemical, or biological conditions. Proteins are essential cellular components whose function is closely linked to its structure and functional groups (Shah and Misra 2011).

The genotypic characteristics of an organism can be predicted through a study of its genome, while the phenotypic characteristics are usually a result of changed protein expression. A study on *C. reinhardtii* has focused on its centrioles and eyespot apparatus (Jamers et al. 2009), while Vener (2007) has reviewed the reversible phosphorylation of the thylakoid proteins in plants and *C. reinhardtii* in response to ambient light, CO_2, and redox conditions (Vener 2007). Together, these reports have advanced our understanding of the proteins involved in the light harvesting, flagella, and eyespot apparatus formation in *C. reinhardtii*.

More recently, due to the importance being attached to algal-derived third-generation biofuels, researchers have shifted their focus and tried to study quantitative proteomics in different algal species under stress conditions. A proteome study was reported in *Nannochloropsis oculata* under nitrogen starvation (Tran et al. 2016), and several heat shock proteins in *C. reinhardtii* were down-regulated during short duration, high light stress (Mahong et al. 2012). Nitrogen starvation stimulus helps to study the various responses of *C. reinhardtii* in the transcriptome, proteome, and photosynthetic metabolism (Schmollinger et al. 2014) and similarly in *Tisochrysis lutea* and *Neochloris oleoabundans* (Garnier et al. 2014; Morales-Sánchez et al. 2016). Recently, lipid droplet surface protein was identified in selected microalgae strains, and this new identification could develop a better understanding of organelle dynamics (Sirikhachornkit et al. 2018). The response of microalgae to N starvation and the corresponding changes happening in the proteome are very useful for the prediction of metabolic pathways, and knowledge is very crucial if we are to be able to stimulate desired fatty acid or triacylglycerol (TAG) accumulation.

LIPIDOMICS AND METABOLOMICS

Metabolomics and lipidomics can be targeted or untargeted, where untargeted metabolomics covers the global metabolite levels in the cell lying under a certain mass range, while targeted metabolomics quantifies a few selected metabolites (Cajka and Fiehn 2016; Vinayavekhin and Saghatelian 2010). Metabolomics is parallel to transcriptome and proteome, in which complete identification and characterization of cellular metabolites with highly accurate quantifications is achieved for an organism ((Lee and Fiehn 2008). Metabolomics offers an inside-out understanding of molecular interactions inside the cells and depicts the overall progression from RNA and protein analysis at the molecular level (Bino et al. 2004). At present, mass spectroscopy (MS) and nuclear magnetic resonance (NMR)-based approaches are dominant in exploring global untargeted metabolomics studies (Dunn and Hankemeier 2013). The most common challenge in metabolomics is deciphering the features such as mass signals into singlet metabolites when an MS-based approach is employed (Zamboni et al. 2015).

Lipidomics can be used to quantify different lipid classes along with their molecular species (Brügger 2014). A lipidome gives insights into lipid remodeling during altered environmental conditions such as nitrogen starvation. A study on *Chlorella* sp. (Trebouxiophyceae) and *Nannochloropsis* sp. (Eustigmatophyceae) has shown that in nitrogen depletion, phosphoglycerolipids tend to increase while long-chain fatty acids in TAGs were broken down (Martin et al. 2014). Another study to assess the variation in lipidomes due to heat stress on *C. reinhardtii* found that at 42°C, cells produce higher polyunsaturated TAGs and diacylglycerols (DAGs), while major chloroplastic monogalactosyldiacyl glycerol sn1-18:3/sn2-16:4 was decreased, triggering an increase in accumulation of DAG sn1-18:3/sn2-16:4 and TAG sn1-18:3/sn2-16:4/sn3-18:3 (Légeret et al. 2016). The study also revealed that TAGs are converted from DAGs via direct conversion from monogalactosyldiacylglycerols (MGDG). The study also finds that the third fatty acid of a TAG is generally originated from a phosphatidyl ethanolamine or a diacylglyceryl-O-4′- (N, N, N, -trimethyl)-homoserine betaine.

A metabolomic analysis of *Phaeodactylum tricornutum* confirms the findings of Légeret et al. (2016)

that betaine lipid diacylglyceroltrimethylhomoserine (DGTS) and MGDG are major contributors to TAGs (Popko et al. 2016). They also found that 16:0 and 16:1(n-7) content is improved in TAGs during nitrogen deprivation along with a metabolite, sedoheptulose, which may have a role in carbon channeling back to the Calvin cycle as it is deregulated. The analysis of lipidomic changes in the non-model green microalga *Ettlia oleoabundans* (a.k.a. *Neochloris oleoabundans*) at different timepoints revealed that the lipid profile under stress consists of a high amount of TAGs, phosphatidylglycerol (PGs), DGGDs, SQDGs, and chlorophyll are interconnected with photosynthetic efficiency (Matich et al. 2018).

ADVANCED MICROSCOPY AND CELL DYNAMICS

Microfluorescence and electron microscopy are used to analyze the morphology and elemental composition of microalgae, whereas new techniques like confocal Raman microscopy, X-ray spectroscopy (EDX), secondary ion mass spectrometry (SIMS) and their combinations are explored to study microalgae (Moudříková et al. 2017). Chiu et al. (2017) has compared the conventional methods for quantification of lipids and carbohydrates in *Chlamydomonas* sp. with Raman spectroscopy and also observed their concentration maps at subcellular resolution. Another study has simultaneously quantified TAGs, protein, starch, and lipids using single-cell Raman spectra; they have used 13 Raman peak markers which can characterize the temporal dynamics of these biological macromolecules (He et al. 2017).

There has also been a study revealing the carotenoid and lipid dynamics in microalga *Haematococcus pluvialis* under high light, elucidating the movement of astaxanthin towards the cellular periphery (Ota et al. 2018). This study has explained that the dynamics of astaxanthin are different from other pigments like lutein, β-carotene, and chlorophylls during light stress and their movement is helped by lipid droplets inside the cell.

INTEGRATION OF MULTIOMICS DATA

The availability of high throughput datasets like genome, proteome, transcriptome, lipidome and metabolome and their interactions provide an in-depth insight into molecular dynamics, while integrating all of the datasets along with the microscopic datasets lets us learn more about the proper functioning of cells along with their mechanics and dynamics. The modeling of such datasets is still a big challenge. Lutz and co-workers have performed integrated omics analyses of arctic snow algae's functionality and adaptability. They showed that green and red snow have different physicochemical properties and hence the associated algae are different. The green snow is wet carbon and nutrient rich and is dominated by the alga *Microglena* sp., and its metabolic profile shows that involved metabolites are good for growth, while red snow is nutrient poor and dry therefore it has various *Chloromonas* spp. with more reserved metabolites (Lutz et al. 2015). Such multiple omics studies could be of greater meaning for understanding the ecosystems and its variables.

TRANSCRIPTIONAL ENGINEERING—A NOVEL PERSPECTIVE

Transcription is the fundamental process of cellular metabolism, wherein the DNA sequence is converted into a complementary mRNA through the action of RNA polymerase. There are a number of proteins, known as transcription factors (TFs), which guide this process. The role of TFs in the cell cycle is to correctly guide the rate of DNA transcription such that the right mRNA is expressed in the right cell at the correct time. TFs regulate the transcription process by binding to specific DNA sequences and interacting with the RNA polymerase to activate or repress the process (Bajhaiya et al. 2017).

Overexpression of TFs or the target gene, or a mutation in the TFs, could lead to overproduction of certain metabolites without affecting the growth profile. Transcriptional engineering in microalgae primarily deals with altering the transcription process by influencing the TFs or the target gene itself (Bajhaiya et al. 2017). PSR1 has been identified as a transcription regulator in *C. reinhardtii*, which was used to modulate starch and lipid biosynthesis under phosphate limitation (Bajhaiya et al. 2017). The overexpression of PSR1 increases triacylglycerol (TAG) accumulation without affecting growth (Ngan et al. 2015). In another approach, a foreign TF, GmDof4 from soybean (*Glycine max*), was overexpressed in *Chlorella ellipsoidea* (mixotrophic growth), resulting in the upregulation of 22 lipid/fatty acid biosynthesis genes, with acetyl coenzyme A carboxylase (ACCase) being particularly affected (Zhang et al. 2014). Another TF, the AtWRI1 from *Arabidopsis*, was expressed in *Nannochloropsis salina*, which increased its total lipid content under both normal and osmotic stressed conditions (Kang et al. 2017). In order to have more carbon diverted towards lipid biosynthesis, a ZnCys TF was identified and downregulated in *N. gaditana* using a CRISPR/Cas9 toolkit which doubled the lipid production by selectively allocating carbon to lipid biosynthesis, but it negatively affected the growth of the cells. However, after further modifications, the mutants were able to achieve 103% lipid productivity compared to control, with only a minor loss of biomass productivity (Ajjawi et al. 2017). Overexpression of a basil leucine zipper (bZIP) type TF in *N. salina* showed enhanced growth coupled with increased lipid content of the cells (Kwon 2018). Another approach is to address the low biomass yields in the case of nutrient-deprived microalgae with a high lipid content. The TF, CHT7, was found to be a repressor of cellular quiescence in *Chlamydomonas* sp. (Tsai et al. 2014). In addition, studies are ongoing to understand more about the algal TFs involved in transcription to have more control over gene expression (Anderson et al. 2017; Thiriet-Rupert et al. 2018). A lack of understanding has been a major bottleneck in the utilization of this technique on a large scale. It is expected that the current focus on genetically engineered algae for biofuels would remove the knowledge gaps to a major extent.

BIG DATA ANALYSIS

Biologists are grappling with huge data sets, and therefore there is an urgent need to tackle new challenges for processing such large volumes of information in the form of genomic snippets. Analysis of such data sets is itself a herculean task, involving various approaches like phylogenomics, subcellular localization, etc.

PHYLOGENOMICS OF METABOLIC PATHWAYS

A comparative understanding of metabolic pathways with a significant amount of information is necessary for developing and enhancing various lipid/by-products from engineered microalgae (Kapase et al. 2018). Genomic data from various genome assembly projects generate an increasing number of sequences which are usually not annotated and express very limited user-defined functions with no detailed pathways, structures or genome information, limiting the construction of biosynthetic pathways (Khozin-Goldberg and Cohen 2011; Reijnders et al. 2014). There is a significant gap between the unknown and non-validated gene/protein functions in algae.

Phylogenomics is the study of the evolutionary background of biological lineages based on the comparative analysis of genome-scale data, which simultaneously allows us to refer to the various biological queries at a scale not possible earlier. Large-scale data mining and analysis of sequential data sets offer a new insight into the development of new metabolic models, providing details of the molecular evolution of metabolic pathways (Misra et al. 2012). A phylogenomic study by Misra et al. (2012) deduced that lipid biosynthetic pathway related genes among Prasinophytes, Chlorophytes, Streptophytes, and Rhodophytes clustered based on exon-intron assembly, conserved motif arrangement and functionality. Also, genomes systematically mined from species belonging to Chlorophyta, Heterokontophyta, Rhodophyta, and Haptophyta identified 289 enzymes involved in lipid metabolic pathways, consequently building the Database of Enzymes of Microalgal Biofuel Feedstock (dEMBF). dEMBF is the first database developed for the enzymes involved and is responsible for lipid synthesis from 15 algal genomes, thereby building an informative platform for enzyme queries and analysis (http://bbprof.immt.res.in/embf, Misra et al. 2016). The Algae Gene Co-expression database (ALCOdb—http://alcodb.jp) is another database providing information on various microalgal gene co-expression at the interspecies level based on comparison and network analysis, facilitating microalgal molecular understanding and evolutionary approach (Aoki et al. 2016) among two model algae, *C. reinhardtii*, and *Cyanidioschyzon merolae*, highlighting the major gene family belonging to higher plants. Phylogenomics study of red algae highlights the evolution of the mevalonate (MVA) pathway for isoprenoid biosynthesis in Rhodophyta and offers in-depth understanding of the origin and evolution of various genes and metabolic pathways involved in 15 red algal species of Rhodophyta (Qiu et al. 2016).

SUB-CELLULAR NETWORKS

Gene regulatory networks (GRNs) are the graphical representation of biological systems data, serving as a network-based model for understanding underlying systems and mechanisms. Why some genes are more active than others during stress (Fan et al. 2012; Macneil and Walhout 2011) is not fully understood as there are undefined underlying regulatory mechanisms involved which are very much interconnected and interdependent. Recent studies have identified and characterized set of genes encoding transcription factors (TFs) and transcription regulators (TRs), which chiefly control lipid accumulation and metabolism (Sardar et al. 2016). Complex networks with underlying transcriptional regulatory hubs that control the lipid accumulation, a set of architecture guided by universal principles, like many networks, are linked to cells' metabolic system, which are interconnected by a large number of nodes and internodes, called hubs. Accordingly, these hubs can become robust or fragile under certain conditions. Lipid accumulation in *Chlamydomonas* during N stress expresses 70 TF and TR genes bringing out metabolic regulation and response to cellular growth in a chronological order. Some novel genes directly involved in TAG metabolism included AP2-15, FHA10, and MYBL13 and two groups of specific and permanent hubs were identified (Gargouri et al. 2015). The GRN modules and the overall topology are analysed in many systems, as observed by the occurrence of TF and gene hubs; these GRNs are not random in nature and can be visualized and studied by computational and mathematical tools (Babu et al. 2004). Identification of various networks, their interconnectivity and dynamics might allow better understanding and characterization of the key nodes involved in various pathway and metabolic analyses (Macneil and Walhout 2011).

PREDICTION OF TRANSCRIPTIONAL FACTORS

Transcriptional factors (TFs) control the expression of a gene at the mRNA synthesis level by interacting physically with the *cis*-regulatory genomic DNA sequences to control the expression of target genes by a highly regulated differential gene expression. TFs can either repress or activate transcription, and some can control/regulate both according to the cellular content. Engineering TFs in algae for augmenting TAG accumulation is a potential and emerging field, and to date around 147 putative TFs and 87 putative TRs have been reported in *C. reinhardtii* (Courchesne et al. 2009). Transcription factor-SNO3 overexpression in *C. reinhardtii* increased total lipids among wild-type from 25% to 36% in SNO3 strain during nitrogen stress conditions; likewise, overexpressed lipogenesis TF in *Chlorella ellipsoids* increased lipid accumulation by 52% (Arora et al. 2018; Gimpel et al. 2015).

Zinc-finger protein transcription factors have been reported for enhanced metabolite production by TF engineering. There are many types of zinc finger proteins grouped according to the number and order of their Cys and His residues, which bind with the zinc ion. Knocking out the fungal Zn(II)2 Cys6-encoding genes improved the partitioning of total carbon to lipids by 20% in wild-type, which was enhanced to 40–55% in mutants with a two-fold increase in lipids (~5.0 g m^{-2} d^{-1}) (Ajjawi et al. 2017). There is a long way to go to determine and identify TFs involved in lipid modulation in microalgae. Almost 20 TFs were identified by using RNA-seq analysis of *N. gaditana* during nitrogen stress conditions by employing CRISPR-Cas9 insertional mutagenesis of 18 of these TFs.

GENOME-SCALE METABOLIC MODELS (GSMMS)

The metabolism of a living organism is the complete set of chemical reactions required for life; numerous enzymes efficiently play the role of catalysts in these reactions. There are usually two main points to be considered while studying these reactions: first by kinetics, i.e., unknown for most of the reactions and other is through determination of stoichiometry (Baart and Martens 2012). Genome-scale metabolic models (GSMMs) can be constructed and modeled once, after gathering enough of the annotated algal genome or transcriptome data available and topology of the metabolic network is analyzed. Initial draft models are generated directly from the available genome annotation data and finalized simultaneously by adding various experimental datasets, literature review and gap-filling steps; the final model evolves with the inclusion of all the reactions alga performs and various associated genes and constraints (Reijnders et al. 2014).

GSMMs for *Ostreococcus tauri* and *O. Lucimarinus* were developed based on the available sequenced annotation data, and a lot of gap-filling designs were required for reactions for the production of the various metabolites before this model could be accounted for the production of the biomass constituents. Most comprehensively studied algal metabolic models are iRC1080 (Chang et al. 2014) and AlgaGEM (Dal'Molin 2011), accounting for the varying degrees of cellular compartments; in iRC1080, half (865/1730) of the non-transport reactions occur in different compartments except cytosol, while this figure is about 12% (201/1617) for AlgaGEM. The first GSMM, iNS934 for *N. salina*, with advanced version reported 2345 reactions and 934 genes from lipid and nitrogen metabolism (Loira et al. 2017); the *Chlorella vulgaris* UTEX 395 reconstruction showed 2294 reactions, 843 genes, and 1770 metabolites from the validation process from transcriptomics data during photoautotrophic, mixotrophic and heterotrophic growth conditions, successfully demonstrating an increased growth rate by altering the culture medium (Zuñiga et al. 2016). There are various online tools available, and GREAT (Genome Regulatory Architecture Tools) is one of the web portals (http://absynth.issb.genopole.fr/GREAT) for studying and analyzing genome architecture and visualization (Bouyioukos et al. 2016). GSMMs for various microalgae will be able to reconstruct the metabolic network pathway design by engineering algal species for the enhanced production of various metabolites (Reijnders et al. 2014; Tomar and De 2013; Zomorrodi et al. 2012).

RECONSTRUCTION OF MICROALGAL CELL FACTORIES

Systems biology and high-throughput technologies are playing a key role in enhancing the production of fine chemicals in algae (Fu et al. 2016). More specifically, algae as a carbon capture mechanism for the production of net carbon negative biofuels as an approach has been dealt with in a related study (Raslavičius et al. 2018). The development of algae as cellular green factories will rely significantly on our understanding of the algal metabolism with its various interconnected pathways and branches. Furthermore, to fully understand algal metabolism, it is necessary to integrate a computational biology approach to understand the working of an algal cell. A recent study has focused on a platform-independent software, Cameo, which aims to optimize the reconstruction of cellular factories through computer-aided metabolic engineering (Cardoso et al. 2018). Another flux balance analysis–based platform, Redirector, has been reported for targeting specific metabolites in complex pathways (Rockwell et al. 2013).

Genome editing technologies such as CRISPR/Cas9, zinc finger nucleases (ZFNs) and transcription activator-like effector nucleases (TALENs) have led to the development of genetically robust microalgal strains which have resulted in altered gene expressions. Taken together, these technologies have resulted in various different microalgal strains designed to produce a particular product in excess (Maeda et al. 2018).

CONCLUSIONS

The development of microalgal cell factories is not possible without an understanding of their physiology and metabolism. The tools and techniques used previously are not competent enough to let us understand the cell dynamics of microalgae. There is a need to elaborate our understanding in microalgae by using omics-based approaches, including genomics, metabolomics, transcriptomics, and proteomics, using next-generation advancements along with microscopic analysis. Data analysis using phylogeny and subcellular localization would help us determine key limiting factors in the functionality of metabolic pathways among microalgae.

REFERENCES

Ajjawi, I., Verruto, J., Aqui, M., Soriaga, L. B., Coppersmith, J., Kwok, K., Peach, L., Orchard, E., Kalb, R., Xu, W., Carlson, T. J., Francis, K., Konigsfeld, K., Bartalis, J., Schultz, A., Lambert, W., Schwartz, A. S., Brown, R., and Moellering, E. R. Lipid production in *Nannochloropsis gaditana* is doubled by decreasing expression of a single transcriptional regulator. *Nat. Biotechnol.* 2017. 35: 647–652.

Anderson, M. S., Muff, T. J., Georgianna, D. R., and Mayfield, S. P. Towards a synthetic nuclear transcription system in green algae: Characterization of *Chlamydomonas reinhardtii* nuclear transcription factors and identification of targeted promoters. *Algal Res.* 2017. 22: 47–55.

Aoki, Y., Okamura, Y., Ohta, H., Kinoshita, K., and Obayashi, T. ALCOdb: Gene coexpression database for microalgae. *Plant Cell Physiol.* 2016. 57: e3.

Arora, N., Pienkos, P. T., Pruthi, V., Poluri, K. M., and Guarnieri, M. T. Leveraging algal omics to reveal potential targets for augmenting TAG accumulation. *Biotechnol. Adv.* 2018. 36: 1274–1292.

Baart, G. J. E., and Martens, D. E. Genome-scale metabolic models: Reconstruction and analysis. In: M. Christodoulides (Ed.), *Neisseria meningitidis: Methods in Molecular Biology* (Vol. 799). 2012. 107–126. Totowa, NJ: Humana Press.

Babu, M. M., Luscombe, N. M., Aravind, L., Gerstein, M., and Teichmann, S. A. Structure and evolution of transcriptional regulatory networks. *Curr. Opin. Struct. Biol.* 2004. 14: 283–291.

Bajhaiya, A. K., Ziehe Moreira, J., and Pittman, J. K. Transcriptional engineering of microalgae: Prospects for high-value chemicals. *Trends Biotechnol.* 2017. 35: 95–99.

Beauvais-Flück, R., Slaveykova, V. I., and Cosio, C. Transcriptomic and physiological responses of the green microalga *Chlamydomonas reinhardtii* during short-term exposure to subnanomolar methylmercury concentrations. *Environ. Sci Technol.* 2016. 50: 7126–7134.

Bino, R. J., Hall, R. D., Fiehn, O., Kopka, J., Saito, K., Draper, J., Nikolau, B. J., Mendes, P., Roessner-Tunali, U., Beale, M. H., Trethewey, R. N., Lange, B. M., Wurtele, E. S., and Sumner, L. W. Potential of metabolomics as a functional genomics tool. *Trends Plant Sci.* 2004. 9: 418–425.

Bouyioukos, C., Bucchini, F., Elati, M., and Képès, F. GREAT: A web portal for genome regulatory architecture tools. *Nucleic Acids Res.* 2016. 44: W77–W82.

Brodie, J., Chan, C. X., De Clerck, O., Cock, J. M., Coelho, S. M., Gachon, C., Grossman, A. R., Mock, T., Raven, J. A., Smith, A. G., Yoon, H. S., and Bhattacharya, D. The algal revolution. *Trends Plant Sci.* 2017. 22: 726–738.

Brügger, B. Lipidomics: Analysis of the lipid composition of cells and subcellular organelles by electrospray ionization mass spectrometry. *Annu. Rev. Biochem.* 2014. 83: 79–98.

Cadoret, J.-P., Garnier, M., and Saint-Jean, B. Microalgae, functional genomics and biotechnology. *Adv. Bot. Res.* 2012. 64: 285–341.

Cajka, T., and Fiehn, O. Toward merging untargeted and targeted methods in mass spectrometry-based metabolomics and lipidomics. *Anal. Chem.* 2016. 88: 524–545.

Cardoso, J. G. R., Jensen, K., Lieven, C., Lærke Hansen, A. S., Galkina, S., Beber, M., Özdemir, E., Herrgård, M. J., Redestig, H., and Sonnenschein, N. Cameo: A Python library for computer aided metabolic engineering and optimization of cell factories. *ACS Synth. Biol.* 2018. 7: 1163–1166.

Chang, R. L., Ghamsari, L., Manichaikul, A., Hom, E. F. Y., Balaji, S., Fu, W., Shen, Y., Hao, T., Palsson, B. O., Salehi-Ashtiani, K., and Papin, J. A. Metabolic network reconstruction of *Chlamydomonas* offers insight into light-driven algal metabolism. *Mol. Syst. Biol.* 2014. 7: 518–518.

Chen, B., Wan, C., Mehmood, M. A., Chang, J. S., Bai, F., and Zhao, X. Manipulating environmental stresses and stress tolerance of microalgae for enhanced production of lipids and value-added products—A review. *Bioresour. Technol.* 2017. 244: 1198–1206.

Chiu, L. D., Ho, S. H., Shimada, R., Ren, N. Q., and Ozawa, T. Rapid in vivo lipid/carbohydrate quantification of single microalgal cell by Raman spectral imaging to reveal salinity-induced starch-to-lipid shift. *Biotechnol. Biofuels.* 2017. 10: 1–9. doi: 10.1186/s13068-016-0691-y.

Courchesne, N. M. D., Parisien, A., Wang, B., and Lan, C. Q. Enhancement of lipid production using biochemical, genetic and transcription factor engineering approaches. *J. Biotechnol.* 2009. 141: 31–41.

Dal'Molin, C. G., Quek, L.-E., Palfreyman, R. W., and Nielsen, L. K. AlgaGEM—A genome-scale metabolic reconstruction of algae based on the *Chlamydomonas reinhardtii* genome. *BMC Genom.* 2011. 12 Suppl 4: S5.

Dunn, W. B., and Hankemeier, T. Mass spectrometry and metabolomics: Past, present and future. *Metabolomics.* 2013. 9: 1–3.

Fan, J., Yan, C., Andre, C., Shanklin, J., Schwender, J., and Xu, C. Oil accumulation is controlled by carbon precursor supply for fatty acid synthesis in *Chlamydomonas reinhardtii.* *Plant Cell Physiol.* 2012. 53: 1380–1390.

Fu, W., Chaiboonchoe, A., Khraiwesh, B., Nelson, D. R., Al-Khairy, D., Mystikou, A., Alzahmi, A., and Salehi-Ashtiani, K. Algal cell factories: Approaches, applications, and potentials. *Mar. Drugs.* 2016. 14: 1–19.

Gao, C., Wang, Y., Shen, Y., Yan, D., He, X., Dai, J., and Wu, Q. Oil accumulation mechanisms of the oleaginous microalga *Chlorella protothecoides* revealed through its genome, transcriptomes, and proteomes. *BMC Genomics.* 2014. 15: 582. doi: 10.1186/1471-2164-15-582.

Gargouri, M., Park, J. J., Holguin, F. O., Kim, M. J., Wang, H., Deshpande, R. R., ... Gang, D. R. Identification of regulatory network hubs that control lipid metabolism in *Chlamydomonas reinhardtii.* *J. Exp. Bot.* 2015. 66: 4551–4566.

Garnier, M., Carrier, G., Rogniaux, H., Nicolau, E., Bougaran, G., Saint-Jean, B., and Cadoret, J. P. Comparative proteomics reveals proteins impacted by nitrogen deprivation in wild-type and high lipid-accumulating mutant strains of *Tisochrysis lutea.* *J. Proteomics.* 2014. 105: 107–120.

Gimpel, J. A., Henríquez, V., and Mayfield, S. P. In metabolic engineering of eukaryotic microalgae: Potential and challenges come with great diversity. *Front. Microbiol.* 2015. 6: 1376.

He, Y., Zhang, P., Huang, S., Wang, T., Ji, Y., and Xu, J. Label-free, simultaneous quantification of starch, protein and triacylglycerol in single microalgal cells. *Biotechnol. Biofuels.* 2017. 10: 275.

Jamers, A., Blust, R., and De Coen, W. Omics in algae: Paving the way for a systems biological understanding of algal stress phenomena? *Aquat. Toxicol.* 2009. 92: 114–121.

Jansson, J. K., Neufeld, J. D., Moran, M. A., and Gilbert, J. A. Omics for understanding microbial functional dynamics. *Environ. Microbiol.* 2012. 14: 1–3.

Jutur, P. P., Nesamma, A. A., and Shaikh, K. M. Algae-derived marine oligosaccharides and their biological applications. *Front. Mar. Sci.* 2016. 3: 83.

Kang, N. K., Kim, E. K., Kim, Y. U., Lee, B., Jeong, W. J., Jeong, B. R., and Chang, Y. K. Increased lipid production by heterologous expression of AtWRI1 transcription factor in *Nannochloropsis salina.* *Biotechnol. Biofuels.* 2017. 10: 231.

Kapase, V. U., Nesamma, A. A., and Jutur, P. P. Identification and characterization of candidates involved in the production of OMEGAs in microalgae: A gene mining and phylogenomic approach. *Prep. Biochem. Biotechnol.* 2018 0: 1–10.

Keasling, J. D. Synthetic biology and the development of tools for metabolic engineering. *Metab. Eng.* 2012. 14: 189–195.

Khozin-Goldberg, I., and Cohen, Z. Unraveling algal lipid metabolism: Recent advances in gene identification. *Biochimie.* 2011. 93: 91–100.

Kim, H., Jang, S., Kim, S., Yamaoka, Y., Hong, D., Song, W. Y., Nishida, I., Li-Beisson, Y., and Lee, Y. The small molecule fenpropimorph rapidly converts chloroplast membrane lipids to triacylglycerols in *Chlamydomonas reinhardtii.* *Front. Microbiol.* 2015. 6: 54.

Kwon, S., Kang, N. K., Koh, H. G., Shin, S. E., Lee, B., Jeong, B. R., and Chang, Y. K. Enhancement of biomass and lipid productivity by overexpression of a bZIP transcription factor in *Nannochloropsis salina.* *Biotechnol. Bioeng.* 2018. 115: 331–340.

Lee, D. Y., and Fiehn, O. High quality metabolomic data for *Chlamydomonas reinhardtii.* *Plant Methods.* 2008. 4: 7.

Légeret, B., Schulz-Raffelt, M., Nguyen, H. M., Auroy, P., Beisson, F., Peltier, G., Blanc, G., and Li-Beisson, Y. Lipidomic and transcriptomic analyses of *Chlamydomonas reinhardtii* under heat stress unveil a direct route for the conversion of membrane lipids into storage lipids. *Plant Cell Environ.* 2016. 39: 834–847.

Loira, N., Mendoza, S., Cortés, M. P., Rojas, N., Travisany, D., Di Genova, A., Gajardo, N., Ehrenfeld, N., and Maass, A. Reconstruction of the microalga *Nannochloropsis salina* genome-scale metabolic model with applications to lipid production. *BMC Syst. Biol.* 2017. 11: 66.

Lu, H., Giordano, F., and Ning, Z. Oxford nanopore MinION sequencing and genome assembly. *Genom. Proteom. Bioinform.* 2016. 14: 265–279.

Lutz, S., Anesio, A. M., Field, K., and Benning, L. G. Integrated "Omics", targeted metabolite and single-cell analyses of arctic snow algae functionality and adaptability. *Front. Microbiol.* 2015. 6: 1323.

Macneil, L. T., and Walhout, A. J. M. Gene regulatory networks and the role of robustness and stochasticity in the control of gene expression. *Genome Res.* 2011. 21: 645–657.

Maeda, Y., Yoshino, T., Matsunaga, T., Matsumoto, M., and Tanaka, T. Marine microalgae for production of biofuels and chemicals. *Curr. Opin. Biotechnol.* 2018. 50: 111–120.

Mahong, B., Roytrakul, S., Phaonaklop, N., Wongratana, J., and Yokthongwattana, K. Proteomic analysis of a model unicellular green alga, *Chlamydomonas reinhardtii*, during short-term exposure to irradiance stress reveals significant down regulation of several heat-shock proteins. *Planta.* 2012. 235: 499–511.

Martin, G. J. O., Hill, D. R. A., Olmstead, I. L. D., Bergamin, A., Shears, M. J., Dias, D. A., Kentish, S. E., Scales, P. J., Botté, C. Y., and Callahan, D. L. Lipid profile remodeling in response to nitrogen deprivation in the microalgae *Chlorella* sp. (Trebouxiophyceae) and *Nannochloropsis* sp. (Eustigmatophyceae). *PLOS ONE.* 2014. 9: e103389.

Martin, J. A., and Wang, Z. Next-generation transcriptome assembly. *Nat. Rev. Genet.* 2011. 12: 671–682.

Matich, E. K., Ghafari, M., Camgoz, E., Caliskan, E., Pfeifer, B. A., Haznedaroglu, B. Z., and Atilla-Gokcumen, G. E. Time-series lipidomic analysis of the oleaginous green microalga species *Ettlia oleoabundans* under nutrient stress. *Biotechnol. Biofuels.* 2018. 11: 29.

Misra, N., Panda, P. K., Parida, B. K., and Mishra, B. K. Phylogenomic study of lipid genes involved in microalgal biofuel production—Candidate gene mining and metabolic pathway analyses. *Evol. Bioinform. Online.* 2012. 8: 545–564.

Morales-Sánchez, D., Kyndt, J., Ogden, K., and Martinez, A. Toward an understanding of lipid and starch accumulation in microalgae: A proteomic study of *Neochloris oleoabundans* cultivated under N-limited heterotrophic conditions. *Algal Res.* 2016. 20: 22–34.

Moudříková, Š., Nedbal, L., Solovchenko, A., and Mojzeš, P. Raman microscopy shows that nitrogen-rich cellular inclusions in microalgae are microcrystalline guanine. *Algal Res.* 2017. 23: 216–222.

Ngan, C. Y., Wong, C. H., Choi, C., Yoshinaga, Y., Louie, K., Jia, J., Chen, C., Bowen, B., Cheng, H., Leonelli, L., Kuo, R., Baran, R., García-Cerdán, J. G., Pratap, A., Wang, M., Lim, J., Tice, H., Daum, C., Xu, J., Northen, T., Visel, A., Bristow, J., Niyogi, K. K., and Wei, C. L. Lineage-specific chromatin signatures reveal a regulator of lipid metabolism in microalgae. *Nat. Plants.* 2015. 1: 15107.

Ota, S., Morita, A., Shinsuke, O., Hirata, A., Sek, S., and Okuda, K. Carotenoid dynamics and lipid droplet containing astaxanthin in response to light in the green alga *Haematococcus pluvialis*. *Sci. Rep.* 2018. 8: 1.

Paliwal, C., Mitra, M., Bhayani, K., Bharadwaj, S.V.V., Ghosh, T., Dubey, S., and Mishra, S. Abiotic stresses as tools for metabolites in microalgae. *Bioresour. Technol.* 2017. 244: 1216–1226.

Popko, J., Herrfurth, C., Feussner, K., Ischebeck, T., Iven, T., Haslam, R., Hamilton, M., Sayanova, O., Napier, J., Khozin-Goldberg, I., and Feussner, I. Metabolome analysis reveals betaine lipids as major source for triglyceride formation, and the accumulation of sedoheptulose during nitrogen-starvation of *Phaeodactylum tricornutum*. *PLOS ONE.* 2016. 11: e0164673.

Qiu, H., Yoon, H. S., and Bhattacharya, D. Red algal phylogenomics provides a robust framework for inferring evolution of key metabolic pathways. *PLoS Curr. Tree Life.* 2016: 1–15.

Rai, V., Karthikaichamy, A., Das, D., Noronha, S., Wangikar, P. P., and Srivastava, S. Multi-omics frontiers in algal research: Techniques and progress to explore biofuels in the postgenomics world. *Omics.* 2016. 20: 387–399.

Raslavičius, L., Striūgas, N., and Felneris, M. New insights into algae factories of the future. *Renew. Sust. Energ. Rev.* 2018. 81: 643–654.

Reijnders, M. J. M. F., van Heck, R. G. A., Lam, C. M. C., Scaife, M. A., dos Santos, V. A. P. M., Smith, A. G., and Schaap, P. J. Green genes: Bioinformatics and systems-biology innovations drive algal biotechnology. *Trends Biotechnol.* 2014. 32: 617–626.

Rismani-Yazdi, H., Haznedaroglu, B. Z., Bibby, K., and Peccia, J. Transcriptome sequencing and annotation of the microalgae *Dunaliella tertiolecta*: Pathway description and gene discovery for production of next-generation biofuels. *BMC Genom.* 2011. 12: 148.

Rismani-Yazdi, H., Haznedaroglu, B. Z., Hsin, C., and Peccia, J. Transcriptomic analysis of the oleaginous microalga *Neochloris oleoabundans* reveals metabolic insights into triacylglyceride accumulation. *Biotechnol. Biofuels.* 2012. 5: 74.

Rockwell, G., Guido, N. J., and Church, G. M. Redirector: Designing cell factories by reconstructing the metabolic objective. *PLOS Comput. Biol.* 2013. 9: e1002882.

Sardar, R., Shaikh, K. M., Jutur, P. P., Group, I. B., Asaf, A., and Marg, A. Lipid metabolism and carbon concentrating reinhardtii using regulatory networks. In: International Conference on Bioinformatics Systems Biology 2016. 1–4. Allahabad, India: IEEE.

Sasso, S., Pohnert, G., Lohr, M., Mittag, M., and Hertweck, C. Microalgae in the postgenomic era: A blooming reservoir for new natural products. *FEMS Microbiol. Rev.* 2012. 36: 761–785.

Schmollinger, S., Mühlhaus, T., Boyle, N. R., Blaby, I. K., Casero, D., Mettler, T., Moseley, J. L., Kropat, J., Sommer, F., Strenkert, D., Hemme, D., Pellegrini, M., Grossman, A. R., Stitt, M., Schroda, M., and Merchant, S. S. Nitrogen-sparing mechanisms in *Chlamydomonas* affect the transcriptome, the proteome, and photosynthetic metabolism. *Plant Cell.* 2014. 26: 1410–1435.

Shah, T. R., and Misra, A. Proteomics. In: *Challenges in Delivering Therapeutic Genomics and Proteomics.* 2011. 387–427. London, UK: Elsevier.

Singh, R., Mattam, A. J., Jutur, P., and Yazdani, S. S. Synthetic biology in biofuels production. In: Meyers, D. R. A. (Ed.), *Reviews in Cell Biology and Molecular Medicine.* 2016. 144–176. Wiley-VCH Verlag GmbH & Co. KGaA.

Sirikhachornkit, A., Suttangkakul, A., Vuttipongchaikij, S., and Juntawong, P. De novo transcriptome analysis and gene expression profiling of an oleaginous microalga *Scenedesmus acutus* TISTR8540 during nitrogen deprivation-induced lipid accumulation. *Sci. Rep.* 2018. 8: 3668.

Skjånes, K., Rebours, C., and Lindblad, P. Potential for green microalgae to produce hydrogen, pharmaceuticals and other high value products in a combined process. *Crit. Rev. Biotechnol.* 2013. 33: 172–215.

Sun, Z., Chen, Y. F., and Du, J. Elevated CO2 improves lipid accumulation by increasing carbon metabolism in *Chlorella* sorokiniana. *Plant Biotechnol. J.* 2016. 14: 557–566.

Thiriet-Rupert, S., Carrier, G., Trottier, C., Eveillard, D., Schoefs, B., Bougaran, G., Cadoret, J.-P., Chénais, B., and Saint-Jean, B. Identification of transcription factors involved in the phenotype of a domesticated oleaginous microalgae strain of *Tisochrysis lutea*. *Algal Res.* 2018. 30: 59–72.

Tomar, N., and De, R. K. Comparing methods for metabolic network analysis and an application to metabolic engineering. *Gene.* 2013. 521: 1–14.

Tran, N.-A. T., Padula, M. P., Evenhuis, C. R., Commault, A. S., Ralph, P. J., and Tamburic, B. Proteomic and biophysical analyses reveal a metabolic shift in nitrogen deprived *Nannochloropsis oculata*. *Algal Res.* 2016. 19: 1–11.

Tsai, C. H., Warakanont, J., Takeuchi, T., Sears, B. B., Moellering, E. R., and Benning, C. The protein Compromised Hydrolysis of Triacylglycerols 7 (CHT7) acts as a repressor of cellular quiescence in *Chlamydomonas*. *Proc. Natl Acad. Sci. U.S.A.* 2014. 111: 15833–15838.

Vener, A. V. Environmentally modulated phosphorylation and dynamics of proteins in photosynthetic membranes. *Biochim. Biophys. Acta.* 2007. 1767: 449–457.

Vinayavekhin, N., and Saghatelian, A. Chapter 30—Untargeted metabolomics. In: *Current Protocols in Molecular Biology.* 2010. 1–24.

Wu, G., Hufnagel, D. E., Denton, A. K., and Shiu, S. H. Retained duplicate genes in green alga *Chlamydomonas reinhardtii* tend to be stress responsive and experience frequent response gains. *BMC Genom.* 2015. 16: 149.

Yao, L., Tan, T. W., Ng, Y. K., Ban, K. H. K., Shen, H., Lin, H., and Lee, Y. K. RNA-Seq transcriptomic analysis with Bag2D software identifies key pathways enhancing lipid yield in a high lipid-producing mutant of the non-model green alga *Dunaliella tertiolecta*. *Biotechnol. Biofuels.* 2015. 8: 191.

Zamboni, N., Saghatelian, A., and Patti, G. J. Defining the metabolome: Size, flux, and regulation. *Mol. Cell.* 2015. 58: 699–706. doi: 10.1016/j.molcel.2015.04.021.

Zhang, J., Hao, Q., Bai, L., Xu, J., Yin, W., Song, L., Xu, L., Guo, X., Fan, C., Chen, Y., and Ruan, J. Overexpression of the soybean transcription factor GmDof4 significantly enhances the lipid content of *Chlorella ellipsoidea*. *Biotechnol. Biofuels.* 2014. 7: 128.

Zomorrodi, A. R., Suthers, P. F., Ranganathan, S., and Maranas, C. D. Mathematical optimization applications in metabolic networks. *Metab. Eng.* 2012. 14: 672–686.

Zuñiga, C., Li, C. T., Huelsman, T., Levering, J., Zielinski, D. C., McConnell, B. O., Long, C. P., Knoshaug, E. P., Guarnieri, M. T., Antoniewicz, M. R., Betenbaugh, M. J., and Zengler, K. Genome-scale metabolic model for the green alga *Chlorella vulgaris* UTEX 395 accurately predicts phenotypes under autotrophic, heterotrophic, and mixotrophic growth conditions. *Plant Physiol.* 2016. 172: 589–602.

24 Bioactive Metabolites
Genetic Regulation and Potential Market Implications

Ayse Kose, Claire Remacle, Young-Woo Kim, Suphi S. Oncel, and Murat Elibol

CONTENTS

Introduction 273
Microalgal Physiology for Production of Bioactive Metabolites 274
   Lipid Synthesis 274
Strain Engineering for Yield Efficiency and Enhanced Lipid Production 274
   Transformation Methods 274
   Genetic Engineering 277
Downstream Processes 277
Size of the Global Market for Algal Products 277
Conclusions 279
Acknowledgments 279
References 279

BOX 24.1 SALIENT FEATURES

Nature-derived therapeutics are currently attracting more attention than chemical synthesis and synthetic ones. Today the expansion of the natural bioactive molecule market is starting to dominate, and it seems that generic drug and therapeutic understanding will give way to new products. Thus, controlled production conditions for naturally originating therapeutics are of importance. The public is also paying more attention to natural molecules and the demand ofor bioactive molecules is also increasing dramatically. Bioactive molecules from microalgae show antioxidant, antibacterial, antiviral, anticancer, skin regenerative, sunscreen, antihypertensive, neuroprotective and immunostimulatory effects which are favorable for the pharmaceutical, nutraceutical and cosmetics industries. Microalgal bioactive metabolites are emerging in the market for pharmaceutical, nutraceuticals, cosmetics, food and feed purposes. With recent developments, new products have come on to the market as a combinatory result of molecular biology and bioprocesses. Since we have started to understand more about algal physiology, gene regulation and metabolic networks, it has been slightly easier to increase the productivity of certain products such as fatty acids, pigments, proteins and recombinant therapeutics. In this chapter, certain fundamental bioactive pathways, their market potential and some essential information on bioprocesses are discussed.

INTRODUCTION

Bioactive metabolites are certain metabolites that are synthesized inside cells or secreted in the extra-cellular environment (Kiuru et al. 2014), and they show certain activities on various sources which make them important compounds in biotechnology. Bioactive metabolites can be synthesized by using the tools available in industrial biotechnology or the recent techniques developed in genetic engineering (Westerhoff and Palsson 2004; Hlavova et al. 2015).

Microalgae, being one of the complex single celled organisms of metabolic networks, are great model species to study development, metabolic differentiation, and stress physiology-related bioactive metabolite production. Yet, to date, more than 1,000 bioactive compounds have been defined, most of which have antibacterial, antifungal, antiviral, antioxidant, anticancer and immune-enhancing activities (Table 24.1). Bioactive metabolites such as proteins, peptides, fatty acids, and carbohydrates produced through primary metabolism or secondary metabolism compounds, such as carotenoids, amino acids, enzymes, are gaining more attraction lately.

In this chapter, we will discuss the promising bioactive metabolites of microalgae as a result of their metabolic network and regulation, and we will highlight a selection methodology to study a commercialization roadmap considering metabolic engineering as a target tool in the bioprocess scheme.

MICROALGAL PHYSIOLOGY FOR PRODUCTION OF BIOACTIVE METABOLITES

Microalgae have significant potential for the production of biomass and bioproducts, including lipids that can be transformed into biodiesel, long-chain polyunsaturated fatty acids for human health and high-value pigments with antioxidant properties. Most production processes are currently based on light-dependent assimilation of CO_2.

LIPID SYNTHESIS

Fatty acids, the building blocks for triacylglycerides (TAG) and membrane lipids, are synthesized in the chloroplasts. The first two enzymes catalyzing their synthesis from acetyl-CoA units are the acetyl-CoA carboxylase (ACCase) and the fatty acid synthase (FAS) complex (Fan et al. 2011; Merchant et al. 2012). The resulting fatty acids (acyl-CoA) can be used directly in the chloroplast to sequentially acylate glycerol-3-phosphate (G-3-P) by chloroplast resident acyl-transferases (glycerol-3-phosphate acyltransferase and lysophosphatidyl acyltransferase) to produce lysophosphatidic acid (LysoPa) and phosphatidic acid (PA). The PA and its dephosphorylated product diacylglycerol (DAG) generated in the chloroplast serve primarily as precursors for structural lipids of the photosynthetic membranes (Fan et al. 2011). Fatty acids can also be exported into the cytosol and used to acylate G-3-P in the endoplasmic reticulum (ER) by ER-resident acyltransferase isoforms. The resultant PA and DAG can be used to synthesize both membrane lipids and storage TAG (reviewed in Fan et al. 2011; Bellou et al. 2014). The saturated and monounsaturated C16 and C18 fatty acids can be further processed in the ER for diversification (elongation, additional unsaturations) and acyl lipid synthesis, depending on the species (Murphy 2001), in a way which has not yet been elucidated (Bellou et al. 2014). In adverse conditions such as nitrogen limitation, TAG accumulation is increased and in some oleaginous microalgae (ones that store carbon in lipids rather than in starch), the lipid fraction of the biomass can reach more than 40% when the carbon and nitrogen supplies are properly manipulated (Guccione et al. 2014). The increased TAG content can be attributed to a re-routing of the central carbon metabolism, including reallocation of fixed carbon towards fatty acids and lipids as well as the conversion of non-lipid cell components to neutral lipids (Roessler 1990; Jaeger et al. 2017). Biodiesel production requires transesterification of TAG in the presence of methanol, resulting in methyl esters of fatty acids (biodiesel) and glycerol. Transesterification is catalyzed by acids, alkalis and lipase enzymes, with alkali-catalyzed transesterification being 4000 times faster than acid-catalyzed reaction (reviewed in Chisti 2007).

Long-chain polyunsaturated fatty acids (LC-PUFAs) from 20 to 22 carbon length (20–22) represent the most valuable products due to increased human consumption, especially docosahexaenoic acid (DHA, 22:6n-3) (Doughman et al. 2007). About 10% of LC-PUFAs produced today come from cultured microalgae, with an increasing trend due to the limitations of fish-oil production (Khozin-Goldberg et al. 2011). Synthesis of LC-PUFAs is present in various classes of marine phytoplankton such as Bacillariophyta (diatoms like *Phaeodactylum* sp), Eustigmatophyta (*Nannochloropsis* sp.) or Dinoflagellata (*Pyrocystis fusiformis*) and in the freshwater microalgae class of Chlorophyta (*Ostreococcus tauri*) (reviewed in Khozin-Goldberg et al. 2011).

STRAIN ENGINEERING FOR YIELD EFFICIENCY AND ENHANCED LIPID PRODUCTION

TRANSFORMATION METHODS

Genetic transformation is compartmentalized into three types of organelles: mitochondria, chloroplast and nucleus. The only microalga where the three genomes can be selectively transformed is the green microalga *Chlamydomonas reinhardtii*, where chloroplastic and mitochondrial genomes can be transformed using particle bombardment (Boynton et al. 1988; Remacle et al. 2006) while the nuclear genome is transformed by electroporation or glass-bead agitation (Kindle et al. 1989; reviewed in Mussnug 2015). Homologous replacement is achieved in the organellar genomes (Boynton et al. 1988; Larosa et al. 2006).

TABLE 24.1
Bioactive Ingredients from Microalgae Species and Their Mechanism of Action

Pigments
- Astaxanthin
- Cantaxanthin
- Chlorophyll
- Fucoxanthin
- Lutein
- Lycopene
- Phycocyanin
- Phycoerytrin
- Zeaxanthin
- B-carotene

Pharmaceuticals and Cosmeceuticals
- Amino acids
- Bioactive peptides
- CoenzymeQ10
- Immunotoxins
- Polysaccharides
- Recombinant proteins
- Vaccines
- Vitamins

Fatty Acids
- PUFAs

Single Cell Proteins
- Spirulina
- Chlorella

Nutraceuticals
- CoenzymeQ10
- Fatty acids
- Pigments
- Polysaccharides
- Proteins and peptides
- Single cell proteins
- Vitamins

Food Ingredients
- Aromatics
- Colorant
- Gelling and thickening agents
- Hydrocolloids
- Polysaccharides
- Smart packaging

Others
- Alkaloids
- Flavanoids
- Phenolics
- Phytols
- Sterols
- Terpenoids

Mechanism of Action

- Antiadhesive
- antiatherogenicity
- Antiaging
- Antibacterial
- Anticancer
- Anticellullite
- Anticoagulant
- Antidiabetic
- Antifungal
- Antihelmintic
- Antihyperlipidemia
- Antimutagenic
- Antiobesity
- Antioxidant

- Antiprotozoan
- Antiviral
- Antiwrinkle
- Brightening
- Collagen boosting
- Drug-carriers
- Hepatoprotective
- Hydrating
- Immunostimulant
- Neuroprotective
- Osmoregulators
- Skin whitening
- Sunscreen

(*Continued*)

TABLE 24.1 (CONTINUED)
Bioactive Ingredients from Microalgae Species and Their Mechanism of Action

Emerging Species

- *Anabaena sp.*
- *Anabaena variabilis*
- *Botryococcus braunii*
- *Chaetoceros calcitrans*
- *Chlamydomonas reinhardtii*
- *Chlorella ellipsoidea*
- *Chlorella minutissima*
- *Chlorella protothecoides*
- *Chlorella sp.*
- *Chlorella vulgaris*
- *Chlorella zofingiensis*
- *Crypthecodinium*
- *Dunaliella salina*
- *Dunaliella tertiolecta*
- *Euglena gracilaris*
- *Galdieria sulphuraria*
- *Haematococcus pluvialis*
- *Isochrysis sp.*
- *Monodus subterraneus,*
- *Nannochloropsis salina*
- *Nostoc commune*
- Nos*toc muscorum*
- *Nostoc punctiforme*
- *Nostoc sp.*
- *Nostoc flagelliforme*
- *Oedogonium*
- *Pavlova*
- *Pavlova lutheri*
- *Phaeodactylum tricornutum*
- *Porphyridium cruentum*
- *Porphyridium sp.*
- *Scenedesmus quadricauda*
- *Schizochytrium*
- *Spirogyra*
- *Spirulina sp.*
- *Stichococcus bacillaris*
- *Synecococcus*
- *Synococystis*
- *Tetraselmis*
- *Thalassiosira*
- *Tolypothrix sp.*
- *Ulkenia*

However, DNA integration into the nuclear genome occurs by random integration. Transgene expression in the nuclear genome has been reported to be difficult in *Chlamydomonas*, with potential reasons for this being biased codon usage, epigenetic transgene silencing, positional effects and chromatin structure (reviewed in Mussnug 2015). The hybrid promoter HSP70A-RBCS2 promoter (Schroda et al. 2000) is the most common one for transgene expression. Recently, the expression level of transgene of interest has been shown to be significantly increased if an intron is added in the coding sequence of the gene of interest every 200–300 nucleotides (Baier et al. 2018). Different selection markers are available for *C. reinhardtii*, the most commonly used mediating antibiotic/herbicide resistance by expression of a foreign transgene (reviewed in Mussnug 2015).

Apart from *Chlamydomonas*, on which transformation has been routinely performed in laboratories for nearly 30 years, efforts have been made to transform other microalgae with biotechnological potential over the last 10 years. In diatoms such as *Phaeodactylum tricornutum* and *Thalassiosira pseudonana*, the nuclear genome is transformed by particle bombardment. Several promoters are available for transgene expression such as the lhcf1 promoter, which is widely used and derived from the gene-encoding chlorophyll A/C-binding light-harvesting complex protein. The most common selection marker confers resistance to zeocin. Chloroplast genome transformation has also been reported (reviewed in Velmurugan and Deka 2018).

In haptophytes, transformation is difficult due to the complexity of the haptophyte cell membrane coccosphere. Stable transformation systems have been established for a few species such as *Emiliania huxleyi* and *Isochrysis galbana* using particle bombardment and agrobacterium-mediated stable DNA transfer, respectively (reviewed in Velmurugan and Deka 2018).

The genus *Nannochloropsis* belonging to the eustigmatophyceae family has been also shown to be transformed and used for overexpression of proteins involved in lipid metabolism. Various promoters and selection markers are used (reviewed in Poliner et al. 2018).

In addition, genome editing methods, such as the CRISPR/Cas9 system, have been recently established in microalgae such as *C. reinhardtii* (Shin et al. 2016), *P. tricornutum* (Kroth et al. 2018) and *Nannochloropis* (Poliner et al. 2018).

Genetic Engineering

Both growth rates and lipid content should be improved on order to reduce the cost of lipid production in microalgae (Davis et al. 2011). To improve growth rates, efforts have been made to improve photosynthetic efficiency. Molecular targets showing improved growth rates include the sedoheptulose-1,7-bisphosphatase (SBPase) of the Calvin cycle and the phosphoenolpyruvate carboxylase. The SBPase of *C. reinhardtii* was expressed in a glycerol-producing halotolerant, *Dunaliella bardawil*. The *D. bardawill* transformant showed improved photosynthetic performance with increased total organic carbon content and osmoticum glycerol production (Fang et al. 2012). The downregulation of the *C. reinhardtii* phosphoenolpyruvate carboxylase (PEPCase), which catalyzes the transformation of phosphoenolpyruvate (PEP) into oxaloacetate while fixing CO_2, led to a 20% increase in triacylglycerol (TAG) content and overall lower growth rates. On the contrary, overexpression of the enzyme led to increased growth rates (Deng et al 2014).

Another option relies on the improvement of light utilization. By decreasing the antenna size (Tetali et al. 2007; Beckmann et al. 2009), it is possible to obtain higher biomass densities under high light in mass cultivation systems due to lower shading effects (reviewed in Gommaa et al. 2016). In a similar way, reducing non-photochemical quenching also contributes to higher biomass densities (Berteotti et al. 2016).

Manipulating fatty acids synthesis is still in its infancy because of incomplete knowledge of the regulation of carbon partitioning and detailed fatty-acid synthesis and catabolism. For example, engineering the Fatty Acid Synthesis complex (FAS) has been done, leading to modified composition of fatty acids but not increased content (Blatti et al. 2012). Similarly, manipulation of expression of DGAT genes did not contribute to modifying lipid content (Deng et al. 2012; La Russa et al. 2012). Other targets include a multifunctional enzyme (lipase/phospholipase/acyltransferase) whose downregulation in *P. tricornutum* increased lipid content without affecting growth rate in the diatom *Thalasossiera pseudonana* (Trentacoste et al. 2013) (reviewed in Bellou et al 2014).

DOWNSTREAM PROCESSES

After cultivation in either ponds or PBRs, microalgal biomass need to be separated from suspension culture, which is called the *harvesting* stage (Wang et al., 2012); this is actually a key process with a special emphasis on the product cost (Christenson and Sims 2011). Due to its economy-related effect, projected to be over 20% of the total cost, the harvesting method should be selected with specific attention without underestimating the interaction between all the other process steps from upstream to downstream (Milledge and Heaven 2013). Dewatering is the critical step in the harvesting process. The main approach to this step involves first separating the bulk water from the culture and then concentrating the microalgae later. This approach targets the efficient use of both energy and time. Keeping this strategy in mind, the main harvesting methods such as centrifugation, sedimentation, flocculation, electrocoagulation, flotation or filtration can be classified according to the driving force that is incorporated during the process, which can be mechanical, electrical, chemical, biological or a combination of these (Pragya et al. 2013).

A novel approach for microalgal harvesting is magnetic microalgal harvesting, which uses magnetic particles and an external magnetic field for the separation of cells. The trick in this method lies in the fact that both the microalgal cells and the magnetic particles have negatively charged surfaces. This negative charge between the cell and the particle, connected by the addition of cationic polyelectrolytes in the culture, results in the direct linking of the cells with the magnetic particles, which are then harvested using an external magnetic field (Wang et al. 2015).

SIZE OF THE GLOBAL MARKET FOR ALGAL PRODUCTS

Microalgae, along with the other taxonomical group cyanobacteria, has enormous potential to produce cosmeceuticals, pharmaceuticals, food additives and colorants, aromatic compounds and certain other fine chemicals. The point here is the balance between the financial reliability of the developed products and the scientific impact, and contribution to the scientific area, to build the industrial biotechnological infrastructure if the marketing is final desire (Figure 24.1).

The market for algal products is estimated to be approximately 4 billion US dollars, and the major factor in this is the increasing tendency to use natural foods, cosmetics and hygiene products, as growing numbers of people substitute their nutritional habits with vegetarian and/or vegan products (http://www.algaeindustrymagazine.com/

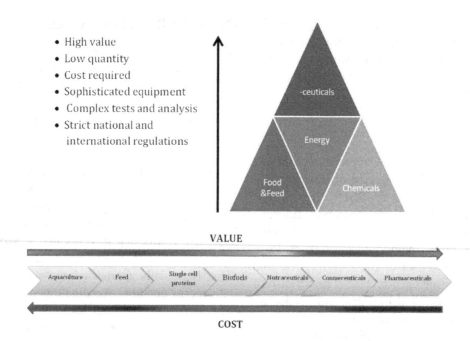

FIGURE 24.1 (See color insert.) Value and cost relation of fundamental microalgal compounds.

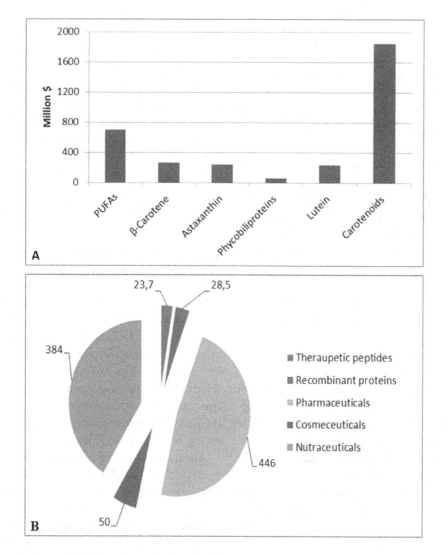

FIGURE 24.2 (See color insert.) Market share of microalgal bioactive compounds: (A) fine chemicals, (B) % main sector share.

algae-products-market-growth-projections-2018-2023). Health consciousness along with environmental consciousness represent a new gap in the market that entrepreneurships are trying to fill (Gouveia et al. 2008; Garcia et al. 2017). Over 75% of the algal market is comprised of single-cell protein capsules used as a dietary supplement (Becker 2007). The market is mostly dominated by these compounds, which are produced in China, Japan and Taiwan and distributed all over the globe. There are also local producers on a small scale as a fashion brand or designer production. The global market contribution of *Spirulina* produced is between 800 and 1200 tons annually, and it is expected to be worth 2000 million US Dollars by 2026, up from around 700 Million Dollars back in 2016 (https://www.persistencemarketresearch.com/market-research/spirulina-market.asp). The colorful pigments market, mostly for carotenoids, is another increasing source of global attention to which microalgae could contribute. The size of this market is estimated to reach more than 1.53 billion US dollars by 2021 (www.marketsandmarkets.com). Certain microalgae and cyanobacteria species are candidates for carotenoid production: *Haematococcus pluvialis* is the lead species for astaxanthin production, *Duneliella salina* for β-carotene and *Spirulina* for phycocyanin, and many more species are used for the production of pigments. Only astaxanthin itself has a market size of 814 million US dollars, and it is estimated to expand more (www.marketsandmarkets.com). The current prices for 1,000 mg of lutein, astaxanthin, β-carotene and zeaxanthin are 2.5, 1.8, 0.6 and 10 $, respectively (Bhalamurugan et al. 2018). Figure 24.2 shows the sizes of the global markets for certain microalgal products, mostly carotenoids, and their share in the total carotenoids market along with the major global markets where microalgal products could find a place.

CONCLUSIONS

Microalgae as novel industrial microorganisms have great potential for the development of novel bioactive ingredients for several industries. However, most of the compounds are not determined in clinical trial as purified form; crude extracts are preferred due to the complexity of the downstream processes. Although microalgae have some limitations, recent trends in both bioprocesses and genetic engineering techniques highlight that large-scale production of microalgae products for the health, food and cosmetics industries will be realized in the near future.

ACKNOWLEDGMENTS

The authors acknowledge the support of KONNECT (Strengthening STI Cooperation between the EU and Korea, Promoting Innovation and the Enhancement of Communication for Technology-related Policy Dialogue). The chapter is written with the contribution of the partners from ALGACTIVE-KONNECT Project.

REFERENCES

Baier, T., Wichmann, J., Kruse, O., Lauersen, K. J. Intron-containing algal transgenes mediate efficient recombinant gene expression in the green microalga *Chlamydomonas reinhardtii*. *Nucleic Acids Res.* 2018. 46:6909–6919.

Becker, E. W. Micro-algae as a source of protein. *Biotechnol Adv.* 2007. 25(2):207–210.

Beckmann, J., Lehr, F., Finazzi, G., Hankamer, B., Posten, C., Wobbe, L., Kruse, O. Improvement of light to biomass conversion by de-regulation of light-harvesting protein translation in *Chlamydomonas reinhardtii*. *J Biotechnol.* 2009. 142:70–77.

Bellou, S., Baeshen, M. N., Elazzazy, A. M., Aggeli, D., Sayegh, F., Aggelis, G. Microalgal lipids biochemistry and biotechnological perspectives. *Biotechnol Adv.* 2014. 32:1476–1493.

Berteotti, S., Ballottari, M., Bassi, R. Increased biomass productivity in green algae by tuning non-photochemical quenching. *Sci Rep.* 2016. 6:21339.

Berteotti, S., Ballottari, M., Bassi, R. Increased biomass productivity in green algae by tuning non-photochemical quenching. *Sci Rep.* 2016. 6:21339.

Bhalamurugan, G.L., Valerie, O., Mark, L. Valuable bioproducts obtained from microalgal biomass and their commercial applicatiobs: A review. *Environ. Eng.* 2018.

Blatti, J. L., Beld, J., Behnke, C. A., Mendez, M., Mayfield, S. P., Burkart, M. D. Manipulating fatty acid biosynthesis in microalgae for biofuel through protein–protein interactions. *PLOS ONE.* 2012. 7(9):e42949.

Boynton, J. E., Gillham, N. W., Harris, E. H., Hosler, J. P., Johnson, A. M., Jones, A. R., Randolph-Anderson, B. L., Robertson, D., Klein, T. M., Shark, K. B. Chloroplast transformation in Chlamydomonas with high velocity microprojectiles. *Science.* 1988. 240: 1534–1538.

Chisti, Y. Biodiesel from microalgae. *Biotechnol Adv.* 2007. 25:294–306.

Christenson, L., Sims, R. Production and harvesting of microalgae for wastewater treatment, biofuels, and bioproducts. *Biotechnol Adv.* 2011. 29:686–702.

Davis, R., Aden, A., Pienkos, P. T. Techno-economic analysis of autotrophic microalgae for fuel production. *Appl Energy.* 2011. 88:3524–3531.

Deng, X., Cai, J., Li, Y., Fei, X. Expression and knockdown of the PEPC1 gene affect carbon flux in the biosynthesis of triacylglycerols by the green alga *Chlamydomonas reinhardtii*. *Biotechnol Lett.* 2014. 36:2199–2208.

Deng, X. D., Gu, B., Li, Y. J., Hu, X. W., Guo, J. C., Fei, X. W. The roles of acyl-CoA: Diacylglycerol acyltransferase 2 genes in the biosynthesis of triacylglycerols by the green algae *Chlamydomonas reinhardtii*. *Mol Plant.* 2012. 5:945–947.

Doughman, S. D., Krupanidhi, S., Sanjeevi, C. B. Omega-3 fatty acids for nutrition and medicine: Considering microalgae oil as a vegetarian source of EPA and DHA. *Curr Diabetes Rev.* 2007. 3:198–203.

Fan, J., Andre, C., Xu, C. A chloroplast pathway for the de novo biosynthesis of triacylglycerol in *Chlamydomonas reinhardtii*. *FEBS Lett*. 2011. 585:1985–1991.

Fang, L., Lin, H. X., Low, C. S., Wu, M. H., Chow, Y., Lee, Y. K. Expression of the *Chlamydomonas reinhardtii* sedoheptulose-1,7-bisphosphatase in *Dunaliella bardawil* leads to enhanced photosynthesis and increased glycerol production. *Plant Biotechnol J.* 2012. 10:1129–1135.

García, J. L., de Vicente, M., Galán, B. Microalgae, old sustainable food and fashion nutraceuticals. *Microb Biotechnol*. 2017. 10:1017–1024.

Gomaa, M. A., Al-Haj, L., Abed, R. M. M. Metabolic engineering of Cyanobacteria and microalgae for enhanced production of biofuels and high-value products. *J Appl Microbiol*. 2016. 121:919–931.

Gouveia, L., Batista, A. P., Sousa, I., Raymundo, A., Bandarra, N. M. Microalgae in novel food products. In: *Food Chemistry Research Developments* (ed. Papadopoulos, K. N.). 2008. 75–111. New York, NY: Nova Publishers.

Guccione, A., Biondi, N., Sampietro, G., Rodolfi, L., Bassi, N., Tredici, M. R. Chlorella for protein and biofuels: From strain selection to outdoor cultivation in a Green Wall Panel photobioreactor. *Biotechnol Biofuels*. 2014. 7:84.

Hlavova, M., Turoczy, Z., Bisova, K. Improving microalgae for biotechnology: From genetics to synthetic biology. *Biotech Adv*. 2015. 33(6):1194–1203.

Jaeger, D., Winkler, A., Mussgnug, J. H., Kalinowski, J., Goesmann, A., Kruse, O. Time-resolved transcriptome analysis and lipid pathway reconstruction of the oleaginous green microalga *Monoraphidium neglectum* reveal a model for triacylglycerol and lipid hyperaccumulation. *Biotechnol Biofuels*. 2017. 10:197.

Khozin-Goldberg, I., Iskandarov, U., Cohen, Z. LC-PUFA from photosynthetic microalgae: Occurrence, biosynthesis, and prospects in biotechnology. *Appl Microbiol Biotechnol*. 2011. 91:905–915.

Kindle, K. L., Schnell, R. A., Fernández, E., Lefebvre, P. A. Stable nuclear transformation of Chlamydomonas using the Chlamydomonas gene for nitrate reductase. *J Cell Biol*. 1989. 109:2589–2601.

Kiuru, P., D'Auria, M. V., Muller, C. D., Tammela, P., Vuorela, H., Yli-Kauhaluoma, J. Exploring marine resources for bioactive compounds. *Planta Med*. 2014. 80:1234–1246.

Kroth, P. G., Bones, A. M., Daboussi, F., Ferrante, M. I., Jaubert, M., Kolot, M., Nymark, M., Río Bártulos, C., Ritter, A., Russo, M. T., Serif, M., Winge, P., Falciatore, A. Genome editing in diatoms: Achievements and goals. *Plant Cell Rep*. 2018. 37:1401–1408.

Larosa, V., Coosemans, N., Motte, P., Bonnefoy, N., Remacle, C. Reconstruction of a human mitochondrial complex I mutation in the unicellular green alga Chlamydomonas. *Plant J*. 2006. 70:759–768.

La Russa, M., Bogen, C., Uhmeyer, A., Doebbe, A., Filippone, E., Kruse, O., Mussgnug, J. H. Functional analysis of three type-2 DGAT homologue genes for triacylglycerol production in the green microalga *Chlamydomonas reinhardtii*. *J Biotechnol*. 2012. 162:13–20.

Merchant, S. S., Kropat, J., Liu, B., Shaw, J., Warakanont, J. TAG, you're it! Chlamydomonas as a reference organism for understanding algal triacylglycerol accumulation. *Curr Opin Biotechnol*. 2012. 23:352–363.

Milledge, J. J., Heaven, S. A review of the harvesting of micro-algae for biofuel production. *Rev Environ Sci Biotechnol*. 2013. 12:165–178.

Murphy, D. J. The biogenesis and functions of lipid bodies in animals, plants and microorganisms. *Progr Lip Res*. 2001. 40:325–438.

Mussgnug, J. H. Genetic tools and techniques for *Chlamydomonas reinhardtii*. *Appl Microbiol Biotechnol*. 2015. 99:5407–5418.

Poliner, E., Farré, E. M., Benning, C. Advanced genetic tools enable synthetic biology in the oleaginous microalgae *Nannochloropsis sp*. *Plant Cell Rep*. 2018. 37:1383–1399.

Pragya, N., Pandey, K. K., Sahoo, P. K. A review on harvesting, oil extraction and biofuels production technologies from microalgae. *Renew Sustain Energy Rev*. 2013. 24:159–171.

Remacle, C., Cardol, P., Coosemans, N., Gaisne, M., Bonnefoy, N. High-efficiency biolistic transformation of Chlamydomonas mitochondria can be used to insert mutations in complex I genes. *Proc Natl Acad Sci USA*. 2006. 103:4771–4776.

Roessler, P. G. Environmental control of glycerolipid metabolism in microalgae: Commercial implications and future research directions. *J Phycol*. 1990. 26:393–399.

Schroda, M., Blöcker, D., Beck, C. F. The HSP70A promoter as a tool for the improved expression of transgenes in Chlamydomonas. *Plant J*. 2000. 21:121–131.

Shin, S. E., Lim, J. M., Koh, H. G., Kim, E. K., Kang, N. K., Jeon, S., Kwon, S., Shin, W. S., Lee, B., Hwangbo, K., Kim, J., Ye, S. H., Yun, J. Y., Seo, H., Oh, H. M., Kim, K. J., Kim, J. S., Jeong, W. J., Chang, Y. K., Jeong, B. R. CRISPR/Cas9-induced knockout and knock-in mutations in *Chlamydomonas reinhardtii*. *Sci Rep*. 2016. 6:27810.

Tetali, S. D., Mitra, M., Melis, A. Development of the light-harvesting chlorophyll antenna in the green alga *Chlamydomonas reinhardtii* is regulated by the novel Tla1 gene. *Planta*. 2007. 225:813–829.

Trentacoste, E. M., Shrestha, R. P., Smith, S. R., Glé, C., Hartmann, A. C., Hildebrand, M., Gerwick, W. H. Metabolic engineering of lipid catabolism increases microalgal lipid accumulation without compromising growth. *Proc Natl Acad Sci USA*. 2013. 110:19748–19753.

Velmurugan, N., Deka, D. Transformation techniques for metabolic engineering of diatoms and haptophytes: Current state and prospects. *Appl Microbiol Biotechnol*. 2018. 102:4255–4267.

Wang, B., Lan, C. Q., Horsman, M. Closed photobioreactors for production of microalgal biomasses. *Biotechnol Adv*. 2012. 30:904–912.

Wang, S., Stiles, A. R., Guo, C., Liu, C. Harvesting microalgae by magnetic separation: A review. *Algal Res*. 2015. 9:178–185.

Westerhoff, H. V., Palsson, B. O. The evolution of molecular biology into systems biology. *Nat Biotechnol*. 2004. 22:1249–1252.

25 Leveraging Genome Sequencing Strategies for Basic and Applied Algal Research, Exemplified by Case Studies

Ariana A. Vasconcelos and Vitor H. Pomin

CONTENTS

Abbreviations 281
Introduction 281
Phylogeny and Evolution 282
Biofuels 284
Production of Recombinant Proteins 285
Production of Bioproducts – the Case of Astaxanthin 286
Bioremediation 287
Perspectives and Concluding Remarks 288
References 288

BOX 25.1 SALIENT FEATURES

Algae constitute a diverse and important group of organisms widely distributed on our planet. In addition to their ecological role, algae are exploited as a source of active biomolecules and are the basis in the large-scale food chain. Genome studies of various species of algae provided insight into many important aspects, including adaptation in environments, evolution of the genome through horizontal gene transfer and molecular diversity. These discoveries in the field of genomics have boosted other fields such as genetic engineering, both with regard to the production of transgenic algae and the use of these algae for the production of recombinant proteins that are very useful for the manufacture of vaccines, insecticides, biohydrogen and others metabolites that can be successfully used in medicine or industry. In this chapter, we discuss some examples of algae that have had their genomes sequenced and how the knowledge can be generally employed to benefit society in terms of the exploitation of algae and their related biotechnology.

ABBREVIATIONS

BKT: beta-ketolase
CTB: cholera toxin B subunit
DNA: deoxyribonucleic acid
EST: expressed sequence tag
H_2: molecular hydrogen
HPV: human papillomavirus
IgG: immunoglobulin
mAb: monoclonal antibody
NGS: next-generation sequencing
O_2: molecular oxygen
PA83: anthrax protective antigen 83
Pfs25: *Plasmodium falciparum* surface protein 25
Pfs28: *Plasmodium falciparum* surface protein 28
ROS: reactive oxygen species

INTRODUCTION

Algae constitute a diverse and important group of organisms widely distributed on our planet. These organisms range from unicellular to multicellular forms (Barsanti and Gualtieri 2014) and they play a fundamental ecological role as photosynthesizers, being responsible for the production of 50% of all the oxygen produced on Earth

(Cardozo et al. 2007; Thornton 2012). In addition to the ecological role they play, algae are exploited as a source of active biomolecules and the basis in the large-scale food chain. In recent years, algae have been prominent in the food and pharmaceutical industries and served as a rich source for the production of cosmetics and, more recently, of biofuels (Raja et al. 2013).

Although algae have valuable importance in evolutionary studies, our knowledge about this subject is still limited (Hlavova et al. 2015). This is because there are just very few available genetic and molecular tools to evaluate the relationship of functional genes with their potencies in algae activities (Tirichine and Bowler 2011). With the opening up of research on algal genomics and full genome sequencing, new perspectives or understanding of basic aspects and industrial relevance is emerging (Walker et al. 2005).

The first algal genome sequenced was achieved in 2004 for the red alga *Cyanidioschyzon merolae* (Matsuzaki et al. 2004). In the same year, a sketch of the genome of the marine diatom *Thalassiosira pseudonana* was also generated (Armbrust et al. 2004). In 2005, the introduction of next-generation sequencing (NGS), a technology that enables DNA sequencing on platforms capable of generating information about millions of base pairs in a single run, together with the growing interest in algae as heating mitigators and as an alternative source of biofuel, has further boosted studies on the sequencing of the genome of various algae (Kim et al. 2014).

In this chapter, we will discuss some examples of algae that have had their genomes sequenced and how this knowledge can be generally employed to benefit society in terms of the exploitation of algae and their related biotechnology (Table 25.1).

A good example is the green alga *Chlamydomonas reinhardtii*, which had its complete genome sequenced (Pröschold et al. 2005; Merchant et al. 2007) and today is a model organism to study photosynthesis and origin of flagella. Other examples of species that have had their genomes decoded in the last decade are the green algae *Ostreococcus tauri* (Derelle et al. 2006), *Ostreococcus lucimarinus* (Palenik et al. 2007), *Micromonas pusilla* (Worden et al. 2009), *Chlorella variabilis* (Blanc et al. 2010), *Volvox carteri* (Prochnik et al. 2010), the red algae *Chondrus crispus* (Colle'n et al. 2013), *Porphyridium purpureum* (Bhattacharya et al. 2013) and *Cyanophora paradoxa* (Price et al. 2012).

Genome data provided insights on many key issues, such as adaptation in environments, the evolution of the genome through horizontal gene transfer, and the study of molecular diversity. These discoveries in the field of genomics have propelled the field of genetic engineering, both with regards to the production of transgenic algae, and the use of these algae for the production of recombinant proteins which are very useful for manufacture of vaccines, insecticides, biohydrogen and other metabolites that can be successfully employed in medicine or industry (Hallmann 2007).

PHYLOGENY AND EVOLUTION

Throughout many decades we have sought to better understand the origin of the millions of species that inhabit our planet. In recent years, the increasing availability of genomic and transcriptomic data from organisms has facilitated the discoveries in this particular area (Tomato Genome Consortium 2012; Rands et al. 2013; Hori et al. 2014). The advances made in the field have contributed to advance phylogenetic studies and correlations involving algal species (Delsuc et al. 2005; Kim et al. 2014; Yokono et al. 2018).

Turmel and coworkers (2006) have sequenced the chloroplast genome of *Chara vulgaris* and compared the sequence with previously known *Mesostigma* (Mesostigmatales), *Chlorokybus* (Chlorokybales), *Staurastrum* and *Zygnema* (Zygnematales), *Chaetosphaeridium* (Coleochaetales) and some terrestrial plants. It was inferred that chloroplast genome remained largely unchanged in terms of genetic content, gene order and intron composition during the transition from charophyte green algae to terrestrial plants, indicating thus an upcoming evolutionary relationship between these two plants. From the analyses of inversion-based genomic rearrangements, no changes were observed in the order of genes during the transition from charophyte green algae to terrestrial plants (Turmel et al. 2006). In order to elucidate the evolutionary history of the main characteristics of algae and terrestrial plants, a group led by Bhattacharya et al. (2013) sequenced the 70 million-base pair nuclear genome of the unicellular algal *Cyanophora paradoxa* CCMP329 (Pringsheim strain). Price et al. (2012) have shown that plastids studied could be used to trace their origin to a single ancestral, supporting thus the hypothesis of monophyly in Plantae. This has helped to clarify the permanent question in eukaryotic evolution about monophyly in the Plantae kingdom (McFadden and van Dooren 2004; Rodríguez-Ezpeleta et al. 2005; Chan et al. 2011). With this study, it became clear that the ancestral Plantae contained many of the major innovations that can serve as an initial element to constitute the plant genomes and terrestrial algae.

C. paradoxa was indeed a great model of study. This species has shown characteristics for retaining ancestral mechanisms of starch biosynthesis, fermentation and

TABLE 25.1
Main Species of Algae that had their Genomes Characterized and Application of these Studies

Organism	Year	Relevance	References
Cyanidioschyzon merolae	2004	First complete algal genome. Provides a model system for the study of the origin, evolution and fundamental mechanisms of eukaryotic cells.	Matsuzaki et al. (2004)
Thalassiosira pseudonana	2004	The use of genome sequence to infer ocean ecology.	Armbrust et al. (2004)
Chlamydomonas reinhardtii	2005; 2007	Organism well studied. Its genomics provide data for theevolutionary study, use as a system of expression of human therapeutic proteins and vaccines.	Pröschold et al. (2005), Merchant et al. (2007)
Ostreococcus tauri	2006	Good candidate for biological models, such as cell division and/or genome evolution studies.	Derelle et al. (2006)
Chara vulgaris	2006	Studies on the origin and evolution of algae.	Turmel et al. (2006)
Ostreococcus lucimarinus	2007	Provided insights into the unique metal metabolism of these organisms.	Palenik et al. (2007)
Micromonas pusilla	2009	It provided valuable insights into ecological differentiation and the dynamic nature of early plant evolution.	Worden et al. (2009)
Chlorella variabilis		The genome reveals adaptation to photosymbiosis, coevolution with viruses.	Blanc et al. (2010)
Volvox carteri	2010	Important to investigate the evolution of multicellularity and development.	Prochnik et al. (2010)
Cyanophora paradoxa	2012	Very important in evolution studies. Contributed to the understanding of photosynthesis and plant evolution.	Price et al. (2012)
Chondrus crispus	2013	Provides insight into marine red algae metabolism and adaptations to the marine environment.	Colle'n et al. (2013)
Porphyridium purpureum	2013	Contributed to the study of evolution in algae.	Bhattacharya et al. (2013)
Klebsormidium flaccidum	2014	Important in the study of evolution, it contributed to elucidate the initial transition step from aquatic algae to terrestrial plants.	Hori et al. (2014)
Botryococcus braunii		Important in the research of development of biofuels from algae.	Browne et al. (2017)
Chromochloris zofingiensis		Provides information on the improvement of carotenoid production (example: Astaxanthin).	Roth et al. (2017)
Chara braunii	2018	Important in the study of evolution. It allowed the understanding of innovations that characterize the genomes of plants and terrestrial algae.	Nishiyama et al. (2018)

translocation of plastid proteins. In this case, the presence of protein-conducting channels in the outer and inner membranes of *C. paradoxa* plastid casings was a reference to the system of the photosynthetic eukaryotes. The analyses have shown traces of an ancient binding with parasites such as *Chlamydiae* in the genomes of *C. paradoxa* and other plants. Apparently, chlamydia-like bacteria seemed to donate genes related to plastid photosynthesis and polysaccharide polymerization to be stored in the cytosol (Chan et al. 2011).

The colonization of the soil by plants was a key event in the evolution of life (Pires and Dolan 2012). In this regard, in the study of Hori et al. (2014), the authors have reported the genome sequence of the algae *Klebsormidium flaccidum*. Comparison of the genome sequences with those from other algae and terrestrial plants have shown that *K. flaccidum* had many specific genes found also in terrestrial plants (Hori et al. 2014). This alga produces several plant hormones and homologs of the signaling intermediates required for hormonal actions in higher plants. The genome of *K. flaccidum* also encoded a primitive system for protection against the harmful effects of high intensity light. The presence of this plant-related system in *K. flaccidum* has suggested that during evolution, this alga had acquired the fundamental machinery necessary for adaptation to the terrestrial environments (Hori et al. 2014).

A recent study also concerned with the mechanisms involved during the early terrestrification of plants was recently published (Nishiyama et al. 2018). In this study, besides reporting the genome of *Chara braunii*, comparative genomic analyses were performed and demonstrated that *C. braunii* was phylogenetically close to terrestrial plants (Figure 25.1).

The genome of *C. braunii* is very rich in terms of phylogenetic information and points related to the evolutionary timing of the major steps that allowed life to progress on the terrestrial habitat (Nishiyama et al. 2018).

Important parts in the gradual evolutionary origin of plants would come from investigations concerning the general and specific characteristics of enzymes involved in cell wall biosynthesis, underlying events during cell division, correlations involving phytohormones, and more elaborated transcriptional regulations (Martin and Allen 2018). In the recent study of Nishiyama et al. (2018), phylogenetic analyses of shared proteins and genes, as well as of independent sets of structural genomic data have provided robust support for the notion that the charophyte order *Charales* can occupy a basal position relative to *Coleochaetales* e *Zygnematales*. This conception has challenged the current view on the phylogeny of charophyte green algae. Comparative analyses of structural genomic data have also indicated that the chloroplast genome remained largely unchanged in terms of a sequence of genes, gene content and composition levels of the introns during the transition from charophyte green algae to terrestrial plants (Nishiyama et al. 2018). In all, knowledge about the genome of algal species can indeed help us to understand how terrestrial plants have moved to the terrestrial habitat and how, after occupying this new environment, these autotrophs participated to increase oxygen content into the Earth atmosphere, and how some of the evolutionary steps have contributed to the current conditions of the planet.

BIOFUELS

Nowadays fossil fuels meet the growing energy needs of the world. However, its burning promotes climate change and depletes natural reserves (Singh and Singh 2012). Given these impacts, biofuels produced from renewable resources can be a more sustainable alternative (Radakovits et al. 2010; Simas-Rodrigues et al. 2015). Photosynthetic algae, both microalgae and macroalgae, have been considered as potential resources for the production of biofuels (Sheehan et al. 1998). In this regard, initial attempts to produce fuel from microalgae date from the late 1970s because of the oil crisis (U.S. DOE 2010). At that time, the price for production of fuel from fossil was considerably high (Hannon et al. 2010). Hence, it was necessary to develop a process to make algae-derived biofuels more practical and economically viable.

The production of biofuels from algae on a high and required scale requires an understanding of the algal biology, genetics, gene expression, as well as the development of molecular tools to modify these algae in order to fully direct the biotechnology related to these marine organisms to the fuel industry, since biofuels are becoming one of the biggest commodities in the world (Radakovits et al. 2010). This possibility can be amplified with the advancement in molecular techniques for genome sequencing. Knowledge and genetic

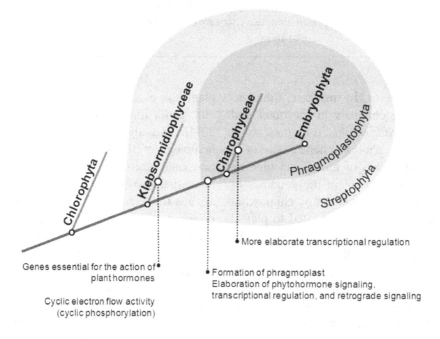

FIGURE 25.1 Cladogram symbolizing the evolution of charophytic algae to terrestrial plants, highlighting some characteristics that allowed this terrestrification. (Adapted from Nishiyama et al. 2018.)

improvements are increasingly being used to enhance the efficiency of this production (Hannon et al. 2010; Radakovits et al. 2010).

Recently the genome of the green fuel-producing microalga, *Botryococcus braunii*, was sequenced by a team of researchers. This alga has been the focus of studies because its cells produce large amounts of hydrocarbons, which can be converted into fuels (Browne et al. 2017). It was the growing use of the hydrocarbons of this alga as biofuel that encouraged the scholars to unravel the genome of *B. braunii*. This is expected to hasten our understanding of the role of genes and enzymes involved in the production of the hydrocarbons (Browne et al. 2017; Texas A&M AgriLife Communications 2017).

Another practical example that has attracted the sectors of biotechnology and bioenergy is related to the metabolism of molecular hydrogen (H_2), which has been studied extensively in *C. reinhardtii* (Hankamer et al. 2007; Hemschemeier et al. 2009). H_2 is believed to be the ideal fuel for the future because of its high energy content and clean combustion since its burning produces only water (Smith et al. 1992). Photoproduction of H_2 is possible in microalgae, both eukaryotic and prokaryotes because these species have hydrogenase enzymes that allow them to produce hydrogen under certain conditions (Das and Veziroğlu 2001). However, one of the major challenges of H_2 production by microalgae is the incompatibility between oxygen, photosynthesis and anaerobic H_2 production due to the high sensitivity of the enzyme hydrogenase to O_2 (Benemann 2000; Kruse and Hankamer 2010).

In this way, the research in photosynthesis has been expanded aiming for the development of methods for the production of photobiological hydrogen in green microalgae such as *C. reinhardtii* (Melis 2007; Mus et al. 2007; Nguyen et al. 2008). Some studies have demonstrated that it is possible to produce photobiological H_2, reducing the size of the light sensing antenna, and inhibiting state transitions and hydrogenase engineering (Hankamer et al. 2007; Melis 2007). In combination with physiological and biochemical approaches, these studies allowed a greater understanding of H_2 metabolism and maturation of hydrogenase enzymes in green microalgae. In this context, knowledge of metabolism coupled with genetic tools provides the emergence strategies to optimize the production of H_2 eukaryotic photosynthetic organisms (Radakovits et al. 2010).

Based on the aforementioned statements, we can clearly point out that advances in the field of genomics (such as sequenced whole genomes) have brought a new perspective to this area of production of biofuels using algae as the main resources.

PRODUCTION OF RECOMBINANT PROTEINS

Besides the extraction of hydrocarbons to obtain biofuels (Browne et al. 2017) and the production of photobiological hydrogen as an energy source (Melis 2007), the expression of recombinant proteins has also attracted the biotech sector in the algal field (Fuhrmann 2004). This is due to the projected application of transgenic microalgae as systems for the production of recombinant proteins with therapeutic and industrial action. In this way, the use of transgenic microalgae becomes promising in the scope of cost reduction and simplicity in the production of recombinant proteins (Gong et al. 2011). The rapid vegetative growth and high cellular densities of algae growing in low-cost environments – saline – contribute to the simplicity of this process (Franklin and Mayfield 2004). In addition to these advantages, the identification of genes and the availability of new genomic tools, such as the creation of databases of expressed sequence tags (ESTs), promote the growth of research in the area of expression of recombinant proteins in algae (Fuhrmann 2004).

To date, recombinant proteins, such as monoclonal antibodies, vaccines, hormones, therapeutic proteins, among others, have been produced experimentally from the nuclear genome or transgenic chloroplast of *C. reinhardtii* microalgae (Franklin and Mayfield 2005). Both the nuclear genome (Merchant et al. 2007) and the chloroplast (Maul et al. 2002) were sequenced, and the databases are available for search or cloning. One of the main reasons for using *C. reinhardtii* as an expression system is the wide range of molecular genetics tools available for this organism. Among the available molecular tools are various elements (such as sequences) of well characterized nuclear chloroplasts and promoters, and even more important, means for nuclear and chloroplast transformations (Mayfield and Franklin 2005).

An example of the production of human therapeutic proteins in transgenic algae has been shown in the work of Tran et al. (2009), where it was possible to synthesize and assemble a human monoclonal antibody (mAb) in chloroplasts of eukaryotic green alga *C. reinhardtii*. The antibody, 83K7C, is derived from a human immunoglobulin (IgG1) that has its action against the protective antigen of anthrax 83 (PA83). Studies have shown that this antibody blocks the effects of anthrax toxin in animal models (Tran et al. 2009). The 83K7C expressed in the alga was bound to PA83 *in vitro* with similar affinity to the 83K7C antibody expressed in mammalian cell culture. This work showed that *C. reinhardtii* chloroplasts have the ability to synthesize and assemble complex and functional human antibodies (Tran et al. 2009).

In another study, the expression model of the chloroplast genome *C. reinhardtii* was used for the expression of the fibronectin D2 binding domain of *Staphylococcus aureus* fused with the B subunit of cholera toxin. The objective was to produce an oral vaccine, which can be stored in ambient temperature without the need for refrigeration (Dreesen et al. 2010). In the tests using mice models it was shown that an oral immunization program with an algae-based vaccination significantly reduced pathogen burden in the spleen and gut of treated mice, and further promoted protection in 80% of them against lethal doses of *S. aureus*. It is important to note that the seaweed vaccine remained stable for more than 1.5 years at room temperature (Dreesen et al. 2010).

Another study has investigated the possibility of producing vaccines that blocked the transmission of malaria using the chloroplast expression model of *C. reinhardtii* (Gregory et al. 2012). For this study, the researchers selected antibodies that recognize the surface protein *Plasmodium falciparum* 25 (Pfs25) and 28 (Pfs28) and interrupt the sexual development of parasites inside the midgut of the mosquito. These proteins are difficult to produce in traditional recombinant systems because they have structurally complex and non-glycosylated domains. Already the production in chloroplasts of algae presents an advantage; characterized by double eukaryotic complex proteins, do not glycosyl these proteins (Gregory et al. 2012). The authors were able to successfully produce Pfs25 or Pfs28 and found that the antigens were structurally similar to native proteins. In addition, the antibodies designed for these recombinant proteins were able to recognize Pfs25 and Pfs28 from *P. falciparum*. Thus, algae showed promise for the production of candidate vaccine proteins (Gregory et al. 2012).

The work of Demurtas et al. (2013) reports the production of a vaccine based on Human Papillomavirus (HPV) E7 protein, using as system *C. reinhardtii*. This protein represents a key antigen for the development of therapeutic vaccines against HPV related cancers and lesions (Dermutas et al. 2013). The expression cassette which contained the protein encoding sequences was introduced into the algae chloroplast genome by homologous recombination. After the protein was obtained, *in vivo* tests were performed showing the induction of specific anti-E7 immunoglobulins (IgGs) and the proliferation of E7-specific T cells in C57BL/6 mice vaccinated with total extract of *C. reinhardtii* and with purified protein (Dermutas et al. 2013). In addition to the *in vivo* tests, *in vitro* tests were also performed with a tumor cell line expressing the E7 protein. These tests showed high levels of tumor protection, indicating successful expression of the E7GGG protein in soluble, immunogenic form (Dermutas et al. 2013). Obtaining E7 protein from microalgae enables detailed biochemical and physicochemical studies to confirm its structure and biological activity, which until now have been performed only on proteins expressed in bacterial systems (Dermutas et al. 2013).

Research on the production of recombinant proteins in algae expression systems continues in search of further advances. The knowledge that is available today is due to the availability of databases of genomes, especially information about *Chlamydomonas* (Furhrlan 2004). The studies cited here indicated that *C. reinhardtii* can be used as an alternative organism for the expression of recombinant proteins and important for future biotechnological applications. Although levels of expression need to be improved to make commercial interest possible (Franklin and Mayfield 2005; Eichler-Stahlberg et al. 2009).

PRODUCTION OF BIOPRODUCTS – THE CASE OF ASTAXANTHIN

One of the alternatives to make feasible the use of microalgae as components of biofuels is to obtain valuable co-products which would lead to economically feasible commercial production of microalgae (Roth et al. 2017). Some studies are being carried out in this direction and some species of algae have been used for the production of supplements, polyunsaturated fatty acids and carotenoids (Khan et al. 2018; Vaşas 2018). A well-known product by the industry is the carotenoid named astaxanthin (3,3'-dihydroxy-β, β-carotene-4,4'-dione) (Figure 25.2) (Yuan et al. 2011; Ambati et al. 2014). Carotenoids are a group of molecules that play a significant role in the photosynthetic pathway of microalgae (Ranga Rao et al. 2018). These molecules have the ability to prevent the formation of reactive oxygen species, and so are known as excellent antioxidants (Ranga Rao et al. 2013).

Some studies have highlighted the antioxidant (Naguib 2000) and anti-inflammatory (Park et al. 2010) benefits of astaxanthin for human health applications, including anticancer actions (Palozza et al. 2009), cardiovascular disease prevention (Fassett and Combes 2011), neurodegenerative diseases (Ryu et al. 2012), inflammatory diseases (Suzuki et al. 2006) and diabetes prevention (Uchiyama et al. 2002).

The first ever cloning and expression of β-carotene ketolase from *Haematococcus pluvialis* into the same organism has been reported by Kathiresan et al (2015). In this study, the β-carotene ketolase (bkt) gene was isolated from *H. pluvialis* and cloned in a vector pRT100

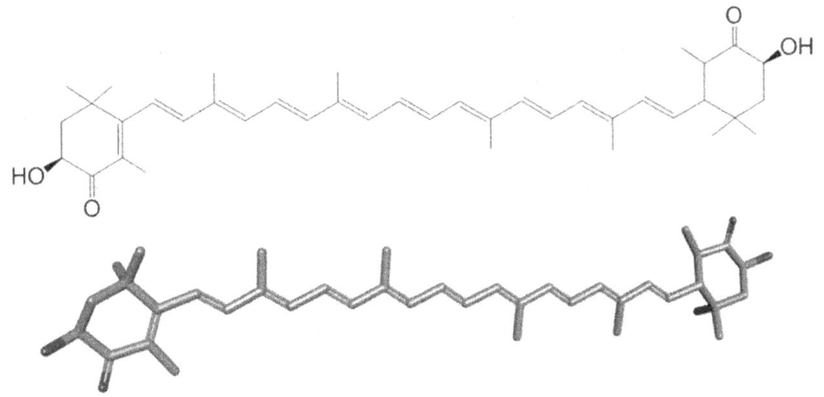

FIGURE 25.2 Chemical structure of astaxanthin. On the bottom, carbon backbone and oxygen atoms are shown in light and dark grey, respectively. Hydrogen atoms were omitted for simplification.

and further mobilized to a binary vector pCAMBIA 1304. The T-DNA of pCAMBIA 1304, which consists of cloned bkt, was successfully transformed to *H. pluvialis* through Agrobacterium mediation. Total carotenoids and astaxanthin content in the transformed cells were found to be 2-3-fold higher, while the intermediates like echinenone and canthaxanthin were found to be 8-10-fold higher than in the control cells. The expression level of carotenogenic genes like phytoene synthase (psy), phytoene desaturase (pds), lycopene cyclase (lcy), bkt and β-carotene hydroxylase (bkh) were found to be higher in transformed cells compared to the non-transformed (NT) *H. pluvialis*.

Similarly, the first ever metabolically engineered *Dunaliella*, an alga known for β-carotene production, was reported by Anila et al (2015). *Dunaliella* is a commercially important marine alga producing high amount of β-carotene. In their study *Dunaliella salina* was transformed for ketocarotenoid production by pathway modification including a gene for bkt from *H. pluvialis*. Furthermore, it was engineered for encoding β-carotene ketolase (4,4'β-oxygenase) along with chloroplast targeting for the production of ketocarotenoids. The bkt under the control of the *Dunaliella* Rubisco smaller subunit promoter along with its transit peptide sequence was introduced into the alga through standardized Agrobacterium-mediated transformation procedure. A notable upregulation of the endogenous hydroxylase level of transformants was observed where the BKT expression was higher in nutrient-limiting conditions. Carotenoid analysis of the transformants through HPLC and MS analysis showed the presence of astaxanthin and canthaxanthin with maximum content of 3.5 and 1.9 μg/g DW, respectively. This study for the first time reported the feasibility of using *D. salina* for the production of ketocarotenoids including astaxanthin.

Until recently the focus for astaxanthin production from algae was centric to *Haematococcus pluvialis*. However, a recent work has brought the genome assembly of the green algae *Chromochloris zofingiensis* and highlighted the importance of this information to boost the production of astaxanthin (Roth et al. 2017). The genome of *C. zofingiensis* provided insight into the astaxanthin pathway and brought a relationship between carotenoid biosynthesis in this algae and *C. reinhardtii*. The genome study also revealed that there is a broad distribution of carotenoid biosynthesis genes in many chromosomes, as is typical in eukaryotes (Roth et al. 2017). Two genes encode beta-ketolase (BKT), a key enzyme that synthesizes astaxanthin, BKT1 and BKT genome analysis allowed the identification of the BKT1 gene as a determinant for the production of astaxanthin; it was still possible to identify genes that could be incomplete in biosynthesis and astaxanthin accumulation pathways. Understanding the genomics and transcriptomics of *C. zofingiensis* contributes to this becoming an attractive model not only for fundamental studies of its biology but also for the financially viable and environmentally sustainable production of biofuels and bioproducts (Roth et al. 2017). Thus the exploitation of algae for the production of useful metabolites is another example of the successful usage of algae for the benefits of human society.

BIOREMEDIATION

Bioremediation is a process that uses living organisms to perform an ecological cleaning. These organisms will degrade, transform or remove waste and pollutants from environments such as water and soil. Species of algae, mainly microalgae, have been used to clean effluent pollutants, including having a high capacity to remove heavy metals from wastewater (Wilde and Benemann 1993;

Rawat et al. 2011). The use of different algae for wastewater treatment has been the subject of research and development for several decades. A number of species of algae of different genus such as *Botryococcus*, *Chlamydomonas*, *Chlorella* and *Phormidium* have been proven for different phytoremediation purposes. However, the selection of algal species more suitable for the removal of certain pollutants requires more studies (Rawat et al. 2011).

A good model would be *Chlamydomonas*, this alga has its genes sequenced and well characterized, and can be useful as a source of new genes for the transformation of plants or other algae. An example is the genes involved in metal transport, which were described in the work of Rubinelli et al. (2002). In addition, this alga presents in the structure of its cell wall sulfated oligosaccharides that play a role in the binding of heavy metals (Xue et al. 1988). This feature opens the possibility of manipulation of these algal genes to increase the synthesis of cell wall polysaccharides or other metal-binding sites on the cell surface, providing greater metal removal (Mehta and Gaur 2005).

The exploration of biological mechanisms at the molecular level to produce manipulated organisms with greater capacity of biosorption and selectivity for specific metallic ions can be used to develop new biosorbents. The high cost of conventional technologies to reduce concentrations of toxic metal ions in wastewater at acceptable rates has led to the exploration of genetic and protein engineering approaches to producing more profitable "green" biosorbents (Zeraatkar et al. 2016). Although innumerable possibilities can be found in this field, transgenic applications of algae for use as bioremediation agents are still scarce and more studies must be done in this direction.

PERSPECTIVES AND CONCLUDING REMARKS

Knowledge about genomes allows the understanding of events occurring in organisms and genetic manipulation. This knowledge has many advantages, especially for the biotechnology industry. The comprehension of the complexity in metabolic pathways along with genome data can lead to the development of transformation protocols and genetic engineering that can allow production and exploitation of natural and high-value products such as antibodies, hormones and vaccines at economically and industrially viable levels.

Algal species may be more productive with the use of new cultivation protocols, harvesting and processing can be improved, leading to greater stability and minimized extraction costs. The applications of genomics, proteomics and metabolomics would introduce numerous possibilities in the areas of systems and synthetic biology. This represents an enormous potential for solutions to environmental problems, for example, finding alternatives to fossil fuels or improving the metabolism of microalgae to heavy metals.

Furthermore, the elucidation of specific regulatory networks may also contribute to increase the levels of recombinant protein production by the manipulation of additional loci that may contribute to protein folding, translation efficiency, extra-cellular protein secretion among other possibilities. The enhanced understanding of genome has tremendously advanced the applications of algal biotechnology and will continue to provide exciting opportunities in the future.

REFERENCES

Ambati, R. R., Phang, S. M., Ravi, S., Aswathanarayana, R. G. Astaxanthin: Sources, extraction, stability, biological activities and its commercial applications – A review. *Mar. Drugs*. 2014, 12: 128–152.

Anila, N., Simon, D. P., Chandrashekar, A., Ravishankar, G. A., Sarada, R. Metabolic engineering of *Dunaliella salina* for the production of ketocarotenoids. *Photosynth. Res.* 2015, 127(3): 321–333.

Armbrust, E. V., Berges, J. A., Bowler, C., Green, B. R., Martinez, D., Putnam, N. H., Zhou, S., Allen, A. E., Apt, K. E., Bechner, M., Brzezinski, M. A., Chaal, B. K., Chiovitti, A., Davis, A. K., Demarest, M. S., Detter, J. C., Glavina, T., Goodstein, D., Hadi, M. Z., Hellsten, U., Hildebrand, M., Jenkins, B. D., Jurka, J., Kapitonov, V. V., Kröger, N., Lau, W. W., Lane, T. W., Larimer, F. W., Lippmeier, J. C., Lucas, S., Medina, M., Montsant, A., Obornik, M., Parker, M. S., Palenik, B., Pazour, G. J., Richardson, P. M., Rynearson, T. A., Saito, M. A., Schwartz, D. C., Thamatrakoln, K., Valentin, K., Vardi, A., Wilkerson, F. P., Rokhsar, D. S. The genome of the diatom *Thalassiosira pseudonana*: Ecology, evolution, and metabolism. *Science*. 2004, 306: 79–86.

Barsanti, L., Gualtieri, P. *Algae: Anatomy, Biochemistry, and Biotechnology*. 2014, 361. CRC Press.

Benemann, J. R. Hydrogen production by microalgae. *J. Appl. Phycol.* 2000, 12: 291–300.

Bhattacharya, D., Price, D. C., Chan, C. X., Qiu, H., Rose, N., Ball, S., Weber, A. P. M., Arias, M. C., Henrissat, B., Coutinho, P. M., Krishnan, A., Zäuner, S., Morath, S., Hilliou, F., Egizi, A., Perrineau, M. M., Yoon, H. S. Genome of the red alga *Porphyridium purpureum*. *Nat. Commun.* 2013, 4: 1941.

Blanc, G., Duncan, G., Agarkova, I., Borodovsky, M., Gurnon, J., Kuo, A., Lindquist, E., Lucas, S., Pangilinan, J., Polle, J., Salamov, A., Terry, A., Yamada, T., Dunigan, D. D., Grigoriev, I. V., Claverie, J. M., Van Etten, J. L. The *Chlorella variabilis* NC64A genome reveals adaptation to photosymbiosis, coevolution with viruses, and cryptic sex. *Plant Cell*. 2010, 22: 2943–2955.

Browne, D. R., Jenkins, J., Schmutz, J., Shu, S., Barry, K., Grimwood, J., Chiniquy, J., Sharma, A., Niehaus, T. D., Weiss, T. L., Koppisch, A. T., Fox, D. T., Dhungana, S., Okada, S., Chappell, J., Devarenne, T. P. Draft nuclear genome sequence of the liquid hydrocarbon – Accumulating green microalga *Botryococcus braunii* race B (Showa). *Genome Announc.* 2017, 5: e00215-17.

Cardozo, K. H. M., Guaratini, T., Barros, M. P., Falcão, V. R., Tonon, A. P., Lopes, N. P., Campos, S., Torres, M. A., Souza, A. O., Colepicolo, P., Pinto, E. Metabolites from algae with economical impact. *Comp. Biochem. Physiol. C Toxicol. Pharmacol.* 2007, 146: 60–78.

Chan, C. X., Yang, E. C., Banerjee, T., Yoon, H. S., Martone, P. T., Estevez, J. M., Bhattacharya, D. Red and green algal monophyly and extensive gene sharing found in a rich repertoire of red algal genes. *Curr. Biol.* 2011, 21: 328–333.

Collén, J., Porcel, B., Carré, W., Ball, S. G., Chaparro, C., Tonon, T., Barbeyron, T., Michel, G., Noel, B., Valentin, K., Elias, M., Artiguenave, F., Arun, A., Aury, J. M., Barbosa-Neto, J. F., Bothwell, J. H., Bouget, F. Y., Brillet, L., Cabello-Hurtado, F., Capella-Gutiérrez, S., Charrier, B., Cladière, L., Cock, J. M., Coelho, S. M., Colleoni, C., Czjzek, M., Da Silva, C., Delage, L., Denoeud, F., Deschamps, P., Dittami, S. M., Gabaldón, T., Gachon, C. M., Groisillier, A., Hervé, C., Jabbari, K., Katinka, M., Kloareg, B., Kowalczyk, N., Labadie, K., Leblanc, C., Lopez, P. J., McLachlan, D. H., Meslet-Cladiere, L., Moustafa, A., Nehr, Z., Nyvall Collén, P., Panaud, O., Partensky, F., Poulain, J., Rensing, S. A., Rousvoal, S., Samson, G., Symeonidi, A., Weissenbach, J., Zambounis, A., Wincker, P., Boyen, C. Genome structure and metabolic features in the red seaweed *Chondrus crispus* shed light on evolution of the *Archaeplastida*. *Proc. Natl Acad. Sci. U.S.A.* 2013, 110: 5247–5252.

Das, D., Veziroğlu, T. N. Hydrogen production by biological processes: A survey of literature. *Int. J. Hydrog. Energy.* 2001, 26: 13–28.

Delsuc, F., Brinkmann, H., Philippe, H. Phylogenomics and the reconstruction of the tree of life. *Nat. Rev. Genet.* 2005, 6: 361–375.

Demurtas, O. C., Massa, S., Ferrante, P., Venuti, A., Franconi, R., Giuliano, G. A *Chlamydomonas*-derived human papillomavirus 16 E7 vaccine induces specific tumor protection. *PLOS ONE.* 2013, 8: e61473.

Derelle, E., Ferraz, C., Rombauts, S., Rouzé, P., Worden, A. Z., Robbens, S., Partensky, F., Degroeve, S., Echeynié, S., Cooke, R., Saeys, Y., Wuyts, J., Jabbari, K., Bowler, C., Panaud, O., Piégu, B., Ball, S. G., Ral, J. P., Bouget, F. Y., Piganeau, G., De Baets, B., Picard, A., Delseny, M., Demaille, J., Van de Peer, Y., Moreau, H. Genome analysis of the smallest free-living eukaryote *Ostreococcus tauri* unveils many unique features. *Proc. Natl Acad. Sci. U.S.A.* 2006, 103: 11647–11652.

Dreesen, I. A., Charpin-El Hamri, G., Fussenegger, M. Heat-stable oral alga-based vaccine protects mice from *Staphylococcus aureus* infection. *J. Biotechnol.* 2010, 145: 273–280.

Eichler-Stahlberg, A. E., Weisheit, W., Ruecker, O., Heitzer, M. Strategies to facilitate transgene expression in *Chlamydomonas reinhardtii*. *Planta.* 2009, 229: 873–883.

Fassett, R. G., Coombes, J. S. Astaxanthin: A potential therapeutic agent in cardiovascular disease. *Mar. Drugs.* 2011, 9: 447–465.

Franklin, S. E., Mayfield, S. P. Prospects for molecular farming in the green alga *Chlamydomonas*. *Curr. Opin. Plant Biol.* 2004, 7: 159–165.

Franklin, S. E., Mayfield, S. P. Recent developments in the production of human therapeutic proteins in eukaryotic algae. *Expert Opin. Biol. Ther.* 2005, 5: 225–235.

Fuhrmann, M. Production of antigens in *Chlamydomonas reinhardtii*: Green microalgae as a novel source of recombinant proteins. *Methods Mol. Med.* 2004, 94: 191–195.

Gong, Y., Hu, H., Gao, Y., Xu, X., Gao, H. Microalgae as platforms for production of recombinant proteins and valuable compounds: Progress and prospects. *J. Ind. Microbiol. Biotechnol.* 2011, 38: 1879–1890.

Gregory, J. A., Li, F., Tomosada, L. M., Cox, C. J., Topol, A. B., Vinetz, J. M., Mayfield, S. Algae-produced Pfs25 elicits antibodies that inhibit malaria transmission. *PLOS ONE.* 2012, 7: e37179.

Hallmann, A. Algal transgenics and biotechnology. *Transgenic Plant J.* 2007, 1: 81–98.

Hankamer, B., Lehr, F., Rupprecht, J., Mussgnug, J. H., Posten, C., Kruse, O. Photosynthetic biomass and H2 production by green algae: From bioengineering to bioreactor scale-up. *Physiol. Plant.* 2007, 131: 10–21.

Hannon, M., Gimpel, J., Tran, M., Rasala, B., Mayfield, S. Biofuels from algae: Challenges and potential. *Biofuels.* 2010, 1: 763–784.

Hemschemeier, A., Melis, A., Happe, T. Analytical approaches to photobiological hydrogen production in unicellular green algae. *Photosynth. Res.* 2009, 102: 523–540.

Hlavova, M., Turoczy, Z., Bisova, K. Improving microalgae for biotechnology – From genetics to synthetic biology. *Biotechnol. Adv.* 2015, 33: 1194–1203.

Hori, K., Maruyama, F., Fujisawa, T., Togashi, T., Yamamoto, N., Seo, M., Sato, S., Yamada, T., Mori, H., Tajima, N., Moriyama, T., Ikeuchi, M., Watanabe, M., Wada, H., Kobayashi, K., Saito, M., Masuda, T., Sasaki-Sekimoto, Y., Mashiguchi, K., Awai, K., Shimojima, M., Masuda, S., Iwai, M., Nobusawa, T., Narise, T., Kondo, S., Saito, H., Sato, R., Murakawa, M., Ihara, Y., Oshima-Yamada, Y., Ohtaka, K., Satoh, M., Sonobe, K., Ishii, M., Ohtani, R., Kanamori-Sato, M., Honoki, R., Miyazaki, D., Mochizuki, H., Umetsu, J., Higashi, K., Shibata, D., Kamiya, Y., Sato, N., Nakamura, Y., Tabata, S., Ida, S., Kurokawa, K., Ohta, H. *Klebsormidium flaccidum* genome reveals primary factors for plant terrestrial adaptation. *Nat. Commun.* 2014, 5: 3978.

Kathiresan, S., Chandrashekar, A., Ravishankar, G. A., Sarada, R. Regulation of astaxanthin and its intermediates through cloning and genetic transformation of β–carotene ketolase in *Haematococcus pluvialis*. *J. Biotechnol.* 2015, 196–197: 33–41.

Khan, M. I., Shin, J. H., Kim, J. D. The promising future of microalgae: Current status, challenges, and optimization of a sustainable and renewable industry for biofuels, feed, and other products. *Microb. Cell Fact.* 2018, 17: 36.

Kim, K. M., Park, J. H., Bhattacharya, D., Yoon, H. S. Applications of next-generation sequencing to unravelling the evolutionary history of algae. *Int. J. Syst. Evol. Microbiol.* 2014, 64: 333–345.

Kruse, O., Hankamer, B. Microalgal hydrogen production. *Curr. Opin. Biotechnol.* 2010, 21: 238–243.

Martin, W. F., Allen, J. F. An algal greening of land. *Cell.* 2018, 174: 256–258.

Matsuzaki, M., Misumi, O., Shin-I, T., Maruyama, S., Takahara, M., Miyagishima, S. Y., Mori, T., Nishida, K., Yagisawa, F., Nishida, K., Yoshida, Y., Nishimura, Y., Nakao, S., Kobayashi, T., Momoyama, Y., Higashiyama, T., Minoda, A., Sano, M., Nomoto, H., Oishi, K., Hayashi, H., Ohta, F., Nishizaka, S., Haga, S., Miura, S., Morishita, T., Kabeya, Y., Terasawa, K., Suzuki, Y., Ishii, Y., Asakawa, S., Takano, H., Ohta, N., Kuroiwa, H., Tanaka, K., Shimizu, N., Sugano, S., Sato, N., Nozaki, H., Ogasawara, N., Kohara, Y., Kuroiwa, T. Genome sequence of the ultrasmall unicellular red alga *Cyanidioschyzon merolae* 10D. *Nature.* 2004, 428: 653–657.

Maul, J. E., Lilly, J. W., Cui, L., dePamphilis, C. W., Miller, W., Harris, E. H., Stern, D. B. The *Chlamydomonas reinhardtii* plastid chromosome: Islands of genes in a sea of repeats. *Plant Cell.* 2002, 14: 2659–2679.

Mayfield, S. P., Franklin, S. E. Expression of human antibodies in eukaryotic micro-algae. *Vaccine.* 2005, 23: 1828–1832.

McFadden, G. I., van Dooren, G. G. Evolution: Red algal genome affirms a common origin of all plastids. *Curr. Biol.* 2004, 14: R514–R516.

Mehta, S. K., Gaur, J. P. Use of algae for removing heavy metal ions from wastewater: Progress and prospects. *Crit. Rev. Biotechnol.* 2005, 25: 113–152.

Melis, A. Photosynthetic H2 metabolism in *Chlamydomonas reinhardtii* (unicellular green algae). *Planta.* 2007, 226: 1075–1086.

Merchant, S. S., Prochnik, S. E., Vallon, O., Harris, E. H., Karpowicz, S. J., Witman, G. B., Terry, A., Salamov, A., Fritz-Laylin, L. K., Maréchal-Drouard, L., Marshall, W. F., Qu, L. H., Nelson, D. R., Sanderfoot, A. A., Spalding, M. H., Kapitonov, V. V., Ren, Q., Ferris, P., Lindquist, E., Shapiro, H., Lucas, S. M., Grimwood, J., Schmutz, J., Cardol, P., Cerutti, H., Chanfreau, G., Chen, C. L., Cognat, V., Croft, M. T., Dent, R., Dutcher, S., Fernández, E., Fukuzawa, H., González-Ballester, D., González-Halphen, D., Hallmann, A., Hanikenne, M., Hippler, M., Inwood, W., Jabbari, K., Kalanon, M., Kuras, R., Lefebvre, P. A., Lemaire, S. D., Lobanov, A. V., Lohr, M., Manuell, A., Meier, I., Mets, L., Mittag, M., Mittelmeier, T., Moroney, J. V., Moseley, J., Napoli, C., Nedelcu, A. M., Niyogi, K., Novoselov, S. V., Paulsen, I. T., Pazour, G., Purton, S., Ral, J. P., Riaño-Pachón, D. M., Riekhof, W., Rymarquis, L., Schroda, M., Stern, D., Umen, J., Willows, R., Wilson, N., Zimmer, S. L., Allmer, J., Balk, J., Bisova, K., Chen, C. J., Elias, M., Gendler, K., Hauser, C., Lamb, M. R., Ledford, H., Long, J. C., Minagawa, J., Page, M. D., Pan, J., Pootakham, W., Roje, S., Rose, A., Stahlberg, E., Terauchi, A. M., Yang, P., Ball, S., Bowler, C., Dieckmann, C. L., Gladyshev, V. N., Green, P., Jorgensen, R., Mayfield, S., Mueller-Roeber, B., Rajamani, S., Sayre, R. T., Brokstein, P., Dubchak, I., Goodstein, D., Hornick, L., Huang, Y. W., Jhaveri, J., Luo, Y., Martínez, D., Ngau, W. C., Otillar, B., Poliakov, A., Porter, A., Szajkowski, L., Werner, G., Zhou, K., Grigoriev, I. V., Rokhsar, D. S., Grossman, A. R. The *Chlamydomonas* genome reveals the evolution of key animal and plant functions. *Science.* 2007, 318: 245–250.

Mus, F., Dubini, A., Seibert, M., Posewitz, M. C., Grossman, A. R. Anaerobic acclimation in *Chlamydomonas reinhardtii* anoxic gene expression, hydrogenase induction, and metabolic pathways. *J. Biol. Chem.* 2007, 282: 25475–25486.

Naguib, Y. M. A. Antioxidant activities of astaxanthin and related carotenoids. *J. Agric. Food Chem.* 2000, 48: 1150–1154.

Nguyen, A. V., Thomas-Hall, S. R., Malnoë, A., Timmins, M., Mussgnug, J. H., Rupprecht, J., Kruse, O., Hankamer, B., Schenk, P. M. Transcriptome for photobiological hydrogen production induced by sulfur deprivation in the green alga *Chlamydomonas reinhardtii*. *Eukaryot. Cell.* 2008, 7: 1965–1979.

Nishiyama, T., Sakayama, H., de Vries, J., Buschmann, H., Saint-Marcoux, D., Ullrich, K. K., Haas, F. B., Vanderstraeten, L., Becker, D., Lang, D., Vosolsobě, S., Rombauts, S., Wilhelmsson, P. K. I., Janitza, P., Kern, R., Heyl, A., Rümpler, F., Villalobos, L. I. A. C., Clay, J. M., Skokan, R., Toyoda, A., Suzuki, Y., Kagoshima, H., Schijlen, E., Tajeshwar, N., Catarino, B., Hetherington, A. J., Saltykova, A., Bonnot, C., Breuninger, H., Symeonidi, A., Radhakrishnan, G. V., Van Nieuwerburgh, F., Deforce, D., Chang, C., Karol, K. G., Hedrich, R., Ulvskov, P., Glöckner, G., Delwiche, C. F., Petrášek, J., Van de Peer, Y., Friml, J., Beilby, M., Dolan, L., Kohara, Y., Sugano, S., Fujiyama, A., Delaux, P. M., Quint, M., Theißen, G., Hagemann, M., Harholt, J., Dunand, C., Zachgo, S., Langdale, J., Maumus, F., Van Der Straeten, D., Gould, S. B., Rensing, S. A. The *Chara* genome: Secondary complexity and implications for plant terrestrialization. *Cell.* 2018, 174: 448. e24–464.e24.

Palenik, B., Grimwood, J., Aerts, A., Rouzé, P., Salamov, A., Putnam, N., Dupont, C., Jorgensen, R., Derelle, E., Rombauts, S., Zhou, K., Otillar, R., Merchant, S. S., Podell, S., Gaasterland, T., Napoli, C., Gendler, K., Manuell, A., Tai, V., Vallon, O., Piganeau, G., Jancek, S., Heijde, M., Jabbari, K., Bowler, C., Lohr, M., Robbens, S., Werner, G., Dubchak, I., Pazour, G. J., Ren, Q., Paulsen, I., Delwiche, C., Schmutz, J., Rokhsar, D., Van de Peer, Y., Moreau, H., Grigoriev, I. V. The tiny eukaryote *Ostreococcus* provides genomic insights into the paradox of plankton speciation. *Proc. Natl Acad. Sci. U.S.A.* 2007, 104: 7705–7710.

Palozza, P., Torelli, C., Boninsegna, A., Simone, R., Catalano, A., Mele, M. C., Picci, N. Growth-inhibitory effects of the astaxanthin-rich alga *Haematococcus pluvialis* in human colon cancer cells. *Cancer Lett.* 2009, 283: 108–117.

Park, J. S., Chyun, J. H., Kim, Y. K., Line, L. L., Chew, B. P. Astaxanthin decreased oxidative stress and inflammation and enhanced immune response in humans. *Nutr. Metab.* 2010, 7: 18.

Pires, N. D., Dolan, L. Morphological evolution in land plants: New designs with old genes. *Philos. Trans. R. Soc. Lond. B Biol. Sci.* 2012, 367: 508–518.

Price, D. C., Chan, C. X., Yoon, H. S., Yang, E. C., Qiu, H., Weber, A. P. M., Schwacke, R., Gross, J., Blouin, N. A., Lane, C., Reyes-Prieto, A., Durnford, D. G., Neilson, J. A. D., Lang, B. F., Burger, G., Steiner, J. M., Löffelhardt, W., Meuser, J. E., Posewitz, M. C., Ball, S., Arias, M. C., Henrissat, B., Coutinho, P. M., Rensing, S. A., Symeonidi, A., Doddapaneni, H., Green, B. R., Rajah, V. D., Boore, J., Bhattacharya, D. Cyanophora paradoxa genome elucidates origin of photosynthesis in algae and plants. *Science.* 2012, 335: 843–847.

Prochnik, S. E., Umen, J., Nedelcu, A. M., Hallmann, A., Miller, S. M., Nishii, I., Ferris, P., Kuo, A., Mitros, T., Fritz-Laylin, L. K., Hellsten, U., Chapman, J., Simakov, O., Rensing, S. A., Terry, A., Pangilinan, J., Kapitonov, V., Jurka, J., Salamov, A., Shapiro, H., Schmutz, J., Grimwood, J., Lindquist, E., Lucas, S., Grigoriev, I. V., Schmitt, R., Kirk, D., Rokhsar, D. S. Genomic analysis of organismal complexity in the multicellular green alga *Volvox carteri. Science.* 2010, 329: 223–226.

Pröschold, T., Harris, E. H., Coleman, A. W. Portrait of a species: *Chlamydomonas reinhardtii. Genetics.* 2005, 170: 1601–1610.

Radakovits, R., Jinkerson, R. E., Darzins, A., Posewitz, M. C. Genetic engineering of algae for enhanced biofuel production. *Eukaryot. Cell.* 2010, 9: 486–501.

Raja, A., Vipin, C., Aiyappan, A. Biological importance of marine algae – An overview. *Int. J. Curr. Microbiol. App. Sci.* 2013, 2: 222–227.

Rands, C. M., Darling, A., Fujita, M., Kong, L., Webster, M. T., Clabaut, C., Emes, R. D., Heger, A., Meader, S., Hawkins, M. B., Eisen, M. B., Teiling, C., Affourtit, J., Boese, B., Grant, P. R., Grant, B. R., Eisen, J. A., Abzhanov, A., Ponting, C. P. Insights into the evolution of Darwin's finches from comparative analysis of the *Geospiza magnirostris* genome sequence. *BMC Genom.* 2013, 12: 14–95.

Ranga Rao, A., Deepika, G., Ravishankar, G. A., Sarada, R., Narasimharao, B. P., Bo, L., Su, Y. Industrial potential of carotenoid pigments from microalgae: Current trends and future prospects. *Crit. Rev. Food Sci. Nutr.* 2018: 1–22.

Ranga Rao, A., Sindhuja, H. N., Dharmesh, S. M., Udaya Sankar, K., Sarada, R., Ravishankar, G. A. Effective inhibition of skin cancer, tyrosinase and antioxidative properties by astaxanthin and astaxanthin estersfromgreen alga *Haematococcus pluvialis. J. Agri. Food Chem.* 2013, 61: 3842–3851.

Rawat, I., Ranjith Kumar, R., Mutanda, T., Bux, F. Dual role of microalgae: Phycoremediation of domestic wastewater and biomass production for sustainable biofuels production. *Appl. Energy.* 2011, 88: 3411–3424.

Rodríguez-Ezpeleta, N., Brinkmann, H., Burey, S. C., Roure, B., Burger, G., Löffelhardt, W., Bohnert, H. J., Philippe, H., Lang, B. F. Monophyly of primary photosynthetic eukaryotes: Green plants, red algae, and glaucophytes. *Curr. Biol.* 2005, 15: 1325–1330.

Roth, M. S., Cokus, S. J., Gallaher, S. D., Walter, A., Lopez, D., Erickson, E., Endelman, B., Westcott, D., Larabell, C. A., Merchant, S. S., Pellegrini, M., Niyogi, K. K. Chromosome-level genome assembly and transcriptome of the green alga *Chromochloris zofingiensis* illuminates astaxanthin production. *Proc. Natl Acad. Sci. U. S. A.* 2017, 114: E4296–E4305.

Rubinelli, P., Siripornadulsil, S., Gao-Rubinelli, F., Sayre, R. T. Cadmium- and iron-stress-inducible gene expression in the green alga *Chlamydomonas reinhardtii*: Evidence for H43 protein function in iron assimilation. *Planta.* 2002, 215: 1–13.

Ryu, S. K., King, T. J., Fujioka, K., Pattison, J., Pashkow, F. J., Tsimikas, S. Effect of an oral astaxanthin prodrug (CDX-085) on lipoprotein levels and progression of atherosclerosis in LDLR and ApoE mice. *Atherosclerosis.* 2012, 222: 99–105.

Sheehan, J., Dunahay, T., Benemann, J., Roessler, P. *Look Back at the U.S. Department of Energy's Aquatic Species Program: Biodiesel from Algae, Close-Out Report.* United States, 1998.

Simas-Rodrigues, C., Villela, H. D. M., Martins, A. P., Marques, L. G., Colepicolo, P., Tonon, A. P. Microalgae for economic applications: Advantages and perspectives for bioethanol. *J. Exp. Bot.* 2015, 66: 4097–4108.

Singh, B. R., Singh, O. Global trends of fossil fuel reserves and climate change in the 21st Century. In: *Fossil Fuel and the Environment*, Shahriar Khan (editor). 2012. 167–192. InTech.

Smith, G. D., Ewart, G. D., Tucker, W. Hydrogen production by cyanobacteria. *Int. J. Hydrogen Energy.* 1992, 17: 695–698.

Suzuki, Y., Ohgami, K., Shiratori, K., Jin, X. H., Ilieva, I., Koyama, Y., Yazawa, K., Yoshida, K., Kase, S., Ohno, S. Suppressive effects of astaxanthin against rat endotoxin induced uveitis by inhibiting the NF-kB signaling pathway. *Exp. Eye Res.* 2006, 82: 275–281.

Texas A&M AgriLife Communications. Genome sequence of fuel-producing alga announced. *ScienceDaily.* 2007. <www.sciencedaily.com/releases/2017/05/170510174850.htm> (accessed August 13, 2018).

Thornton, D. C. O. Primary production in the ocean. In: *Advances in Photosynthesis – Fundamental Aspects*, Mohammad Najafpour (editor). 2012. 563–587. InTech.

Tirichine, L., Bowler, C. Decoding algal genomes: Tracing back the history of photosynthetic life on Earth. *Plant J.* 2011, 66: 45–57.

Tomato Genome Consortium. The tomato genome sequence provides insights into fleshy fruit evolution. *Nature.* 2012, 485: 635–641.

Tran, M., Zhou, B., Pettersson, P. L., Gonzalez, M. J., Mayfield, S. P. Synthesis and assembly of a full-length human monoclonal antibody in algal chloroplasts. *Biotechnol. Bioeng.* 2009, 104: 663–673.

Turmel, M., Otis, C., Lemieux, C. The chloroplast genome sequence of *Chara vulgaris* sheds new light into the closest green algal relatives of land plants. *Mol. Biol. Evol.* 2006, 23: 1324–1338.

U.S. DOE. 2010. National Algal Biofuels Technology Roadmap. U.S. Department of Energy, Office of Energy Efficiency and Renewable Energy, Biomass Program.

Uchiyama, K., Naito, Y., Hasegawa, G., Nakamura, N., Takahashi, J., Yoshikawa, T. Astaxanthin protects β-cells against glucose toxicity in diabetic db/db mice. *Redox Rep.* 2002, 7: 290–293.

Vasas, G. Microalgae as the source of natural products. *Orv. Hetil.* 2018, 159: 703–708.

Walker, T. L., Collet, C., Purton, S. Algal transgenics in the genomic era. *J. Phycol.* 2005, 41: 1077–1093.

Wilde, E. W., Benemann, J. R. Bioremoval of heavy metals by the use of microalgae. *Biotechnol. Adv.* 1993, 11: 781–812.

Worden, A. Z., Lee, J. H., Mock, T., Rouzé, P., Simmons, M. P., Aerts, A. L., Allen, A. E., Cuvelier, M. L., Derelle, E., Everett, M. V., Foulon, E., Grimwood, J., Gundlach, H., Henrissat, B., Napoli, C., McDonald, S. M., Parker, M. S., Rombauts, S., Salamov, A., Von Dassow, P., Badger, J. H., Coutinho, P. M., Demir, E., Dubchak, I., Gentemann, C., Eikrem, W., Gready, J. E., John, U., Lanier, W., Lindquist, E. A., Lucas, S., Mayer, K. F., Moreau, H., Not, F., Otillar, R., Panaud, O., Pangilinan, J., Paulsen, I., Piegu, B., Poliakov, A., Robbens, S., Schmutz, J., Toulza, E., Wyss, T., Zelensky, A., Zhou, K., Armbrust, E. V., Bhattacharya, D., Goodenough, U. W., Van de Peer, Y., Grigoriev, I. V. Green evolution and dynamic adaptations revealed by genomes of the marine picoeukaryotes Micromonas. *Science.* 2009, 324: 268–272.

Xue, H. B., Stumm, W., Sigg, L. The binding of heavy metals to algal surfaces. *Water Res.* 1988, 22: 917–926.

Yokono, M., Satoh, S., Tanaka, A. Comparative analyses of whole genome protein sequences from multiple organisms. *Sci. Rep.* 2018, 8: 6800.

Yuan, J. P., Peng, J., Yin, K., Wang, J. H. Potential health-promoting effects of astaxanthin: A high-value carotenoid mostly from microalgae. *Mol. Nutr. Food Res.* 2011, 55: 150–165.

Zeraatkar, A. K., Ahmadzadeh, H., Talebi, A. F., Moheimani, N. R., McHenry, M. P. Potential use of algae for heavy metal bioremediation, a critical review. *J. Environ. Manage.* 2016, 181: 817–831.

Index

α-tocopherol, 212

β-carotene, 111, 114, 139–145, 169, 195–196, 249, 279
    carotenoid pigments from algae, 140–141
    Dunaliella salina, 142–144
        health and nutraceutical products, 143
        processing and manufacturing requirements, 143–144
    and human health, 141–142
    isoprenoid-carotenoids, 140
    in marine organisms, 142
    market price, 143
    overview, 139–140
β-cyclocitral, 94
β-ionone, 94
β-sitosterol, 53

1,1-diphenyl-2-picrylhydrazyl (DPPH), 18
2′,7′-dichlorofluorescin diacetate (DCFH-DA), 18
7,12-dimethylbenz[α]anthracene (DMBA), 194
2-keto-3-methylvalerate (2-KMV), 94
2-ketoacid decarboxylase, 94
2-ketoisocaproate (2-KIC), 94
2-KIC, *see* 2-ketoisocaproate (2-KIC)
2-KMV, *see* 2-keto-3-methylvalerate (2-KMV)
2-methylisoborneol (2-MIB), 94
5′-deoxyadenosylcobalamin (AdoB$_{12}$), 105
5-methoxydec-9-ynoic acid, 86
13-hydroxy dechlorofontonamide, 192
83K7C antibody, 285

AAD, *see* Aldehyde decarbonylase (AAD)
AAR, *see* Acylacyl carrier protein reductase (AAR)
Abysmal light utilization efficiency, 26
ACCase, *see* Acetyl-CoA carboxylase (ACCase)
Accelerated solvent extraction (ASE), 237
Acetyl-CoA carboxylase (ACCase), 274
Acylacyl carrier protein reductase (AAR), 94
Adenosine triphosphate (ATP), 75, 90
Adiponectin, 131
AdoB$_{12}$, *see* 5′-deoxyadenosylcobalamin (AdoB$_{12}$)
Adsorption technique, 95
Advanced microscopy and cell dynamics, 265
Agaran-type polysaccharide (GFP08), 226
Agardhiella subulata, 65

AIP, *see* Atherogenic index of plasma (AIP)
Alcaligenes, 9
ALCOdb, *see* Algae Gene Co-expression database (ALCOdb)
Alcoholic precipitation, 29–30
Alcohols, 94
Aldehyde decarbonylase (AAD), 94
Algae, 207–215
    benefits to industry, 214
        feed, 214
        food, 214
    chemical composition, 209–212
        amino acids and proteins, 210
        antioxidants, 212
        carbohydrates/polysaccharides, 210
        lipids and sterols, 210–212
        minerals and vitamins, 212
    health benefits as functional food, 212–214
        anti-diabetic activity, 213
        anti-obesity activity, 212–213
        growth promoter of gut bacteria, 214
        growth promoting activity in animals, 213
        source of dietary fiber, 213
    in Mascarene Islands, 209
        macroalgae, 209
        microalgae, 209
    overview, 208
    types, 208
Algae Gene Co-expression database (ALCOdb), 266
AlgaGEM model, 267
Alginates, 33–40, 72
    biological properties, 39–40
    gel properties, 39
    overview, 34–35
    physicochemical characterization, 38
        critical overlap concentrations, 38
        macromolecular magnitudes, 38
        rheological measurements, 38–39
            dynamic oscillatory, 39
            steady shear, 38–39
    structural characterization, 35–37
        footprints and main functional groups identification, 36
        global composition, 35–36
        M/G ratios determination, 36
        sequence and blocks distribution, 36–37
Alginic acid, 72
Aliphatic alcohols, 95
Aliphatic ketones, 94
Alkaloids, 192
Allophycocyanin, 116, 212
Alteromonas macleodii, 9

Amansia multifida, 226
Amarouciaxanthin A, 247
Amino acid, 152, 210
Amphidinolides (APDNs), 196
Amphidinols (APDLs), 196
Anabaena variabilis, 29
Angiogenesis, 75, 223, 224–226
Anionic polysaccharides, 221–227
    action on angiogenesis, 224–226
    angiogenesis and tubulogenesis, 223
    overview, 222
    structure and pharmacological actions, 222–223
Antiaging effects, 154
Antiangiogenic effect, 250
Anticancer compounds, 185–198
    alkaloids, 192
    bioactive peptides, 186–187, 190–191
    carotenoids, 193–196
        β-carotene, 195–196
        astaxanthin, 194–195
        fucoxanthin, 195
    C-phycocyanin, 193
    dinoflagellate toxins, 196–197
    lipid compounds, 191–192
    overview, 185–186
    polyketides, 191
Anticancer effect, 251
Antidiabetic activity, 213
Antidiabetic effect, 252
Anti-inflammatory activity, 250–251
Anti-obesity activity, 212–213
Anti-obesity effect, 251–252
Antioxidant, 212
    activity, 248–249
Antiretroviral therapy (ART), 74
APDLs, *see* Amphidinols (APDLs)
APDNs, *see* Amphidinolides (APDNs)
Aplanospores, 123
Apollo program, 182
Apratoxins, 190
Arginine, 152
Arsenic toxicity, 103
ART, *see* Antiretroviral therapy (ART)
Arthosporia, 238
Arthrospira platensis, 181, 182
Arthrospira (Spirulina), 162, 169, 170
Ascophyllum nodosum, 33, 34, 103, 212, 213, 214, 224
ASE, *see* Accelerated solvent extraction (ASE)
Astaxanthin (ATX), 111, 113–114, 194–195, 286–287
Atherogenic index of plasma (AIP), 101
ATP, *see* Adenosine triphosphate (ATP)
ATX, *see* Astaxanthin (ATX)

ATX and its esters, 121–132
  bioavailability, 127, 129
  cardiovascular disease prevention, 130
  commercial production, 124
  for diabetics, 130–131
  Haematococcus pluvialis
    breaking aplanospores, 123–124
    carotenoid composition, 124
    life cycle and carotenoid accumulation, 123
  health and nutraceutical applications, 129
  human clinical studies, 129
  inflammatory disorders prevention, 130
  inhibiting cancer, 131
  origin and occurrence, 122–123
  overview, 122
  preventing neurodegenerative diseases, 131
  production through photobioreactor cultivation, 123
  protection from ultraviolet radiation, 129–130
  safety, 126
  for salmon and trout feeds, 124–126
  supporting immune system, 130
  ulcers and gastric injury prevention, 130

BAT, *see* Brown adipose tissues (BAT)
Bead milling, 123
Beta-cell regeneration, 103
Beta-ketolase (BKT), 287
Bifidobacteria, 213
Bifurcaria bifurcata, 65
Big data analysis, 262, 266
Bioactive compounds extraction techniques, 230–238
  conventional extraction, 230
  modern extraction techniques, 230–238
    enzyme-assisted extraction, 238
    microwave-assisted extraction, 237–238
    pressurized fluid extraction, 237
    supercritical CO_2 extraction, 231, 237
    ultrasound-assisted extraction, 237
Bioactive metabolites, 273–279
  downstream processes, 277
  genetic engineering, 277
  global market for algal products, 277–279
  microalgal physiology for production, 274
    lipid synthesis, 274
  overview, 273–274
  transformation methods, 274–277
Bioactive peptides, 86, 186–187, 190–191
Biodiesel production, 274
Biofuels, 284–285
Biological life support systems (BLSS), 178–179
Biomedical applications of fucoidan, 14–20
  drug delivery, 17–20
  tissue engineering, 15–17
  wound dressing, 20
Bioproducts, 286–287
Bioremediation, 287–288
Biosynthetic pathways, 91, 94–95
Bis (trimethysilyl) trifluoroacetamide (BSTFA), 36
BKT, *see* Beta-ketolase (BKT)
BLSS, *see* Biological life support systems (BLSS)
Bone-protection, 250
Botryococcus braunii, 285
Brentuximabvedotin, 187
Brown adipose tissues (BAT), 75
Brown seaweed, *see Sargassum polycystum*
BSTFA, *see* Bis (trimethysilyl) trifluoroacetamide (BSTFA)

CAGR, *see* Compound annual growth rate (CAGR)
Calcium alginates, 39, 40
Calvin cycle, 90, 265
CAM, *see* Chorioallantoic membrane (CAM)
Cameo software, 268
Cancer, 185; *see also* Anticancer compounds; Anticancer effect
Candida albicans, 64, 238
Canistrocarpus cervicornis, 65
Canthaxanthin (CTX), 194
Carbohydrates, 162, 165, 210
Carbon dioxide induce gelation, 39
Carotenogenesis, 143
Carotenoids, 90, 112–113, 151, 153, 167, 169, 193–196, 212, 245–246, 247, 279, 286–287
  β-carotene, 195–196
  astaxanthin, 194–195
  fucoxanthin, 195
Catechins, 212
Caulerpa cylindracea, 62
Cavitation, 83, 84
CDER, *see* Center for Drug Evaluation and Research (CDER)
Cellular polysaccharides, 6–7
Center for Drug Evaluation and Research (CDER), 104
Center for Veterinary Medicine (CVM), 54
Central nervous system (CNS), 76
Centre for Food Safety and Applied Nutrition (CFSAN), 54
Centrifugation, 113
Cerebrovascular protection, 250
CFSAN, *see* Centre for Food Safety and Applied Nutrition (CFSAN)
Chara braunii, 284–285
Chemical harvesting methods, 113
Chitosan-alginate fucoidan composites, 17–18
Chlamydomonas reinhardtii, 172, 197, 263, 264, 265, 274, 276, 282, 285–286

Chlorella, 5, 7, 43–47, 162, 169, 170, 183, 186, 239
  antimicrobial activity, 53
  commercial production, 47–49
  industrial potential, 49–51, 53
  market value, 53
  safety regulations, 53–54
Chlorella ellipsoidea, 150, 265
Chlorella gracilis, 86
Chlorella marina, 194
Chlorella protothecoides, 263
Chlorella pyrenoidosa, 53, 150, 179, 213
Chlorella sorokiniana, 49, 115
Chlorella vulgaris, 46, 47, 49, 50, 86, 115, 154, 182, 191, 194, 210, 238, 267
Chlorella zofingiensis, 114, 115
Chlorophyceae, 209
Chlorophyll, 90, 115, 151, 153, 167, 169
Chlorophyta, 82
Cholesterol, 53
Chorioallantoic membrane (CAM), 221, 226
Chromochloris zofingiensis, 287
Chromophyta, 82
Circular and raceway ponds, 27
Circular basins, 27
Cladosiphon okamuranus, 224
Closed reactor systems, 27
$CN-B_{12}$, *see* Cyanocobalamin ($CN-B_{12}$)
CNS, *see* Central nervous system (CNS)
Coccomyxa onubensis, 115
Coelastrella sp., 115
Column photobioreactors, 48
Commercial-scale algae cultivation, 26–28
  circular and raceway ponds, 27
  closed reactor systems, 27
  cultivation tanks, 26
  open cultivation systems, 26
  shallow ponds, 26
  sloped open systems, 27
  tubular reactors, 27–28
Compound annual growth rate (CAGR), 143
Condensation-based recovery system, 95
Confocal Raman microscopy, 265
Continuous cultivation regime, 28
Conventional solvent extraction, 113
Corallina pilulifera, 76
C-phycocyanin, 116, 193, 198
CRISPR/Cas9 technology, 268
Critical overlap concentrations, 38
Cryotopic gelation, 39
Crypthecodinium cohnii, 238
Cryptophycins, 190
CTX, *see* Canthaxanthin (CTX)
Cultivation tanks, 26
Curacin A, 191
CVM, *see* Center for Veterinary Medicine (CVM)
Cyanidioschyzon merolae, 282
Cyanobacteria, 26, 82, 178, 180, 186
  anticancer compounds, 185–198
    alkaloids, 192

bioactive peptides, 186–187, 190–191
carotenoids, 193–196
C-phycocyanin, 193
dinoflagellate toxins, 196–197
lipid compounds, 191–192
overview, 185–186
polyketides, 191
Cyanobacterial exopolysaccharide, see Exopolysaccharide (EPS)
Cyanocobalamin (CN-B$_{12}$), 105
Cyanophora paradoxa, 282–283
Cyanophyceae, 86
Cyclic depsipeptides, 187
Cyclic lipopeptides, 187
Cyclic peptides, 187

DAG, see Diacylglycerol (DAG)
Database of Enzymes of Microalgal Biofuel Feedstock (dEMBF), 266
DCFH-DA, see 2′,7′-dichlorofluorescin diacetate (DCFH-DA)
dEMBF, see Database of Enzymes of Microalgal Biofuel Feedstock (dEMBF)
Deschlorohapalindole3, 192
Desmodesmus sp., 115
Dewatering, 277
DGTS, see Diacylglyceroltrimethylhomoserine (DGTS)
DHA, see Docosahexaenoic acid (DHA)
Diabetes mellitus, 99, 100, 130
Diabetic nephropathy, 103, 131
Diacylglycerol (DAG), 264, 274
Diacylglyceroltrimethylhomoserine (DGTS), 265
Diafiltration, 6, 29, 30
Dialysis/diffusion setting method, 39
Dictyota pfaffii, 65
Dietary Supplement Health and Education Act (DSHEA), 54
Dietetic Products Nutrition and Allergies (NDA), 54
Dimethylallyl diphosphate (DMADP), 94
Dimethyldisulfide (DMDS), 94
Dimethylsulfide (DMS), 94, 95
Dimethylsulphoniopropionate (DMSP), 94–95
Dimethyltrisulfide (DMTS), 94
Dinoflagellate toxins, 196–197
Dissolvable liberated polysaccharide (DLP), 28
Diterpenes, 73
DLP, see Dissolvable liberated polysaccharide (DLP)
DMADP, see Dimethylallyl diphosphate (DMADP)
DMBA, see 7,12-dimethylbenz[α]anthracene (DMBA)
DMDS, see Dimethyldisulfide (DMDS)
DMS, see Dimethylsulfide (DMS)
DMSP, see Dimethylsulphoniopropionate (DMSP)

DMTS, see Dimethyltrisulfide (DMTS)
Docosahexaenoic acid (DHA), 7, 54, 131, 151, 153, 230, 238, 249
Dolastatin17, 187
Downstream processes, 113, 277
DPPH, see 1,1-diphenyl-2-picrylhydrazyl (DPPH)
Dried nori products, 106–107
DSHEA, see Dietary Supplement Health and Education Act (DSHEA)
Dunal, Felix, 142
Dunaliella, 43, 45, 238
Dunaliella bardawil, 139, 140, 143, 144, 277
Dunaliella salina, 111, 112, 114, 151, 196, 239, 287
    and β-carotene, 142–143
    health and nutraceutical products, 143
    processing and manufacturing requirements, 143–144
Dunaliella tertiolecta, 263
Duneliella salina, 279
Durvillea, 33, 34
Dynamic oscillatory measurements, 39

Ecklonia, 33, 34
Eckstolonol, 75
ECM, see Extracellular matrix (ECM)
Edible seaweeds
    vitamin B12 from, 105–108
        bioavailability, 108
        characterization, 107–108
        content, 106–107
        overview, 105–106
EDTA, see Ethylenediaminetetraacetic acid (EDTA)
EDX, see X-ray spectroscopy (EDX)
EFSA, see European Food Safety Authority (EFSA)
Egg-box system, 33, 39
Eicosapentaenoic acid (EPA), 7, 54, 131, 151, 153, 230
Eklonia stolonifera, 75, 103
Electrospun nanofibers, 17
Endoplasmic reticulum (ER), 274
Enterobacteria, 214
Enteromorpha, 210
Enzyme-assisted extraction, 238
EPA, see Eicosapentaenoic acid (EPA)
EPS, see Exopolysaccharide (EPS)
ER, see Endoplasmic reticulum (ER)
ESA, see European Space Agency (ESA)
Escherichia coli, 62, 108
Ethylenediaminetetraacetic acid (EDTA), 28, 29
Ettlia oleoabundans, 265
European Algal Biomass Association, 172
European Commission, 33, 40, 122, 171
European Community Regulation on Food Safety, 53

European Food Safety Authority (EFSA), 54, 171
European Medicines Agency, 6
European Space Agency (ESA), 45, 181
Exopolysaccharide (EPS), 25–30
    extraction and purification, 28
    overview, 26
    production, 26–28
        commercial-scale algae cultivation, 26–28
        laboratory-scale bioreactors, 28
    treatments for extraction, 28–30
        isolation, 29
        peripheral ultrafiltration, 30
        ultrasound/microwave-assisted technique, 30
        via alcoholic precipitation, 29–30
Extracellular matrix (ECM), 152
Extra-cellular polysaccharides, 4–7, 9
Extraction kinetics, 84
Extrinsic aging, 154

FA, see Formaldehyde (FA)
FAFF, see Future of Algae for Food & Feed (FAFF)
Farnesyl diphosphate (FDP), 94
FAS, see Fatty-acid synthesis (FAS)
Fatty-acid synthesis (FAS), 274, 277
FD&C, see Federal Food, Drug and Cosmetic Act (FD&C)
FDP, see Farnesyl diphosphate (FDP)
Federal Food, Drug and Cosmetic Act (FD&C), 54
FGF, see Fibroblast growth factors (FGF)
FGFR, see Fibroblast growth receptor (FGFR)
Fibroblast growth factors (FGF), 221, 223
Fibroblast growth receptor (FGFR), 190
Filtration, 6, 113
Flat plate photobioreactor, 48
Flavonoids, 169
Flavonols, 212
Flocculation, 113
Folkman, Judah, 222
Formaldehyde (FA), 28–29
Fourier transform infrared (FT-IR) spectroscopy, 14, 20, 36
Fraction rich in fucans (FRF), 225
FT-IR, see Fourier transform infrared (FT-IR) spectroscopy
Fucans, 72
Fucoidan, 13–20
    biomedical applications, 14–20
        drug delivery, 17–20
        tissue engineering, 15–17
        wound dressing, 20
    characterization, 14
    isolation from marine macroalgae, 14
    overview, 13
    sulfated polysaccharides, 14
Fucoidans, 73
Fucosterol, 65

Fucoxanthin (FX), 195, 198, 245–252
   bioavailability, 247
   biological activity, 248–252
      antiangiogenic effect, 250
      anticancer effect, 251
      antidiabetic effect, 252
      anti-inflammatory activity, 250–251
      antiobesity effect, 251–252
      antioxidant activity, 248–249
      bone-protection, 250
      cerebrovascular protection, 250
      hepatoprotective effect, 249
      ocular protection, 250
      skin protection, 249–250
   metabolites, 247
   overview, 245–246
   safety aspects, 248
   sources, 246
   structure, 246–247
Fucoxanthinol, 247
Fucus serratus, 72
Fucus spiralis, 224
Fucus vesiculosus, 222, 224, 225, 248
Functional omics, 263
Fusarium oxysporum, 64
Future of Algae for Food & Feed (FAFF), 172
FX, *see* Fucoxanthin (FX)

Gaz Chromatography coupled with Mass Spectrometry and Electron Ionization (GC/MS-EI), 36
GDP, *see* Geranyl diphosphate (GDP)
Gelation mechanism, 39
Gelonium multiform, 197
Gel permeation chromatography, 7
Gemcitabine-resistant human pancreatic cancer cells (GR-HPCCs), 194
Gemini missions, 182
Generally recognized as safe (GRAS), 54, 230, 231
Gene regulatory networks (GRNs), 266, 267
Genetic engineering, 262, 277
Genome Regulatory Architecture Tools (GREAT), 267
Genome-scale metabolic models (GSMMs), 267
Genome sequencing strategies, 281–288
   biofuels, 284–285
   bioproducts, 286–287
   bioremediation, 287–288
   overview, 281–282
   phylogeny and evolution, 282–284
   recombinant proteins, 285–286
Genomics, 263, 282, 285, 287, 288
Geosmin, 94
Geranyl diphosphate (GDP), 94
Geranylgeranyl pyrophosphate, 140
Glenn, John, 181
Glutaraldehyde (GTA), 29
Glycoproteins, 9

Glycosylated haemoglobin (HbA1C) levels, 100
Good Manufacturing Practice (GMP), 240
Gracilaria edulis, 213
GRAS, *see* Generally recognized as safe (GRAS)
Grateloupia filicinia, 66
Grateloupian elliptica, 103
Gravity sedimentation, 113
GREAT, *see* Genome Regulatory Architecture Tools (GREAT)
Green microalgae, 131, 150, 285
Green seaweed, 151–152, 211
GR-HPCCs, *see* Gemcitabine-resistant human pancreatic cancer cells (GR-HPCCs)
GRNs, *see* Gene regulatory networks (GRNs)
GSMMs, *see* Genome-scale metabolic models (GSMMs)
GTA, *see* Glutaraldehyde (GTA)
Gymocin-A, 196

Haematococcus, 43, 45, 113–114, 186
Haematococcus pluvialis, 111, 112, 113–114, 121, 151, 194, 197, 265, 279, 286–287
   breaking aplanospores, 123–124
   carotenoid composition, 124
   life cycle and carotenoid accumulation, 123
Hamburger, Clara, 142
Hapalindole X, 192
Harvesting stage, 277
HbA1C, *see* Glycosylated haemoglobin (HbA1C) levels
HCMV, *see* Human cytomegalovirus (HCMV)
Helicobacter pylori, 130
Hematococcus, 238, 239
Hematocysts, 123
Hepatocytes fatty degeneration, 103
Hepatoprotective effect, 249
Herpes simplex viruses (HSV), 66
Heterotrophic growth, 90
High-fat diet (HFD), 249
High intensity ultrasound (HIU), 82
High Performance Anion Exchange Chromatography with Pulsed Amperometric Detection (HPAEC-PAD), 36
Himantothallus grandifolius, 64
HIU, *see* High intensity ultrasound (HIU)
HIV, *see* Human immunodeficiency virus (HIV)
Hizikia fusiformis, 75, 76, 213
Homing device, 190
HPAEC-PAD, *see* High Performance Anion Exchange Chromatography with Pulsed Amperometric Detection (HPAEC-PAD)

HPV E7 protein, Human Papillomavirus (HPV) E7 protein
HSV, *see* Herpes simplex viruses (HSV)
Human cytomegalovirus (HCMV), 74
Human immunodeficiency virus (HIV), 66, 74, 75
Human Papillomavirus (HPV) E7 protein, 286
Human umbilical vein endothelial cells (HUVEC), 76, 224, 225, 226
Hydrocarbons, 94, 95
Hyperglycemia, 99, 100, 130
Hypnea cornuta, 65
Hypnea musciformis, 103

ICGB, *see* International Cooperative Biodiversity Group (ICGB)
IL, *see* Interleukin (IL)
Immunoglobulin production, 130
Inorganic carbon (CO_2), 90
Insulin sensitivity, 100
Integrated omics (iOMICS), 261
Interleukin (IL), 40
Internal setting method, 39
International Cooperative Biodiversity Group (ICGB), 186–187
Intra-cellular polysaccharides, 4–5
Intrinsic aging, 154
iOMICS, *see* Integrated omics (iOMICS)
Ion-exchange chromatography, 7
IPP, *see* isopentenyl diphosphate (IPP)
iRC1080model, 267
Isolation techniques, 155–156
Isomalyngamide, 191
isopentenyl diphosphate (IPP), 94
Isoprenoid-carotenoids, 140–141
Isoprenoid pathways, 94

Joint Research Centre, 171

Karenia mikimotoi, 196
Karlodinium veneficum, 197
Karlotoxins (KTX), 197
Ketones, 94
Klebsiella pneumoniae, 63
Klebsormidium flaccidum, 283
Korabl-Sputnik4, 179
KTX, *see* Karlotoxins (KTX)
Kupffer cells, 103

Laboratory-scale bioreactors, 28
   algae cultivation regimes, 28
   optimization strategies for yield improvement, 28
Lactobacilli spp., 214
Lactobacteria, 213
Laminarans, 72
Laminaria, 33, 34, 40
Laminaria japonica, 225
Laminaria saccharina, 224
Laucysteinamide A, 192
Laurencia dendroidea, 65

Index

Laxaphycins, 190–191
LC/MS-MS, see Liquid chromatography/electrospray ionization-tandem mass spectrometry (LC/MS-MS)
LC-PUFAs, see Long-chain polyunsaturated fatty acids (LC-PUFAs)
Leishmania amazonensis, 65
Leuconostoc, 9
Life support system (LSS), 178–179, 180, 183
Limnothrix redekei, 29
Lipid, 152, 162, 165, 210–212
    compounds, 191–192
    metabolites, 131
    synthesis, 274
Lipidomics and metabolomics, 264–265
Lipophilicity, 247
Liquid chromatography/electrospray ionization-tandem mass spectrometry (LC/MS-MS), 105
Liquid–liquid extraction, 155
LIU, see Low intensity ultrasound (LIU)
LMF, see Low molecular weight Fucoidan (LMF)
LMW, see Low molecular weight (LMW)
Lobophora variegata, 225
Long-chain polyunsaturated fatty acids (LC-PUFAs), 274
Low intensity ultrasound (LIU), 82
Low molecular weight Fucoidan (LMF), 252
Low molecular weight (LMW), 225
LSS, see Life support system (LSS)
Lutein, 114–115
Lycopene, 155
Lyophilization method, 17

Macroalgae, see Seaweeds
Macrocystis genus, 40
Macrocystis pyrifera, 33, 34
Macrophomina phaseolina, 64
Macrozooids, 123
MAE, see Microwave-assisted extraction (MAE)
Marine algal bioactives, 71–77
    anti-angiogenic, 75–76
    anti-cancer, 73
    anti-obesity, 75
    antioxidant, 75
    anti-viral, 73–75
    health claims, 73
    neuroprotective, 76
    overview, 71–73
    photoprotection, 76–77
MeB_{12}, see Methylcobalamin (MeB_{12})
Mechanical methods, 113
Melanin, 155
MELiSSA, see Micro-Ecological Life Support System Alternative (MELiSSA)
Membrane filtration systems, 6, 7
MEP, see Methylerythritol phosphate (MEP) pathway
Mercury projects, 182
Metabolic pathways, 266
Meta-hydroxydiphenyl (MHDP), 35–36
Methionine, 95
Methylcobalamin (MeB_{12}), 105
Methylerythritol phosphate (MEP) pathway, 94
Mevalonate pathway (MVA), 94
MGDG, see Monogalactosyldiacylglycerols (MGDG)
M/G ratios determination, 36
MHDP, see Meta-hydroxydiphenyl (MHDP)
Micelles formation, 247
Microalgae, 3, 161–173, 186, 205, 209
    anticancer compounds, 185–198
        alkaloids, 192
        bioactive peptides, 186–187, 190–191
        carotenoids, 193–196
        C-phycocyanin, 193
        dinoflagellate toxins, 196–197
        lipid compounds, 191–192
        overview, 185–186
        polyketides, 191
    extracting bioactive compounds, 229–240
        biochemical composition, 230
        commercial and industrial applications, 238–239
        economic potential, 239
        overview, 230
        techniques, 230–231, 237–238
    on market, 169–170
        Arthrospira (Spirulina) and *Chlorella*, 170
    molecular tools, 261–268
        advanced microscopy and cell dynamics, 265
        big data analysis, 266
        functional omics, 263
        genome-scale metabolic models (GSMMs), 267
        genomics, 263
        lipidomics and metabolomics, 264–265
        microalgal cell factories reconstruction, 268
        multiomics data integration, 265
        overview, 262
        phylogenomics of metabolic pathways, 266
        proteomics, 264
        sub-cellular networks, 266–267
        transcriptional engineering, 265–266
        transcriptional factors prediction, 267
        transcriptomics, 263–264
    nutraceutical aspects, 177–183
        designing space food, 181–182
        future scenarios, 182–183
        life support systems (LSS), 178–179
        overview, 177–178
        in space, 179–181
        terraforming, 179
    nutrients for human consumption, 162–169
        pigments and polyphenols, 167, 169
        proteins, carbohydrates and lipids, 162, 165
        vitamins and minerals, 165, 167
    overview, 161–162
    perspectives and constraints, 171–173
    pigments, 111–116
        β-carotene, 114
        astaxanthin, 113–114
        carotenoids, 112–113
        chlorophylls, 115
        lutein, 114–115
        overview, 111
        phycobiliproteins, 116
    polysaccharides, 3–9
        extraction, 4–7
        industrial applications and prospects, 7, 9
        overview, 3–4
        purification, 7
    volatile organic compounds (VOC), 89–96
        biosynthesis, 91, 94–95
        commercial perspective, 95
        factors affecting, 90
        overview, 89–90
        techniques for recovery, 95
Micro- and macroalgae in cosmetics, 149–157
    anticellulite and slimming effects, 155
    biological activities, 154
        antiaging effects, 154
        moisturizing/hydration action, 154
    as functional ingredients, 150–153
    isolation techniques, 155–156
    marketing potential, 156–157
    overview, 149–150
    photo-protective action, 154–155
    whitening/melanin-inhibiting effects, 155
Micro-Ecological Life Support System Alternative (MELiSSA), 181, 183
Microwave-assisted extraction (MAE), 30, 155, 156, 237–238
Microwave technologies, 6
Microzooids, 123
Minerals, 165, 167, 212
Molecular tools in microalgae, 261–268
    advanced microscopy and cell dynamics, 265
    big data analysis, 266
    functional omics, 263
    genome-scale metabolic models (GSMMs), 267
    genomics, 263
    lipidomics and metabolomics, 264–265

microalgal cell factories
    reconstruction, 268
multiomics data integration, 265
overview, 262
phylogenomics of metabolic
    pathways, 266
proteomics, 264
sub-cellular networks, 266–267
transcriptional engineering, 265–266
transcriptional factors prediction, 267
transcriptomics, 263–264
Monilini alaxa, 64
Monogalactosyldiacylglycerols (MGDG), 264, 265
Monosaccharides, 210, 222
Monounsaturated fatty acids, 165
Multiomics data integration, 265
Muriellopsis sp., 115
MVA, *see* Mevalonate pathway (MVA)

NAMC, *see* Nonyl 8-acetoxy-6-methyloctanoate (NAMC)
Nannochloropsis gaditana, 263
Nannochloropsis oculata, 238, 264
Nannochloropsis salina, 265
NASA, *see* National Aeronautics and Space Administration (NASA)
National Aeronautics and Space Administration (NASA), 45, 178, 180, 181
Natural moisturizing factors (NMFs), 152, 154
Natural (NF) fucoidan, 225
NDA, *see* Dietetic Products Nutrition and Allergies (NDA)
Neochlorisoleo abundans, 263
Newtonian fluid, 38
Next-generation sequencing (NGS), 282
NF, *see* Natural (NF) fucoidan
NGS, *see* Next-generation sequencing (NGS)
NIDDM, *see* Non-insulin-dependent diabetes mellitus (NIDDM)
Nizimuddinia zanardini, 38
NMFs, *see* Natural moisturizing factors (NMFs)
NMR, *see* Nuclear magnetic resonance (NMR) spectra
Non-insulin-dependent diabetes mellitus (NIDDM), 99
Non-mechanical methods, 113
Non-Newtonian fluid, 38
Nonpolar compounds, 238
Nonpolar lipids, 165
Nonribosomal peptide synthetase (NRPS), 186, 191
Non-small-cell lung cancer (NSCLC), 190, 193, 195, 198
Non-solvent induce phase separation, 39
Nonyl 8-acetoxy-6-methyloctanoate (NAMC), 192
Nori, *see Porphyra yezoensis*

Nostoc calcicola, 29
NRPS, *see* Nonribosomal peptide synthetase (NRPS)
NSCLC, *see* Non-small-cell lung cancer (NSCLC)
Nuclear magnetic resonance (NMR) spectra, 36, 37

Ocular protection, 250
Open cultivation, 26, 47
Open-pond methods, 112, 229
Organic carbon substrates, 90
Oscillatory rheology, 39
OSF, *see* Oversulfated (OSF) fucoidan
Ostreococcus Lucimarinus, 267
Ostreococcus tauri, 267
Outdoor cultivation, 45
Oversulfated (OSF) fucoidan, 225
Oxidative stress, 251

Palmaria palmata, 210
Palmella, 123
Parallel streak method, 62
PCL, *see* Polycaprolactone (PCL)
PDT, *see* Photodynamic therapy (PDT)
Pectenotoxins (PTX), 196
Pelvetia babingtonii, 103
Penicillium digitatum, 64
Peripheral ultrafiltration, 30
Pertraction, 95
Pervaporation, 95
Petalonia binghamiae, 103, 252
Phaeodactylum tricornutum, 172, 264, 276
Phaeophyceae, 209
Phenolic acids, 169
Pheophorbide, 72
Pheophytin, 72, 76
Phlorotannins, 73, 75, 76, 103, 212
Phospholipids, 247
Photobioreactors, 112, 229
Photobioreactor tubes, 27–28, 48
Photodynamic therapy (PDT), 193
Photo-protective action, 154–155
Phycobilins, 151, 169
Phycobiliproteins, 90, 116, 153, 167, 169, 212, 238
Phycocyanin, 116, 212
Phycoerythrins, 116, 153, 212
Phycoerythrocyanin, 116
Phylogenomics, 266
Phylogeny and evolution, 282–284
Phytochemicals, 167, 169
Pigments, 167, 169
PKS, *see* Polyketide synthase hybrid (PKS) pathways
Plasmodium falciparum, 286
PLE, *see* Pressurized liquid extraction (PLE)
Polar compounds, 238
Polar lipids, 165
Polycaprolactone (PCL), 17
Polyelectrolyte complexation, 17

Polyketides, 191
Polyketide synthase hybrid (PKS) pathways, 186, 191
Polyphenols, 103, 167, 169, 212
Polysaccharides, 72, 150, 154, 162, 210, 224
Polysiphonis spp., 213
Polyunsaturated aldehydes (PUA), 192
Polyunsaturated fatty acids (PUFAs), 165, 192, 211, 212, 230, 240, 286
Porphyra spp., 105, 106–107
Porphyra tenera, 210
Porphyra yezoensis, 214
Porphyridium, 4, 5, 7, 151, 214, 238
Porphyridium cruentum, 6, 29
PPARγ, *see* Proliferator-activated receptor gamma (PPARγ)
Pressurized fluid extraction, 237
Pressurized liquid extraction (PLE), 237
Pressurized solvent extraction (PSE), 237
Proliferator-activated receptor gamma (PPARγ), 103
Proline methyl ester, 86
Proteins, 162, 165, 210
Proteomics, 264
PSE, *see* Pressurized solvent extraction (PSE)
Pseudoalteromonas sp., 9
PTX, *see* Pectenotoxins (PTX)
PUA, *see* Polyunsaturated aldehydes (PUA)
PUFAs, *see* Polyunsaturated fatty acids (PUFAs)
Pyropheophytin, 72

Quasi-continuous cultivation regime, 28

Rabbit aortic endothelial cells (RAEC), 224
Raceway ponds, 27, 47, 48, 112
RAEC, *see* Rabbit aortic endothelial cells (RAEC)
Reactive oxygen species (ROS), 130, 248, 249, 250
Recombinant DNA technology, 28
Recombinant proteins, 285–286
Red macroalgae, 152
Regulation (EC) No. 262/99, 171
Regulation (EU) No. 2015/2283, 171
Rhizoctonia solani, 64
Rhodophyceae, 209
Rhodophyta, 82
ROS, *see* Reactive oxygen species (ROS)

Saccharina japonica, 154, 251
Salmonella typhimurium, 62
Sargasssum tenerrimum, 64
Sargassum filipendula, 63
Sargassum fulvellum, 76
Sargassum integerrimum, 224
Sargassum latifolium, 62, 63
Sargassum linearifolium, 65
Sargassum platycarpum, 62
Sargassum polycystum, 99–104, 103, 213, 246

action mechanisms of extracts, 100–102
antidiabetic effects, 103–104
effects on diabetic pancreas, liver, and kidneys, 102–103
overview, 99–100
Sargassum stenophyllium, 224
Sargassum vulgare, 222
Saturated fatty acids, 165
Scenedesmus, 45, 183
Scenedesmus acutus, 263
Scenedesmus almeriensis, 115
Scenedesmus obliquus, 115
Schizochytrium, 214, 238
SDF-1, *see* Stem cell derived factor 1 (SDF-1)
Seaweeds, 61–66, 150, 205, 209; *see also* Marine algal bioactives
alginates from, 33–40
biological properties, 39–40
gel properties, 39
overview, 34–35
physicochemical characterization, 38
rheological measurements, 38–39
structural characterization, 35–37
anionic polysaccharides from, 221–227
action on angiogenesis, 224–226
angiogenesis and tubulogenesis, 223
overview, 222
structure and pharmacological actions, 222–223
antibacterial activity, 62–64
antifungal activity, 64–65
antiprotozoal activity, 65
antiviral activity, 65–66
bioactive compounds, 62
fucoidan from, 14
overview, 61–62
SEC-MALLS, *see* Size exclusion chromatography coupled with refractometer and/or Multi-Angle Light Scattering analysis (SEC-MALLS)
Secondary ion mass spectrometry (SIMS), 265
Secondary metabolites, 153
Selenium-enriched phycocyanin (Se-PC), 193
Semi-continuous cultivation regime, 28
Se-PC, *see* Selenium-enriched phycocyanin (Se-PC)
SFE, *see* Supercritical fluid extraction (SFE)
Shallow ponds, 26
Silver nanoparticles (AgNPs), 64
SimrisAlg AB, 172
SIMS, *see* Secondary ion mass spectrometry (SIMS)
Siphonoxanthin, 76
Size exclusion chromatography coupled with refractometer and/or Multi-Angle Light Scattering analysis (SEC-MALLS), 38

Skin protection, 249–250
Skylab program, 182
Sloped open systems, 27
Small molecular mass (SMM), 29
Sodium alginate, 39
Sodium hydroxide (NaOH), 28
Solid–liquid extraction, 155
Solieria chordalis, 211
Sonication, 83–84
Soxhlet apparatus, 155, 230
Spatoglossum scroederi, 222
Spectroscopic methods, 36–37
Sphingomonas, 9
Spirulina, 7, 43, 44, 151, 186, 210, 279
antimicrobial activity, 53
commercial production, 45–46
industrial potential, 49–51, 53
market value, 53
microalgae market, 44–45
safety regulations, 53–54
Spirulina maxima, 46
Spirulina platensis, 46, 53, 193, 213, 214
Staphylococcus aureus, 62, 63, 286
Staphylococcus epidermidis, 63
Staphylococcus saprophyticus, 63
Steady shear measurements, 38–39
Stem cell derived factor 1 (SDF-1), 18
Sterols, 210–212
Stypopodium zonale, 65
Sub-cellular networks, 266–267
Subcritical water extraction (SWE), 155–156
Sulfated polysaccharides, 7, 14, 73, 222
Sulfur compounds, 94
Supercritical CO_2 extraction, 231, 237
Supercritical fluid extraction (SFE), 155
Supercritical water extraction (SWE), 237
Surface tension, 84
SWE, *see* Subcritical water extraction (SWE); Supercritical water extraction (SWE)
Symplostatin3, 187

TAGs, *see* Triacylglycerides (TAGs); Triacylglycerols (TAGs)
TALENs, *see* Transcription activator-like effector nucleases (TALENs)
Tasco®, 214
TBARS, *see* Thiobarbituric acid reactive substances (TBARS)
T2DM, *see* Type 2 diabetes mellitus (T2DM)
Terpenoid pathways, 94
Terraforming, 179
TFs, *see* Transcription factors (TFs)
TGF-β1, *see* Transforming growth factor (TGF-β1)
Thalassiosira pseudonana, 276, 282
Thiobarbituric acid reactive substances (TBARS), 250
Tissue engineering, 15–17
Titov, Gherman, 181

TMCS, *see* Trimethylchlorosilane (TMCS)
TNF-α, *see* Tumor necrosis factor (TNF-α)
Transcription activator-like effector nucleases (TALENs), 268
Transcriptional engineering, 265–266
Transcription factors (TFs), 265–267
Transcription regulators (TRs), 266
Transcriptomics, 263–264
Transformation methods, 274–277
Transforming growth factor (TGF-β1), 20
Triacylglycerides (TAGs), 274
Triacylglycerols (TAGs), 247, 262, 264, 265
Trichomonas vaginalis, 65
Trichomoniasis, 65
Trimethylchlorosilane (TMCS), 36
TRs, *see* Transcription regulators (TRs)
Tubular photobioreactor, 48
Tubular reactors, 27–28
Tubulogenesis, 223
Tumor necrosis factor (TNF-α), 40
Two membrane-based techniques, 95
Type 2 diabetes mellitus (T2DM), 99, 100, 131
Tyrosinase inhibitors, 155

UAE, *see* Ultrasound-assisted extraction (UAE)
UCP1, *see* Uncoupling protein 1 (UCP1)
Ultrafiltration, 6, 30
Ultrasonication, 6
Ultrasound-assisted extraction (UAE), 30, 81–86, 155, 156, 237
bioactive compounds
extraction, 84
from fresh water and marine microalgae, 86
proteins, peptides and amino acids, 86
microalgae
application, 86
definition and classification, 82
overview, 81–82
physical parameters, 82–84
principle, 82
solvent parameters, 84
Ultrasound cleaning bath, 84
Ultrasound probe system, 84
Ultrasound waves, 82
Ultraviolet radiation, 76, 154–155, 249–250
Ulva armoricana, 211
Ulva fasciata, 62
Ulva lactuca, 62
Ulva pertuse, 210
Ulva reticulata, 62
Uncoupling protein 1 (UCP1), 75
Undaria pinnatifida, 213, 250, 251, 252
United States Food and Drug Administration (USFDA), 6, 35, 40, 104, 122, 125, 143, 172, 187, 230, 231
US Space Shuttle, 182

Vapour pressure, 84
Vascular endothelial growth factor (VEGF), 221, 223, 224, 225, 226
Vasculogenesis, 223
Vegetative phases, 123
VEGF, *see* Vascular endothelial growth factor (VEGF)
Viridamides A and B, 86
Viscosity, 84
Vitamin B12 from edible seaweeds, 105–108
    bioavailability, 108
    characterization, 107–108
    content, 106–107
    overview, 105–106
Vitamin E, *see* α-tocopherol
Vitamins, 165, 167, 212
VOC, *see* Microalgae, volatile organic compounds (VOC)

White adipose tissues (WAT), 75
World Health Organization (WHO), 64, 185

Xanthomonas, 9
X-ray spectroscopy (EDX), 265

Yessotoxin (YTX), 196

Zeaxanthin, 114
Zinc finger nucleases (ZFNs), 268
Zinc-finger protein transcription factors, 267
Zond spacecraft, 179
Zoospores, 123